Elliptic Systems in the Plane

Editorial Board

Elliptic Systems in the Plane

W. L. Wendland

Technische Hochschule Darmstadt

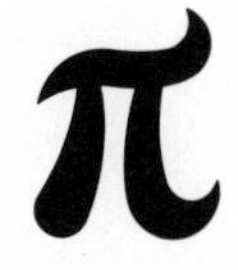

Pitman

London San Francisco Melbourne

Dedicated to

Prof. Dr. Dr. h.c. W. Haack

Pitman Publishing Limited
39 Parker Street, London WC2B 5PB

Fearon-Pitman Publishers Inc.
6 Davis Drive, Belmont, California 94002, USA

Associated Companies
Copp Clark Ltd, Toronto
Pitman Publishing New Zealand Ltd, Wellington
Pitman Publishing Pty Ltd, Melbourne

First Published 1979

AMS Subject Classifications: (main) 35J55, 35A92, 30A97, 65N15, 65N30
(subsidiary) 35Q15, 35J65, 45E05

Library of Congress Cataloging in Publication Data

Wendland, Wolfgang, 1936–
Elliptic systems in the plane.

(Monographs and studies in mathematics; 3)
Includes bibliographies.
1. Differential equations, Elliptic. 2. Boundary value problems. I. Title.
QA377.W45 515′.353 78–6499

ISBN 0–273–01013–1

Typeset in The Universities Press (Belfast) Ltd
Reproduced and printed by photolithography and bound in Great Britain at The Pitman Press, Bath

Preface

This book treats both theoretical and computational aspects of elliptic systems of first-order partial differential equations in two variables. The topics include: normal forms, representation theorems, boundary value problems, function theoretic properties and constructive and computational methods. While the problems treated are mainly linear, some semilinear and nonlinear problems are included.

The general theory for elliptic systems has been developed around the *a priori* estimates of Agmon, Douglis and Nirenberg, the Fourier transformation and the Trace theorem and already there has been a vast amount of work done. Concise presentations of this approach can be found in the important books by L. Hörmander [6] and J. Lions and E. Magenes [7].

Systems in the plane, however, allow a more concise and simpler approach by means of function theoretic methods. For two unknown functions this approach is extensively represented in the books by L. Bers [1], I. N. Vekua [9], [10] and in part II of the book by W. Haack and the author [5]. The Fredholm theory for second-order systems with the Laplacian as principal part can already be found in A. Bitsadze's book [2] and for strongly elliptic equations in G. Fichera's book [3]. Function theoretic properties are treated by R. Gilbert [4] and W. Tutschke [8].

In the present book more general systems with more than two unknowns are treated and here some new difficulties arise. For more simplicity the presentation of boundary value problems is restricted to simply connected domains although most of the methods can be extended to multiply connected domains in a similar way as in [9]. Also the class of solutions is mainly restricted to Hölder continuously differentiable functions. The book consists of two parts.

In Part A the solvability theory for Riemann Hilbert problems and also constructive aspects are presented. It begins with systems for two real valued unknowns reviewing the already classical work of L. Bers, W. Haack and I. N. Vekua. Based on the corresponding Fredholm properties and *a priori* estimates, the solvability theory for general systems in Hölder spaces is developed by using an elementary homotopy technique which perturbs the Cauchy–Riemann equations into the general equations with and without normal form. The corresponding semilinear and nonlinear problems are presented together with results on existence and regularity. For the construction and approximation of solutions we present equivalent singular integral equations and in the analytic case Vekua's generating operator. Some properties of the solution such as the similarity principle, zeros of the solution, unique continuation and reflection principles are discussed.

Part B is devoted to three different computational methods. It begins with the integral equation method for the fundamental problems presenting the asymptotic error analysis for the corresponding discretized Fredholm integral equations by the use of collectively compact operator theory and approximation of conformal mappings. The corresponding discretized and continuous equations are solved by using a successive approximation in connection with analytic continuation of the resolvent operator. Then we present Fredholm integral equations of the first kind in connection with layer methods and their treatment with Galerkin's method including asymptotic error analysis. The same problems are also treated with finite differences. Using the Green and Neumann functions on the grids and the continuum we obtain asymptotic error estimates. The method is based on equivalent discrete and nondiscrete Fredholm integral equations of the second kind defining a family of collectively compact operators on piecewise continuous functions. For the general problem with general boundary conditions we minimize the defect of the mapping defined by the differential operator and boundary conditions relative to different norms. Here the least squares method with finite elements as trial functions converges asymptotically of optimal order in Sobolev spaces. The corresponding discrete equations defined by numerical quadratures yield approximations of the same order if the quadrature formulas provide sufficient accuracy. If the problems are specialized to the simplest case, the generalized analytic functions and their first boundary value problem then we can also show pointwise asymptotic error estimates.

One of the aims of this book is to attempt a unified presentation of theoretical and computational aspects for plane elliptic problems which have many applications in flow problems, elasticity, acoustics and many other fields. Some problems are collected in an Appendix.

The view of this representation is on one hand to show the fineness of the theory of the general elliptic boundary value problems by the use of elementary methods of functional analysis and complex function theory, and on the other hand to give a treatment of the most common computational methods for solving these first-order problems.

The references are collected at the end of each chapter. Some of the chapters end with additional references on some recent results which are briefly classified.

The manuscript of this book arose in connection with seminars which the author has given at the University of Delaware, the Oregon State University and at the Technische Hochschule Darmstadt and was supported by these universities and the Exchange-Visitor-Program G-I-5 of the Fulbright Commission. My sincere appreciation goes to a number of colleagues, especially J. Asvestas, D. Colton, M. Costabel, R. P. Gilbert, B. Goldschmidt, H. Grabmüller, G. Gray, G. C. Hsiao, B. Kawohl, P. Kopp, H. Löffler, E. Meister, K. Merten, W. Moss, B. Rüpprich, W. Tutschke, C. Waid and R. Weinacht for advice, encouragement and for valuable criticisms of the manuscript. I am particularly indebted to David Colton for his many comments and constant help with the manuscript. Special mention is due to Mrs. B. Heinzl who greatly helped my English, to Mrs. A. Chandler, Mrs. D. Templeton, and especially Mrs. U. Abou El-Seoud for typing the manuscript and to the staff of Pitman for preparing the manuscript for printing. I express my appreciation to the publishers for their patience and accommodation to my wishes. Last but not least I express my thanks to my academic teachers W. Haack and G. Hellwig who founded my interest in the presented field and to my wife Gisela and my children Katrin and David who provided the peaceful and understanding atmosphere which I needed for the completion of this book.

W. L. Wendland
Darmstadt, Germany
August 1978

References

1 Bers, L. Function-theoretical properties of solutions of partial differential equations of elliptic type. *Ann. Math. Studies* No. 33, 1954.
2 Bitsadze, A. V. *Boundary value problems.* North Holland, Amsterdam, 1968.
3 Fichera, G. Linear elliptic differential systems and eigenvalue problems. *Lecture Notes in Mathematics* No. 8, Springer, Berlin, 1965.
4 Gilbert, R. P. Constructive methods for elliptic equations. *Lecture Notes in Mathematics* No. 365, Springer, Berlin, 1974.
5 Haack, W. and Wendland, W. *Lectures on partial and Pfaffian differential equations.* Pergamon Press, Oxford, 1972.
6 Hörmander, L. *Linear partial differential operators.* Springer, Berlin, New York, Heidelberg, 1964.
7 Lions, J. L. and Magenes, E. *Non-homogeneous boundary value problems and applications.* Springer, Berlin, New York, Heidelberg, 1972.

8 Tutschke, W. *Partielle komplexe Differentialgleichungen in einer und in mehreren komplexen Variablen.* VEB Deutscher Verlag d. Wiss., Berlin, 1977.

9 Vekua, I. N. *Generalized analytic functions.* Pergamon Press, Oxford, 1962.

10 Vekua, I. N. *New methods for solving elliptic equations.* North Holland, Amsterdam, 1967.

Contents

A

Solvability Properties of Riemann Hilbert Problems

1

Review of the Riemann–Hilbert Problem in Hilbert's Normal Form

This chapter gives a review on the basic results for problems in normal form with two unknowns

$$\begin{aligned} u_x - v_y &= Au + Bv + c, \\ u_y + v_x &= \tilde{A}u + \tilde{B}v + \tilde{c} \end{aligned} \quad \text{in } \mathsf{G}, \tag{1.0.1}$$

$$u \cos \tau(s) - v \sin \tau(s) = \phi \quad \text{on } \dot{\mathsf{G}}.$$

The presentation follows essentially the book by Haack and Wendland [3]. Other presentations of the theory to (1.0.1) can also be found in the books [2], [4]. A very extensive study of these systems in L_p-spaces and Hölder spaces is presented in the book by Vekua [9].

Here only the basic properties of the boundary value problem (1.0.1) are considered.

At the end of Section 1.3, the problem (1.0.1) is formulated as a Fredholm mapping. Noether's theorem for (1.0.1) and its formal adjoint is formulated by the use of regular bilinear forms in suitable pairs of Hölder-spaces according to Jörgens [6] and Heuser [5].

In Section 1.4, new results on a simple class of semilinear equations are presented generalizing the special boundary value problems considered in [1] and [15].

There the problems

$$\begin{aligned} u_x - v_y &= f_1(x, y, u, v), \\ u_y + v_x &= f_2(x, y, u, v) \end{aligned} \quad \text{in } \mathsf{G},$$

$$u \cos \tau(s) - v \sin \tau(s) = \phi \quad \text{on } \dot{\mathsf{G}}$$

are solved by the use of an embedding and the Schauder–Leray technique.

1.1 The First Boundary Value Problem (see also [3, Section 9.4])

Consider the boundary value problem (BVP)

$$\begin{aligned} &u_x - v_y = Au + Bv + C, \\ &u_y + v_x = \tilde{A}u + \tilde{B}v + \tilde{C}, \\ &u|_{\dot{G}} = \psi. \end{aligned} \tag{1.1.1}$$

Throughout this chapter, G shall denote a bounded, simply connected domain in R^2 with a Hölder continuously differentiable, positively oriented boundary, $\dot{G} \in C^{1+\alpha}$. Sometimes, we shall use the isotropic coordinates

$$z = x + iy, \qquad \bar{z} = x - iy, \tag{1.1.2}$$

and identify R^2 with the complex plane. Demanding that the total differential of a function ϕ satisfies

$$d\phi = \phi_x\, dx + \phi_y\, dy = \phi_z\, dz + \phi_{\bar{z}}\, d\bar{z}, \tag{1.1.3}$$

we obtain the complex directional derivatives

$$\phi_z = \tfrac{1}{2}(\phi_x - i\phi_y), \qquad \phi_{\bar{z}} = \tfrac{1}{2}(\phi_x + i\phi_y), \qquad \phi_x = \phi_z + \phi_{\bar{z}}, \\ \phi_y = i(\phi_z - \phi_{\bar{z}}). \tag{1.1.4}$$

Setting $w = u + iv$, the BVP (1.1.1) can be written in the complex form

$$w_{\bar{z}} = (u + iv)_{\bar{z}} = \bar{a}w + b\bar{w} + c, \qquad \operatorname{Re} w|_{\dot{G}} = \psi, \tag{1.1.5}$$

where

$$a = \tfrac{1}{4}[A + \tilde{B} + i(B - \tilde{A})], \qquad b = \tfrac{1}{4}[A - \tilde{B} + i(\tilde{A} + B)], \qquad c = \tfrac{1}{2}(C + i\tilde{C}). \tag{1.1.6}$$

The method used here for solving this boundary value problem is to convert it to an equivalent integral equation by using the Green functions of the first and second kind (the latter is often called the Neumann function). The Green function of the first kind is determined by the following equations

$$\begin{aligned} &G^I(z, \zeta) = \frac{1}{2\pi} \log \frac{1}{|z - \zeta|} + Z^I(z, \zeta), \quad z, \zeta \in G. \\ &[d, d_n G^I] = 0, \quad z \neq \zeta, \\ &G^I|_{z \in \dot{G}} = 0, \quad \zeta \in G, \end{aligned} \tag{1.1.7}$$

where Z^I denotes a suitable smooth function.

Here, the Pfaffian form d_n is defined by

$$d_n\phi = \phi_x\, dy - \phi_y\, dx = i(\phi_{\bar{z}}\, d\bar{z} - \phi_z\, dz), \tag{1.1.8}$$

and the wedge product of the total differential form d by $d_n\phi$ is

$$[d, d_n\phi] = (\phi_{xx} + \phi_{yy})[dx, dy] = 2i\phi_{z\bar{z}}[dz, d\bar{z}]. \tag{1.1.9}$$

It is well known that G^I exists and is uniquely determined by the above properties. If $\phi(z)$ is a conformal mapping of G onto the unit disk, G^I can be explicitly expressed by

$$G^I(z, \zeta) = -\frac{1}{2\pi}\log\left|\frac{\phi(z)-\phi(\zeta)}{1-\phi(z)\overline{\phi(\zeta)}}\right|. \tag{1.1.10}$$

Warschawski showed in an elegant way that ϕ is in $C^{1+\alpha}(\bar{G})$ and $\phi' \neq 0$ on $\bar{G}$ [13]. Green's function of the second kind satisfies the following equations

$$\begin{aligned} &G^{II}(z, \zeta) = \frac{1}{2\pi}\log\frac{1}{|z-\zeta|} + Z^{II}, \quad z, \zeta \in G, \\ &[d, d_n G^{II}] = 0, \quad z \neq \zeta \\ &d_n G^{II}|_{\dot{G}} = -\Sigma^{-1}\sigma(s)\, ds, \quad \zeta \in G \end{aligned} \tag{1.1.11}$$

where Z^{II} is a smooth function and σ is some positive normalizing function with $\oint_{\dot{G}}\sigma\, ds = \Sigma \neq 0$. It is well known that Green's function of the second kind corresponding to a given normalizing function σ exists and is uniquely determined if we demand the normalizing condition

$$\oint_{\dot{G}} \sigma(s) G^{II}(z(s), \zeta)\, ds = 0, \quad \zeta \in \bar{G}. \tag{1.1.12}$$

The Green function of the second kind corresponding to the special weight function

$$\sigma = \left|\frac{d\phi}{dz}\right|\Big|_{\dot{G}} > 0, \quad \Sigma = 2\pi, \tag{1.1.13}$$

can be expressed explicitly by

$$G^{II}(z, \zeta) = -\frac{1}{2\pi}\log|\{\phi(z)-\phi(\zeta)\}\{1-\phi(z)\overline{\phi(\zeta)}\}|. \tag{1.1.14}$$

But this G^{II} is only one of infinitely many corresponding to different choices of the normalizing function σ (see [3] Section 4.7).

The symmetry of G^I is evident from (1.1.10). Similarly, the symmetry of G^{II} can be seen from an explicit formula (see [3] Section 4.7, formula 4.7.24). The following estimates which hold for G^I as well as G^{II} follow from the properties of ϕ

$$|G| \leq C(1+|\log r|), \quad |G_x| \leq C/r, \quad |G_{xx}| \leq C/r^2,$$
$$rG, \quad r^{1+\alpha}G_x \in C^\alpha(\bar{\mathsf{G}}) \quad (r=|x-y|) \tag{1.1.15}$$

and corresponding estimates for the other derivatives.

Now we find the equivalent integral equations. Let ζ be a fixed point of G and let K be a disk with radius $\rho>0$ and centre ζ which is completely contained in G. Using Gauss' integral theorem, we have

$$\oint_{\dot{\mathsf{G}}\cup(-\dot{\mathsf{K}})} \{u(\mathrm{d}_n G^I - \mathrm{i}\,\mathrm{d}G^{II}) + \mathrm{i}v(\mathrm{d}_n G^{II} - \mathrm{i}\,\mathrm{d}G^I)\}$$
$$= \iint_{\mathsf{G-K}} [\mathrm{d}u, \mathrm{d}_n G^I - \mathrm{i}\,\mathrm{d}G^{II}] + \mathrm{i}[\mathrm{d}v, \mathrm{d}_n G^{II} - \mathrm{i}\,\mathrm{d}G^I]$$
$$= \iint_{\mathsf{G-K}} [u_z\,\mathrm{d}z + u_{\bar{z}}\,\mathrm{d}\bar{z}, -\mathrm{i}G^I_z\,\mathrm{d}z + \mathrm{i}G^I_{\bar{z}}\,\mathrm{d}\bar{z} - \mathrm{i}G^{II}_z\,\mathrm{d}z - \mathrm{i}G^{II}_{\bar{z}}\,\mathrm{d}\bar{z}]$$
$$+\mathrm{i}[v_z\,\mathrm{d}z + v_{\bar{z}}\,\mathrm{d}\bar{z}, -\mathrm{i}G^{II}_z\,\mathrm{d}z + \mathrm{i}G^{II}_{\bar{z}}\,\mathrm{d}\bar{z} - \mathrm{i}G^I_z\,\mathrm{d}z - \mathrm{i}G^I_{\bar{z}}\,\mathrm{d}\bar{z}]$$
$$= \mathrm{i} \iint_{\mathsf{G-K}} \{w_{\bar{z}}(G^I_z + G^{II}_z) + \bar{w}_z(G^I_{\bar{z}} - G^{II}_{\bar{z}})\}[\mathrm{d}z, \mathrm{d}\bar{z}]. \tag{1.1.16}$$

Let us examine the terms in (1.1.16). On the circle $\dot{\mathsf{K}}$

$$G^{I/II}|_{\dot{\mathsf{K}}} = -\frac{1}{4\pi}\log(z-\zeta)(\overline{z-\zeta}) + Z^{I/II} \tag{1.1.17}$$

holds with functions $Z^{I/II}$ harmonic in K. Therefore, the integrand in $\oint_{\dot{\mathsf{K}}}$ becomes

$$\mathrm{d}_n G^{I/II} - \mathrm{i}\,\mathrm{d}G^{II/I}|_{\dot{\mathsf{K}}} = \frac{1}{4\pi}\left\{\mathrm{i}\frac{\mathrm{d}z}{z-\zeta} - \mathrm{i}\frac{\mathrm{d}\bar{z}}{\overline{z-\zeta}} + \mathrm{i}\frac{\mathrm{d}z}{z-\zeta} + \mathrm{i}\frac{\mathrm{d}\bar{z}}{\overline{z-\zeta}}\right\}$$
$$+\mathrm{d}_n Z^{I/II} - \mathrm{i}\,\mathrm{d}Z^{II/I}$$
$$= \frac{\mathrm{i}\,\mathrm{d}z}{2\pi(z-\zeta)} + \text{regular terms}. \tag{1.1.18}$$

We have then

$$-\oint_{\dot{\mathsf{K}}} \{u(\mathrm{d}_n G^I - \mathrm{i}\,\mathrm{d}G^{II}) + \mathrm{i}v(\mathrm{d}_n G^{II} - \mathrm{i}\,\mathrm{d}G^I)\} = -\oint_{|z-\zeta|=\rho} w(z)\frac{\mathrm{i}\,\mathrm{d}z}{2\pi(z-\zeta)}$$
$$+\oint u\{\cdots\} + \oint \mathrm{i}v\{\cdots\}$$
$$= w(\zeta) + o(1), \tag{1.1.19}$$

and the line integral over $\dot{\mathsf{K}}$ approaches $w(\zeta)$ as $\rho \to 0$. The line integral over $\dot{\mathsf{G}}$ is of the form

$$\oint_{\dot{\mathsf{G}}} \{u(\mathrm{d}_n G^I - \mathrm{i}\,\mathrm{d}G^{II}) + \mathrm{i}v\,\mathrm{d}_n G^{II}\} = \oint_{\dot{\mathsf{G}}} \psi(\mathrm{d}_n G^I - \mathrm{i}\,\mathrm{d}G^{II}) - \mathrm{i}\Sigma^{-1}\oint_{\dot{\mathsf{G}}} v\sigma\,\mathrm{d}s = -\theta(\zeta) - \mathrm{i}\Sigma^{-1}\oint_{\dot{\mathsf{G}}} v\sigma\,\mathrm{d}s. \tag{1.1.20}$$

Substituting the relation

$$\mathrm{d}_n G^I - \mathrm{i}\,\mathrm{d}G^{II}\Big|_{\dot{\mathsf{G}}} = \frac{\mathrm{i}}{\pi}\frac{\mathrm{d}\phi(z)}{\phi(z)-\phi(\zeta)}\Big|_{\dot{\mathsf{G}}} + \chi(s)\,\mathrm{d}s \tag{1.1.21}$$

where $\chi(s)$ is a suitable boundary function depending on σ,* into (1.1.20) it follows that θ is holomorphic and satisfies

$$\theta_{\bar{\zeta}} = 0 \quad \text{and} \quad \operatorname{Re}\theta|_{\dot{\mathsf{G}}} = \psi. \tag{1.1.22}$$

Given ψ, θ is a known function.

Now, replacing $w_{\bar{z}}$ and $\bar{w}_z$ in (1.1.16) by the right-hand side of (1.1.5) and its conjugate expression respectively and letting ρ tend to zero gives

$$\begin{aligned} w(\zeta) - \theta(\zeta) - \mathrm{i}\Sigma^{-1}\oint_{\dot{\mathsf{G}}} v\sigma\,\mathrm{d}s &= \mathrm{i}\iint_{\mathsf{G}} \{c(G_z^I + G_z^{II}) + \bar{c}(G_{\bar{z}}^I - G_{\bar{z}}^{II})\}[\mathrm{d}z, \mathrm{d}\bar{z}] \\ &\quad + \mathrm{i}\iint_{\mathsf{G}} \{(\bar{a}w + b\bar{w})(G_z^I + G_z^{II}) + (a\bar{w} + \bar{b}w)(G_{\bar{z}}^I - G_{\bar{z}}^{II})\}[\mathrm{d}z, \mathrm{d}\bar{z}] \\ &\equiv F(\zeta) + \mathsf{K}w. \end{aligned} \tag{1.1.23}$$

Given c, the term $F(\zeta)$ is a known function, while the term $\mathsf{K}w$ contains the unknown w. Therefore, (1.1.16) yields the following integral equation for w

$$w(\zeta) = \mathsf{K}w + \theta(\zeta) + F(\zeta) + \mathrm{i}\Sigma^{-1}\oint_{\dot{\mathsf{G}}} v\sigma\,\mathrm{d}s. \tag{1.1.24}$$

The following theorem shows the equivalence of the boundary value problem (1.1.1) and the integral equation (1.1.24).

* For the special case $\sigma = \|\phi'\|_{\dot{\mathsf{G}}}$ one gets $\chi(s)\,\mathrm{d}s = -\mathrm{i}\,\mathrm{d}\phi/\phi|_{\dot{\mathsf{G}}}$ (see [3] pp. 118 and 122).

Theorem 1.1.1 *Let the coefficients a, b and c in* (1.1.1) *be in the class* $C^\alpha(\bar{G})$ *and let* ψ, $\sigma \in C_R^{1+\alpha}(\dot{G})$, $0<\alpha<1$, *with* $\sigma>0$.*

(*i*) *If w is a continuously differentiable solution of the BVP* (1.1.1), *then w satisfies the integral equation* (1.1.24).

(*ii*) *If w is a continuous solution of the integral equation* (1.1.24), *then w belongs to* $C^{1+\alpha}(\bar{G})$ *and satisfies the differential equations and the boundary condition* (1.1.1).

Proof Conclusion (i) was proven above. For (ii), let us separate (1.1.24) into its real and imaginary parts

$$u(\zeta)=-\oint_{\dot{G}} \psi\, d_n G^I + \iint_G [-(Au+Bv+C)\,dy + (\tilde{A}u+\tilde{B}v+\tilde{C})\,dx, dG^I],$$

$$v(\zeta)=\oint_{\dot{G}} \psi\, dG^{II} - \iint_G [(Au+Bv+C)\,dx + (\tilde{A}u+\tilde{B}v+\tilde{C})\,dy, dG^{II}] + \Sigma^{-1}\oint_{\dot{G}} v\sigma\, ds. \quad (1.1.25)$$

The first term $-\oint_{\dot{G}} \psi\, d_n G^I$ is the harmonic extension of the $C^{1+\alpha}$ boundary values ψ; therefore, it belongs to $C^{1+\alpha}(\bar{G})$ according to Privalov's theorem (see e.g. [2] p. 401). If $u+iv$ is a continuous solution of (1.1.25), then

$$|u(\zeta_1)-u(\zeta_2)| \leq \text{const}\,|\zeta_1-\zeta_2| + \text{const}\iint_G \{|G_x^I(z,\zeta_1)-G_x^I(z,\zeta_2)| + |G_y^I(z,\zeta_1)-G_y^I(z,\zeta_2)|\}[dx, dy]. \quad (1.1.26)$$

The explicit representation (1.1.10) leads to the estimate

$$\begin{aligned}|G_x^I(z,\zeta_1)-G_x^I(z,\zeta_2)| &\leq \text{const}\,(\phi)\cdot\left\{|\zeta_1-\zeta_2|^\alpha\left(\frac{1}{|z-\zeta_1|}+\frac{1}{|z-\zeta_2|}\right) + \left|\frac{1}{|z-\zeta_1|}-\frac{1}{|z-\zeta_2|}\right|\right\}\\ &\leq \text{const}\,(\phi)\left\{|\zeta_1-\zeta_2|^\alpha\left(\frac{1}{|z-\zeta_1|}+\frac{1}{|z-\zeta_2|}\right) + \frac{|\zeta_1-\zeta_2|}{|z-\zeta_1||z-\zeta_2|}\right\}. \quad (1.1.27)\end{aligned}$$

* $C_R^{1+\alpha}(\dot{G})$ denotes the Banach space of real functions in the class $C^{1+\alpha}(\dot{G})$ with scalar field R.

Integrating (1.1.27) over G yields

$$\iint_{G} |G_x^I(z,\zeta_1)-G_x^I(z,\zeta_2)|[dx, dy] \leq \text{const}\,(\phi, G)\cdot|\zeta_1-\zeta_2|^{\alpha}$$

$$+\text{const}\,(\phi, G)|\zeta_1-\zeta_2||\log|\zeta_1-\zeta_2|| \quad (1.1.28)$$

(for the integration of the last term see e.g. [7] Section 15). Hence, we have the inequality

$$\iint_{G} \{|G_x^I(z,\zeta_1)-G_x^I(z,\zeta_2)|+|G_y^I(z,\zeta_1)-G_y^I(z,\zeta_2)|\}[dx, dy]$$

$$\leq \text{const}\,(\phi, G, \alpha)|\zeta_1-\zeta_2|^{\alpha}. \quad (1.1.29)$$

Using (1.1.29) in (1.1.26), we obtain

$$|u(\zeta_1)-u(\zeta_2)| \leq \text{const}\,(u, v, a, \ldots,)\cdot|\zeta_1-\zeta_2|^{\alpha}. \quad (1.1.30)$$

A similar estimate holds for v. Thus, a continuous solution u, v of the integral equation (1.1.25) is Hölder continuous in $\bar{G}$.

Using similar methods, it follows that u and v are Hölder continuously differentiable (see e.g. [3] Section 4.4, Lemma 2). To show that u and v satisfy the differential equations, we substitute (1.1.25) into the right-hand side of Gauss' theorem

$$\iint_{F} (u_x - v_y)[dx, dy] = \oint_{\dot{F}} u\, dy + v\, dx, \quad (1.1.31)$$

for arbitrary subdomains $F \subseteq G$, change the orders of integration, and use some relations found on p. 97 of [3]. Theorem 1.1.1 shows that the solvability properties of the BVP (1.1.1) follow from the solvability properties of the integral equation.

In general, K is *not* linear in the vector space of complex-valued functions over the field of complex numbers. To avoid this difficulty, we consider the space of complex-valued functions over the scalar field of *real* numbers, i.e. only linear combinations $\alpha w_1 + \beta w_2$ with real α, β are allowed. We supply this vector space with the topology of uniform convergence and denote this Banach space by $C^0(\bar{G})$ furnished with the norm

$$\|w\|_0 = \max_{\bar{G}} \sqrt{(u^2+v^2)}. \quad (1.1.32)$$

We now prove that K is a compact mapping of $C^0(\bar{G})$ into itself,

$$K: C^0(\bar{G}) \xrightarrow{\text{compact}} C^0(\bar{G}). \quad (1.1.33)$$

To this end, let $\|w\|_0 \leq 1$. Then

$$|\mathrm{Re}\,[\mathsf{K}w(\zeta_1) - \mathsf{K}w(\zeta_2)]| = \left| \iint_{\mathsf{G}} \{[Au + Bv][G_x^I(z, \zeta_1) - G_x^I(z, \zeta_2)] + [\tilde{A}u + \cdots +]\}[\mathrm{d}x, \mathrm{d}y] \right|$$

$$\leq \mathrm{const}\,(A, B, \tilde{A}, \tilde{B}) \iint_{\mathsf{G}} \{|G_x^I(z, \zeta_1) - G_x^I(z, \zeta_2)| + |G_y^I(z, \zeta_1) - G_y^I(z, \zeta_2)|\}[\mathrm{d}x, \mathrm{d}y]. \quad (1.1.34)$$

Using the estimate (1.1.29) and the same estimates for the imaginary part we obtain the inequality

$$|\mathsf{K}w(\zeta_1) - \mathsf{K}w(\zeta_2)| \leq \mathrm{const}\,(A, \ldots, \tilde{B}, \phi, \mathsf{G}, \alpha) \cdot |\zeta_1 - \zeta_2|^\alpha, \qquad (1.1.35)$$

where the constant is independent of w, $\|w\|_0 \leq 1$. Therefore, K maps the unit ball in $C^0(\bar{\mathsf{G}})$ into a set of uniformly bounded and equi-Hölder-continuous functions. It follows from the Arzelà–Ascoli theorem that this set is compact in $C^0(\bar{\mathsf{G}})$ which proves (1.1.33). *Thus,* (1.1.24) *turns out to be a Fredholm equation of the second kind.*

Let us replace $\Sigma^{-1} \oint_{\dot{\mathsf{G}}} v\sigma\, \mathrm{d}s$ in (1.1.25) by an arbitrary real constant κ and suppose for the moment that (1.1.25) is uniquely solvable for each κ (this is the content of Theorem 1.1.2 below). Denote this solution by u_κ, v_κ. Multiplying the second equation of (1.1.25) by σ and integrating over $\dot{\mathsf{G}}$, we obtain

$$\oint_{\dot{\mathsf{G}}} v_\kappa \sigma\, \mathrm{d}s = \kappa \Sigma, \qquad (1.1.36)$$

by using the normalizing condition (1.1.12) for G^{II}. Hence, by Theorem 1.1.1 u_κ, v_κ is a solution of BVP (1.1.1). This means that there exists a one-dimensional affine subspace of $C^0(\bar{\mathsf{G}})$ consisting of solutions $w_\kappa = u_\kappa + \mathrm{i}v_\kappa$ of BVP (1.1.1).

Theorem 1.1.2 *For any given* $c \in C^\alpha(\bar{\mathsf{G}})$, $\psi \in C^\alpha(\dot{\mathsf{G}})$ (c *complex-valued and* ψ *real-valued*) *and any given real constant* κ, *there exists exactly one continuous solution* $w = u + \mathrm{i}v$ *of the integral equation* (1.1.25) *which satisfies the side condition*

$$\Sigma^{-1} \oint_{\dot{\mathsf{G}}} v\sigma\, \mathrm{d}s = \kappa.$$

In order to prove Theorem 1.1.2, we need the following theorem, which is one statement of the so-called *similarity principle* and goes partly back to Carleman.

Theorem 1.1.3 *If* $w \in C^1(\mathsf{G}) \cap C^\alpha(\bar{\mathsf{G}})$ *is a solution in* G *of the homogeneous system*

$$w_{\bar{z}} = \bar{a}w + b\bar{w}, \tag{1.1.37}$$

then there exists a pair of functions $f, \tilde{w} \in C^\alpha(\bar{\mathsf{G}})$ *with* f *holomorphic in* G, *such that*

$$w(z) = f(z)\mathrm{e}^{\tilde{w}(z)}. \tag{1.1.38}$$

The pair f, $\tilde{w}$ *is uniquely determined, e.g. by the side conditions*

$$\operatorm{Im}\,\tilde{w}\Big|_{\dot{\mathsf{G}}} = \chi \quad \text{and} \quad c_0\Sigma = \oint_{\dot{\mathsf{G}}} \operatorname{Re}\,\tilde{w}\sigma\,\mathrm{d}s, \tag{1.1.39}$$

where $\sigma > 0$ *is a given normalizing function with* $\Sigma = \oint_{\dot{\mathsf{G}}} \sigma\,\mathrm{d}s$, $\chi \in C^\alpha(\dot{\mathsf{G}})$ *is some given function and* $c_0 \in R$ *is some given real constant.* (The side conditions have to be changed if G is multiply connected.)

Remarks For real analytic a and b, this theorem was proved first by Vekua, see [10] p. 83. The similarity principle holds even for rather singular coefficients a, b, as Vekua showed in [11].

Proof The proof, which shall be sketched here, is a slight variation of one given in [2] p. 378 ff. due to Bers and Nirenberg. Hence we omit the details, noting only the different formulations.

Let us define

$$g(z) = \begin{cases} \bar{a} + b\bar{w}/w & \text{if} \quad w \neq 0, \\ 0 & \text{if} \quad w = 0. \end{cases} \tag{1.1.40}$$

Then g is a bounded, Lebesgue measurable function because w is assumed to be continuous. Using G^I and G^{II}, the desired function $\tilde{w}$ is given by

$$\tilde{w} = \mathrm{i}\oint_{\dot{\mathsf{G}}} \chi(-\mathrm{d}_n G^I + \mathrm{i}\,\mathrm{d}G^{II}) + c_0 - \iint_{\mathsf{G}} \{g(G_z^I + G_z^{II}) + \bar{g}(G_{\bar{z}}^I - G_{\bar{z}}^{II})\} \times [\mathrm{d}z, \mathrm{d}\bar{z}] \tag{1.1.41}$$

Using inequality (1.1.29) and the same estimate for G^{II}, (1.1.7) and (1.1.11) respectively, it follows that

$$\tilde{w} \in C^\alpha(\bar{\mathsf{G}}), \qquad \operatorname{Im}\,\tilde{w}|_{\dot{\mathsf{G}}} = \chi, \qquad c_0 = \Sigma^{-1}\oint_{\dot{\mathsf{G}}} \operatorname{Re}\,\tilde{w}\sigma\,\mathrm{d}s. \tag{1.1.42}$$

If we define

$$f(z)=w(z)e^{-\tilde{w}(z)}, \tag{1.1.43}$$

then f is Hölder continuous in $\bar{G}$ and holomorphic in G.
This can be proven exactly as in [2, p. 378].

The uniqueness of f, $\tilde{w}$ can easily be shown by using function theoretic arguments. We omit the proof.

Proof of Theorem 1.1.2 Using the Fredholm alternative (see e.g. [6] Section 5.17), it suffices to show that a solution of the homogeneous integral equation ($\kappa=0$, $\psi\equiv 0$, $c\equiv 0$) vanishes identically. By Theorem 1.1.1, a solution $w=u+iv$ of the homogeneous integral equation satisfies

$$w_{\bar{z}}=\bar{a}w+b\bar{w}, \qquad \mathrm{Re}\, w|_{\dot{G}}=0, \qquad \oint_{\dot{G}} v\sigma\, ds=0. \tag{1.1.44}$$

Using Carleman's Theorem 1.1.3 with $\chi\equiv 0$ and $c_0=0$, w can be represented in the form

$$w(z)=f(z)e^{\tilde{w}(z)}, \tag{1.1.45}$$

where

$$\mathrm{Im}\, \tilde{w}|_{\dot{G}}=0 \text{ (and e.g. } \oint_{\dot{G}} \mathrm{Re}\, \tilde{w}\sigma\, ds=0). \tag{1.1.46}$$

But then

$$\mathrm{Re}\, w|_{\dot{G}}=\mathrm{Re}\,(fe^{\tilde{w}})|_{\dot{G}}=e^{\mathrm{Re}\,\tilde{w}}\, \mathrm{Re}\, f|_{\dot{G}}=0, \tag{1.1.47}$$

and therefore f must be an imaginary constant iv and we have

$$w(z)=ive^{\tilde{w}(z)}. \tag{1.1.48}$$

Finally, it follows that $v=0$ and $w\equiv 0$ since

$$0=\oint_{\dot{G}} v\sigma\, ds=v\oint_{\dot{G}} \mathrm{Im}\,(ie^{\mathrm{Re}\tilde{w}})\sigma\, ds=v\oint_{\dot{G}} e^{\mathrm{Re}\tilde{w}}\sigma\, ds. \tag{1.1.49}$$

Combining Theorems 1.1.2 and 1.1.1, we obtain *Hellwig's existence theorem:*

Theorem 1.1.4 *All solutions of the homogeneous first BVP* (1.1.1) *are of the form*

$$w_0=\nu w_h, \quad \nu \in R, \tag{1.1.50}$$

with $w_h\neq 0$ *in* $\bar{G}$. *For any given* $c \in C^{\alpha}(\bar{G})$ *and* $\psi \in C^{1+\alpha}(\dot{G})$ *there exists a one-dimensional affine set of solutions. A unique solution is obtained by prescribing the boundary norm* κ.

Proof Except for (1.1.50), the proof follows immediately from Theorems 1.1.1 and 1.1.2. To prove (1.1.50), let w_h be the solution of integral equation (1.1.25) with κ in place of $\Sigma^{-1} \oint_{\dot{G}} v\sigma \, ds$ corresponding to $c \equiv 0$, $\psi \equiv 0$, $\kappa = 1$. By Theorem 1.1.1, w_h is a solution of the homogeneous BVP (1.1.1). Again using Theorem 1.1.3 with $\chi \equiv 0$ and $c_0 = 0$, w_h can be represented by

$$w_h(z) = f(z) \, e^{\tilde{w}}, \tag{1.1.51}$$

which leads to

$$w_h(z) = i\nu \, e^{\tilde{w}} \tag{1.1.52}$$

with $\nu \neq 0$ because $\kappa = 1$. Therefore, $w_h \neq 0$ in $\bar{G}$.

Remarks The assumptions on a and b can be weakened without restrictions on the theorems. If a and b are L^p-functions then all theorems remain valid except for statements about the regularity of w_0. If we use generalized derivatives in the sense of Pompeiu as Vekua did in his book [9], then the integral equations (1.1.25) can be used in a similar way for developing the L^p-theory.

1.2 The Boundary Value Problem with Characteristic $\leqslant 0$

First, we remove the function c from

$$w_{\bar{z}} - \bar{a}w - b\bar{w} = c, \qquad \operatorname{Re} e^{i\tau} w|_{\dot{G}} = \psi \tag{1.2.1}$$

by a well-known procedure. If g satisfies

$$g_{\bar{z}} - \bar{a}g - b\bar{g} = c, \qquad \operatorname{Re} g|_{\dot{G}} = 0, \tag{1.2.2}$$

then it suffices to consider the equation for the new unknown function

$$v = w - g: \tag{1.2.3}$$

$$v_{\bar{z}} - \bar{a}v - b\bar{v} = 0, \qquad \operatorname{Re} e^{i\tau} v|_{\dot{G}} = \psi - \operatorname{Re} e^{i\tau} g|_{\dot{G}} \equiv \hat{\psi}. \tag{1.2.4}$$

Hence, without loss of generality, we can restrict the considerations to a homogeneous equation

$$w_{\bar{z}} - \bar{a}w - b\bar{w} = 0, \qquad \operatorname{Re} e^{i\tau} w|_{\dot{G}} = \psi. \tag{1.2.5}$$

Because we are looking for $C^{1+\alpha}(\bar{G})$ solutions, we must assume that

$$e^{i\tau} \in C^{1+\alpha}(\dot{G}). \tag{1.2.6}$$

We shall see that the solvability properties depend in an essential way on the so-called 'characteristic' or 'index' which is defined by the number of revolutions made by the vector $e^{i\tau}$ when the boundary point moves around the boundary in the positive sense

$$n \equiv \frac{1}{2\pi} \oint_{\dot{G}} d\tau = \frac{1}{2\pi} \oint_{\dot{G}} d \arg e^{i\tau}.$$

Using Carleman's Theorem from Section 1.1, we see that a solution w to (1.2.5) can be represented by

$$w = f(z)\hat{w}, \tag{1.2.7}$$

where f is a holomorphic function. For the new unknown function $\hat{w}$, we get the new problem

$$\hat{w}_{\bar{z}} = \bar{a}\hat{w} + b\frac{\bar{f}}{f}\bar{\hat{w}}, \qquad \operatorname{Re} e^{i\tau} f\hat{w}|_{\dot{G}} = \psi. \tag{1.2.8}$$

It was Hilbert's idea to use a factorization of the form (1.2.7) to simplify the boundary condition by a suitable choice of the holomorphic function f.

First, let us look at the simplest case.

(1) $n=0$ Since $n=0$ if and only if $\tau(L)=\tau(0)$, it follows that $\tau \in C^{1+\alpha}(\dot{G})$. Let us define the holomorphic function f by

$$f(z) = e^{i\{\oint \tau(d_n G^I - i\, dG^{II})\}} \equiv e^{\phi(z)}. \tag{1.2.9}$$

Then ϕ and f are holomorphic and $f \in C^{1+\alpha}(\bar{G})$ by Privalov's Theorem. Moreover,

$$f|_{\dot{G}} = e^{\phi|_{\dot{G}}} \equiv e^{-i\tau + s}$$

which implies that

$$\operatorname{Re} \hat{w}|_{\dot{G}} = \hat{\psi} \equiv \psi\, e^{-s}. \tag{1.2.10}$$

Using the results of Section 1.1, we have

Theorem 1.2.1 *If* $n=0$, *the set of solutions to* (1.2.5) (*with* $\psi \in C_R^{1+\alpha}(\dot{G})$) *is a one-dimensional affine subspace* (*of* $C^{1+\alpha}(\bar{G})$) *and of the form*

$$w = w_p + \nu w_h, \quad \nu \in R,$$

and $w_h \neq 0$ *in* $\bar{G}$.

Next, we consider the case of negative characteristics.

(2) $n<0$ For motivation, let us first look at the homogeneous problem $\psi=0$. Then the boundary condition

$$\operatorname{Re} e^{i\tau} w|_{\dot{G}} = 0$$

means for $w|_{\dot{G}} \neq 0$, that $w \perp e^{-i\tau}$ on $\dot{G}$ and, therefore, $w|_{\dot{G}}$ must perform $|n|$ revolutions. Because $w = fe^{\tilde{w}}$, f also performs $|n|$ revolutions. By the well-known function theoretic argument principle, f must have $|n|$ zeros in G. This suggests the factorization

$$w = \prod_{j=1}^{|n|} (z - z_j)\hat{w}, \qquad z_1, \ldots, z_{|n|} \in G. \tag{1.2.11}$$

The new differential equation for $\hat{w}$ is

$$\hat{w}_{\bar{z}} = \bar{a}\hat{w} + b \prod_{j=1}^{|n|} \frac{(\bar{z} - \bar{z}_j)}{(z - z_j)} \bar{\hat{w}}. \tag{1.2.12}$$

The coefficient of $\bar{\hat{w}}$ has bounded singularities at the points z_j, but according to the remark at the end of the first section, all theorems of that section maintain their validity with such coefficients. The new boundary condition is

$$\operatorname{Re} e^{i\hat{\tau}} \hat{w}|_{\dot{G}} \equiv \operatorname{Re} e^{i\tau} \prod_{j=1}^{|n|} \frac{(z - z_j)}{|z - z_j|} \hat{w}|_{\dot{G}} = \hat{\psi} \equiv \psi \prod_{j=1}^{|n|} \frac{1}{|z - z_j|_{\dot{G}}} \tag{1.2.13}$$

with characteristic zero, since

$$\frac{1}{2\pi} \oint_{\dot{G}} d\hat{\tau} = \frac{1}{2\pi} \oint_{\dot{G}} d\tau + \frac{1}{2\pi} \sum_{j=1}^{|n|} \oint_{\dot{G}} d \arg (z - z_j) = -|n| + |n| = 0.$$

It follows from Theorem 1.2.1 that

Theorem 1.2.2 *For any given $\psi \in C_R^{1+\alpha}(\dot{G})$ and any chosen zeros $z_1, \ldots, z_{|n|}$, there exists a set of solutions to* (1.2.5) *of the form*

$$w = \prod_{j=1}^{|n|} (z - z_j)\{\hat{w}_p + \nu \hat{w}_h\}, \qquad \nu \in R, \ \hat{w}_h \neq 0 \quad \text{in} \quad \bar{G}.$$

Obviously, this family of solutions depends on $2|n| + 1 = -2n + 1$ real constants.

Let $X = C^{1+\alpha}(\bar{G})$ and $Y = C^{\alpha}(\bar{G}) \times C_R^{1+\alpha}(\dot{G})$.
Consider the linear operator $R: X = C^{1+\alpha}(\bar{G}) \to C^{\alpha}(\bar{G}) \times C_R^{1+\alpha}(\dot{G}) = Y$ defined by

$$Rw = (w_{\bar{z}} - \bar{a}w - b\bar{w}, \qquad \operatorname{Re} e^{i\tau} w|_{\dot{G}}). \tag{1.2.14}$$

Obviously, R is continuous.

The solution set of $Rw = \chi$ must be an affine subspace of X. Theorem 1.2.2 suggests that $\dim N(R) = -2n + 1$.

To prove this, we first establish

Theorem 1.2.3 *In case $n<0$, for any set of points $z_1, \ldots, z_{|n|} \in \mathrm{G}$, there exist $2|n|+1$ nontrivial solutions $w_0, \ldots, w_{2|n|}$ of the homogeneous boundary value problem* (1.2.5) *with*

$$w_0(z_j)=0$$
$$w_{2k}(z_j)=\delta_{jk}, \qquad w_{2k-1}(z_j)=\mathrm{i}\delta_{jk}, \qquad j, k=1, \ldots, |n|. \tag{1.2.15}$$

Proof We obtain w_0 from Theorem 1.2.2 with $\psi=0$, $\nu \neq 0$. Now let us consider w_1 and w_2. To find them, we need two solutions $\tilde{w}_1$, $\tilde{w}_2$ of the form

$$\tilde{w}_{1,2}=\prod_{j=2}^{|n|}(z-z_j)v_{1,2} \tag{1.2.16}$$

which are linearly independent at z_1, i.e. $\lambda\tilde{w}_1(z_1)+\mu\tilde{w}_2(z_1)=0$, λ, $\mu \in R$ implies that $\lambda=\mu=0$. For v, it follows that

$$v_{\bar{z}}-\bar{a}v-\prod_{j=2}^{|n|}\frac{\overline{(z-z_j)}}{(z-z_j)}b\bar{v}=0 \tag{1.2.17}$$

and

$$\mathrm{Re}\left(\mathrm{e}^{\mathrm{i}\tau}\prod_{j=2}^{|n|}\frac{(z-z_j)}{|z-z_j|}v|_{\dot{\mathrm{G}}}\right)=0 \tag{1.2.18}$$

which is a BVP with characteristic -1. We now try to find two linearly independent solutions to (1.2.17) and (1.2.18). Let v_a, $v_b \neq 0$ be solutions to (1.2.17) which satisfy the boundary conditions of the first kind

$$\begin{aligned} &\mathrm{Re}\, v_a|_{\dot{\mathrm{G}}}=0, \qquad \mathrm{Im}\, v_b|_{\dot{\mathrm{G}}}=0, \\ &\mathrm{Im}\, v_a|_{\dot{\mathrm{G}}}>0, \qquad \mathrm{Re}\, v_b|_{\dot{\mathrm{G}}}>0. \end{aligned} \tag{1.2.19}$$

It follows that v_a and v_b are linearly independent at z_1. To show this, let

$$\mu v_a(z_1)+\lambda v_b(z_1)=0 \quad \text{with} \quad \mu, \lambda \in R,$$

and let

$$v_0=\mu v_a+\lambda v_b.$$

Then v_0 has a zero at z_1. On the other hand, v_0 solves a BVP with characteristic zero because

$$\mathrm{Re}\left(\frac{(i\bar{v}_0)}{|v_0|}v_0\right)|_{\dot{\mathrm{G}}}=0$$

provided $\mu^2+\lambda^2\neq 0$. This follows because

$$\mathrm{i}\bar{v}_0|_{\dot{G}}=\mu \operatorname{Im} v_a+\mathrm{i}\lambda \operatorname{Re} v_b$$

is restricted to a half-plane and consequently makes no revolutions. But the nontrivial solution of a homogeneous problem with characteristic 0 has no zeros. Therefore, only $\lambda=\mu=0$ is possible.

Since the functions v_a, v_b do not fulfill the boundary condition (1.2.18), we add solutions v_α, v_β to (1.2.17) with zeros at z_1:

$$v_\alpha : v_\alpha(z_1)=0, \quad \operatorname{Re}\left(e^{\mathrm{i}\tau}\prod_{j=2}^{|n|}\frac{(z-z_j)}{|z-z_j|}\, v_\alpha|_{\dot{G}}\right)=-\operatorname{Re}\left(e^{\mathrm{i}\tau}\prod_{j=2}^{|n|}\frac{(z-z_j)}{|z-z_j|}\, v_a|_{\dot{G}}\right), \tag{1.2.20}$$

$$v_\beta : v_\beta(z_1)=0, \quad \operatorname{Re}\left(e^{\mathrm{i}\tau}\prod_{j=2}^{|n|}\frac{(z-z_j)}{|z-z_j|}\, v_\beta|_{\dot{G}}\right)=-\operatorname{Re}\left(e^{\mathrm{i}\tau}\prod_{j=2}^{|n|}\frac{(z-z_j)}{|z-z_j|}\, v_b|_{\dot{G}}\right).$$

Then, $v_1=v_a+v_\alpha$ and $v_2=v_b+v_\beta$ are two solutions of (1.2.17) and (1.2.18), which are linearly independent at z_1. Consequently, the functions

$$\tilde{w}_{1,2}\equiv\prod_{j=2}^{|n|}(z-z_j)v_{1|2} \tag{1.2.21}$$

are also linearly independent at z_1. Then each of the complex equations

$$\lambda\tilde{w}_1(z_1)+\mu\tilde{w}_2(z_1)=\begin{cases}1\\ \mathrm{i}\end{cases}$$

has exactly one real pair $\{{\lambda,\mu \atop \hat\lambda,\hat\mu}\}$ of solutions and

$$w_2=\lambda\tilde{w}_1+\mu\tilde{w}_2,\ w_1=\hat{\lambda}\tilde{w}_1+\hat{\mu}\tilde{w}_2 \tag{1.2.22}$$

are the desired functions. Theorem 1.2.3 is proved, if we interchange z_1 and z_2 and so on.

Now, let us show that the above constructed set of solutions of (1.2.5) *forms a basis for the null space of* R. First, we examine the linear independence of this set.

Suppose that there exist $\lambda_0,\ldots,\lambda_{2|n|}\in R$ such that

$$\sum_{j=0}^{2|n|}\lambda_j w_j=0, \quad \text{for all} \quad z\in\bar{G}.$$

Then setting $z=z_k$, it follows from (1.2.15) that $\lambda_{2k}+\mathrm{i}\lambda_{2k-1}=0$ which implies that $\lambda_1=\lambda_2=\cdots=\lambda_{2|n|}=0$. Therefore,

$$\lambda_0 w_0 \equiv 0,$$

but w_0 vanishes only at $z=z_k$, $k=1,\ldots,|n|$. Hence, $\lambda_0=0$.

Now, let $w(z)$ be any solution of the homogeneous problem. Then we calculate real constants $\lambda_1,\ldots,\lambda_{2|n|}$ from the equations

$$\lambda_{2k}+\mathrm{i}\lambda_{2k-1}=\lambda_{2k}w_{2k}(z_k)+\lambda_{2k-1}w_{2k-1}(z_k)=w(z_k), \qquad k=1,\ldots,|n|.$$

Then

$$\tilde{w}(z)=w(z)-\sum_{j=1}^{2|n|}\lambda_j w_j(z)$$

is again a solution of the homogeneous problem, but now with the zeros z_j, $j=1,\ldots,|n|$. From Carleman's Theorem 1.1.3, we know that $\tilde{w}$ can be represented by

$$\tilde{w}=\sum_{j=1}^{|n|}(z-z_j)\hat{w},$$

where $\hat{w}$ solves the BVP (1.2.12) and (1.2.13) with $\psi\equiv 0$ which has characteristic zero and for which Theorem 1.2.1 holds. This means that

$$\hat{w}=\lambda\hat{w}_h.$$

On the other hand,

$$w_0=\nu\prod_{j=1}^{|n|}(z-z_j)\hat{w}_h, \qquad \nu\neq 0.$$

It follows that

$$\hat{w}=\lambda_0 w_0, \qquad \lambda_0=\lambda/\nu.$$

Altogether, we have the identity

$$w(z)=\sum_{j=0}^{2|n|}\lambda_j w_j(z) \quad \text{for all} \quad z\in\bar{\mathsf{G}}. \tag{1.2.23}$$

Hence $w_0,\ldots,w_{2|n|}$ form a basis for $\mathsf{N}(\mathsf{R})$.

We have proven that

$$\alpha\equiv\dim N(\mathsf{R})=1-2n<\infty, \tag{1.2.24}$$

and for all (c,ψ) there exists $w\in C^{1+\alpha}(\bar{\mathsf{G}})$ such that $\mathsf{R}w=(c,\psi)$. This means that the range of R is $C^{\alpha}(\bar{\mathsf{G}})\times C^{1+\alpha}_{\mathsf{R}}(\dot{\mathsf{G}})=Y$. Hence $R(\mathsf{R})$ is closed and

$$\beta\equiv\operatorname{codim} R(\mathsf{R})=0. \tag{1.2.25}$$

Thus R is a 'Noethernian' or 'Fredholm' operator. The difference

$$\nu = \alpha - \beta = -2n + 1 = 1 - \frac{1}{\pi} \oint_{\dot{\mathsf{G}}} \mathrm{d} \arg \mathrm{e}^{\mathrm{i}\tau} \tag{1.2.26}$$

is the so-called *index* of the linear operator R. Atkinson showed that the index is stable for arbitrary linear compact perturbations and for small continuous perturbations. Also, an equation involving a Fredholm operator is normally solvable, that is

$$\mathsf{R}w = (c, \psi)$$

has a solution if and only if the right-hand side is orthogonal to the null space of the adjoint operator (see e.g. [5]). But, in our case we have

$$\beta = \dim N(\mathsf{R}^*) = 0$$

and, therefore, no solvability conditions are needed.

If we assume that the index formula (1.2.26) also holds for positive characteristic, then for $n > 0$, $\nu < 0$, and the number of solvability conditions β must be greater than zero. We show this in the next section.

Since in the case $n \leqslant 0$ the solution exists but is nonunique and since the homogeneous problem has a basis characterized by Theorem 1.2.3, one may add $-2n+1$ suitable side conditions to obtain a uniquely solvable well posed problem.

Theorem 1.2.4 *Let the points $z_1, \ldots, z_{|n|} \in \mathsf{G}$ be chosen as well as the weight function $\sigma > 0$ on $\dot{\mathsf{G}}$ and fixed in the following. Then the boundary value problem*

$$\begin{aligned} w_{\bar{z}} - \bar{a}w - b\bar{w} &= c \quad \text{in} \quad \mathsf{G}, \\ \operatorname{Re} \mathrm{e}^{\mathrm{i}\tau} w &= \psi \quad \text{on} \quad \dot{\mathsf{G}}, \\ w(z_j) &= \gamma_j \quad \text{for} \quad j = 1, \ldots, |n|, \\ \oint \operatorname{Im} \mathrm{e}^{\mathrm{i}\tau} w \sigma \, \mathrm{d}s &= \kappa \end{aligned} \tag{1.2.27}$$

is uniquely solvable for every given right-hand side

$$(c, \psi, \gamma_1, \ldots, \gamma_{|n|}, \kappa) \in C^\alpha(\bar{\mathsf{G}}) + C^{1+\alpha}(\dot{\mathsf{G}}) \times \mathbb{C}^{|n|} \times R \equiv Z.$$

The proof follows directly from Theorem 1.2.3 and the representation for w_0,

$$w_0 = \prod_{j=1}^{|n|} (z - z_j) \exp\left\{\oint_{\dot{\mathsf{G}}} \tilde{\tau}(\mathrm{i}\, \mathrm{d}_n G^I + \mathrm{d}G^{II})\right\} \hat{w}_0 = f(z)\hat{w}_0 \tag{1.2.28}$$

where $\tilde{\tau}$ is defined by

$$e^{i\tilde{\tau}} = e^{i\tau} \prod_{j=1}^{|n|} \frac{(z - z_j)}{|z - z_j|} \tag{1.2.29}$$

and $\hat{w}_0$ solves a homogenous system and the boundary condition

$$\operatorname{Re} \hat{w}_0 = 0 \quad \text{on} \quad \dot{\mathsf{G}}. \tag{1.2.30}$$

The general solution of the first two equations in (1.2.27) takes the form

$$w = w_p + \sum_{j=0}^{2|n|} \lambda_j w_j$$

with w_j satisfying (1.2.15). The point conditions in (1.2.27) yield

$$\lambda_{2k} + i\lambda_{2k-1} = \gamma_k - w_p(z_k), \qquad k = 1, \ldots, |n|;$$

and from (1.2.28–30) the last condition in (1.2.27) determines λ_0 according to Theorem 1.1.4,

$$\lambda_0 \oint_{\dot{\mathsf{G}}} \operatorname{Im} \hat{w}_0 \prod_{j=1}^{|n|} |z - z_j| \exp\left\{\oint_{\dot{\mathsf{G}}} \tilde{\tau}\, dG^{II}\right\} \sigma\, ds$$
$$= \kappa - \oint_{\dot{\mathsf{G}}} \operatorname{Im} e^{i\tau}\left(w_p + \sum_{j=1}^{n} \lambda_j w_j\right) \sigma\, ds$$

where for $\hat{w}_0$ the weight function $\prod_{j=1}^{|n|} |z - z_j|\, \sigma \exp\{\oint_{\dot{\mathsf{G}}} \tilde{\tau}\, dG^{II}\}$ has to be used.

The problem (1.2.27) is also *well posed*. Since the mapping from $w \in C^{1+\alpha}(\bar{\mathsf{G}})$ into Z, defined by the left-hand side of (1.2.27), is a linear continuous bijective mapping, the closed graph theorem ([5] p. 321, Theorem 80.1) implies that the inverse mapping is *continuous*, too.

Thus, for the solution w of (1.2.27) the following estimate is valid:

$$\|w\|_{C^{1+\alpha}(\bar{\mathsf{G}})} \leq k\{\|c\|_{C^\alpha(\bar{\mathsf{G}})} + |\psi|_{C^{1+\alpha}(\dot{\mathsf{G}})} + \sum_{j=1}^{|n|} |\gamma_j| + |\kappa|\} \tag{1.2.31}$$

Here, the constant k depends on G, a, b, τ, σ and z_j, $j = 1, \ldots, |n|$.

In the following Theorem we shall prove another *a priori* estimate where the constant depends only on an upper bound of $|a|$ and $|b|$. This *a priori* estimate will prove to be rather useful for some semilinear problems in Section 1.4.

Theorem 1.2.5 *For all $h \in C^1(\bar{\mathrm{G}})$ satisfying the homogeneous boundary and side conditions*

$$\begin{aligned} &\operatorname{Re} \mathrm{e}^{\mathrm{i}\tau} h|_{\dot{\mathrm{G}}} = 0, \\ &h(z_j) = 0, \quad j = 1, \ldots, |n|, \\ &\oint_{\dot{\mathrm{G}}} \operatorname{Im} \mathrm{e}^{\mathrm{i}\tau} h\sigma \, \mathrm{d}s = 0, \end{aligned} \tag{1.2.32}$$

the a priori *estimate*

$$\|h\|_0 \leq \gamma \|h_{\bar{z}} - \bar{a}h - b\bar{h}\|_0 \tag{1.2.33}$$

holds for any a, b with $\|a\|_0$, $\|b\|_0 \leq K$ and where the constant γ depends only on K, G, *σ, τ and z_j but not on a, b or h.*

For the special case $\tau \equiv 0$ and $n = 0$ this *a priori* estimate was proven in [15].

Using $\|a\|_\alpha$, $\|b\|_\alpha \leq K$ instead of the sup-norms, one can even prove an *a priori* estimate

$$\|h\|_{C^{1+\alpha}(\bar{\mathrm{G}})} \leq \gamma' \|h_{\bar{z}} - \bar{a}h - b\bar{h}\|_{C^\alpha(\bar{\mathrm{G}})} \tag{1.2.34}$$

of the same kind. But here the estimate (1.2.31) for $a = b = 0$ would be used where k was not found in a constructive way. Therefore let us present another approach where the constant γ is found in a constructive manner.

Proof First let us prove (1.2.33) in the special case of $a = b = 0$. To this end let us write h in the form

$$h = h_1 + h_2 \tag{1.2.35}$$

where

$$h_2 = \sum_{j=1}^{|n|} h_{\bar{z}}(z_j) P_j(\bar{z}) \tag{1.2.36}$$

and the P_j are polynomials in $\bar{z}$ of order $2|n|$ satisfying

$$P_j(\bar{z}_k) = 0, \qquad \frac{\mathrm{d}}{\mathrm{d}\bar{z}} P_j(\bar{z}_k) = \delta_{jk}; \qquad j, k = 1, \ldots, |n|. \tag{1.2.37}$$

Then the function h_1 can be written as

$$h_1 = \prod_{j=1}^{|n|} (z - z_j) \mathrm{e}^{\Phi(z)} \tilde{h} \tag{1.2.38}$$

where the holomorphic function $\Phi(z)$ is given by

$$\Phi = \mathsf{T} + \mathrm{i}\mathsf{F} \equiv \oint_{\dot{\mathsf{G}}} \tilde{\tau}(\mathrm{i}\,\mathrm{d}_n G^I + \mathrm{d}G^{II}), \tag{1.2.39}$$

with $\tilde{\tau}$ from (1.2.29).

The function $\tilde{h}$ satisfies

$$\begin{aligned} &\operatorname{Re} \tilde{h}|_{\dot{\mathsf{G}}} = -\prod_{j=1}^{|n|} |z - z_j|^{-1} \mathrm{e}^{-T} \operatorname{Re} \mathrm{e}^{\mathrm{i}\tau} h_2|_{\dot{\mathsf{G}}}, \\ &\oint_{\dot{\mathsf{G}}} \operatorname{Im} \tilde{h} \left\{ \prod_{j=1}^{|n|} |z - z_j|\, \mathrm{e}^{T} \sigma \right\} \mathrm{d}s = -\oint_{\dot{\mathsf{G}}} \operatorname{Im} \mathrm{e}^{\mathrm{i}\tau} \sigma h_2 \,\mathrm{d}s \end{aligned} \tag{1.2.40}$$

and is continuously differentiable. Hence $\tilde{h}$ can be represented by (1.1.23),

$$\begin{aligned} \tilde{h}(\zeta) = {} & \mathrm{i} \iint_{\mathsf{G}} \{\tilde{h}_{\bar{z}}(G^I_z + G^{II}_z) + \bar{\tilde{h}}_z(G^I_{\bar{z}} - G^{II}_{\bar{z}})\}[\mathrm{d}z, \mathrm{d}\bar{z}] \\ & + \oint_{\dot{\mathsf{G}}} \prod_{j=1}^{|n|} |z - z_j|^{-1} \mathrm{e}^{-T} \left\{ \operatorname{Re} \mathrm{e}^{\mathrm{i}\tau} \sum_{k=1}^{|n|} h_{\bar{z}}(z_k) P_k(\bar{z}) \right\} (\mathrm{d}_n G^I - \mathrm{i}\,\mathrm{d}G^{II}) \\ & - \mathrm{i} \left\{ \oint_{\dot{\mathsf{G}}} \prod_{j=1}^{n} |z - z_j|\, \mathrm{e}^{T} \sigma \,\mathrm{d}s \right\}^{-1} \oint_{\dot{\mathsf{G}}} \operatorname{Im} \mathrm{e}^{\mathrm{i}\tau} \sigma \sum_{k=1}^{|n|} h_{\bar{z}}(z_k) P_k(\bar{z}) \,\mathrm{d}s \end{aligned} \tag{1.2.41}$$

where G^{II} corresponds to the weight function $\prod_{j=1}^{|n|} |z - z_j|\, \mathrm{e}^{T} \sigma$.

Inserting (1.2.38), (1.2.35) and (1.2.36) we end up with the representation formula

$$\begin{aligned} h(\zeta) = {} & \sum_{j=1}^{n} h_{\bar{z}}(z_j) P_j(\bar{\zeta}) \\ & + \mathrm{i} \iint_{\mathsf{G}} \left\{ (h_{\bar{z}} - h_{2\bar{z}}) \prod_{j=1}^{|n|} \frac{(\zeta - z_j)}{(z - z_j)} \exp\,(\phi(\zeta) - \phi(z))(G^I_z + G^{II}_z) \right. \\ & \left. + (\bar{h}_z - \bar{h}_{2z}) \prod_{j=1}^{|n|} \frac{\overline{(\zeta - z_j)}}{\overline{(z - z_j)}} \exp\,(\overline{\phi(\zeta)} - \overline{\phi(z)})(G^I_{\bar{z}} - G^{II}_{\bar{z}}) \right\} [\mathrm{d}z, \mathrm{d}\bar{z}] \\ & + \prod_{j=1}^{|n|} (\zeta - z_j) \exp \phi(\zeta) \left\{ \oint_{\dot{\mathsf{G}}} \prod_{l=1}^{|n|} |z - z_l|^{-1}\, \mathrm{e}^{-T} \{ \operatorname{Re} \mathrm{e}^{\mathrm{i}\tau} \sum_{k=1}^{|n|} h_{\bar{z}}(z_k) P_k(\bar{z}) \} \right. \\ & \qquad \times (\mathrm{d}_n G^I - \mathrm{i}\,\mathrm{d}G^{II}) \\ & \left. - \mathrm{i} \left\{ \oint_{\mathsf{G}} \prod_{l=1}^{|n|} |z - z_l|\, \mathrm{e}^{T} \sigma \,\mathrm{d}s \right\}^{-1} \oint_{\dot{\mathsf{G}}} \operatorname{Im} \mathrm{e}^{\mathrm{i}\tau} \sum_{k=1}^{|n|} h_{\bar{z}}(z_k) P_k(\bar{z}) \,\mathrm{d}s \right\}. \end{aligned} \tag{1.2.42}$$

Although $G_z^I + G_z^{II}$ has a singularity of the form $|z-\zeta|^{-1}$, for the integrals the inequality

$$\prod_{j=1}^{|n|} |\zeta - z_j| \iint_G |z-z_j|^{-1}|\zeta - z|^{-1}[dx, dy] \leq k' \prod_{j=1}^{|n|} |\zeta - z_j|\{1 + |\log|\zeta - z_j||\} \tag{1.2.43}$$

can be used (see Michlin [7] Satz 2, p. 83). Hence (1.2.42) yields the inequality

$$\|h\|_0 \leq \hat{\gamma}\|h_{\bar{z}}\|_0. \tag{1.2.44}$$

For the more general proposition (1.2.33) let us introduce a function v defined by

$$v_{\bar{z}} = \begin{cases} \bar{a} + b\frac{\bar{h}}{h} & \text{where} \quad h \neq 0 \\ \bar{a} & \text{where} \quad h = 0 \end{cases} \quad \text{and}$$

$$\operatorname{Im} v|_{\dot{G}} = 0, \qquad \oint_{\dot{G}} \operatorname{Re} v\sigma \, ds = 0. \tag{1.2.45}$$

Then inequality (1.2.44) in the special case $n = 0$, $\tau \equiv 0$ implies for v by the use of $\|a\|_0$, $\|b\|_0 \leq K$ the estimate

$$\|v\|_0 \leq 2\gamma' K. \tag{1.2.46}$$

Setting

$$f \equiv h \exp(-v) - \alpha\phi_0 \tag{1.2.47}$$

where ϕ_0 is the holomorphic solution of the boundary value problem

$$\phi_{0\bar{z}} = 0, \qquad \operatorname{Re} e^{i\tau}\phi_0|_{\dot{G}} = 0, \qquad \phi_0(z_j) = 0, \qquad \oint_{\dot{G}} \operatorname{Im} e^{i\tau}\phi_0\sigma \, ds = 1 \tag{1.2.48}$$

and α is chosen as

$$\alpha \equiv \oint_{\dot{G}} \operatorname{Im}(h \exp(-v) e^{i\tau}\sigma) \, ds \tag{1.2.49}$$

one can easily show that f satisfies the boundary value problem

$$f_{\bar{z}} = c \exp(-v), \qquad \operatorname{Re} e^{i\tau} f|_{\dot{G}} = 0, \tag{1.2.50}$$

$$f(z_j) = 0, \quad j = 1, \ldots, |n|, \qquad \oint_{\dot{G}} \operatorname{Im}(e^{i\tau} f\sigma) \, ds = 0$$

where

$$c = h_{\bar{z}} - \bar{a}h - b\bar{h}. \tag{1.2.51}$$

Then it follows from (1.2.42) for f and with (1.2.46) that

$$\|f\|_0 \leq \hat{\gamma}\|c\|_0 \exp(2\gamma' K) \tag{1.2.52}$$

holds. (1.2.47) yields

$$\begin{aligned}\|h\|_0 &\leq \exp(2\gamma' K)\{\|f\|_0 + |\alpha|\, \|\phi_0\|_0\} \\ &\leq \hat{\gamma} \exp(4\gamma' K)\|c\|_0 + \|\phi_0\|_0 \exp(2\gamma' K)|\alpha|.\end{aligned} \tag{1.2.53}$$

On the other hand, (1.2.47) implies for α the relation

$$\alpha = -\left\{\oint (\operatorname{Im} \phi_0 e^{i\tau})(\exp v)\sigma \, ds\right\}^{-1} \oint \operatorname{Im} e^{i\tau} f(\exp v)\sigma \, ds. \tag{1.2.54}$$

The nominator cannot vanish since $\phi_0 e^{i\tau}|_{\dot{G}} = \operatorname{Im} \phi_0 e^{i\tau}|_{\dot{G}} \neq 0$ according to Theorem 1.2.2.

Inserting (1.2.54) into (1.2.53) and estimating $\|f\|_0$ by (1.2.52) we find the desired inequality

$$\|h\|_0 \leq \{\hat{\gamma} \exp(4\gamma' K) + \|\phi_0\|_0 \exp(10\gamma' K)\gamma(\phi_0)\}\|c\|_0. \tag{1.2.55}$$

1.3 The Boundary Value Problem of Positive Characteristic

We now consider BVP (1.2.5) in case

$$n = \frac{1}{2\pi} \oint_{\dot{G}} d\tau > 0.$$

First, let us look for solutions with poles. If $z_1, \ldots, z_n \in G$ and the factorization

$$w = \prod_{j=1}^{n} \frac{1}{(z - z_j)} \hat{w} \tag{1.3.1}$$

is substituted into (1.2.5) we obtain the BVP

$$\hat{w}_{\bar{z}} = \bar{a}\hat{w} + \left\{b \prod_{j=1}^{n} \frac{\overline{(z - z_j)}}{(z - z_j)}\right\} \bar{\hat{w}},$$

$$\operatorname{Re}\left(e^{i\tau} \prod_{j=1}^{n} \frac{|z - z_j|}{(z - z_j)} \hat{w}\bigg|_{\dot{G}}\right) = \hat{\psi}$$

for $\hat{w}$ which has characteristic 0. Hence, all solutions of this BVP are of the form

$$\hat{w} = \hat{w}_p + \nu \hat{w}_h, \quad \nu \in R, \quad \hat{w}_h \neq 0,$$

and all solutions of (1.2.5) with $n>0$ are of the form

$$w=\left(\prod_{j=1}^{n}\frac{1}{(z-z_j)}\right)\{\hat{w}_p+\nu\hat{w}_h\},\ \nu\in R. \tag{1.3.2}$$

Let us now consider the question of uniqueness for regular solutions to $\mathsf{R}w=0$. Then w has the form (1.3.2) with $\hat{w}_p=0$.

In order for w to be in class C^1, it is necessary that $\nu\hat{w}_h(z_j)=0$ for all j. Hence $\nu=0$ and $w\equiv 0$ in $\bar{\mathsf{G}}$.

Next, we consider solvability conditions for regular solutions. For a moment let us look at the inhomogeneous equation $\mathsf{R}w=(c,\psi)$ where R is the operator defined by (1.2.14). If w is a regular solution and g is regular, then by Gauss' theorem

$$\begin{aligned}2\mathrm{Re}\oint_{\dot{\mathsf{G}}} wg\,\mathrm{d}z &= \oint_{\dot{\mathsf{G}}} wg\,\mathrm{d}z+\bar{w}\bar{g}\,\mathrm{d}\bar{z}=\iint_{\mathsf{G}}[\mathrm{d},\, wg\,\mathrm{d}z+\bar{w}\bar{g}\,\mathrm{d}\bar{z}]\\
&=\iint_{\mathsf{G}}\{-w_{\bar{z}}g-wg_{\bar{z}}+\bar{w}_z\bar{g}+\bar{w}\bar{g}_z\}[\mathrm{d}z,\mathrm{d}\bar{z}]\\
&=\iint_{\mathsf{G}}\{-(\bar{a}w+b\bar{w}+c)g-wg_{\bar{z}}+(a\bar{w}+\bar{b}w+\bar{c})\bar{g}+\bar{w}\bar{g}_z\}\\
&\qquad\times[\mathrm{d}z,\mathrm{d}\bar{z}]\\
&=\iint_{\mathsf{G}}\{-w(g_{\bar{z}}+\bar{a}g-\bar{b}\bar{g})+\bar{w}(\bar{g}_z+a\bar{g}-bg)\}[\mathrm{d}z,\mathrm{d}\bar{z}]\\
&\quad+\iint_{\mathsf{G}}(-cg+\bar{c}\bar{g})[\mathrm{d}z,\mathrm{d}\bar{z}].\end{aligned}\tag{1.3.3}$$

Hence we have

Proposition 1.3.1 *If w is a regular solution to $\mathsf{R}w=(c,\psi)$ and g is a regular solution to the adjoint equation*

$$g_{\bar{z}}=-\bar{a}g+\bar{b}\bar{g}, \tag{1.3.4}$$

then

$$2\mathrm{Re}\oint_{\dot{\mathsf{G}}} wg\,\mathrm{d}z=\iint_{\mathsf{G}}(-cg+\bar{c}\bar{g})[\mathrm{d}z,\mathrm{d}\bar{z}].$$

Now if g satisfies the homogeneous adjoint boundary condition

$$\operatorname{Re}\left(-\mathrm{i}e^{-\mathrm{i}\tau}\frac{\mathrm{d}z}{\mathrm{d}s}g\right)\Big|_{\dot{G}}=0, \tag{1.3.5}$$

then

$$\begin{aligned}2\operatorname{Re}\oint_{\dot{G}} wg\,\mathrm{d}z &= 2\operatorname{Re}\oint_{\dot{G}} e^{\mathrm{i}\tau}w\cdot e^{-\mathrm{i}\tau}g\,\mathrm{d}z\\ &= 2\oint_{\dot{G}}\Big\{\operatorname{Re}(e^{\mathrm{i}\tau}w)\cdot\operatorname{Re}\left(e^{-\mathrm{i}\tau}\frac{\mathrm{d}z}{\mathrm{d}s}g\right)\mathrm{d}s\\ &\qquad\qquad -\operatorname{Im}(e^{\mathrm{i}\tau}w)\cdot\operatorname{Im}\left(e^{-\mathrm{i}\tau}\frac{\mathrm{d}z}{\mathrm{d}s}g\right)\mathrm{d}s\Big\}\\ &= 2\oint_{\dot{G}}\operatorname{Re}(e^{\mathrm{i}\tau}w)\operatorname{Re}\left(e^{-\mathrm{i}\tau}\frac{\mathrm{d}z}{\mathrm{d}s}g\right)\mathrm{d}s.\end{aligned}$$

Let us define the (formal) adjoint operator $\mathsf{R}^*\colon X\to Y$ by

$$\mathsf{R}^*g\equiv\left\{g_{\bar{z}}+\bar{a}g-\bar{b}\bar{g},\ \operatorname{Re}\left(-\mathrm{i}e^{-\mathrm{i}\tau}\frac{\mathrm{d}z}{\mathrm{d}s}g\right)\Big|_{\dot{G}}\right\}. \tag{1.3.6}$$

Then by the above, we have

Theorem 1.3.2 *A necessary condition for the existence of a regular solution to* $\mathsf{R}w=(c,\psi)$ *with* $n>0$ *is that*

$$\oint_{\dot{G}}\psi\operatorname{Im}\left(-\mathrm{i}e^{-\mathrm{i}\tau}\frac{\mathrm{d}z}{\mathrm{d}s}g\right)\mathrm{d}s=\mathrm{i}\iint_{G}(-cg+\bar{c}\bar{g})[\mathrm{d}x,\mathrm{d}y] \tag{1.3.7}$$

for all g *such that* $\mathsf{R}^*g=(0,0)$.

In the next theorem, we show how to construct regular solutions to $\mathsf{R}w=(c,\psi)$.

Theorem 1.3.3 *A sufficient condition for the existence of a regular solution to* $\mathsf{R}w=(c,\psi)$ *with* $n>0$ *is that* (1.3.7) *holds for all regular solution to* $\mathsf{R}^*g=(0,0)$.

Proof As above, without loss of generality, we can assume that $c\equiv 0$. Let us choose $z_0, z_1, \ldots, z_{n-1}\in G$. Associated with the n poles $z_0,\ldots,z_{n-1}$ we have the solutions

$$w=\left\{\prod_{j=0}^{n-1}\frac{1}{(z-z_j)}\right\}\{\hat{w}_p+\nu\hat{w}_h\},\quad \nu\in R. \tag{1.3.8}$$

We show that, for a particular value of ν, w will be continuous at z_j, $j=0,\ldots,n-1$. In order that w be continuous at z_j, $j=0,\ldots,n-1$, we must have

$$\hat{w}_p(z_j)+\nu\hat{w}_h(z_j)=0, \qquad j=0,\ldots,n-1. \tag{1.3.9}$$

Now (1.3.9) represents $2n$ real linear equations for only *one* real unknown ν. Let us assume for the moment that the rank of the algebraic system represented by (1.3.9) is 1. Then $\nu \in R$ is uniquely determined. From Carleman's theorem, we have

$$\hat{w}:=\hat{w}_p+\nu\hat{w}_h=\prod_{j=0}^{n-1}(z-z_j)\tilde{w}$$

with $\tilde{w}\in C^{1+\alpha}(\bar{\mathsf{G}}-(\bigcup_{j=0}^{n-1}\{z_j\}))\cap C^{\alpha}(\bar{\mathsf{G}})$. This implies $w\in C^{\alpha}(\bar{\mathsf{G}})$ and with the equations (1.2.5) even that $w\in C^{1+\alpha}(\bar{\mathsf{G}})$.

To complete the proof we have only to show that the rank of the system of $2n$ real linear equations represented by (1.3.10) is 1, provided that (1.3.7) with $c\equiv 0$ holds for all regular solution to $\mathsf{R}^*g=(0,0)$. Since the BVP $\mathsf{R}^*g=(0,0)$ has characteristic

$$n^*=\frac{1}{2\pi}\oint_{\dot{\mathsf{G}}}\mathrm{d}\arg\left(-\mathrm{i}e^{-\mathrm{i}\tau}\frac{\mathrm{d}z}{\mathrm{d}s}\right)=-\frac{1}{2\pi}\oint_{\dot{\mathsf{G}}}\mathrm{d}\tau+\frac{1}{2\pi}\oint_{\dot{\mathsf{G}}}\mathrm{d}\arg\frac{\mathrm{d}z}{\mathrm{d}s}$$
$$=-n+1\leq 0,$$

according to Section 1.2 it has $1-2(1-n)=2n-1$ solutions $g_0,\ldots,g_{2n-2}$ which form a basis for $N(\mathsf{R}^*)$ satisfying

$$\begin{aligned} &g_0(z_j)=0,\\ &g_{2k}(z_j)=\delta_{jk}, \qquad j,k=1,\ldots,n-1,\\ &g_{2k-1}(z_j)=i\delta_{jk}, \end{aligned} \tag{1.3.10}$$

in the case $n>1$. (For $n=1$ the latter equations are not necessary.) Let $\varepsilon>0$ be sufficiently small so that the disk K_r with radius ε and centre z_r is completely contained in G for all $r=0,\ldots,n-1$.
By using the solvability conditions

$$\oint_{\dot{\mathsf{G}}}\psi\rho_l\,\mathrm{d}s=0, \qquad l=0,\ldots,2n-2$$

with

$$\rho_l=\mathrm{Im}\left(-\mathrm{i}e^{-\mathrm{i}\tau}\frac{\mathrm{d}z}{\mathrm{d}s}g_l\right),$$

it follows from Proposition 1.3.1 that

$$\mathrm{Re}\sum_{r=0}^{n-1}\oint_{\dot{K}_r}\left(\prod_{j=0}^{n-1}\frac{1}{(z-z_j)}\right)\cdot\{\hat{w}_p(z)+\nu\hat{w}_h(z)\}g_l(z)\,\mathrm{d}z=0,$$
$$l=0,\ldots,2n-2, \quad \nu\in R. \tag{1.3.11}$$

Since $g_0(z_1)=g_0(z_2)=\cdots=g_0(z_{n-1})=0$ for $n>1$, we have, letting $\varepsilon\to 0$ in (1.3.11) with $l=0$,

$$\operatorname{Re}\,(2\pi i\{\hat{w}_p(z_0)+\nu\hat{w}_h(z_0)\}g_0(z_0))=0, \qquad \nu\in R$$

and the same relation for $n=1$ also.
Hence,

$$\operatorname{Im}\{\hat{w}_p(z_0)g_0(z_0)\}=0,$$
$$\operatorname{Im}\{\hat{w}_h(z_0)g_0(z_0)\}=0,$$

or

$$\begin{aligned}&\operatorname{Im}\hat{w}_p(z_0)\operatorname{Re}g_0(z_0)+\operatorname{Re}\hat{w}_p(z_0)\operatorname{Im}g_0(z_0)=0,\\&\operatorname{Im}\hat{w}_h(z_0)\operatorname{Re}g_0(z_0)+\operatorname{Re}\hat{w}_h(z_0)\operatorname{Im}g_0(z_0)=0.\end{aligned}\tag{1.3.12}$$

Since $g_0(z_0)\neq 0$, the determinant of the homogeneous system (1.3.12) must vanish, i.e., we must have

$$\operatorname{Re}\hat{w}_p(z_0)\operatorname{Im}\hat{w}_h(z_0)-\operatorname{Im}\hat{w}_p(z_0)\operatorname{Re}\hat{w}_h(z_0)=0.$$

This implies that the system of two real equations in one real unknown ν represented by

$$\hat{w}_p(z_0)+\nu\hat{w}_h(z_0)=0$$

has a unique solution $\tilde{\nu}\in R$. It follows from (1.3.10) and (1.3.11) for $n>1$ that

$$\operatorname{Re}2\pi i\{\hat{w}_p(z_k)+\tilde{\nu}\hat{w}_h(z_k)\}=0,$$
$$\operatorname{Re}2\pi i\cdot i\{\hat{w}_p(z_k)+\tilde{\nu}\hat{w}_h(z_k)\}=0, \qquad k=1,\ldots,n-1,$$

which together imply that

$$\hat{w}_p(z_k)+\tilde{\nu}\hat{w}_h(z_k)=0, \qquad k=1,\ldots,n-1.$$

Hence, the rank of the system represented by (1.3.9) is 1 and the proof is complete.

Let us examine the question how many linearly independent solvability conditions are implied by (1.3.7). Since the BVP $\mathsf{R}^*g=(0,0)$ has characteristic $-n+1$, the solutions $g_0,\ldots,g_{2n-2}$ are linearly independent in X, i.e.

$$\sum_{j=0}^{2n-2}\lambda_j g_j=0, \qquad \lambda_j\in R, \quad z\in\bar{\mathsf{G}}$$

implies that $\lambda_j=0$, $0\leqslant j\leqslant 2n-2$. Now consider the linear functionals on

Y defined by

$$f_{g_j}[(c, \psi)] = \mathrm{i} \iint_{\mathsf{G}} (cg_j - \bar{c}\bar{g}_j)[\mathrm{d}x, \mathrm{d}y] + \oint_{\dot{\mathsf{G}}} \psi \operatorname{Im}\left(-\mathrm{i}e^{-\mathrm{i}\tau} \frac{\mathrm{d}z}{\mathrm{d}s} g_j\right) \mathrm{d}s.$$

We want to show that $\{f_{g_0}, \ldots, f_{g_{2n-2}}\}$ is a linearly independent set of functionals on Y. Suppose

$$\sum_{j=0}^{2n-2} \lambda_j f_{g_j} = 0, \qquad \lambda_j \in R.$$

Then for all $(c, \psi) \in Y$ we have

$$\mathrm{i} \iint_{\mathsf{G}} (c\Sigma\lambda_j g_j - \bar{c}\Sigma\lambda_j \bar{g}_j)[\mathrm{d}x, \mathrm{d}y] + \oint_{\dot{\mathsf{G}}} \psi\Sigma\lambda_j \operatorname{Im}\left(-\mathrm{i}e^{-\mathrm{i}\tau} \frac{\mathrm{d}z}{\mathrm{d}s} g_j\right) \mathrm{d}s = 0. \tag{1.3.13}$$

Setting $\psi = 0$ in (1.3.13) we have that for all c

$$\iint_{\mathsf{G}} (c\Sigma\lambda_j g_j - \bar{c}\Sigma\lambda_j \bar{g}_j)[\mathrm{d}x, \mathrm{d}y] \doteq 0. \tag{1.3.14}$$

For arbitrary real c, (1.3.14) implies that $\operatorname{Im} \Sigma\lambda_j g_j = 0$ in G and for arbitrary imaginary c, that $\operatorname{Re} \Sigma\lambda_j g_j = 0$ in G. Hence $\Sigma\lambda_j g_i = 0$ in G and by continuity in $\bar{\mathsf{G}}$. But this implies that $\lambda_j = 0$, $0 \leq j \leq 2n-2$.

This proof shows that $f_{g_0}, \ldots, f_{g_{2n-2}}$ are linearly independent already on the subspace $C^\alpha(\bar{\mathsf{G}})$ of Y. Similarly, it can be shown that these functionals are independent already on $C_R^{1+\alpha}(\dot{\mathsf{G}})$ too (see [3], Section 12.4).

Now, let us summarize the results obtained so far. The operator R defined by (1.2.14) is a bounded, linear transformation from X to Y. Let $n>0$. Then $\alpha = \dim N(\mathsf{R}) = 0$. Since we have shown that there are $2n-1$ linearly independent solvability conditions for $\mathsf{R}x = y$, it follows that $\beta = \operatorname{codim} R(\mathsf{R}) = 2n-1$. Let us define a bilinear form on $Y \times X$ by

$$\langle (c, \psi), g \rangle = \mathrm{i} \iint_{\mathsf{G}} (cg - \bar{c}\bar{g})[\mathrm{d}x, \mathrm{d}y] + \oint_{\dot{\mathsf{G}}} \psi \operatorname{Im}\left(-\mathrm{i}e^{-\mathrm{i}\tau} \frac{\mathrm{d}z}{\mathrm{d}s} g\right) \mathrm{d}s.$$

It is easily seen that $\langle \cdot, \cdot \rangle$ is continuous on $Y \times X$ and *regular*. By *regular*, we mean $\langle y, x \rangle = 0$ for all $y \in Y$ implies $x = 0$ and $\langle y, x \rangle = 0$ for all $x \in X$ implies $y = 0$. Now let

$$Y_0 = \{(c, \psi) \in Y : \langle (c, \psi), g \rangle = 0 \quad \text{for all} \quad g \in N(\mathsf{R}^*)\}.$$

Then because the $2n-1$ linearly independent solvability conditions are necessary, $R(\mathsf{R}) \subseteq Y_0$, and because they are sufficient, $Y_0 \subseteq R(\mathsf{R})$. Hence, $R(\mathsf{R}) = Y_0$. Since Y_0 is a closed subspace of Y, so is $R(\mathsf{R})$. Thus R is a *Fredholm operator* (see e.g. [8]) with index

$$\nu(\mathsf{R}) = \alpha - \beta = 1 - 2n.$$

The operator R^* defined by (1.3.6) is a bounded, linear transformation from X to Y. The index of its boundary condition is $-n+1 \leq 0$. As we have shown in Section 1.2, R^* is also Fredholm; in particular, $\alpha^* = \dim N(\mathsf{R}^*) = 2n-1$, $\beta^* = \operatorname{codim} R(\mathsf{R}^*) = 0$ and $R(\mathsf{R}^*) = Y$. Thus R^* has the Fredholm index

$$\nu(\mathsf{R}^*) = \alpha^* - \beta^* = 2n - 1 = -\nu(\mathsf{R}).$$

Now let us define a bilinear form on $X \times Y$ by

$$\langle\langle w, (d, \chi)\rangle\rangle = -\oint_{\dot{G}} \operatorname{Im}(e^{i\tau} w)\chi \, ds - i \iint_{G} (wd - \bar{w}\bar{d})[dx, dy].$$

Again, it is easily seen that $\langle\langle \cdot, \cdot \rangle\rangle$ is continuous on $X \times Y$ and regular. Moreover, using Gauss' formula, we obtain

$$\langle \mathsf{R}w, g\rangle = \langle\langle w, \mathsf{R}^* g\rangle\rangle, \qquad w, g \in X.$$

It seems natural in this case to say that R and R^* are adjoint with respect to these bilinear forms.

These results are a special case of the following theorem which will prove to be useful in what follows (see [5], [6], [14]).

Theorem 1.3.4 *Suppose*

(*i*) *X_1, Y_1, X_2 and Y_2 are Banach spaces;*

(*ii*) *A and B are Fredholm operators mapping X_1 into Y_1 and X_2 into Y_2 respectively;*

(*iii*) *$\langle \cdot, \cdot \rangle$ and $\langle\langle \cdot, \cdot \rangle\rangle$ are continuous, regular bilinear forms on $Y_1 \times X_2$ and $X_1 \times Y_2$, respectively;*

(*iv*) *A and B are adjoint with respect to these bilinear forms; i.e. $\langle Ax_1, x_2\rangle = \langle\langle x_1, Bx_2\rangle\rangle$ for all $(x_1, x_2) \in X_1 \times X_2$.*

Then $\nu(A) = -\nu(B)$ if and only if A and B are normally solvable with respect to $\langle \cdot, \cdot \rangle$ and $\langle\langle \cdot, \cdot \rangle\rangle$;
i.e. $\nu(A) = -\nu(B)$ if and only if

$Ax_1 = y_1$ is solvable if and only if $\langle y_1, x_2\rangle = 0$
for all $x_2 \in N(B)$, and
$Bx_2 = y_2$ is solvable if and only if $\langle\langle x_1, y_2\rangle\rangle = 0$
for all $x_1 \in N(A)$.

Proof If A and B are normally solvable with respect to $\langle\cdot,\cdot\rangle$ and $\langle\langle\cdot,\cdot\rangle\rangle$, then because $\langle\cdot,\cdot\rangle$ and $\langle\langle\cdot,\cdot\rangle\rangle$ are regular $\alpha(A)=\beta(B)$ and $\alpha(B)=\beta(A)$ so that $\nu(A)=\alpha(A)-\beta(A)=\beta(B)-\alpha(B)=-\nu(B)$.

Now suppose $\nu(A)=-\nu(B)$. Let A^* and B^* denote the topological adjoints of A and B, respectively. For $x_2\in N(B)$, let $\xi_{x_2}(\cdot)=\langle\cdot,x_2\rangle$. We claim that the mapping Φ defined by $\Phi(x_2)=\xi_{x_2}$ is an injection of $N(B)$ into $N(A^*)$. Because $\langle\cdot,\cdot\rangle$ is regular $\Phi: X_2\to Y_1^*$ is well-defined and 1-1. Since $\langle\cdot,\cdot\rangle$ is continuous, ξ_{x_2} belongs to the topological dual space Y_1^*. Now $\langle\langle x_1,Bx_2\rangle\rangle=\langle Ax_1,x_2\rangle=\xi_{x_2}(Ax_1)=0$ for all $x_1\in X_1$ implies that $A^*\xi_{x_2}=0$.
Hence $\xi_{x_2}\in N(A^*)$. Thus

$$\alpha(B)=\dim N(B)\leqslant\dim N(A^*)=\beta(A).$$

Similarly, we obtain

$$\alpha(A)=\dim N(A)\leqslant\dim N(B^*)=\beta(B).$$

Hence

$$\alpha(A)\leqslant\beta(B)=\alpha(B)-\nu(B)=\alpha(B)+\nu(A)\leqslant\beta(A)+\nu(A)=\alpha(A),$$

so that $\alpha(A)=\beta(B)$ and $\alpha(B)=\beta(A)$.

Hence, the mapping Φ introduced above is also onto. Thus, $Ax_1=y_1$ is solvable if and only if $y_1^*(y_1)=0$ for all $y_1^*\in N(A^*)$—that means—if and only if $\xi_{x_2}(y_1)=\langle y_1,x_2\rangle=0$ for all $x_2\in N(B)$. Similarly, $Bx_2=y_2$ is solvable if and only if $\langle\langle x_1,y_2\rangle\rangle=0$ for all $x_1\in N(A)$, and the proof is complete.

1.4 Semilinear First-Order Problems

Let us consider the semilinear boundary value problem

$$\begin{aligned} &w_{\bar z}=H(z,\bar z,w,\bar w) \quad \text{in} \quad G,\\ &\mathrm{Re}\,(e^{i\tau(s)}w)=\psi \qquad \text{on} \quad \dot G;\ \tau,\psi\in C^{1+\alpha}(\dot G). \end{aligned} \tag{1.4.1}$$

For convenience, we restrict ourselves to the case of non-positive characteristic

$$n\equiv\frac{1}{2\pi}\oint_{\dot G} d\tau\leqslant 0. \tag{1.4.2}$$

Such problems with $\tau\equiv 0$ were investigated by Begehr and Gilbert in [1] by the use of Schauder's fixed point theorem and by Wendland in [15] with the Schauder–Leray technique. Although the assumptions on the smoothness of H in [15] are stronger than in [1] we shall use this method because it works without difficulties for negative n also, whereas those in [1] seem to be more involved. Let us assume that

$$H,\ H_w,\ H_{\bar{w}},\ H_{ww},\ H_{w\bar{w}},\ H_{\bar{w}\bar{w}} \in C^0(\bar{G}\times\mathbb{C}) \tag{1.4.3}$$

and *that all these functions are bounded by a constant* K where the derivatives correspond to (1.1.4). In addition, for *each fixed value* of $w \in \mathbb{C}$ we assume

$$H(z, \bar{z}, w, \bar{w}) \in C^\alpha(\bar{G}). \tag{1.4.4}$$

The smoothness properties and boundedness in (1.4.3) seem to be very strong. But in the case that a solution to (1.4.1) exists or an *a priori* bound for the solution is known, the boundedness condition is not necessary because then H can be multiplied by a smooth cut-off function $\chi(w)$ which has a sufficiently large compact support. We shall come back to this point in the Corollary 1.4.2.

According to the linear case (1.2.27) let us choose $|n|$ fixed points $z_j \in G$, $j=1,\dots,|n|$ and let us modify (1.4.1) by imposing the additional conditions

$$\begin{aligned} &w(z_j)=\gamma_j, \qquad j=1,\dots,|n|=\tilde{n} \\ &\oint_{\dot{G}} \operatorname{Im}(e^{i\tau}w)\sigma\, ds=\kappa \end{aligned} \tag{1.4.5}$$

where the γ_j are given complex numbers and κ real. Then we can state:

Theorem 1.4.1 *The semilinear boundary value problem* (1.4.1) *with* (1.4.5) *has exactly one solution* $w \in C^{1+\alpha}(\bar{G})$.

Proof The main tool of the proof will be the *a priori* estimate (1.2.33) applied to the following embedding. Let us consider the family of problems

$$w_{\bar{z}}=tH \quad \text{in } G \text{ with} \quad t\in[0,1] \tag{1.4.6}$$

and with the boundary and side conditions

$$\begin{aligned} &\operatorname{Re} e^{i\tau}w|_{\dot{G}}=\psi, \qquad \oint_{\dot{G}} \operatorname{Im} e^{i\tau}w\sigma\, ds=\kappa, \\ &w(z_j)=\gamma_j, \qquad j=1,\dots,|n|. \end{aligned} \tag{1.4.7}$$

For $t=0$ the solution can easily be constructed by making use of the method in Section 1.2. Using this solution as the initial approximation, we solve (1.4.6, 7) for $t=t_1>0$ by *Newton's method.* Such combinations between embedding and Newton's method were investigated e.g. by Wacker [12].

The known $w(z, t_1)$ defines the initial approximation for Newton's method corresponding to (1.4.6) with $t = t_2 > t_1$. We shall show that after finitely many steps $t_1, t_2, \ldots, t_N$ the solution for $t = 1$, the original problem, can be found.

To this end let us consider Newton's method for a fixed value of $t = t_j$:

$$w_{n+1,\bar{z}} = t_j H_w(z, \bar{z}, w_n, \bar{w}_n)[w_{n+1} - w_n] + t_j H_{\bar{w}}(z, \bar{z}, w_n, \bar{w}_n)[\bar{w}_{n+1} - \bar{w}_n] + t_j H(z, \bar{z}, w_n, \bar{w}_n) \quad (1.4.8)$$

and (1.4.7) for w_{n+1}; $n = 0, 1, \ldots$; $w_0 \equiv w(t_{j-1}, z)$.
For *each* interation is this a *linear* problem for w_{n+1} which can be solved by the method of Section 1.2. For the convergence of (1.4.8) we consider $w_{n+1} - w_n$, which satisfies the equation

$$w_{n+1,\bar{z}} - w_{n\bar{z}} = t_j \{H_w(z, \bar{z}, w_n, \bar{w}_n)[w_{n+1} - w_n] + H_{\bar{w}}[\bar{w}_{n+1} - \bar{w}_n] + \tfrac{1}{2} H_{(w,\,w)}[w_n - w_{n-1}]^2 + H_{(w,\,\bar{w})}|w_n - w_{n-1}|^2 + \tfrac{1}{2} H_{(\bar{w},\,\bar{w})}[\bar{w}_n - \bar{w}_{n-1}]^2\} \quad (1.4.9)$$

and

$$\operatorname{Re} e^{i\tau}(w_{n+1} - w_n)|_{\dot{G}} = 0, \qquad \oint_{\dot{G}} \operatorname{Im} e^{i\tau}(w_{n+1} - w_n)\sigma \, ds = 0,$$
$$(w_{n+1} - w_n)(z_j) = 0, \qquad j = 1, \ldots, \tilde{n}, \quad (1.4.10)$$

where $H_{(w,\,w)}$ is defined by the mean value formula,

$$H_{(w,\,w)} \equiv \int_0^1 H_{ww}(z, \bar{z}, \tau w_n + (1-\tau) w_{n-1}, \tau \bar{w}_n + (1-\tau)\bar{w}_{n-1}) \tau \, d\tau, \quad (1.4.11)$$

and $H_{(w,\,\bar{w})}$, $H_{(\bar{w},\,\bar{w})}$ are defined similarly.

The difference $w_{n+1} - w_n = h$ fulfills (1.2.32), hence, the *a priori* estimate (1.2.33) with (1.4.9) implies

$$\|w_{n+1} - w_n\|_0 \leq 2Kt_j\gamma(K)\|w_n - w_{n-1}\|_0^2. \quad (1.4.12)$$

Thus (1.4.9) converges if the initial approximation satisfies

$$2Kt_j\gamma(K)\|w_1 - w_0\|_0 < 1. \quad (1.4.13)$$

Since we use $w_0 = w(t_{j-1}, z)$, we obtain for the difference $w_1 - w_0$ the differential equation

$$w_{1\bar{z}} - w_{0\bar{z}} = t_j H_w(z, \bar{z}, w_0, \bar{w}_0)[w_1 - w_0] + t_j H_{\bar{w}}[\bar{w}_1 - \bar{w}_0] + (t_j - t_{j-1}) H(z, \bar{z}, w_0, \bar{w}_0) \quad (1.4.14)$$

and the homogeneous conditions (1.4.5). Again, the *a priori* estimate (1.2.33) yields

$$\|w_1-w_0\|_0 \leq (t_j-t_{j-1})K\gamma(K)\|H\|_0 \leq (t_j-t_{j-1})K^2\gamma(K). \tag{1.4.15}$$

Thus, according to (1.4.15), t_j has to be chosen such that

$$t_j(t_j-t_{j-1})<[2K^3\gamma^2(K)]^{-1}, \qquad j=1,2,\ldots. \tag{1.4.16}$$

Letting $t_0=0$, we choose t_1 and solve (1.4.6) by the use of (1.4.8), then choose t_2 and solve (1.4.6) by the use of (1.4.8). Repeating finitely many times we end up with $t_N=1$ and $w(z,1)$, the solution of the desired problem (1.4.1), (1.4.5).

For the proposed uniqueness let w_1, w_2 be any two solutions of (1.4.1), (1.4.5). Then the difference $w^*\equiv w_1-w_2$ satisfies a homogeneous boundary value problem

$$w^*_{\bar z}=H_{(w)}w^*+H_{(\bar w)}\bar w^*,$$

$$\operatorname{Re} e^{i\tau}w^*|_{\dot G}=0, \qquad \oint_{\dot G} \operatorname{Im} e^{i\tau}w^*\sigma\, ds=0, \qquad w^*(z_j)=0, j=1,\ldots,\tilde n, \tag{1.4.17}$$

where $H_{(w)}$, $H_{(\bar w)}$ are defined similarly to (1.4.11) depending on w_1, w_2. Considering $H_{(w)}$, $H_{(\bar w)}$ as given functions, (1.4.17) is a homogeneous *linear* problem for w^* which has only the trivial solution $w^*\equiv 0$ according to Theorem 1.2.4.

Now let us consider the problem

$$w_{\bar z}=\bar a w+b\bar w+g(x,y,w,\bar w)+c(x,y) \tag{1.4.18}$$

with homogeneous conditions

$$\operatorname{Re} e^{i\tau}w|_{\dot G}=0,\ w(z_j)=0, \qquad \oint_{\dot G} \operatorname{Im} e^{i\tau}w\sigma\, ds=0. \tag{1.4.19}$$

Let us suppose that g fulfills the smoothness conditions (1.4.3) and also satisfies a growth condition

$$|g(x,y,w,\bar w)|\leq \omega(|w|) \tag{1.4.20}$$

where ω is a monotonic increasing function with

$$\omega(|w|)=0\,(|w|) \quad \text{for} \quad |w|\to 0. \tag{1.4.21}$$

If (1.4.18, 19) has a solution w then it fulfills the *a priori* estimate

$$\|w\|_0\leq \gamma(\|a\|_0+\|b\|_0)\{\|c\|_0+\omega(\|w\|_0)\}. \tag{1.4.22}$$

Let us plot the graph of the function

$$P(r)\equiv r-\gamma\omega(r), (\textit{Fig. 1}) \tag{1.4.23}$$

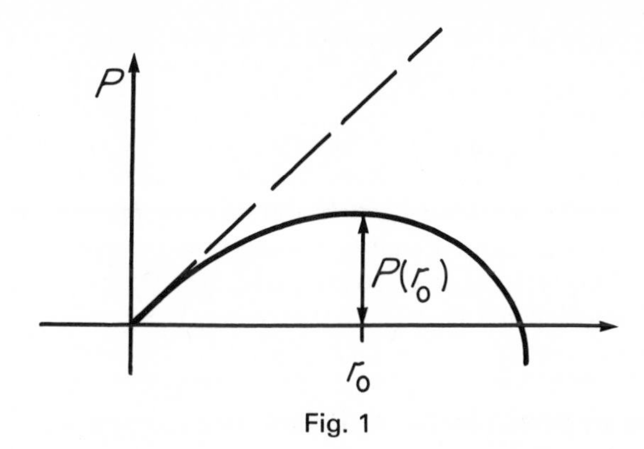

Fig. 1

Let r_0 be chosen such that $P(r)$ is still monotonic increasing for all $0 \leq r \leq r_0$. Then we can prove with Theorem 1.4.1 the following

Corollary 1.4.2 *For* $\gamma \,\|c\|_0 < P(r_0)$ *there exists exactly one solution of* (1.4.18, 19) *with* $\|w\|_0 < r_0$.

Proof Let $\chi(w, \bar{w})$ be a C^∞ cut-off function with compact support and

$$\chi \equiv 1 \quad \text{for} \quad |w| \leq r_0. \tag{1.4.24}$$

Then the family of problems

$$\begin{aligned} &v_{\bar{z}} = [tc + \bar{a}v + b\bar{v} + tg(x, y, v, \bar{v})]\chi(v, \bar{v}), \\ &\operatorname{Re} e^{i\tau} v|_{\dot{G}} = 0, \qquad \oint_{\dot{G}} \operatorname{Im} e^{i\tau} v\sigma \, ds = 0, \; v(z_j) = 0, \; j = 1, \ldots, \tilde{n} \end{aligned} \tag{1.4.25}$$

with $0 \leq t \leq 1$ has exactly one solution v for each t according to Theorem 1.4.1. Of course, $\|v\|_0$ depends continuously on t fulfilling

$$\|v\|_0 \leq \gamma[t\,\|c\|_0 + \omega(\|v\|_0)]$$

or

$$P(\|v\|_0) \leq t\gamma\,\|c\|_0 < P(r_0). \tag{1.4.26}$$

Starting with $t = 0$ and the corresponding $v \equiv 0$, (1.4.26) together with the monotonicity of P yields for this family of solutions

$$\|v\|_0 < r_0. \tag{1.4.27}$$

In the special case of $t = 1$ we have for $w = v$

$$w_{\bar{z}} = c + \bar{a}w + b\bar{w} + g(x, y, w, \bar{w}) \tag{1.4.28}$$

since $\chi(x, y, w(x, y), \bar{w}(x, y)) = 1$.

Remark Begehr and Gilbert investigated in [1]

$$w_{\bar{z}} = f(z, \bar{z}, w, \bar{w})$$

with conditions (1.4.19) and $\tau \equiv 0$. Their assumption ((12) in [1]) assures the *a priori* estimate $\|w\|_0 \leq K$ if the solution exists. Using a cut-off function $\chi(w, \bar{w})$ with $\chi \equiv 1$ for $|w| \leq K$ and defining $H \equiv \chi f$, the Theorem 1.4.1 provides (for smooth enough f) a solution v. By using the estimates in [1] it follows similarly as in Corollary 1.4.1 that $\|v\|_0 \leq K$. Hence $w = v$ is the desired solution in [1].

1.5 Remarks on the Additional References

Generalizations of the Vekua systems to several complex variables and connections with holomorphic functions are studied in [30] and [31]. Some generalizations of the similarity principle for the system (1.1.5) with singular coefficients can be found in [18] and [25]. The behaviour of solutions and representation for coefficients in L^p_{loc} generalizing Vekua's L^p theory for (1.1.5) was investigated in [20]. In [19] the coefficients become infinite at the whole boundary implying a degenerate system. There also the solvability theory for boundary value problems was completely investigated. For another class of boundary conditions including the Poincaré problem the solvability theory can be found in [28]. The appropriate norms and *a priori* estimates for exterior problems with (1.1.5) have been investigated in [17]. Nonlinear Dirichlet and Neumann problems for second-order equations with Laplacian can be found in [21], [22], [23], [26], [27] and [29]. Boundary value problems with nonlinear boundary conditions for holomorphic or generalized analytic functions have been investigated with the Banach fixed point principle and Schauder's fixed point principle in [16] and [32–36], a corresponding uniqueness theorem in [24] and corresponding existence and bifurcation with applications to fluid mechanical problems in [37–40].

References

1 Begehr, H. and Gilbert, R. P., Über das Randwert-Normproblem für ein nichtlineares elliptisches System. In *Proc. Conf. Function Theoretic Methods Part. Diff. Eqn.* Lecture Notes in Maths. No. 561, 112–122, Springer, Berlin, 1976.

2 Courant, R. and Hilbert, D., *Methods of mathematical physics* II. Interscience, New York, 1962.

3 Haack, W. and Wendland, W., *Lectures on Pfaffian and partial differential equations.* Pergamon Press, Oxford, 1972.

4 Hellwig, G., *Partielle Differentialgleichungen.* Teubner, Leipzig, 1962.
5 Heuser, H., *Funktionalanalysis.* Teubner, Leipzig, 1975.
6 Jörgens, K., *Lineare Integraloperatoren.* Teubner, Leipzig, 1970.
7 Michlin, S. G., *Vorlesungen über lineare Integralgleichungen.* Verl. d. Dt. Wiss., Berlin, 1962.
8 Schechter, M., *Principles of functional analysis.* Academic Press, New York, 1971.
9 Vekua, I. N., *Generalized analytic functions.* Pergamon Press, Oxford, 1962.
10 Vekua, I. N., *New methods for solving elliptic equations.* North Holland, Amsterdam, 1967.
11 Vekua, I. N., On one class of elliptic systems with singularities. *Proc. Internat. Conf. on Functional Analysis and Rel. Topics* (Tokyo 1969), pp. 142–147, University Tokyo Press, 1970.
12 Wacker, H. J., Eine Lösungsmethode zur Behandlung nichtlinearer Randwertprobleme. *In* Iterationsverfahren, Numerische Mathematik, Approximationstheorie. *ISNM,* **15,** Birkhäuser, 1970, pp. 245–257.
13 Warschawski, W., On differentiability at the boundary in conformal mapping. *Proc. Am. Math. Soc.* **12,** 614–620, 1961.
14 Wendland, W., Bemerkungen über die Fredholmschen Sätze. *Methoden und Verfahren d. Math. Physik,* **3,** BI Mannheim 722, pp. 141–176, 1970.
15 Wendland, W., An integral equation method for generalized analytic functions. *Proc.* in *Lecture Notes* No. 430, pp. 414–452, Springer, Berlin, 1974.

Additional references

16 Beyer, K., Nichtlineare Randwertprobleme für elliptische Systeme 1. Ordnung für zwei Funktionen von zwei Variablen. *Beiträge zur Analysis* **4,** 31–34, 1972.
17 Beyer, K., Abschätzungen für elliptische Systeme erster Ordnung für zwei Funktionen in Außengebieten des R^2. *Math. Nachr.* **57,** 1–13, 1973.
18 Bliev, N. K., On necessary and sufficient conditions for the existence of analytic solutions of some degenerate system. *Izv. Akad. Nauk Kazakhskoi SSR, Mat. i. Mekh.* **21,** 93–95, 1967.
19 Dzurajev, A., On properties of some degenerate elliptic systems of first order in the plane. *Dokl. Akad. Nauk SSR* **223,** 1975; *Sov. Math. Dokl.* **16,** 1975.
20 Goldschmidt, B., Funktionentheoretische Eigenschaften der Lösungen der Vekuaschen Differentialgleichung mit Koeffizienten in $L_{p,\,loc}(G)$, to appear.
21 Haf, H., Ein nichtlineares Schwingungsproblem mit gemischter linearer Randbedingung. *Monatshefte für Mathematik* **76,** 419–427, 1972.
22 Haf, H., Über ein nichtlineares Schwingungsproblem mit nichtlinearer Dirichletbedingung. *J. f. d. Reine u. Angew. Math.* **258,** 211–220, 1973.
23 Levinson, N., Dirichlet problem for $\Delta u = f(x, y, u)$. *J. Math. Mech.* **12,** 567–575, 1963.
24 Martin, M. H., Local uniqueness in boundary problems. *Proc. Math. Soc. Edinburgh,* II, Ser. A, **17,** 23–26, 1970.

25 Mikhailov, L. G., Special cases in the theory of generalized analytic functions. *Revue Roumaine Math. pur. appl.* **13,** hommage au Professeur N. Theodorescu pour sou 60^e anniversaire, 1403–1408, 1968.
26 Nitsche, J. and Nitsche, J. J., Bemerkungen zum zweiten Randwertproblem der Differentialgleichung $\Delta\phi = \phi_x^2 + \phi_y^2$. *Math. Ann.* **126,** 69–74, 1953.
27 Nitsche, J., Nitsche, J. J., Das zweite Randwertproblem der Differentialgleichung $\Delta u = e^u$. *Arch. Math.* **3,** 460–464, 1952.
28 Schleiff, M., Über geschlossene Lösungen des Poincaréschen Randwertproblems mit Hilfe einer komplexen Differentialgleichung. *Beiträge zur Analysis* **6,** 73–85, 1974.
29 Simader, C., Another approach to the Dirichlet problem for very strongly nonlinear elliptic equations. *In Ordinary and Partial Differential Equations, Proc. Conf.* 1976 Dundee; Lecture Notes No. 564, pp. 423–437, Springer, Berlin, 1976.
30 Tutschke, W., Morphe Funktionen einer und mehrerer Variabler, to appear.
31 Tutschke, W., Pseudoholomorphe Exponentialfunktionen. *Beiträge zur Analysis* **1,** 116–123, 1971.
32 Wolska-Bochenek, J., Probléme non-linéaire à derivée oblique. *Ann. Polon. Math.* **9,** 1961.
33 Wolska-Bochenek, J., Sur un problème non-lineaire d'Hilbert dans le théorie des fonctions pseudo-analytiques. *Zeszyty Naukowe Polytechniki Warszawskiej* 172 *Mat.* **11,** 145–157, 1968.
34 Wolska-Bochenek, J., Sur un problème aux limites discontinues dans la théorie des fonctions pseudo-analytiques. *Zeszyty Naukowe Polytechniki Warszawskiej* 173 *Mat.* **12,** 23–37, 1968.
35 Wolska-Bochenek, J., On some generalized nonlinear problem of the Hilbert type. *Zeszyty Naukowe Polytechniki Warszawskiej* 183 *Mat.* **14,** 15–32, 1968.
36 Wolska-Bochenek, J., A componed non-linear boundary value problem in the theory of pseudo-analytic functions. *Demonstratio Mathematica* **4,** 105–117, 1972.
37 Zeidler, E., Über eine Klasse nichtlinearer singulärer Randwertaufgaben. *Math. Nachr.* **43,** 1–6, 1970.
38 Zeidler, E., Topologische Existenzbeweise für funktionentheoretische Randwertaufgaben mit Anwendungen auf permanente Wellenbewegungen. *Beiträge zur Analysis* **1,** 55–74, 1971.
39 Zeidler, E., Existenz einer Gasblase in einer Parallel- und Zirkulationsströmung unter Berüchsichtigung der Schwerkraft. *Beiträge zur Analysis* **3,** 67–95, 1972.
40 Zeidler, E., Existenzbeweise für asympotische Wirbelwellen. *Beiträge zur Analysis* **3,** 109–134, 1972.

2

The Normal Form of Elliptic Systems with Two Unknowns

Let a system be given by

$$\begin{pmatrix} a_{11} & a_{12} \\ a_{21} & a_{22} \end{pmatrix}\begin{pmatrix} u_x \\ v_x \end{pmatrix}+\begin{pmatrix} b_{11} & b_{12} \\ b_{21} & b_{22} \end{pmatrix}\begin{pmatrix} u_y \\ v_y \end{pmatrix}+\begin{pmatrix} c_{11} & c_{12} \\ c_{21} & c_{22} \end{pmatrix}\begin{pmatrix} u \\ v \end{pmatrix}+\begin{pmatrix} f_1 \\ f_2 \end{pmatrix}=\begin{pmatrix} 0 \\ 0 \end{pmatrix}. \quad (2.0.1)$$

The coefficients in $\mathsf{A}=(a_{jk})$, $\mathsf{B}=(b_{jk})$, $\mathsf{C}=(c_{jk})$ are assumed to be functions of (x, y), which are smooth enough in a suitable domain. The transformation of (2.0.1) into a normal form, which in the elliptic case is the normal form (1.1.1), consists of two parts: an algebraic part depending on the point (x, y) and on the type of (2.0.1), and an analytic part depending on a suitable transformation of the variables.

For the algebraic evaluations, we shall use two different versions. One, using Pfaffian forms, leads to a type-independent normal form. (This is the method of Schneider [17] and Haack [10, p. 195]). The second uses function theoretic methods. It involves complex-valued functions and uses the ellipticity to obtain the normal form (1.1.5) [6].

According to the terminology in Hörmander's book [11] the symbol is introduced and the transformation of the system is presented by means of the eigenvalues and eigenvectors belonging to the coefficients of the principal part. Their dependence on (x, y) is formulated by means of simple Dunford integrals. These formulations prepare the more complicated case of more unknowns.

The analytic evaluations are concerned with finding a suitable solution of a Beltrami-system, which defines a one-to-one mapping on to the new coordinates. If A and B consist of analytic coefficients, then this transformation was given by Gauss [8].

In this case, the existence of Beltrami's transformation ζ with

$$\zeta_{\bar{z}}=q\zeta_z$$

is proved by solving this first-order equation in $\mathbb{C}^2$. The solution is found by successive approximation [13] and the representation of implicit functions in the analytic case [9]. This is a function theoretic method which hinges on the analyticity of the coefficients.

If q is nonanalytic but only L_∞, we present an existence proof for ζ which is a combination of Bojarski's proof in [20] in L_p-spaces, the local proof by Bers and Chern in [5] and the proof by Ahlfors [2] who used the Schauder–Leray technique. This shows the existence of the Hölder continuously differentiable transform ζ. The existence of ζ for $q \in L_\infty$ goes back to Morrey [16]. The whole transformation fails if a_{jk}, b_{jk} in (2.0.1) are not smooth anymore, Vinogradov provided the solvability theory by a suitable approximation for L_∞ coefficients (see *loc. cit* (a) in [20].).

The transformation to the normal form together with the existence theorems in Chapter 1 lead to a simple proof of the *a priori* estimates in Hölder-spaces in accordance with [1] (where the much deeper general *a priori* estimates are obtained for general regular elliptic boundary value problems). The simple estimate

$$\|u+iv\|_{C^{1+\alpha}(\bar{G})} \leq \gamma[|u \cos \tau - v \sin \tau|_{C^{1+\alpha}(\Gamma)} + \|Au+Bv\|_{C^{\alpha}(\bar{G})} + \|u+iv\|_{C^{0}(\bar{G})}] \quad (2.0.2)$$

is basic for the corresponding estimate in the general case.

Bojarski and Iwaniec [3, 4] investigated nonlinear systems

$$\phi_j(u_x, u_y, v_x, v_y, u, v, x, y) = 0, \qquad j = 1, 2$$

which are elliptic in the sense of *Lavrentieff*. For these systems the quasi-conformal mapping ζ is constructed satisfying an equation of the form

$$w_{\bar{z}} = (u+iv)_{\bar{z}} = q(z, w, w_z)w_z, \quad \text{with} \quad |q| \leq q_0 < 1.$$

Finally, the simplest corresponding boundary value problem is investigated.†

2.1 The Distinguishing Pfaffian Form of (2.0.1)

The system (2.0.1) can also be written in the form

$$\begin{pmatrix} a_{11} & b_{11} \\ a_{21} & b_{21} \end{pmatrix}\begin{pmatrix} u_x \\ u_y \end{pmatrix} + \begin{pmatrix} a_{12} & b_{12} \\ a_{22} & b_{22} \end{pmatrix}\begin{pmatrix} v_x \\ v_y \end{pmatrix} + \mathbf{C}\mathbf{u} + \mathbf{f} = \mathbf{0} \quad (2.1.1)$$

where $\mathbf{u} = \begin{pmatrix} u \\ v \end{pmatrix}$. Now let us assume that at each point (x, y) at least one of the determinants

$$a = \begin{vmatrix} a_{11} & b_{11} \\ a_{21} & b_{21} \end{vmatrix}, \qquad b = \begin{vmatrix} a_{12} & b_{12} \\ a_{22} & b_{22} \end{vmatrix}, \quad (2.1.2)$$

† A rather complete investigation of nonlinear problems was given by Bojarski (see *loc. cit* (d) in [20].).

does not vanish. For example, let us suppose

$$b \neq 0 \tag{2.1.3}$$

in a suitable domain. If they both vanish, then either the system degenerates to a first order equation or the new unknowns $u+v$ and $u-v$ satisfy a new system like (2.1.1) with $b \neq 0$.

Because of (2.1.3) the two Pfaffian forms

$$\omega \equiv -b_{12}\,\mathrm{d}x + a_{12}\,\mathrm{d}y \quad \text{and} \quad \tilde{\omega} \equiv -b_{22}\,\mathrm{d}x + a_{22}\,\mathrm{d}y \tag{2.1.4}$$

are linearly independent. (2.1.1) is then equivalent to the Pfaffian equation

$$(-\tilde{\omega}, \omega)\left(\begin{bmatrix} a_{11} & b_{11} \\ a_{21} & b_{21} \end{bmatrix}\begin{pmatrix} u_x \\ u_y \end{pmatrix} + \cdots\right) = \theta$$

or

$$(\mathrm{d}x, \mathrm{d}y)\begin{bmatrix} a_{11}b_{22} - a_{21}b_{12}, & b_{11}b_{22} - b_{12}b_{21} \\ a_{12}a_{21} - a_{22}a_{11}, & -a_{22}b_{11} + a_{12}b_{21} \end{bmatrix}\begin{pmatrix} u_x \\ u_y \end{pmatrix}$$
$$+ b(v_x\,\mathrm{d}x + v_y\,\mathrm{d}y) + (-\tilde{\omega}, \omega)\{\mathbf{C}\mathbf{u} + \mathbf{f}\} = \theta. \tag{2.1.5}$$

If we define these coefficients to be

$$\begin{aligned} g_{12} &= a_{11}b_{22} - a_{21}b_{12}, & g_{22} &= b_{11}b_{22} - b_{12}b_{21}, \\ g_{11} &= -a_{12}a_{21} + a_{22}a_{11}, & g_{21} &= a_{22}b_{11} - a_{12}b_{21}, \end{aligned} \tag{2.1.6}$$

then (2.1.1) becomes equivalent to the Pfaffian equation

$$(-g_{11}u_x - g_{21}u_y)\,\mathrm{d}y + (g_{12}u_x + g_{22}u_y)\,\mathrm{d}x + b\,\mathrm{d}v + (-\tilde{\omega}, \omega)(\mathbf{C}\mathbf{u} + \mathbf{f}) = \theta. \tag{2.1.7}$$

The first Pfaffian form is almost Haack's generalized normal derivative. This normal derivative is defined by

$$\mathrm{d}_n u = \rho[g_{11}u_x + \tfrac{1}{2}(g_{12} + g_{21})u_y]\,\mathrm{d}y - \rho[\tfrac{1}{2}(g_{12} + g_{21})u_x + g_{22}u_y]\,\mathrm{d}x \tag{2.1.8}$$

where ρ is an arbitrarily choosable function with $\rho \neq 0$. This is a generalization of the normal derivative (1.1.8) which corresponded to the Laplacian in (1.1.9). We get the following theorem:

Theorem 2.1.1 *If $b \neq 0$ then the system* (2.0.1) *is equivalent to the Pfaffian equation*

$$-\mathrm{d}_n u + \rho b\,\mathrm{d}v + \rho[\tfrac{1}{2}(g_{12} - g_{21})\,\mathrm{d}u]$$
$$+ \rho[-\tilde{\omega}(c_{11}u + c_{12}v + f_1) + \omega(c_{21}u + c_{22}v + f_2)] = \theta. \tag{2.1.9}$$

This normal form and the d_n-operator (2.1.8) lead to the classification of our system (2.0.1). It is called

$$\left.\begin{matrix}\text{elliptic}\\ \text{hyperbolic}\\ \text{parabolic}\end{matrix}\right\} \text{ if } \quad D = g_{11}g_{22} - \tfrac{1}{4}(g_{12}+g_{21})^2 \begin{cases} > \\ < 0. \\ = \end{cases} \tag{2.1.10}$$

Remark If all coefficients of (2.1.8) vanish, then some parabolic equations with one characteristic behave like hyperbolic equations. Therefore, in [5] these special cases are also called hyperbolic.

The Pfaffian equation (2.1.9) leads to normal forms very easily, even for mixed type equations.

The Beltrami equation can be written as a Pfaffian equation

$$[\mathrm{d}, \mathrm{d}_n\xi] = \theta, \tag{2.1.11}$$

with (2.1.11) being equivalent to

$$\frac{\partial}{\partial x}\left[\rho g_{11}\xi_x + \frac{\rho}{2}(g_{12}+g_{21})\xi_y\right] + \frac{\partial}{\partial y}\left[\frac{\rho}{2}(g_{12}+g_{21})\xi_x + g_{22}\xi_y\right] = 0. \tag{2.1.12}$$

Let us assume that the Beltrami equation (2.1.12) has a solution ξ with the following property:

The functions $\xi(x, y)$ and η where η is given by a solution of $\mathrm{d}\eta = \mathrm{d}_n\xi$ define a homeomorphism in some domain of the x, y plane. (2.1.13)

Using the relation

$$\mathrm{d}_n\eta = -\rho^2 D\,\mathrm{d}\xi \tag{2.1.14}$$

and the new unknowns

$$U := u, \qquad V := \rho b v + \frac{\rho}{2}(g_{12} - g_{21})u, \tag{2.1.15}$$

the Pfaffian equation (2.1.9) becomes

$$-U_\xi\,\mathrm{d}_n\xi - U_\eta\,\mathrm{d}_n\eta + V_\xi\,\mathrm{d}\xi + V_\eta\,\mathrm{d}\eta + \cdots = \theta \tag{2.1.16}$$

or

$$(-U_\xi + V_\eta)\,\mathrm{d}\eta + (\rho^2 D U_\eta + V_\xi)\,\mathrm{d}\xi + \cdots = \theta. \tag{2.1.17}$$

(2.1.17) is equivalent to the system

$$\begin{aligned} U_\xi - V_\eta &= AU + BV + C, \\ \rho^2 D U_\eta + V_\xi &= \tilde{A}U + \tilde{B}V + \tilde{C}, \end{aligned} \tag{2.1.18}$$

with suitable new coefficients. As we shall show later, in the elliptic case

(2.1.11) provides a homeomorphism in any given domain and we can choose

$$\rho^2 = \frac{1}{D}. \tag{2.1.19}$$

Then, (2.1.18) becomes our normal form (1.1.1). Furthermore, the condition $b \neq 0$ is always fulfilled in the elliptic case $D > 0$. This can be seen from the following identity which is also useful in some differential geometric theorems (see [12]).

$$\begin{aligned} D &= ab - \frac{1}{4}\left\{\begin{vmatrix} a_{11} & b_{12} \\ a_{21} & b_{22} \end{vmatrix} - \begin{vmatrix} b_{11} & a_{12} \\ b_{21} & a_{22} \end{vmatrix}\right\}^2 \\ &= \begin{vmatrix} a_{11} & b_{11} \\ a_{12} & b_{12} \end{vmatrix} \cdot \begin{vmatrix} a_{21} & b_{21} \\ a_{22} & b_{22} \end{vmatrix} - \frac{1}{4}\left\{\begin{vmatrix} a_{11} & b_{11} \\ a_{22} & b_{22} \end{vmatrix} + \begin{vmatrix} a_{21} & b_{21} \\ a_{12} & b_{12} \end{vmatrix}\right\}^2 \end{aligned} \tag{2.1.20}$$

2.2 The Transformation Using Complex-Valued Functions

With the real matrix valued functions

$$\mathsf{A} = (a_{jk}), \qquad \mathsf{B} = (b_{jk}), \qquad \mathsf{C} = (c_{jk}), \qquad \mathbf{f} = \begin{pmatrix} f_1 \\ f_2 \end{pmatrix}, \qquad \mathbf{u} = \begin{pmatrix} u \\ v \end{pmatrix} \tag{2.2.1}$$

and the differential operators $\partial/\partial x$, $\partial/\partial y$, the system (2.0.1) becomes

$$\mathsf{A}\frac{\partial}{\partial x}\mathbf{u} + \mathsf{B}\frac{\partial}{\partial y}\mathbf{u} + \mathsf{C}\mathbf{u} + \mathbf{f} = \mathbf{0} \tag{2.2.2}$$

The *symbol* σ of equation (2.2.2) is defined by

$$\sigma(x, y; \tilde{\xi}, \tilde{\eta}) := \mathsf{A}\tilde{\xi} + \mathsf{B}\tilde{\eta} \tag{2.2.3}$$

The symbol is a matrix valued function of the four variables $(x, y; \tilde{\xi}, \tilde{\eta})$. Then (2.2.2) is called

elliptic at (x, y) iff $\sigma(x, y; \tilde{\xi}, \tilde{\eta})$ is a nonsingular matrix for each real pair $(\tilde{\xi}, \tilde{\eta}) \neq (0, 0)$. (See e.g. [11], definition 10.6.1) (2.2.4)

Remark An elementary evaluation shows

$$\det \sigma(x, y; \tilde{\xi}, \tilde{\eta}) = g_{11}\tilde{\xi}^2 + (g_{12} + g_{21})\tilde{\xi}\tilde{\eta} + g_{22}\tilde{\eta}^2. \tag{2.2.5}$$

Therefore the ellipticity means the definiteness of (2.2.5) and is equivalent to the definition (2.1.10).

Now let us consider the elliptic case only. Then, (2.2.4) with $\tilde{\xi}=1$, $\tilde{\eta}=0$ gives

$$g_{11}=\det \mathsf{A}\neq 0. \tag{2.2.6}$$

Without loss of generality, let us therefore assume that

$$a_{jk}=\delta_{jk}, \tag{2.2.7}$$

Thus, we assume that (2.2.2) has the special form

$$\mathbf{u}_x+\mathsf{B}\mathbf{u}_y+\mathsf{C}\mathbf{u}+\mathbf{f}=\mathbf{0}. \tag{2.2.8}$$

Because of (2.2.4) all eigenvalues λ of B are *nonreal*. For transformation into normal form, now let us use the eigenvectors $\mathbf{p}$ of the matrix B^T, the transpose of B,

$$\sum_{j=1}^{2} b_{jk}p^j=\lambda p^k. \tag{2.2.9}$$

In our case, there exist exactly two eigenvalues:

$$\lambda_\pm=\lambda_1(x, y)\pm \mathrm{i}\lambda_2(x, y), \quad (\lambda_2>0) \tag{2.2.10}$$

in each point (x, y). Because B is 2×2, the multiplicity of λ is 1 and we have exactly two complex conjugate eigenvectors:

$$\mathbf{p}_\pm=\left[\begin{pmatrix}p_1\\ p_2\end{pmatrix}\pm \mathrm{i}\begin{pmatrix}\tilde{p}_1\\ \tilde{p}_2\end{pmatrix}\right] \quad \text{with } p_1^2+p_2^2+\tilde{p}_1^2+\tilde{p}_2^2=1, \tag{2.2.11}$$

depending on (x, y). The multiplicity 1 implies:

Theorem 2.2.1 *The eigenvalues $\lambda(x, y)$ and the eigenvectors $\mathbf{p}(x, y)$ have the same regularity as the coefficients of B at all points of ellipticity of* (2.2.8). (See Kato [14] Theorems 5.1, 5.4 and 5.7 pp. 107–114.) *E.g. if $\mathsf{B}\in C^{r+\alpha}(\bar{\mathsf{G}})$ then $\lambda, \mathbf{p}\in C^{r+\alpha}(\bar{\mathsf{G}})$.*

Proof For $\lambda(x, y)$ the conjecture follows immediately from the characteristic equation:

$$\lambda^2-\lambda(b_{11}+b_{22})+\det \mathsf{B}=0 \tag{2.2.12}$$

and the ellipticity.

Because the multiplicity of λ is 1, the eigenvector can locally be represented by the total projection applied to some suitable constant vector $\mathbf{q}_0$ ([14] p. 67 ff.)

$$\mathbf{p}^*(x, y)=\frac{1}{2\pi \mathrm{i}}\oint_{|\zeta-\lambda(x,y)|=\varepsilon} (B^T(x, y)-\zeta I)^{-1}\,\mathrm{d}\zeta\,\mathbf{q}_0,$$

$$\mathbf{p}=(p_1^{*2}+p_2^{*2}+\tilde{p}_1^{*2}+\tilde{p}_2^{*2})^{-1}\mathbf{p}^*(x, y) \tag{2.2.13}$$

with $0 < \varepsilon < \lambda_2(x, y)$ where ε is a constant. (2.2.13) implies the regularity of $\mathbf{p}$.

Scalar multiplication of (2.2.8) by $\mathbf{p}_+$ leads to

$$\mathbf{p}_+^T\mathbf{u}_x + \mathbf{p}_+^T(\mathbf{B}\mathbf{u}_y) + \mathbf{p}_+^T(\mathbf{C}\mathbf{u}) + \mathbf{p}_+^T\mathbf{f} = 0 \tag{2.2.14}$$

and with the unknown complex function

$$U + \mathrm{i}V = w \equiv \mathbf{p}_+^T\mathbf{u}, \tag{2.2.15}$$

to

$$w_x + \lambda_+ w_y + (\mathbf{p}_+^T\mathbf{C} - \mathbf{p}_{+x}^T - \lambda_+\mathbf{p}_{+y}^T)\mathbf{u} + \mathbf{c} = 0 \tag{2.2.16}$$

Lemma 2.2.2 *For an elliptic equation the transformation* (2.2.15) *is regular, i.e.*

$$d \equiv p_1\tilde{p}_2 - p_2\tilde{p}_1 \neq 0 \quad \textit{and} \tag{2.2.17}$$

$$u = \frac{1}{d}(\tilde{p}_2 U - p_2 V), \qquad v = \frac{1}{d}(p_1 V - \tilde{p}_1 U). \tag{2.2.18}$$

Proof Let us assume that d vanishes while $\mathbf{p}$ is given by (2.2.11). Then the eigenvector can be represented by

$$\mathbf{p} = \mu\begin{pmatrix} p_1^* \\ p_2^* \end{pmatrix} \quad \text{with } 0 \neq \mu \in \mathbb{C} \text{ and } p_1^*, p_2^* \text{ real.} \tag{2.2.19}$$

Substituting (2.2.19) into (2.2.9) and dividing by μ we get the equation:

$$\sum_{j=1}^{2} b_{jk}p_j^* = (\lambda_1 + \mathrm{i}\lambda_2)p_k^*, \quad k = 1,2, \tag{2.2.20}$$

which leads to the contradiction $\lambda_2 = 0$ because the b_{jk} are real.

Let

$$\zeta = \xi(x, y) + \mathrm{i}\eta(x, y) \tag{2.2.21}$$

be a complex valued function providing a (local) homeomorphism

$$(x, y) \to (\xi, \eta) \quad \text{or} \quad (z, \bar{z}) \to (\zeta, \bar{\zeta}). \tag{2.2.22}$$

Then the differential operator in (2.2.16) can be transformed into

$$w_x + \lambda_+ w_y = w_\zeta(\zeta_x + \lambda_+\zeta_y) + w_{\bar{\zeta}}(\bar{\zeta}_x + \lambda_+\bar{\zeta}_y). \tag{2.2.23}$$

The first term vanishes if we demand that ζ satisfies the equation

$$\zeta_x + \lambda_+\zeta_y = 0. \tag{2.2.24}$$

This is a first-order system for ξ and η:

$$\begin{aligned} &\xi_x+\lambda_1\xi_y-\lambda_2\eta_y=0,\\ &\eta_x+\lambda_1\eta_y+\lambda_2\xi_y=0. \end{aligned} \tag{2.2.25}$$

Remark *For this first-order system, the relations*

$$g_{11}=1, \qquad g_{12}=b_{22}, \qquad g_{21}=b_{11}, \qquad \lambda_1=\frac{g_{12}+g_{21}}{2}, \qquad \lambda_2=\sqrt{D} \tag{2.2.26}$$

show that (2.2.25) *is equivalent to the Beltrami system*

$$d\eta=d_n\xi \quad \text{with} \quad \rho=\frac{1}{\sqrt{D}}=\frac{1}{\lambda_2} \tag{2.2.27}$$

belonging to (2.1.13).

The Beltrami-system (2.2.24) can be transformed, using

$$\zeta_z+\zeta_{\bar z}+\lambda_+ i(\zeta_z-\zeta_{\bar z})=0 \tag{2.2.28}$$

into the complex version (see [20]II§7)

$$\zeta_{\bar z}=\frac{\lambda_+-i}{\lambda_++i}\zeta_z=q\zeta_z \tag{2.2.29}$$

with

$$q=\frac{\lambda_1+i(\lambda_2-1)}{\lambda_1+i(\lambda_2+1)}=\frac{\lambda_1+i(\sqrt{D}-1)}{\lambda_1+i(\sqrt{D}+1)}. \tag{2.2.30}$$

The ellipticity (i.e. $D>0$) implies that

$$|q|<1. \tag{2.2.31}$$

If we assume that the Beltrami system (2.2.29) provides a solution ζ defining a homeomorphism (2.2.22) in some domain, then (2.2.16) becomes

$$w_{\bar\zeta}(\bar\zeta_x+\lambda_+\bar\zeta_y)+\cdots=\rho w_{\bar\zeta}+\ldots, \tag{2.2.32}$$

where the factor ρ can be written using (2.2.24) as

$$\rho=(\lambda_+-\bar\lambda_+)\bar\zeta_y=2i\lambda_2\bar\zeta_y. \tag{2.2.33}$$

With (2.2.25) the absolute value of ρ^2 becomes

$$|\rho|^2=4\lambda_2(\xi_x\eta_y-\xi_y\eta_x). \tag{2.2.34}$$

It does not vanish if the Jacobian does not vanish and $\lambda_2\neq 0$. The

Jacobian belonging to (2.2.22) where ζ satisfies (2.2.24) can be expressed by

$$\xi_x \eta_y - \xi_y \eta_x = |\zeta_z|^2 - |\zeta_{\bar{z}}|^2 = (1 - |q|^2)\,|\zeta_z|^2, \tag{2.2.35}$$

or using the Pfaffian forms and $d\eta = d_n \xi$, by

$$\begin{aligned}[d\xi, d\eta] &= (\xi_x \eta_y - \xi_y \eta_x)[dx, dy] = [d\xi, d_n \xi] \\ &= \frac{1}{\sqrt{D}} \{g_{11}\xi_x^2 + (g_{12} + g_{21})\xi_x \xi_y + g_{22}\xi_y^2\}[dx, dy]. \end{aligned} \tag{2.2.36}$$

Because we assume ζ to define a homeomorphism, the Jacobian does not vanish, and (2.2.32) becomes the new equation in the complex form

$$w_{\bar{\zeta}} = aw + b\bar{w} + c. \tag{2.2.37}$$

with suitable new complex valued coefficients a, b, c.

2.3 The Solution of Beltrami's System in the Analytic Case

Now let us assume that the coefficient q in

$$\zeta_{\bar{z}} = q(z, \bar{z})\zeta_z, \quad |q| < q_0 < 1 \tag{2.3.1}$$

is an *analytic function* in $(x, y) \in \bar{G}$. Then q as a function of $z, \bar{z}$ also is analytic and possesses an analytic continuation into some domain $D \subset \mathbb{C} \times \mathbb{C}$. All our computations will be made in D. One way for finding a Beltrami homeomorphism is to solve the Cauchy problem in D for

$$\zeta_{z^*} = q(z, z^*)\zeta_z, \qquad \zeta(z, 0) = z. \tag{2.3.2}$$

The Cauchy–Kowalewski Theorem guarantees an analytic solution of (2.3.2). It defines the desired homeomorphism in some neighbourhood of the origin because the Jacobian is continuous in a neighbourhood of the origin and fulfills at the origin the inequality

$$J = (1 - |q|^2)\,|\zeta_z|^2\,|_{z=0, z^*=0} = (1 - |q|^2)|_{(0,0)} \neq 0. \tag{2.3.3}$$

But the Cauchy–Kowalewski proof does not provide a very constructive method for finding ζ.

The linear first-order equation of (2.3.2) can be considered as one equation in the $\mathbb{C} \times \mathbb{C}$ plane. It can be solved by using the method of characteristics. This method leads to a system of Volterra equations and uses successive approximation techniques. The method of characteristics in the complex $\mathbb{C} \times \mathbb{C}$ space was used in the first existence proof for (2.3.1) given by Gauss [8] pp. 193–216. Therefore, the following method is very similar to this old idea.

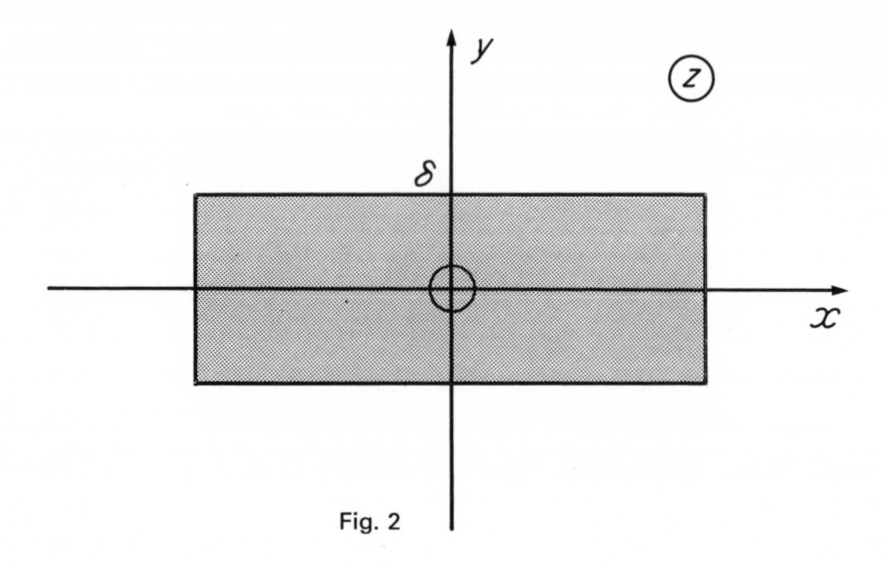

Fig. 2

Instead of (2.3.2) we formulate the Cauchy problem

$$\zeta_{z^*} = q(z, z^*)\zeta_z, \qquad \zeta(x, x) = x. \tag{2.3.4}$$

Now the Cauchy data are particularly given on an interval I of the x-axis (Fig. 2). The characteristic equations to (2.3.4) are given by

$$\frac{dz}{ds} = -q, \qquad \frac{dz^*}{ds} = 1, \qquad \frac{d\zeta}{ds} = 0. \tag{2.3.5}$$

Thus the solution of the system

$$\begin{aligned} z(s, \tau) &= z_0(\tau) - \int_0^s q(z(\sigma, \tau), z^*(\sigma, \tau))\, d\sigma, \qquad z_0(\tau) = \tau \\ z^*(s, \tau) &= z_0^*(\tau) + s, \qquad z_0^*(\tau) = \tau \\ \zeta(s, \tau) &= \zeta_0(\tau), \qquad \zeta_0(\tau) = \tau \end{aligned} \tag{2.3.6}$$

is the parametric representation of the solution to the Cauchy problem (2.3.4). (See [13, pp. 62 ff.] or [10, 20.1.])

Theorem 2.3.1 *Let a constant* $\delta_0 > 0$ *exist such that*

$$\{(z, z^*) \mid |z - x| \leq \delta_0, \quad |z^* - x| \leq \delta_0, \quad x \in I\} \subset \mathsf{D}.$$

Then a strip of width $\delta > 0$ *exists depending on the extension* $q(z, z^*)$ *such that there exists an analytic function*

$$\zeta = \zeta(x + iy, \quad x - iy)$$

solving the equation (2.3.1) *for* $x \in I$, $|y| \leqslant \delta$ *and fulfilling*

$$|\zeta_z| \neq 0. \tag{2.3.7}$$

Remark Later on, we shall see that (2.3.7) suffices to define a homeomorphism of the region $x \in I$, $|y| \leqslant \delta$ on to the image in the ζ-plane.

Using the values of the solutions of (2.3.6) on the new initial curve belonging to $z = x + i\delta$, $z^* = x - i\delta$, we can find a new strip extending the definition domain of ζ. Using Kamke's technique [13], p. 135, it can be shown that this continuation can be extended up to the boundary of the analyticity domain of q in the x, y plane.

Proof We shall see that our proof provides a very simple successive approximation technique for finding ζ. Because of the analyticity of q the iteration

$$Z_{n+1}(s, \tau) = \tau - \int_0^s q[Z_n(\sigma, \tau), \tau + \sigma]\, d\sigma, \qquad Z_0 = \tau \tag{2.3.8}$$

converges uniformly in a suitable region and can be estimated by

$$|Z_{n+1} - Z_n| \leqslant \frac{1}{(n+1)!} |Ms|^{n+1} \tag{2.3.9}$$

with a suitable constant M. The special initial values in (2.3.6) allow another formulation of the iteration (2.3.8) by using z^*, ζ as independent variables. This formulation might be more convenient for computations:

$$Z_{n+1}(z^*, \zeta) = \zeta - \int_0^{z^* - \zeta} q[Z_n(\sigma, \zeta), \sigma + \zeta]\, d\sigma, \qquad Z_0 \equiv \zeta \tag{2.3.10}$$

The limit Z of equation (2.3.8) is analytic since it is a limit of a normal family with respect to (s, τ) or (z, ζ) respectively in some region of $\mathbb{C} \times \mathbb{C}$. It satisfies the integral equation

$$Z(z^*, \zeta) = \zeta - \int_0^{z^* - \zeta} q[Z(\sigma, \zeta), \sigma + \zeta]\, d\sigma \cdot \tag{2.3.11}$$

If we know Z then $\zeta(z, z^*)$ is defined by the implicit equation

$$z = \zeta - \int_0^{z^* - \zeta} q[Z(\sigma, \zeta), \sigma + \zeta]\, d\sigma = Z(z^*, \zeta). \tag{2.3.12}$$

The inverse of an analytic function can be represented by a generalization of the Cauchy integral formula. If we let

$$F(z, \zeta, z^*) = Z(z^*, \zeta) - z \tag{2.3.13}$$

we obtain the representation ([9], p. 18)

$$\zeta=\zeta(z, z^*)=\oint_{|\hat{\zeta}|=\varepsilon>0} \hat{\zeta}\frac{\partial F/\partial \zeta}{F}\,\mathrm{d}\hat{\zeta}. \tag{2.3.14}$$

For computations we see that

$$\zeta_n(z, z^*):=\oint_{|\hat{\zeta}|=\varepsilon>0} \hat{\zeta}\frac{\dfrac{\partial Z_n}{\partial \zeta}(z^*, \hat{\zeta})}{Z_n(z^*, \hat{\zeta})-z}\,\mathrm{d}\hat{\zeta} \tag{2.3.15}$$

defines an approximation converging uniformly to the desired mapping ζ because

$$\frac{\partial Z_n}{\partial \zeta}\to\frac{\partial Z}{\partial \zeta} \tag{2.3.16}$$

converges uniformly since they are derivatives of a normal, convergent family. These partial derivatives fulfill the recursion formulas

$$\left(\frac{\partial Z_{n+1}}{\partial \zeta}\right)(z^*, \zeta)=1+q[Z_n(z^*-\zeta, \zeta), z^*]$$
$$-\int_0^{z^*-\zeta}\left\{q_z[Z_n(\sigma, \zeta), \sigma+\zeta]\cdot\frac{\partial Z_n}{\partial \zeta}+q_{z^*}\right\}\mathrm{d}\sigma. \tag{2.3.17}$$

The $\partial Z_n/\partial \zeta$, needed for (2.3.15), are now computed by (2.3.10) and (2.3.17). For estimates, the functions $\partial Z_n/\partial \zeta$ and Z_n are considered as functions of (s, τ). Then (2.3.17) becomes

$$\left(\frac{\partial Z_{n+1}}{\partial \zeta}\right)(s, \tau)=1+q[Z_n(s, \tau), s+\tau]$$
$$-\int_0^{s}\left\{q_z[Z_n(\sigma, \tau), \sigma+\tau]\frac{\partial Z_n}{\partial \zeta}+q_{z^*}\right\}\mathrm{d}\sigma. \tag{2.3.18}$$

Furthermore, both iterations lead immediately with (2.3.9) to the estimate

$$\left|\frac{\partial Z_{n+1}}{\partial \zeta}-\frac{\partial Z_n}{\partial \zeta}\right|\leqslant m\frac{|ms|^n}{n!} \tag{2.3.19}$$

where m is a suitable constant independent of ζ, s and n. (2.3.19), (2.3.15), and (2.3.9) can be used to obtain an error estimate for $\zeta-\zeta_n$.

For the proof of (2.3.7), we derive from (2.3.12) that

$$1=Z_\zeta\cdot\zeta_z,\qquad \zeta_z=\frac{1}{Z_\zeta}. \tag{2.3.20}$$

Therefore, it suffices to prove $Z_\zeta \neq 0$. Using the transformation

$$\zeta = \tau, \qquad z^* = s + \tau,$$

or (2.3.21)

$$\tau = \zeta, \qquad s = z^* - \zeta$$

the derivative can be written in the form

$$\frac{\partial}{\partial \zeta}[Z(z^*, \zeta)] = \frac{\partial}{\partial \tau}[Z(s, \tau)] - \frac{\partial}{\partial s}[Z(s, \tau)], \tag{2.3.22}$$

where $Z(s, \tau)$ is the solution of

$$Z(s, \tau) = \tau - \int_0^s q[Z(\sigma, \tau), \tau + \sigma]\, d\sigma. \tag{2.3.23}$$

Thus $Z_\zeta \neq 0$ is equivalent to

$$U(s, \tau) \equiv Z_\tau - Z_s \neq 0. \tag{2.3.24}$$

Using (2.3.23), we get by differentiation with respect to s

$$\begin{aligned} U_s &= -\frac{\partial}{\partial \tau}\left\{q[Z(s, \tau), \tau + s] + \frac{\partial}{\partial s} q[Z(s, \tau), \tau + s]\right\} \\ &= -q_z Z_\tau - q_{z^*} + q_z Z_s + q_{z^*} = -q_z \cdot U. \end{aligned} \tag{2.3.25}$$

Since $U(0, \tau) = 1 + q(\tau, \tau)$ from (2.3.24) and (2.3.23), the differential equation (2.3.25) implies

$$U(s, \tau) = Z_\zeta(s, \tau) = [1 + q(\tau, \tau)] \exp\left(-\int_{\sigma=0}^{s} q_z\, d\sigma\right). \tag{2.3.26}$$

Thus the condition $Z_\zeta \neq 0$ is fulfilled if and only if

$$1 + q(\zeta, \zeta) \neq 0. \tag{2.3.27}$$

This is fulfilled in some neighbourhood of the real axis, which proves the theorem.

Example Let us suppose that $q = q_0$ is a *constant*. This case consists of all equations which are equivalent to equations with constant coefficients.

Equation (2.3.11) takes the form

$$Z(z^*, \zeta) = \zeta - q_0(z^* - \zeta) = \zeta(1 + q_0) - q_0 z^*. \tag{2.3.28}$$

The equation (2.3.12) becomes

$$z = \zeta(1 + q_0) - q_0 z^* \tag{2.3.29}$$

and (2.3.14) has the explicit representation

$$\zeta=\frac{1}{1+q_0}(z+q_0z^*), \tag{2.3.30}$$

$$\zeta(z,\bar z)=\frac{z+q_0\bar z}{1+q_0}$$
$$=\frac{1}{1+q_0}(x(1+q_{01})+q_{02}y+\mathrm{i}[y(1-q_{01})+xq_{02}] \tag{2.3.31}$$

where $q_0=q_{01}+\mathrm{i}q_{02}$. Because the linear transformation

$$\hat x=x(1+q_{01})+yq_{02}$$
$$\hat y=xq_{02}+y(1-q_{01})$$

has the Jacobian $1-|q_0|^2$, ζ obviously provides a homeomorphism $(x, y)\to(\xi,\eta)$.

Remark Let us replace (2.3.1) by

$$\zeta_{\bar z}=\frac{P}{Q}\zeta_z \tag{2.3.32}$$

and let us suppose that P, Q are analytic and satisfy

$$|P|\leqslant|Q|q_0,\quad q_0<1\quad\text{and}\quad P=Q=0\Leftrightarrow x=y=0 \tag{2.3.33}$$

in a neighbourhood of $x=y=0$. It is still an open problem whether or not there exists a local *homeomorphism* ζ which has the form

$$\zeta=\frac{\chi_1}{\chi_2} \tag{2.3.34}$$

where χ_1, χ_2 are analytic functions with $\chi_2=0\Leftrightarrow x=y=0$. This problem arises in connection with some rigidity theorems in differential geometry. See [12, Remark to (2.2.8), p. 53].

2.4 Existence Proofs with the Hilbert Transformation

In the existence proofs using the two-dimensional Hilbert transformation the desired local homeomorphism ζ is of the form

$$\zeta(t)=\frac{1}{2\pi\mathrm{i}}\iint\limits_{|z|\leqslant 1}\frac{f(z)}{z-t}[\mathrm{d}z,\mathrm{d}\bar z]+t\equiv Pf+t. \tag{2.4.1}$$

The function ζ has to satisfy

$$\zeta_{\bar{z}} = q\zeta_z \quad \text{where} \quad |q| \leqslant q_0 < 1 \quad \text{with} \quad q \in L_\infty. \tag{2.4.2}$$

Therefore, we need the derivatives of (2.4.1). If f is smooth enough, e.g. $f \in C^\alpha$, $\zeta_{\bar{z}}$ is given by

$$\zeta_{\bar{z}} = f \quad \text{with} \quad f = 0 \quad \text{for} \quad |z| \geqslant 1. \tag{2.4.3}$$

Hence, we assume $q = 0$ for $|z| \geqslant 1$. The other derivative is

$$\zeta_z = \frac{1}{2\pi i} \iint\limits_{|z-t|\leqslant 2} \frac{f(z)-f(t)}{(z-t)^2} [dz, d\bar{z}] + 1. \tag{2.4.4}$$

That follows from

$$\zeta_z(t) = f(t) \frac{1}{2\pi i} \frac{\partial}{\partial t} \iint\limits_{|z-t|\leqslant 2} \frac{1}{(z-t)} [dz, d\bar{z}]$$

$$+ \frac{1}{2\pi i} \iint\limits_{|z-t|\leqslant 2} \frac{f(z)-f(t)}{(z-t)^2} [dz, d\bar{z}] + 1 \tag{2.4.5}$$

with

$$\iint\limits_{|z-t|\leqslant 2} \frac{1}{(z-t)} [dz, d\bar{z}] = 0$$

(see [10] Lemma 2 p. 99).

We denote by

$$Tf \equiv \frac{1}{2\pi i} \iint\limits_{|z-t|\leqslant 2} \frac{f(z)-f(t)}{(z-t)^2} [dz, d\bar{z}] \tag{2.4.6}$$

the Hilbert transformation which can also be expressed in terms of the principal value integral

$$Tf = \text{P.V.} \frac{1}{2\pi i} \iint\limits_{|z-t|\leqslant 2} \frac{f(z)}{(z-t)^2} [dz, d\bar{z}]$$

$$= \lim_{\delta\to 0} \frac{1}{2\pi i} \iint\limits_{\delta\leqslant|z-t|\leqslant 2} \frac{f(z)}{(z-t)^2} [dz, d\bar{z}] \tag{2.4.7}$$

Furthermore, we have the following:

Lemma 2.4.1 *For any $f \in C^\alpha$ with $f=0$ for all $|z| \geq 1$ the following estimates are valid:*

$$\|Tf\|_{L_2(R^2)} = \iint_{R^2} |Tf|^2[dx, dy] = \iint_{|z|\leq 1} |f|^2[dx, dy] = \|f\|_{L_2} \tag{2.4.8}$$

and

$$\|Tf\|_\alpha \leq C_1 \|f\|_\alpha, \tag{2.4.9}$$

where the Hölder-norm $\|\cdot\|_\alpha$ is defined by

$$\|f\|_\alpha \equiv \sup |f| + \sup \frac{|f(z_1)-f(z_2)|}{|z_1-z_2|^\alpha} \equiv \|f\|_0 + [f]_\alpha \tag{2.4.10}$$

Proof ([2]) By Gauss' Theorem, the identity

$$\iint_{|z|\leq R_0} (Tf)\overline{(Tf)}[dz, d\bar{z}] = \iint_{|z|\leq R_0} (Pf)_z \overline{(Pf)_z}[dz, d\bar{z}]$$

$$= \oint_{|z|=R_0} \{(Pf)\overline{(Pf)_z}\, d\bar{z} + (Pf)\overline{(Pf)_{\bar{z}}}\, dz\} + \iint_{|z|\leq R_0} (Pf)_{\bar{z}} \cdot \overline{(Pf)_{\bar{z}}}[dz, d\bar{z}]$$

$$= 0\left(\frac{1}{R_0^2}\right) + \iint_{|z|\leq R_0} f \cdot \bar{f}[dz, d\bar{z}] \tag{2.4.11}$$

holds for any $R_0 > 0$ and any continuous f with compact support. For $R_0 \to \infty$ the boundary integral approaches zero, therefore, (2.4.11) yields (2.4.8).

For the proof of (2.4.9), we use the identity

$$Tf(1) - Tf(-1) = \text{P.V.} \frac{1}{2\pi i} \iint f(z) \left(\frac{1}{(z-1)^2} - \frac{1}{(z+1)^2} \right) [dz, d\bar{z}]$$

$$= \text{P.V.} \frac{1}{2\pi i} \iint f(z) \frac{4z}{(z^2-1)^2} [dz, d\bar{z}]$$

$$= \text{P.V.} \frac{1}{2\pi i} \iint_{x>0} \{f(z)-f(1)\} \frac{4z}{(z^2-1)^2} [dz, d\bar{z}]$$

$$+ \text{P.V.} \frac{1}{2\pi i} \iint_{x<0} \{f(z)-f(-1)\} \frac{4z}{(z^2-1)^2} [dz, d\bar{z}]$$

$$+ \{f(1)-f(-1)\} \text{P.V.} \frac{1}{2\pi i} \iint_{x>0} \frac{4z}{(z^2-1)^2} [dz, d\bar{z}] \tag{2.4.12}$$

with

$$\text{P.V.} \iint_{x>0} \frac{4z}{(z^2-1)^2}[dz, d\bar{z}] + \text{P.V.} \iint_{x<0} \frac{4z}{(z^2-1)^2}[dz, d\bar{z}] = 0. \tag{2.4.13}$$

The identity (2.4.12) implies with $|f(z)-f(\pm 1)| \leqslant |z \mp 1|^\alpha [f]_\alpha$ the inequality

$$\begin{aligned} |Tf(1)-Tf(-1)| &\leqslant \sup \frac{|f(z)-f(t)|}{|z-t|^\alpha} \cdot \left\{ \frac{4}{\pi} \iint_{x>0} \frac{|z-1|^\alpha\, |z|}{|z-1|^2\, |z+1|^2} [dx, dy] \right. \\ &\quad + \frac{4}{\pi} \iint_{x<0} \frac{|z+1|^\alpha\, |z|}{|z-1|^2\, |z+1|^2} [dx, dy] \\ &\quad \left. + \frac{2^\alpha}{2\pi} \left| \text{P.V.} \iint_{x>0} \frac{4z}{(z^2-1)^2} [dz, d\bar{z}] \right| \right\} \\ &\leqslant 2^\alpha \cdot C_1 \sup \frac{|f(z)-f(t)|}{|z-t|^\alpha} = 2^\alpha C_1 [f]_\alpha \end{aligned} \tag{2.4.14}$$

with a suitable constant C_1 depending only on α. Considering instead of f the transformed function

$$\tilde{f}(z) = f\left(\frac{s-t}{2} z + \frac{s+t}{2}\right) \tag{2.4.15}$$

with

$$\sup \frac{|\tilde{f}(z)-\tilde{f}(\tilde{z})|}{|z-\tilde{z}|^\alpha} \leqslant 2^{-\alpha}\, |s-t|^\alpha\, [f]_\alpha \tag{2.4.16}$$

in the inequality (2.4.14) we get the estimate:

$$|Tf(s)-Tf(t)| = |T\tilde{f}(1)-T\tilde{f}(-1)| \leqslant 2^\alpha \cdot C_1 \cdot 2^{-\alpha}\, |s-t|^\alpha\, [f]_\alpha. \tag{2.4.17}$$

This implies the desired inequality (2.4.9).

The equality (2.4.8) shows that T can be extended to the Hilbert space $L_2[(|z| \leqslant 1)]$ such that the extension is a bounded linear operator satisfying (2.4.8).

Theorem 2.4.2 *The functional equation*

$$f = qTf + q \tag{2.4.18}$$

has a unique solution in L_2.

Proof The operator on the right-hand side is a contraction in $L_2[(|z|\leqslant 1)]$:

$$\|qTf-qTg\|_{L_2}\leqslant q_0\|f-g\|_{L_2}, \quad q_0<1. \tag{2.4.19}$$

Therefore, the usual successive approximation

$$f_{n+1}\equiv qTf_n+q, \qquad f_0\equiv 0, \tag{2.4.20}$$

converges in L_2 to the unique solution f.

Remark The solution f of (2.4.18) defines by (2.4.1) a function that can be considered as the desired generalized solution of (2.4.2). For $f\in L_2$ we have only

$$\zeta=Pf+z\in L_p(\Omega) \quad \text{for any compact subset } \Omega\subset R^2 \text{ and any } p<\infty \tag{2.4.21}$$

(see Sobolev [18]). Therefore, ζ is at the first point of view even not continuous. In the following theorems we investigate the regularity of f and ζ.

Theorem 2.4.3 *The solution* $f\in L_2$ *of* (2.4.18) *belongs to* L_p *with a suitable* $p>2$.

This theorem was proven by Bojarski (see also Vekua's book [20], Chapter II, Section 3 ff.). He used the results from Calderon and Zygmund on the Hilbert transformation in L_p-spaces together with the interpolation theorem of Riesz: T can be extended to $L_{p'}$ for any chosen p' such that the extension

$$T=L_{p'}\to L_{p'}[(|z|\leqslant 1)], \tag{2.4.22}$$

is a continuous mapping. Then the convexity theorem of Riesz holds for T (see [7] I, p. 525):

$$\|T\|_p\leqslant 1\cdot(\|T\|_{p'})^{\frac{p-2}{p'-2}\cdot\frac{p'}{p}}, \quad 2\leqslant p\leqslant p'<\infty. \tag{2.4.23}$$

Therefore, the inequality

$$\|qT\|_p\leqslant q_0(\|T\|_{p'})^{\frac{(p-2)p'}{(p'-2)p}}<1 \tag{2.4.24}$$

holds for some $p>2$. With this p the operator qT in (2.4.18) remains a contraction in L_p. Hence, the unique solution f is in L_p.

Theorem 2.4.4 *Let* $q\in C^\alpha$, $|q|\leqslant q_0<1$, $q=0$ *for all* $|z|\geqslant 1$. *Then the solution* f *of* (2.4.18) *is such that* $f\in C^\alpha$.

Proof For the proof we use an idea of Bers and Chern [5, pp. 350 ff.]

who introduced weighted Hölder norms. Let $z_1 \in \mathbb{C}$ be an arbitrarily chosen point which is fixed in the following steps. If we consider

$$q_1 = q(z_1) \tag{2.4.25}$$

instead of q then a homeomorphism is given by (2.3.31)

$$\hat{z} = \frac{z + q_1 \bar{z}}{1 + q_1}. \tag{2.4.26}$$

Now let ζ be the (generalized) solution (2.4.1) of (2.4.2) generated by $f \in L_2$. From Theorem 2.4.3 it follows that $f \in L_p$ for some $p > 2$ and $\zeta \in C^\beta$ with $\beta = 1 - (2/p) > 0$ (see [20]). The (generalized) derivatives of ζ are transformed by

$$1 + T_f = \zeta_z = \zeta_{\hat{z}} \frac{1}{1 + q_1} + \zeta_{\bar{\hat{z}}} \frac{\bar{q}_1}{1 + \bar{q}_1} \tag{2.4.27}$$

and

$$f = \zeta_{\hat{z}} \frac{q_1}{1 + q_1} + \zeta_{\bar{\hat{z}}} \frac{1}{1 + \bar{q}_1}. \tag{2.4.28}$$

Therefore, the equation (2.4.2) is transformed into

$$\zeta_{\bar{\hat{z}}} = \mu \zeta_{\hat{z}} \equiv \frac{1 + \bar{q}_1}{1 - \bar{q}_1 q} \frac{q - q_1}{1 + q_1} \zeta_{\hat{z}}. \tag{2.4.29}$$

Here $\mu \in C^\alpha$, $\mu(z_1) = 0$ with $|\mu| < 1$, and $\zeta_{\hat{z}}$, $\zeta_{\bar{\hat{z}}} \in L_p$, $p > 2$. Now let us choose a function $\chi(s) \in C^\infty$ with a compact support and the properties

$$\left.\begin{array}{rll} \chi = 1 & \text{for all} & 0 \leq s < \delta \\ 0 \leq \chi \leq 1 & \text{for all} & \delta \leq s < 2\delta \\ \chi = 0 & \text{for all} & 2\delta \leq s \end{array}\right\} \tag{2.4.30}$$

where $\delta > 0$ is chosen arbitrarily with $\delta \leq 1$. Then the product

$$\tilde{\zeta} \equiv \chi \zeta \tag{2.4.31}$$

satisfies the equation

$$\tilde{\zeta}_{\bar{\hat{z}}} = \mu \tilde{\zeta}_{\hat{z}} + (\chi_{\hat{z}} - \mu \chi_{\hat{z}}) \zeta. \tag{2.4.32}$$

Because the derivative $\partial / \partial_{\bar{\hat{z}}}$ of

$$P(\mu \tilde{\zeta}_{\hat{z}}) + P[(\chi_{\hat{z}} - \mu \chi_{\hat{z}}) \zeta] \tag{2.4.33}$$

equals the right-hand side of (2.4.32), the function $\tilde{\zeta}$ can be represented by

$$\tilde{\zeta} = P(\mu \tilde{\zeta}_{\hat{z}}) + P[(\chi_{\bar{\hat{z}}} - \mu \chi_{\hat{z}}) \zeta] + \phi(\hat{z}) \tag{2.4.34}$$

where $\phi(\hat{z})$ is a suitable holomorphic function in the $\hat{z}$-plane. Thus, for $\tilde{\zeta}_{\hat{z}}$

the integral equation

$$\tilde{\zeta}_{\hat{z}} = T(\mu\tilde{\zeta}_{\hat{z}}) + T[(\chi_{\bar{\hat{z}}} - \mu\chi_{\hat{z}})\zeta] + \phi_{\hat{z}}$$

$$= \text{P.V.} \frac{1}{2\pi i} \iint\limits_{|\hat{z}|\leq 2\delta} \frac{\mu\tilde{\zeta}_{\hat{z}}}{(\hat{z}-\hat{t})^2}[d\hat{z}, d\bar{\hat{z}}] + T[(\chi_{\bar{\hat{z}}} - \mu\chi_{\hat{z}}]\zeta) + \phi_{\hat{z}} \tag{2.4.35}$$

is fulfilled and $\tilde{\zeta}_{\hat{z}} \in L_p$ for some $p > 0$.

The last two terms are Hölder continuous functions, hence, $\tilde{\zeta}_{\hat{z}}$ will also become Hölder continuous if the principal value integral is a *contraction* operator in the space of Hölder continuous functions. In order to show that, we estimate

$$[T\mu g]_{\beta,\delta} \leq C_1[\mu g]_{\beta,\delta} \leq C_1 \cdot \{\|\mu\|_{0,\delta}[g]_{\beta,\delta} + [\mu]_{\beta,\delta}\, \|g\|_{0,\delta}\} \tag{2.4.36}$$

where

$$\|g\|_{0,\delta} = \max_{|\hat{z}|\leq 2\delta} |g|, \qquad [g]_{\beta,\delta} = \sup_{\substack{|\hat{z}|\leq 2\delta \\ |\hat{t}|\leq 2\delta}} \frac{|g(\hat{z}) - g(\hat{t})|}{|\hat{z}-\hat{t}|^\beta} \tag{2.4.37}$$

and where C_1 is independent of δ for $\delta \leq 1$. Similarly,

$$\|T\mu g\|_{0,\delta} = \sup_{|\hat{t}|\leq 2\delta} \left| \iint\limits_{|\hat{z}|\leq 2\delta} \frac{\mu g(\hat{z}) - \mu g(\hat{t})}{(\hat{z}-\hat{t})^2} [d\hat{z}, d\bar{\hat{z}}] \right|$$

$$\leq [\mu g]_{\beta,\delta} \left| \iint\limits_{|\hat{z}|\leq 2\delta} |\hat{z}-\hat{t}|^{\beta-2}[d\hat{z}, d\bar{\hat{z}}] \right| \leq C_2\delta^\beta[\mu g]_{\beta,\delta} \tag{2.4.38}$$

where C_2 is independent of δ. For proving the contraction property we use the weighted norm

$$|g|_\beta \equiv \|g\|_{0,\delta} + \delta^\beta[g]_{\beta,\delta} \tag{2.4.39}$$

which in $C^\beta\,[(|\hat{z}| \leq 2\delta)]$ is equivalent to $\|g\|_\beta$ for each fixed $\delta > 0$. Then $T\mu g$ satisfies the following inequality,

$$|T\mu g|_\beta = \|T\mu g\|_{0,\delta} + \delta^\beta[T\mu g]_{\beta,\delta} \leq (C_1 + C_2)\delta^\beta[\mu g]_{\delta,\beta}$$

$$\leq (C_1 + C_2)\{\delta^\beta\, \|\mu\|_{0,\delta}\,[g]_{\beta,\delta} + \delta^\beta[\mu]_{\beta,\delta}\, \|g\|_{0,\delta}\}$$

$$\leq (C_1 + C_2)[\max\{\|\mu\|_{0,\delta}, \delta^\beta[\mu]_\beta\}(\|g\|_{0,\delta} + \delta^\beta[g]_{\beta,\delta})],$$

$$|T\mu g|_\beta \leq (C_1 + C_2)\max\{\|\mu\|_{0,\delta}, \delta^\beta[\mu]_\beta\}\, |g|_\beta \leq C\, |\mu|_\beta\, |g|_\beta. \tag{2.4.40}$$

If we choose $\delta > 0$ small enough then $(T\mu\, \cdot)$ becomes a *contraction* in C^β. If we now interpret (2.4.35) as an equation for $\tilde{\zeta}_{\hat{z}}$ where $\phi_{\hat{z}}$ and

$T[(\chi_{\bar{z}}-\mu\chi_{\hat{z}})\zeta]\in C^\beta$ are given, then it has a unique solution in C^β. Thus, the given $\tilde{\zeta}_{\hat{z}}$ equals almost everywhere a C^β-function. Therefore, $f\in C^\beta$ in the whole plane and $\zeta\in C^{1+\beta}$. Now $\tilde{\zeta}_{\bar{z}}\in C^\beta$ satisfies (2.4.35) with $T[(\chi_{\bar{z}}-\mu\chi_{\hat{z}})]\in C^\alpha$ in a neighbourhood of z_1. If δ is again chosen small enough then $T(\mu\,\cdot)$ becomes also a contraction in C^α and $\tilde{\zeta}_{\hat{z}}$ and, therefore, f has to be in C^α.

Corollary 2.4.5 *If $q\in C^\alpha$ and $|q|\leq q_0<1$ then there exists a local homeomorphism in $C^{1+\alpha}$.*

Proof Instead of (2.4.35) we solve

$$\tilde{f}=1+\frac{1}{2\pi\mathrm{i}}\,\mathrm{P.V.}\iint\limits_{|\hat{z}|\leq\delta}\frac{\mu\tilde{f}}{(\hat{z}-\hat{t})^2}[\mathrm{d}_{\hat{z}},\mathrm{d}_{\bar{\hat{z}}}]=1+T(\mu\tilde{f}) \tag{2.4.41}$$

for a suitably small $\delta>0$ in $C^\alpha(\{|\hat{z}|\leq\delta\})$. Then (2.4.40) yields the estimate

$$|\tilde{f}|_\alpha\leq\frac{1}{1-C\,|\mu|_\alpha} \tag{2.4.42}$$

for small enough $\delta>0$. The function

$$\tilde{\zeta}=\hat{t}+\frac{1}{2\pi\mathrm{i}}\iint\limits_{|\hat{z}|\leq\delta}\frac{\mu\tilde{f}}{(\hat{z}-\hat{t})}[\mathrm{d}\hat{z},\mathrm{d}\bar{\hat{z}}] \tag{2.4.43}$$

satisfies the differential equation

$$\tilde{\zeta}_{\bar{\hat{z}}}=\mu\tilde{\zeta}_{\hat{z}} \tag{2.4.44}$$

and furthermore the estimate

$$|\tilde{\zeta}_{\hat{z}}|_\alpha\geq\frac{1}{1+C_1\,|\mu|_\alpha}>0. \tag{2.4.45}$$

Hence, $\hat{\zeta}$ is a local homeomorphism with $\tilde{\zeta}\in C^{1+\alpha}$ in a suitable neighbourhood of $\hat{z}=0$.

Now we are in the position to prove that the mapping ζ (2.4.1) generated by the solution f of (2.4.18) is already a homeomorphism on the whole plane.

Theorem 2.4.6 *Let $q\in C^\alpha$, $|q|\leq q_0<1$, $q=0$ for all $|z|\geq1$. Then the mapping ζ defined by* (2.4.1) *with f from* (2.4.18) *is a homeomorphism of the plane on to itself.* (The proof follows essentially Vekua [20].)

Proof We have to show that for any choice of ζ_0 the complex valued function

$$W \equiv \zeta(x, y) - \zeta_0 \tag{2.4.46}$$

has exactly one zero. That we shall show in two steps.

Step 1 *For any choice of ζ_0, there exists a $R_0 > 0$ such that*

$$W \neq 0 \quad \text{for all} \quad x^2 + y^2 \geqslant R_0^2 \tag{2.4.47}$$

and

$$\frac{1}{2\pi} \oint_{x^2+y^2=R_0^2} \mathrm{d} \arg W = 1. \tag{2.4.48}$$

First we observe that, because the solution f of (2.4.18) has compact support, the function $\zeta(t)$ in (2.4.1) behaves like t for $|t| \to \infty$ and

$$|\zeta - t|, |\zeta_t - 1|, |\zeta_{\bar{t}}| = 0\left(\frac{1}{t}\right). \tag{2.4.49}$$

Hence (2.4.47) follows immediately from the inequality

$$|W| = |\zeta(t) - \zeta_0 + t - t| \geqslant |t| - |\zeta_0| - \frac{k}{|t|} \geqslant R_0 - \frac{k}{R_0} - |\zeta_0|.$$

(2.4.48) is also a consequence of (2.4.49) since

$$\frac{1}{2\pi} \oint_{|t|=R_0} \mathrm{d} \arg W = \frac{1}{2\pi} \oint_{|t|=R_0} \mathrm{d} \arg t + \frac{1}{2\pi} \oint_{|t|=R_0} \mathrm{d} \arg\{1 + \chi\} = 1 + 0$$

with

$$\chi(t, \bar{t}) = \frac{\zeta - t}{t} - \frac{\zeta_0}{t} = 0(R_0^{-1}).$$

Step 2 Now we investigate the zeros of W. Since W satisfies the Beltrami system

$$W_{\bar{z}} = q W_z, \tag{2.4.50}$$

this system can be transformed *locally* into the Cauchy–Riemann system by using the local transformation $\tilde{\zeta}$ of Corollary 2.4.5. Hence, for any z_0, there exists a neighbourhood where W satisfies the Cauchy–Riemann equations,

$$W_{\bar{\tilde{\zeta}}} = 0 \tag{2.4.51}$$

with respect to the new variables $\tilde{\zeta}$ corresponding to that neighbourhood. By Heine–Borel's Theorem the disk $|z| \leqslant R_0$ can be covered by only a finite number of open disks and corresponding local homeomorphisms. If W would have an accumulation point of zeros in one of these disks then W would have an infinite number of zeros in the image domain corresponding to the local transformation where the transformated W is holomorphic. Thus W would vanish identically in this disk. Therefore W has accumulation points of zeros in the nonempty intersections with the neighboured disks. Using for them the same arguments as above, we find, that W would vanish in the neighboured disks also. After a finite number of repetitions we find that W would vanish identically for all $|z| \leq R_0$ in contradiction to (2.4.47). Therefore W can only have isolated zeros z_j. Furthermore, (2.4.47) implies $|z_j| < R_0$, $j = 1, \ldots, N$. Around each of the z_j's, we draw a circle with a radius $\delta > 0$ where δ is chosen small enough such that in each disk $(z \mid |z - z_j| \leq \delta)$ a corresponding local homeomorphism $\tilde{\zeta}_j$ exists. For the holed region $(z \mid |z| \leq R_0 \wedge |z - z_j| \geq \delta, j = 1, \ldots, N)$ we have $W \neq 0$. Corresponding to each hole around z_j let Γ_j be a cut connecting $|z - z_j| = \delta$ with $|z| = R_0$ having the two frontiers Γ_j^+ and Γ_j^-. Then in the cut simply connected domain arg W can be defined as an univalent function and

$$\frac{1}{2\pi} \oint_{|z|=R_0} \mathrm{d} \arg W - \sum_{j=1}^{N} \frac{1}{2\pi} \oint_{|z-z_j|=\delta} \mathrm{d} \arg W + \frac{1}{2\pi} \int_{\Gamma_j} \mathrm{d}[\arg W)|_{\Gamma_j^+} - (\arg W)|_{\Gamma_j^-}] = 0. \quad (2.4.52)$$

Because of $W|_{\Gamma_j^+} = W|_{\Gamma_j^-}$ the value of

$$\frac{1}{2\pi} (\arg W)|_{\Gamma_j^+} - \frac{1}{2\pi} (\arg W)|_{\Gamma_j^-} \quad \text{will equal an integer.}$$

By continuity, this number will be constant along the whole cut Γ_j. That implies

$$\mathrm{d}[(\arg W)|_{\Gamma_j^+} - (\arg W)|_{\Gamma_j^-}] = 0 \quad \text{on} \quad \Gamma_j$$

and (2.4.52) becomes

$$\frac{1}{2\pi} \oint_{|z|=R_0} \mathrm{d} \arg W - \sum_{j=1}^{N} \frac{1}{2\pi} \oint_{|z-z_j|=\delta} \mathrm{d} \arg W = 0. \quad (2.4.53)$$

The value of the first term is known from (2.4.48), whereas in the second term, each circle integral can be interpreted as a contour integral in a suitable $\tilde{\zeta}$-plane with a holomorphic W. Therefore, they count the multiplicities n_j of the zero z_j, and we end up with

$$\sum_{j=1}^{N} n_j = 1. \quad (2.4.54)$$

Consequently, there is exactly one zero, $N=1$, and z_1 has the multiplicity 1.

Note that ζ maps $\mathbb{C}$ bijectively on to $\mathbb{C}$.

Proposition 2.4.7 *The Jacobian of ξ, η never vanishes,* i.e. $\zeta_z \neq 0$.

For this purpose let us suppose there were a zero of $\zeta_z : \zeta_z(z_0)=0$ and with (2.4.2) also $\zeta_{\bar{z}}(z_0)=0$. The Beltrami system (2.4.2) can *locally* be transformed into

$$\zeta_{\bar{\tilde{\zeta}}}=0. \tag{2.4.55}$$

Hence, there is some neighbourhood of z_0 where ζ is a *holomorphic* function of $\tilde{\zeta}$. Since ζ and $\tilde{\zeta}$ are both bijective, there the mapping $\zeta[z(\tilde{\zeta})]$ is conformal providing

$$\zeta_{\tilde{\zeta}}(z_0)\neq 0,$$

in contradiction to

$$\zeta_z(z_0)=\zeta_{\bar{z}}(z_0)=0.$$

Remarks (1) In the case $q\in L_\infty$, $|q|\leqslant q_0<1$, $q=0$ for $|z|\geqslant 1$, Theorem 2.4.6 holds also with $\zeta\in C^\alpha$. It can be proven using an approximation of q by C^α-functions. This is Bojarski's approach and can be found in Vekua's book ([20], pp. 95 ff.).

(2) Ahlfors [2] proved the existence of ζ for $q\in C^\alpha$ with compact support by solving a *family of equations*

$$\zeta_{\bar{z}}=\tau q\zeta_z, \quad \tau\in[0, 1] \quad \text{for} \quad \zeta=\zeta(\tau, z). \tag{2.4.56}$$

He used Schauder's continuation method and showed for the following subset of [0, 1]

$$\gamma=\{\tau\in[0, 1] \mid \zeta(\tau, z) \text{ solves (2.4.56) and defines a } C^{1+\alpha}\text{-homeomorphism}\}, \tag{2.4.57}$$

(i) γ is open with respect to [0, 1] and

(ii) γ is closed.

Therefore $\gamma=[0, 1]$. Furthermore, he found ζ by a finite composition of mappings where each is generated by the solution of a suitable contraction mapping. That γ is open and closed is shown in the following way:

(i) Let $\tau_1\in\gamma$ then $\zeta_1=\zeta(\tau_1, z)$ is a $C^{1+\alpha}$-homeomorphism. If for any τ, ζ is considered to be the composite mapping $\zeta=\zeta[\tau, \zeta_1(z)]$ then (2.4.44) is equivalent to

$$\zeta_{\bar{\zeta}_1} = \frac{\zeta_{1z}}{\bar{\zeta}_{1z}} \frac{(\tau-\tau_1)q}{1-\tau \cdot \tau_1 q} \zeta_{\zeta_1} \equiv \mu(\tau_1, \tau)\zeta_{\zeta_1} \tag{2.4.58}$$

with

$$\|\mu\|_\alpha = |\tau-\tau_1| \left\| \frac{\zeta_{1z} q}{\bar{\zeta}_{1z}(1-\tau \cdot \tau_1 q)} \right\|_\alpha \to 0 \quad \text{for} \quad \tau \to \tau_1. \tag{2.4.59}$$

Therefore, the equation for f,

$$f = \mu \mathrm{T} f + \mu \tag{2.4.60}$$

is uniquely solvable in C^α for all τ in a suitable neighbourhood of τ_1. Thus, γ is open.

(ii) With a careful investigation of f and μ Ahlfors shows that

$$\|\mu\|_\beta \leq |\tau-\tau_1| \cdot C \tag{2.4.61}$$

with a constant C *independent* of τ where $\beta > 0$ is a suitable Hölder exponent. This implies for a family $\zeta(\tau, z)$ that the corresponding densities f_τ converge to a limit f_{τ_0}. Thus, the $\zeta(\tau, z)$ approach $\zeta(\tau_0, z)$ and $\zeta(\tau_0, z)$ becomes also a homeomorphism. The investigation of (2.4.41) shows then that $\zeta \in C^{1+\alpha}$. Thus, γ is closed.

2.5 *A Priori* Estimates

For the investigation of general first-order systems we shall establish *a priori* estimates for the desired solution. They can be obtained from such estimates belonging to a system with only two unknown functions. Such a BVP is given by

$$\mathbf{Du} \equiv \mathbf{u}_x + \mathbf{B}\mathbf{u}_y + \mathbf{C}\mathbf{u} = \mathbf{f} \quad \text{in} \quad \mathrm{G} \tag{2.5.1}$$

and

$$\mathrm{r} \cdot \mathbf{u} = \psi \quad \text{on} \quad \dot{\mathrm{G}}, \qquad \mathrm{r} = (r_1, r_2). \tag{2.5.2}$$

Regarding Section 2.2, an elliptic system can be transformed into Hilbert's normal form for the new unknown function

$$w = \mathbf{p}_+^T \mathbf{u}, \tag{2.5.3}$$

where $\mathbf{p}_+^T$ denotes the transpose of the eigenvector (2.2.11). Then the complex conjugate is

$$\bar{w} = \overline{\mathbf{p}_+^T \mathbf{u}} = \mathbf{p}_-^T \mathbf{u}. \tag{2.5.4}$$

The inverse transformation (2.2.18) is given by the inverse

$$\mathsf{Q} = (\mathbf{q}, \bar{\mathbf{q}}), \tag{2.5.5}$$

of $(\mathbf{p}_+, \mathbf{p}_-)^T$. Q exists satisfying

$$\mathsf{Q}(\mathbf{p}_+, \mathbf{p}_-)^T = (\mathbf{p}_+, \mathbf{p}_-)^T \mathsf{Q} = (\delta_{jk}). \tag{2.5.6}$$

Then the inverse transformation for the unknown functions is given by (2.2.18) or

$$\mathbf{u} = 2 \operatorname{Re} w\mathbf{q}. \tag{2.5.7}$$

The boundary condition becomes

$$\operatorname{Re} Pw \equiv \operatorname{Re}(2\mathbf{q} \cdot \mathsf{r})w = \psi \quad \text{on} \quad \dot{\mathsf{G}}. \tag{2.5.8}$$

After some elementary computations, one gets with (2.2.17)

$$P = 2\mathbf{q} \cdot \mathsf{r} = -\frac{\mathrm{i}}{d}(r_2, -r_1)\mathbf{p}_-. \tag{2.5.9}$$

To get a boundary value problem for w of the form (1.2.1) from (2.5.8), we have to demand $P \neq 0$ on $\dot{\mathsf{G}}$. This condition is called the *Lopatinski condition:*

$$2\mathsf{r} \cdot \mathbf{q} = P \neq 0 \quad \text{on} \quad \dot{\mathsf{G}}. \tag{2.5.10}$$

Remark Here the Lopatinski condition is equivalent to the 'covering condition' ([15], p. 128) and the 'complementing condition' ([1], condition 2).

Now let $\mathsf{B} \in C^{1+\alpha}(\bar{\mathsf{G}}), \mathsf{C}, \mathbf{f} \in C^{\alpha}(\bar{\mathsf{G}})$ and $\mathsf{r}, \psi \in C^{1+\alpha}(\dot{\mathsf{G}})$ and $\dot{\mathsf{G}} \in C^{1+\alpha}$. Referring to Section 1.3 and the properties of the homeomorphism ζ the following theorem holds:

Theorem 2.5.1 *If* (2.5.1) *is elliptic in* $\bar{\mathsf{G}}$ *and the boundary condition* r *in* (2.5.2) *satisfies the Lopatinski condition* (2.5.10) *then the boundary value problem is Fredholm with*

$$\text{nullity} = \max(-2n+1, 0) \quad \text{and} \quad \text{codim (range)} = \max(+2n-1, 0) \tag{2.5.11}$$

where

$$n = \frac{1}{2\pi} \oint_{\dot{\mathsf{G}}} \mathrm{d} \arg P. \tag{2.5.12}$$

Furthermore, the following *a priori* estimate holds, which supports for more general cases the whole solvability theory ([1] and [14]).

Let us denote by $\|\cdot\|_{r+\alpha}$ the $r+\alpha$-Hölder norm belonging to vector functions on G:

$$\|\mathbf{u}\|_{r+\alpha} \equiv \max_{l=1,2} \left\{ \sum_{j+k \leqslant r} \|u^l_{x^j y^k}\|_0 + \sum_{j+k=r} [u^l_{x^j y^k}]_\alpha \right\}$$

where $[\cdot]_\alpha$ denotes the seminorm as in (2.4.10) corresponding to G and by $|\psi|_{r+\alpha}$ the corresponding norm for functions ψ defined on the boundary $\dot{G}$.

Theorem 2.5.2 If (2.5.1) *is elliptic in* $\bar{G}$ *and* (2.5.10) *holds then there exists a constant* γ *independent of* $\mathbf{u}$ *such that the* a priori *estimate*

$$\|\mathbf{u}\|_{1+\alpha} \leqslant \gamma(|\mathbf{ru}|_{1+\alpha} + \|\mathbf{Du}\|_\alpha + \|\mathbf{u}\|_0) \tag{2.5.13}$$

holds.

Proof (1) Let $n \equiv (1/2\pi) \oint \mathrm{d} \arg p > 0$. Then for any $\mathbf{f}$ in the image of D and any ψ in the image of r there exists exactly one solution $w = \mathbf{p}_+^T \mathbf{u}$ of the transformed problem. Referring to Section 1.3, the range of the mapping R belonging to the transformed boundary value problem is a closed subspace of $C^\alpha(\bar{G}) \times C^{1+\alpha}(\dot{G})$. The inverse mapping exists on the range and becomes continuous ([19], p. 180, Theorem 4.2). Therefore, there exists a constant $\tilde{\gamma}$ such that

$$\|w\|_{1+\alpha} \leqslant \tilde{\gamma}(\|\mathbf{p}_+^T \mathbf{f}\|_\alpha + |\psi|_{1+\alpha}) \tag{2.5.14}$$

holds. Thus, we have

$$\|\mathbf{u}\|_{1+\alpha} \leqslant 2\, \|\mathbf{q}\|_{1+\alpha}\, \|w\|_{1+\alpha} \leqslant \gamma(\|\mathbf{Du}\|_\alpha + |\mathbf{r} \cdot \mathbf{u}|_{1+\alpha} + \|\mathbf{u}\|_0). \tag{2.5.15}$$

(2) If $n \leqslant 0$ then the transformed problem is always solvable and has $-2n+1$ eigensolutions $w_0, \ldots, W_{2|n|}$. According to Theorem 1.2.4, any solution satisfies the *a priori* estimate (1.2.31), i.e.

$$\|w\|_{1+\alpha} \leqslant k\Big[\|w_{\bar{\zeta}} - \bar{a}w - b\bar{w}\|_\alpha + |\,|P|^{-1} \operatorname{Re} Pw|_{1+\alpha} + \sum_{j=1}^{|n|} |w(z_j)| + \oint_{\dot{G}} |P|^{-1} \operatorname{Im} Pw\, \sigma\, \mathrm{d}s\Big]. \tag{2.5.16}$$

Inserting (2.5.2), (2.5.3) and (2.5.4) into (2.5.16) yields the desired estimate

$$\|\mathbf{u}\|_{1+\alpha} \leqslant \gamma(\|D\mathbf{u}\|_\alpha + |\mathbf{r} \cdot \mathbf{u}|_{1+\alpha} + \|\mathbf{u}\|_0). \tag{2.5.17}$$

2.6 Nonlinear Elliptic Systems

In connection with quasiconformal mappings, Lavrentieff introduced in 1947 a geometrical concept of ellipticity for general systems of nonlinear equations of the form

$$\Phi(z, \bar{z}, w, \bar{w}, w_z, \overline{w_z}, w_{\bar{z}}, \overline{w_{\bar{z}}}) = 0 \tag{2.6.1}$$

where Φ is a complex valued smooth function of all it's variables. Bojarski and Iwaniec considered these systems again in [3, 4] and showed that for every homeomorphic solution w the system (2.6.1) can be transformed globally to a nonlinear elliptic system

$$z_{\bar{w}} = q(z, \bar{z}, w, \bar{w}, z_w, \overline{z_w})z_w \tag{2.6.2}$$

in the hodograph plane. Then the original solution $w(z)$ satisfies itself an elliptic system of the form

$$w_{\bar{z}} = h(z, \bar{z}, w, \bar{w}, w_z, \overline{w_z}). \tag{2.6.3}$$

Since the transformation is valid for every homeomorphic solution let us consider only system (2.6.3) assuming the Lipschitz condition

$$|h(z, w, \xi_1) - h(z, w, \xi_2)| \leq q_0(z, w)\,|\xi_1 - \xi_2| \tag{2.6.4}$$

with $q_0 < 1$ for all (z, w) in suitable domains.

In the following we shall consider quasiconfomral mappings and also some boundary value problems. The main tool will be the Hilbert transform and the weakly singular integral operator P (2.4.21) with their properties in L_p-spaces. These properties were investigated by Vekua in [20] and we shall use them here without proofs. The basic ideas for these nonlinear problems can already be found in the work of Morrey [16].

First let us consider quasiconformal mappings for which h is more restricted by the assumptions

$$\begin{aligned} &h(z, w, 0) \equiv 0 \text{ for all } z \text{ and } w \text{ and} \\ &h(z, w, \xi) \equiv 0 \text{ for all } |z| \geq R \text{ and all } w \text{ and } \xi \end{aligned} \tag{2.6.5}$$

where R is some positive constant. For convenience let us assume (2.6.4) for *all* z, w. A mapping $\zeta(z)$ is called *quasiconformal* if an inequality

$$|\zeta_{\bar{z}}| \leq q|\zeta_z| \tag{2.6.6}$$

holds with $q < 1$. Under the above assumptions on h the following similarity principle holds:

Theorem 2.6.1 [3] *Any solution w of (2.6.3) in the class $C^{1+\alpha}$ can be represented by*

$$w(z) = F(\zeta(z)) \tag{2.6.7}$$

where $\zeta: \mathbb{C} \to \mathbb{C}$ is a quasiconformal homeomorphism and $F(\cdot)$ is some holomorphic function.

Remark The similarity (2.6.7) holds already for generalized solutions of the Sobolev class $W^{1,2}$.

Proof For a *given* solution w the system (2.6.3) can also be written in the form

$$w_{\bar{z}} = q(z, \bar{z}) w_z \tag{2.6.8}$$

where q is defined by

$$q(z, \bar{z}) \equiv \begin{cases} h[z, \bar{z}, w(z, \bar{z}), \overline{w(z, \bar{z})}, w_z(z, \bar{z}), \overline{w_z(z, \bar{z})}]/w_z(z, \bar{z}) & \text{for } w_z \neq 0, \\ 0 & \text{for } w_z = 0. \end{cases} \tag{2.6.9}$$

This function q is measurable and (2.6.4), (2.6.5) imply

$$|q| \leqslant q_0 < 1, \quad \text{i.e.} \quad q \in L_\infty. \tag{2.6.10}$$

With this function q we construct the homeomorphism ζ (Remark 1 to Proposition 2.4.7). Using generalized derivatives, (2.6.8) implies

$$w_\zeta \zeta_{\bar{z}} + w_{\bar{\zeta}} \bar{\zeta}_{\bar{z}} = q(w_\zeta \zeta_z + w_{\bar{\zeta}} \bar{\zeta}_z),$$
$$w_{\bar{\zeta}}(1 - q\bar{q})\overline{\zeta_z} = 0,$$

i.e. the transformed Cauchy–Riemann equations

$$w_{\bar{\zeta}} = 0. \tag{2.6.11}$$

From (2.6.11) follows the desired representation (2.6.7).

Remark Note that the homeomorphism ζ in (2.6.7) depends on the solution w but does *not* satisfy the nonlinear system (2.6.3) in general.

The following example is due to Goldschmidt (private communication) showing that the above assumptions on h do *not* imply

$$|h_{w_z}| + |h_{\overline{w_z}}| < 1 \tag{2.6.12}$$

for all solutions of (2.6.3) in general. Choose

$$h \equiv \tfrac{3}{4}\phi(\arg w_z)\chi(|z|)(w_z - \overline{w_z})$$

where $\phi \in C^\infty[0, 2\pi]$ is any function with

$$\phi(\alpha) = \begin{cases} 1 & \text{for } |\alpha| \leqslant \dfrac{\pi}{7} \\ 0 & \text{for } |\alpha| \geqslant \dfrac{\pi}{6} \end{cases} \quad \text{satisfying } 0 \leqslant \phi \leqslant 1$$

and $\chi \in C^\infty$ with

$$\chi(|z|)=\begin{cases}1 & \text{for} \quad |z|\leqslant R-1\\ 0 & \text{for} \quad |z|\geqslant R\end{cases} \qquad \text{satisfying} \quad 0\leqslant\chi\leqslant 1.$$

Then h satisfies (2.6.4) and (2.6.5) with $q_0=\frac{3}{4}$ whereas $h_{w_z}=\frac{3}{4}$ and $h_{\overline{w_z}}=-\frac{3}{4}$ in the segment $|\arg w_z|\leqslant \pi/7$. The existence of a solution w for (2.6.3) will be shown in the following.

Among the solutions of (2.6.3) under the assumptions (2.6.4) and (2.6.5) there is even a quasiconformal mapping similar to the Riemann mapping function. Let D_z, D_w denote the unit disks in the z and w plane, respectively. Then the following mapping theorem holds due to Bojarski and Iwaniec:

Theorem 2.6.2 [3, p. 482] *Let* (2.6.4) *and* (2.6.5) *be uniformly satisfied for all* $(z, w, \xi)\in D_z\times D_w\times\mathbb{C}$ *and let h in* (2.6.3) *be measurable in z if w and* w_z *are fixed. Let h be equicontinuous in* $w \in D_w$ *uniformly with respect to* $(z, \xi) \in D_z\times\mathbb{C}$. *Then there exists a quasiconformal homeomorphism w of the closed disk* $D_z\to D_w$ *such that w is a generalized solution of* (2.6.3) *and moreover,* $w(0)=0$, $w(1)=1$.

Proof Using the Green functions (1.1.10), (1.1.14) for the unit disk, the first integral of the right-hand side in the representation formula (1.1.23) plus a suitable constant takes the form

$$P\phi\equiv\frac{1}{2\pi i}\iint\limits_{|t|\leqslant 1}\left[\frac{\phi(t)}{t-z}+\frac{z\overline{\phi(t)}}{1-z\bar{t}}-\frac{\phi(t)}{t-1}-\frac{\overline{\phi(t)}}{1-\bar{t}}\right][dt, d\bar{t}] \tag{2.6.13}$$

defining an operator P which completely continuous in $L_p(D_z)$ for each $p>1$. [20, Theorem 1.26, Theorem 1.27]. P has the properties

$$\operatorname{Re} P\phi|_{|z|=1}=0, \qquad P\phi(1)=0, \qquad (P\phi)_{\bar{z}}=\phi. \tag{2.6.14}$$

For the derivative with respect to z one finds

$$(P\phi)_z=T\phi\equiv\text{P.V.}\frac{1}{2\pi i}\iint\left(\frac{\phi(t)}{(t-z)^2}+\frac{\overline{\phi(t)}}{(1-z\bar{t})^2}\right)[dt, d\bar{t}] \tag{2.6.15}$$

where the operator T has the same properties as T in Section 2.4 which can be proved in exactly the same way. The solution will be determined in the form

$$w=z\exp(P\phi). \tag{2.6.16}$$

Inserting (2.6.16) into (2.6.3) we obtain the nonlinear integral equation for ϕ,

$$\phi=\frac{1}{z}\exp(-P\phi)\cdot h[z, z\exp(P\phi), (1+zT\phi)\exp(P\phi)]. \tag{2.6.17}$$

But, since the right-hand side is singular at $z=0$, one solves a *sequence* of *modified problems* to every $\delta>0$, $\delta<1$:

$$\phi=\chi_\delta \exp(-P\phi)h \qquad \text{with } \chi_\delta\equiv\begin{cases}\dfrac{1}{z} & \text{for } |z|\geqslant\delta>0\\ 0 & \text{for } |z|<\delta.\end{cases} \tag{2.6.18}$$

To solve (2.6.18), let us consider first the equation

$$\psi=\Lambda(\phi,\psi)\equiv\chi_\delta\exp(-P\phi)h[z, \exp(P\phi), \exp(P\phi[1+zT\psi])] \tag{2.6.19}$$

where $\phi\in L_p(D_z)$ is given and $\psi\in L^P$ has to be found. $p>2$ is chosen in such a way that the inequality

$$q_0\|T\psi\|_p\leqslant q_0'\,\|\psi\|_p \qquad \text{with} \qquad q_0'<1 \tag{2.6.20}$$

holds (see Theorem 2.4.3). The assumption (2.6.4) yields for Λ the contraction property

$$\begin{aligned}&\|\Lambda(\phi,\psi)-\Lambda(\phi,\psi')\|_p\\ &\quad\leqslant\left\|\left|\frac{1}{z}\right|\,|\exp(-P\phi)|\exp(P\phi)|\,zT\psi-zT\psi'|\right\|_p\cdot q_0\\ &\quad\leqslant q_0'\,\|\psi-\psi'\|_p.\end{aligned} \tag{2.6.21}$$

Moreover, from (2.6.4) and (2.6.5) it follows that

$$\begin{aligned}\|\Lambda(\phi,0)\|_p&=\|\chi_\delta\exp(-P\phi)h[z, z\exp(P\phi), \exp(P\phi)]\|_p\\ &\leqslant\|\chi_\delta\|_p\, q_0\,|\exp(-P\phi)\exp(P\phi)|<\infty\end{aligned} \tag{2.6.22}$$

for every fixed $\delta>0$. Hence, (2.6.19) can be solved for any given $\phi\in L^p$ by using successive approximation with $\psi_0\equiv0$. The solution $\psi\in L^p$ is unique and satisfies the *a priori* estimate

$$\|\psi\|_p\leqslant\frac{1}{1-q_0'}\|\chi_\delta\|_p\leqslant\frac{1}{1-q_0'}\left(\frac{2\pi}{p-2}\right)^{1/p}\delta^{(2-p)/p}, \quad (p>2). \tag{2.6.23}$$

Since to every $\phi\in L^p$ exists a unique solution ψ of (2.6.19), this defines a mapping

$$\Sigma:\phi\to\psi \quad\text{with}\quad \psi=\Lambda(\phi,\psi). \tag{2.6.24}$$

Since (2.6.23) holds, Σ maps the ball

$$B \equiv \left\{ \phi \mid \|\phi\|_p \leq \frac{1}{1-q_0'} \|\chi_\delta\|_p \right\}$$

into itself. Moreover, Σ is *compact* as will be shown in the following. The mapping P maps B completely continuously into $C^\beta(D)$ for any fixed β with $0<\beta\leq(p-2)/p$ [20, Section I, 6]. Hence every subset of B contains a sequence ϕ_n such that the $P\phi_n$ converge in $C^\beta(D)$. The corresponding $\psi_n = \Sigma(\phi_n)$ satisfy the inequality

$$\begin{aligned} \|\psi_n - \psi_m\|_p &\leq q_0' \|\psi_n - \psi_m\|_p \\ &+\frac{1}{\delta}\{\| \exp(-P\phi_n) h[z, z\exp(P\phi_n), z\exp(P\phi_n)(1+T\psi_n)] \\ &-\exp(-P\phi_m) h[z, z\exp(P\phi_m), z\exp(P\phi_n)(1+T\psi_n)]\|_p \\ &+q_0 |\exp(-P\phi_m)| \, |\exp(P\phi_m)-\exp(P\phi_n)| \, (1+\|T\psi_n\|_p)\}. \end{aligned} \quad (2.6.25)$$

Since h is equicontinuous in the second variable, (2.6.25) ensures that the sequence ψ_n becomes a Cauchy sequence in L^p converging to some ψ in B.

Hence, Schauder's fixed point theorem [5] can be applied to Σ. The fixed point ϕ defines the desired solution,

$$\phi = \Sigma\phi, \qquad \phi = \Lambda(\phi, \phi)$$

of the nonlinear equation (2.6.18) for $\delta>0$. By (2.6.16), ϕ defines a mapping w of the unit disk into itself. w maps $|z|=1$ into $|w|=1$ and $w(0)=0$, $w(1)=1$. The remark to Theorem 2.6.1 and considerations similar to Theorem 2.4.6 show finally that w is a homeomorphism.

For $\delta \to 0$ we obtain a sequence of homeomorphisms. Now we shall show that $w(\delta)$ contains a subsequence tending to the desired homeomorphism as $\delta \to 0$. First we observe that the corresponding ϕ satisfy an *a priori* estimate *independent* of δ,

$$\begin{aligned} \|\phi\|_{p'} &\leq \left\| \left|\frac{1}{z}\right| |\exp(-P\phi)| \, q_0 \, |\exp(P\phi)| \, (1+|z| \, |T\phi|) \right\|_p, \\ &\leq q_0 \left\| \left|\frac{1}{z}\right| \right\|_{p'} + q_0' \|\phi\|_{p'} \end{aligned} \quad (2.6.26)$$

where $p' = p/(p-1) < 2$. But for the exponential nonlinearities in Λ we need estimates in L^p. Therefore let us consider the functions zw for which the inequality

$$|z\phi| \leq q_0(1+|zT\phi|) \quad (2.6.27)$$

holds almost everywhere. Here the last term can be written as

$$zT\phi = \text{P.V.}\frac{1}{2\pi i}\iint_D \frac{t\phi(t)}{(t-z)^2}+\frac{t}{\bar t}\frac{\overline{t\phi(t)}}{(1-z\bar t)^2}[dt, d\bar t]$$

$$+\frac{1}{2\pi i}\iint_D \left(\frac{\phi(t)}{z-t}+\frac{(z-t)}{(1-z\bar t)^2}\overline{\phi(t)}\right)[dt, d\bar t]$$

$$\equiv T^*(t\phi)+P^*\phi. \tag{2.6.28}$$

For T^* one proves exactly in the same manner as in Lemma 2.4.1 that $\|T^*f\|_2=\|f\|_2$ and

$$q_0\|T^*f\|_p \leqslant q_0' \|f\|_p$$

for some $p>2$ which can be identified with p from above. Hence, (2.6.27) implies

$$\|z\phi\|_p \leqslant \frac{q_0}{1-q_0'}(1+\|P^*\phi\|_p). \tag{2.6.29}$$

Since P^* is defined by weakly singular integral operators, it follows from [18, Section 1.27] and (2.6.26) that

$$\|P^*\phi\|_p \leqslant k\|\phi\|_{p'} \leqslant \frac{kq_0}{1-q_0'}\left\|\frac{1}{z}\right\|_{p'} \tag{2.6.30}$$

holds provided $p<3$. (2.6.29) and (2.6.30) imply that $\|z\phi\|_p$ is uniformly bounded independent of δ. This yields the uniform Hölder continuity of $P\phi$ in the ring region $0<\rho\leqslant|z|\leqslant 1$ for fixed ρ. Therefore there exists a subsequence of the homeomorphisms $w(\delta)$ converging uniformly in every ring region to a limit function w. Since every $w(\delta)$ is a homeomorphism and since the convergence holds uniformly in every ring region, the limit function w also becomes a homeomorphism.

Bojarski and Iwaniec remark [4, p. 483] that for the nonlinear equations (2.6.1) the boundary value problems can be solved in a similar way. In the following we shall take up this idea but only for the simplest case of boundary conditions, namely

$$\begin{aligned} &w_{\bar z} = h(z, \bar z, w, \bar w, w_z, \overline{w_z}) \\ &\text{Re } w|_{|z|=1} = \phi, \qquad \text{Im } w(0)=\kappa \end{aligned} \tag{2.6.31}$$

where $\phi \in C^{1+\beta}(\partial D)$, $\beta>0$.
Here, we use a slight modification of P in (2.6.13),

$$P'\omega \equiv \frac{1}{2\pi i}\iint_{|t|\leqslant 1}\left\{\frac{\omega(t)}{t-z}+\frac{z\overline{\omega(t)}}{1-z\bar t}-\frac{1}{2}\frac{\omega(t)}{t}+\frac{1}{2}\frac{\overline{\omega(t)}}{\bar t}\right\}[dt, d\bar t] \tag{2.6.32}$$

which has the properties

$$\operatorname{Re} P'\omega\,\big|_{|z|=1}=0,\qquad \operatorname{Im} P'\omega(0)=0,\qquad (P'\omega)_{\bar z}=\omega,\qquad (P'\omega)_z=T\omega \tag{2.6.33}$$

with T as in (2.6.15). For the inhomogeneous boundary data let us define the holomorphic function θ by

$$\theta(z)\equiv\frac{1}{2\pi\mathrm{i}}\oint_{|t|=1}\phi(t)\frac{t+z}{t-z}\,\mathrm{d}t+\mathrm{i}\kappa-\frac{1}{2\pi\mathrm{i}}\oint_{|t|=1}\phi\,\mathrm{d}\xi,\qquad (t=\xi+\mathrm{i}\eta). \tag{2.6.34}$$

For h let us assume (2.6.4), the equicontinuity in $w\in\mathbb{C}$, and instead of (2.6.5) the following assumption:

To the given boundary values ϕ and κ there exists a constant K such that

$$\|h(z,\theta+P'\chi,\theta_z\|_p\leqslant K(1-q_0') \tag{2.6.35}$$

holds for all χ with $\|\chi\|_p\leqslant K$ and some $p>2$.

In the case of h satisfying (2.6.5), this assumption is fulfilled trivially with $K=(q_0/1-q_0')\,\|\theta_z\|_p$.

Remark The simple example

$$w_{\bar z}=w^2,\quad \operatorname{Re} w\,\big|_{|z|=1}=\operatorname{Re}\frac{1}{\frac{1}{2}-\bar z}\,\Big|_{|z|=1},\quad \operatorname{Im} w(0)=0 \tag{2.6.36}$$

having the unique solution

$$w=\frac{1}{\frac{1}{2}-\bar z} \tag{2.6.37}$$

shows that a growth condition like (2.6.35) is necessary for smooth solutions.

Now we can prove similarly to Theorem 2.6.2 the following:

Theorem 2.6.3 *The boundary value problem (2.6.31) has under the above assumptions a (generalized) solution $w\in C^\beta(D)$, $\beta=(p-2)/p$.*

Proof Inserting

$$w=P'\omega+\theta \tag{2.6.38}$$

into the differential equation (2.6.31) we find the nonlinear integral equation,

$$\omega = h(z, P'\omega + \theta, T\omega + \theta_z), \tag{2.6.39}$$

for the unknown density ω. Let us consider the equation

$$\psi = h(z, P'\chi + \theta, T\psi + \theta_z) \equiv \Lambda(\chi, \psi) \tag{2.6.40}$$

for ψ where χ is any fixed function,

$$\chi \in B \equiv \{\zeta \in L^p \mid \|\chi\|_p \leqslant K\}. \tag{2.6.41}$$

Thanks to (2.6.4) and $\|h(z, P'\chi + \theta, \theta_z)\|_p \leqslant K(1-q_0')$ the equation (2.6.40) can uniquely be solved with $\psi \in L^p$. Thus, the mapping $\Sigma : \chi \to \psi$ is well defined. Assumption (2.6.35) ensures that Σ maps the ball B into itself. The compactness of Σ follows from

$$\begin{aligned}\|\psi_n - \psi_m\|_p &\leqslant q_0' \|\psi_n - \psi_m\|_p \\ &\quad + \|h(z, P'\chi_n + \theta, T\omega_n + \theta_z) - h(z, P'\chi_m + \theta, T\omega_n + \theta_z)\|_p\end{aligned} \tag{2.6.42}$$

and the convergence of $P'\chi_n$ in $C^\beta(D)$, $\beta = (p-2)/p$, where χ_n denotes a suitable sequence from B. Hence Schauder's fixed point theorem provides a solution $\omega \in L^p$ of $\omega = \Sigma\omega$ or $\omega = \Lambda(\omega, \omega)$ or (2.6.39). This ω generates the desired w by (2.6.38).

Remarks If h is given smooth enough then it can be shown that the solution is unique and $w_{\bar z} \in B$. The proof follows from a linear homogeneous system for the difference of two solutions.

For smooth enough h it can also be shown that the solution is in $C^{1+\beta}(D)$. For the proof one can solve (2.6.39) locally similarly to the proof of Theorem 2.4.4.

References

1 Agmon, S., Douglis, A. and Nirenberg, L., Estimates near the boundary for solutions of elliptic partial differential equations satisfying general boundary conditions I, II. *Comm. Pure Appl. Math.* **12,** 623–727, 1959; **17,** 35–92, 1964.

2 Ahlfors, L., Conformality with respect to Riemannian metrics. *Ann. Acad. Sci. Fenn. Ser.* A. I. **206,** 22, 1955.

3 Bojarski, B., Quasiconformal mappings and general structural properties of systems of non linear equations strongly elliptic in the sense of Lavrentieff, to appear.

4 Bojarski, B. and Iwaniec, T., Quasiconformal mappings and non-linear elliptic equations in two variables, I, II. *Bull. de L'Acad. Pol. Sci. ser. math. astr. et phys.* **XXII,** 473–478; 479–484, 1974.

5 Courant, R. and Hilbert, D., *Methods of mathematical physics* II. Interscience, New York, 1962.
6 Douglis, A., A function-theoretic approach to elliptic systems of equations in two variables. *Comm. Pure Appl. Math.* **6,** 259–289, 1953.
7 Dunford, N. and Schwartz, J., *Linear operators.* Interscience, New York, 1972.
8 Gauss, C. F., Gesammelte Werke. Königl. Ges. d. *Wissenschaften Göttigen* **4,** 193–216, 1873.
9 Gilbert, R. P., *Function theoretic methods in partial differential equations.* Academic Press, New York, 1969.
10 Haack, W. and Wendland W., *Lectures of Pfaffian and partial differential equations.* Pergamon Press, Oxford, 1972.
11 Hörmander, L., *Linear partial differential operators.* Springer, Berlin, Heidelberg, New York, 1964.
12 Huck, H., Roitzsch, R., Simon, U., Vortisch, W., Walden, R., Wegner, B. and Wendland, W., Beweismethoden der Differentialgeometrie im Großen. *Lecture Notes* No. 335, Springer, Berlin, Heidelberg, New York, 1973.
13 Kamke, E., *Differentialgleichungen reeller Funktionen.* Akad. Verlagsges, Leipzig, 1930.
14 Kato, T., *Perturbation theory for linear operators.* Springer, Berlin, 1966.
15 Lions, J. L. and Magenes, E., *Non-homogeneous boundary value problems and applications,* I. Springer, Berlin, Heildelberg, New York, 1972.
16 Morrey, C. B., On the solution of quasilinear elliptic partial differential equations. *Trans. Am. Math. Soc.* **43,** 126–166, 1938.
17 Schneider, M., Anfangswertprobleme bei linearen partiellen Differentialgleichungen vom gemischten Typ. *Math. Zeitschr.* **101,** 41–60, 1967.
18 Sobolev, S. L., *Some applications of functional analysis in mathematical physics.* Amer. Math. Soc. Translations. Providence, 1963.
19 Taylor, A., *Functional analysis.* John Wiley, New York, 1967.
20 Vekua, I. N., *Generalized analytic functions.* Pergamon Press, Oxford, 1962.

3

Normal Forms of General First-Order Systems

Whereas the elliptic systems with two unknowns can always be transformed into the Hilbert normal form, the systems for more unknowns may not have a normal form in the case of branching eigenvalues of the coefficient matrix in dependence on (x, y). But 'most' of the systems have a normal form according to the pointwise Jordan form of the coefficients. Therefore first in Section 3.1, the algebraic transformation by means of eigenvalues and generalized eigenvectors of the coefficients is presented and the Lopatinski condition is formulated. Under assumptions on the smooth dependence of the generalized eigenvectors on (x, y), the transformation to normal forms is obtained. The systems with special normal forms are characterized: the so called 'generalized analytic', Pascali [20], as well as the hyperanalytic systems introduced by Douglis [8] and their generalizations [10], [11].

For these systems the Fredholm properties are proved in Section 3.2 by the use of Atkinson's Theorems [2, 23] on Fredholm mappings and the compact embedding of Hölder spaces $H^{\beta} \to H^{\alpha}$ for $\beta > \alpha$ [28]. The case of n separated systems in Hilbert's normal form is homotopically perturbed into the most general *normal form*. To the general system we shall come back in Chapter 4.

In Section 3.3, with an explicit perturbation technique again, the Fredholm index is computed explicitly in terms of a certain winding index belonging to the boundary condition.

In Section 3.4 one adjoint system and the normal solvability are formulated. According to the presentation in Section 3.3, the formal adjoint problem becomes Fredholm again. This together with the index enables the use of Fredholm pairs in dualities for the formulation of solvability conditions and Noether's Theorem. Finally, the *a priori* estimates follow trivially.

It should be pointed out that the first-order systems with normal form are a big class to which belong most of the strongly elliptic second-order systems of mathematical physics. The results coincide partially with those of Vol'pert [27] and for the case of a uniformly triangular normal form with those of Bojarski [4, 5] who used the theory of singular integral equations. The Fredholm properties for systems with C^∞-coefficients can also be found on the last pages of Hörmander's book [14]. For special systems these theorems can already be found in the book by Bitsadze [3]. The homotopy methods in connection with elliptic problems were studied in a rather general manner, e.g. by Labrousse [18]. The Lopatinski–Shapiro condition plays a key role. Hence, in Section 3.5 some results are described on problems with violated Lopatinski–Shapiro condition obtained by Kopp [17]. Other problems by Bitsadze are shortly formulated in connection with results from Szilagyi [24] and Tovmasjan [25, 26]. Certain second-order systems with constant coefficients have been classified with respect to the solvability properties by Tovmasjan [30]. Bitsadze gave recently a survey on some non-Fredholm elliptic problems [29].

3.1 Classification and the Algebraic Normal Forms for Elliptic Systems

The system

$$\mathsf{A}\frac{\partial}{\partial x}\mathbf{u}+\mathsf{B}\frac{\partial}{\partial y}\mathbf{u}+\mathsf{C}\mathbf{u}+\mathbf{f}=\mathbf{0} \tag{3.1.1}$$

with $2n\times 2n$ *real* matrix functions A, B, C and the unknown $2n$ vector function $\mathbf{u}$ is called *elliptic* (in the sense of Petrovski [19, p. 275]) if its *symbol*

$$\sigma(x, y; \tilde{\xi}, \tilde{\eta})\equiv\mathsf{A}\tilde{\xi}+\mathsf{B}\tilde{\eta} \tag{3.1.2}$$

is an invertible $2n\times 2n$ matrix for each real pair $(\tilde{\xi}, \tilde{\eta})\neq(0, 0)$.

Thus, A is nonsingular for an elliptic system (3.1.1) and the elliptic system can always be transformed into the special form

$$\mathbf{u}_x+\mathsf{B}\mathbf{u}_y+\mathsf{C}\mathbf{u}+\mathbf{f}=\mathbf{0}. \tag{3.1.3}$$

For elliptic systems the eigenvalues of B are never real and they form complex conjugated pairs

$$\lambda_\pm^j=\lambda_1^j(x, y)\pm i\lambda_2^j(x, y), \qquad j=1, \ldots, n, \quad \lambda_2^j>0$$

in each point $z=x+iy$. Here $\lambda_\pm^j$ can have a multiplicity higher one which can even change for different points (x, y). Now let us fix (x, y) and let us consider one of the eigenvalues λ_+ in the upper halfplane. Then to this λ_+ in general there belongs a chain of linearly independent generalized eigenvectors $\mathbf{p}_0, \ldots, \mathbf{p}_{r-1}$ of B^T defined by

$$\mathbf{p}_0^T\mathsf{B} - \lambda_+\mathbf{p}_0^T = 0, \tag{3.1.4}$$

$$\mathbf{p}_k^T\mathsf{B} - \lambda_+\mathbf{p}_k^T = \mu_k\mathbf{p}_{k-1}^T, \qquad k = 1, \ldots, r-1, \tag{3.1.5}$$

where $\mu_k \neq 0$ is an arbitrarily chosen function of (x, y) (see e.g. [9, p. 285 ff.]). Then to each eigenvector $\mathbf{p}_0$ there belongs a chain of generalized eigenvectors which may consist only of $\mathbf{p}_0$ itself. If the eigenvectors are suitably chosen then the constructed collection of generalized eigenvectors together with their complex conjugated vectors corresponding to λ_-, will form a basis for C^{2n}. With their help we define new complex valued unknowns

$$w_j = \mathbf{p}_j^T\mathbf{u}, \qquad j = 1, \ldots, n. \tag{3.1.6}$$

where $\mathbf{p}_1, \ldots, \mathbf{p}_n$ denote the generalized eigenvectors belonging to λ_+ in the upper halfplane. Like in Lemma 2.2.2, the ellipticity of (3.1.1) implies that the complex $2n \times 2n$ matrix

$$(\mathbf{p}_1, \bar{\mathbf{p}}_1, \ldots, \mathbf{p}_n, \bar{\mathbf{p}}_n)^T \tag{3.1.7}$$

is nonsingular. The inverse matrix to (3.1.7) can be written as

$$\mathsf{Q} = (\mathbf{q}_1, \bar{\mathbf{q}}_1, \mathbf{q}_2, \bar{\mathbf{q}}_2, \ldots, \mathbf{q}_n, \bar{\mathbf{q}}_n), \tag{3.1.8}$$

and the vectors $\mathbf{p}_j$ and $\mathbf{q}_k$ satisfy the equations

$$\mathbf{p}_j^T\mathbf{q}_l = \delta_{jl} \quad \text{and} \quad \mathbf{p}_j^T\bar{\mathbf{q}}_l = 0 \quad \text{for all} \quad j, l = 1, \ldots, n \tag{3.1.9}$$

and

$$\sum_{j=1}^{n} (p_{j(s)}q_{j(t)} + \bar{p}_{j(s)}\bar{q}_{j(t)}) = \delta_{st} \quad \text{for} \quad s, t = 1, \ldots, 2n, \tag{3.1.10}$$

where $p_{j(s)}$ denote the components of $\mathbf{p}_j$ and $q_{j(t)}$ those of $\mathbf{q}_j$, respectively. If we write (3.1.5) as

$$\mathbf{p}_j^T\mathsf{B} = \lambda_+^j\mathbf{p}_j^T + \mu_j\mathbf{p}_{j-1}^T \quad (\mu_1 = 0), \tag{3.1.11}$$

then (3.1.9) implies

$$\begin{aligned} &\bar{\mathbf{p}}_j^T\mathsf{B}\mathbf{q}_k = 0 \quad \text{and} \\ &\mathbf{p}_j^T\mathsf{B}\mathbf{q}_k = \lambda_+^j\delta_{jk} + \mu_j\delta_{j-1,k}. \end{aligned} \tag{3.1.12}$$

Multiplication of the first equation with $\bar{q}_{j(s)}$ and the second with $q_{j(s)}$ and summation leads with (3.1.10) to

$$\mathsf{B}\mathbf{q}_k = \lambda_+^j\mathbf{q}_k + \mu_{k+1}\mathbf{q}_{k+1} \quad (\mu_{r+1} = 0) \tag{3.1.13}$$

Hence, $\mathbf{q}_k$ is a complete system of generalized eigenvectors belonging to the matrix B.

Now let us return to the boundary value problem. The inverse transformation to (3.1.6) becomes

$$\mathbf{u} = 2 \operatorname{Re} \sum_{j=1}^{n} \mathbf{q}_j w_j. \tag{3.1.14}$$

Multiplying (3.1.3) by $\mathbf{p}_j^T$ and using (3.1.4) or (3.1.5) we get for one eigenvector $\mathbf{p}_0^T$ having the eigenvalue λ_+ the equations

$$\mathbf{p}_0^T \mathbf{u}_x + \lambda_+ \mathbf{p}_0^T \mathbf{u}_y + \mathbf{p}_0^T \mathbf{C} \operatorname{Re} \sum_{j=1}^{n} \mathbf{q}_j^T w_j + \mathbf{p}_0^T \mathbf{f} = 0, \tag{3.1.15}$$

$$\mathbf{p}_k^T \mathbf{u}_x + \lambda_+ \mathbf{p}_k^T \mathbf{u}_y + \mu_k \mathbf{p}_{k-1}^T \mathbf{u}_y + \cdots = 0, \qquad k = 1, \ldots, r-1. \tag{3.1.16}$$

Now let us require the following *very restrictive assumption*:

$$\mathbf{p}_j \in C^{1+\alpha}(\bar{\mathsf{G}}),\ \lambda^j \quad \text{and} \quad \mu_j \in C^{\alpha}(\bar{\mathsf{G}})$$

$$\text{for all} \quad j = 1, \ldots, n,\ 0 < \alpha \leq \frac{1}{n}. \tag{3.1.17}$$

Whereas for $\mathsf{B} \in C^{1+\alpha}$ the unordered set of eigenvalues *always* becomes continuous the eigenvectors may not depend continuously on (x, y) even if $\mathsf{B} \in C^{\infty}$ (see Rellich's example in [16, p. 111]). In Remark 4.2.4 we shall show an example by Hile and Protter for a system without normal form. Another example has been given by Vinogradov. Later we shall omit (3.1.17) for the general case. If B has $2n$ *different* eigenvalues, then Theorem 2.2.1 holds again and (3.1.17) is fulfilled with $\mu_j = 0$, $j = 1, \ldots, n$.

Now (3.1.15) and (3.1.16) become

$$w_{0x} + \lambda_+ w_{0y} + \text{lower order terms} = 0, \tag{3.1.18}$$

$$w_{kx} + \lambda_+ w_{ky} + \mu_k w_{k-1y} + \text{lower order terms} = 0,\ k = 1, \ldots, r-1. \tag{3.1.19}$$

Let $\zeta \in C^{1+\alpha}$ be a Beltrami homeomorphism to λ_+ (see Section 2.4) in $\mathbb{C}$,

$$\zeta_x + \lambda_+ \zeta_y = 0, \tag{3.1.20}$$

and let us choose

$$\mu_k = \mu(\bar{\zeta}_x + \lambda_+ \bar{\zeta}_y) = \mu\rho \quad \text{for} \quad k = 1, \ldots, r-1, \text{ where } \mu \neq 0 \tag{3.1.21}$$

can be chosen arbitrarily i.e. in particular independently of k. Then μ_k does not vanish because of (2.2.34). Note that, according to (3.1.11), this choice of μ_k describes only a scaling of $\mathbf{p}_k$. After division by ρ the system

(3.1.18), (3.1.19) takes the form

$$w_{0\bar{\zeta}}+\sum_{j=0}^{n-1} a_{0j}w_j+\sum_{j=0}^{n-1} b_{0j}\bar{w}_j+c_0=0, \tag{3.1.22}$$

$$w_{k\bar{\zeta}}+\mu\zeta_y w_{k-1\zeta}+\mu\bar{\zeta}_y w_{k-1\bar{\zeta}}+\sum_{j=0}^{n-1} a_{kj}w_j+\sum_{j=0}^{n-1} b_{kj}\bar{w}_j+c_k=0, \quad k=1,\ldots r-1. \tag{3.1.23}$$

The equations (3.1.22) and (3.1.23) belong to just *one* chain of generalized eigenvectors and to *one* eigenvalue. For a *different* eigenvalue we will get a *different homeomorphism* and different integrable directional derivatives $\partial/\partial\bar{\zeta}_j$, $\partial/\partial\zeta_j$.

Remark 3.1.1 If λ_+ is of the multiplicity n but has still n linearly independent eigenvectors $\mathbf{p}_j$, then the normal form becomes D. Pascali's system [20]:

$$\mathbf{w}_{\bar{\zeta}}+\mathbf{a}\mathbf{w}+\mathbf{b}\bar{\mathbf{w}}+\mathbf{c}=\mathbf{0}, \tag{3.1.24}$$

where $\mathbf{w}$ denotes the complex vector valued unknown

$$\mathbf{w}=\text{column}\,(w_1,\ldots,w_n). \tag{3.1.25}$$

Remark 3.1.2 If λ_+ is of the multiplicity n again but has only one eigenvector $\mathbf{p}_0$ and $n-1=r-1$ generalized eigenvectors $\mathbf{p}_k$ where (3.1.4) and (3.1.5) hold in $\bar{G}$, then the normal form becomes the *hyperanalytic system* according to Douglis [8],

$$w_{0\bar{\zeta}}+\mathbf{a}_0^T\mathbf{w}+\mathbf{b}_0^T\bar{\mathbf{w}}+c_0=0, \tag{3.1.26}$$

$$w_{k\bar{\zeta}}+\alpha w_{k-1\zeta}+\beta w_{k-1\bar{\zeta}}+\mathbf{a}_k^T\mathbf{w}+\mathbf{b}_k^T\bar{\mathbf{w}}+c_k=0, \qquad k=1,\ldots,n-1. \tag{3.1.27}$$

If the linking terms a, b have a special triangular shape then (3.1.26) and (3.1.27) become the hyperanalytic system which was investigated by Gilbert and Hile [10, 11]:

$$w_{0\bar{\zeta}}+a_{00}w_0+b_{00}\bar{w}_0+c_0=0, \tag{3.1.28}$$

$$w_{k\bar{\zeta}}+\alpha w_{k-1\zeta}+\beta w_{k-1\bar{\zeta}}+\sum_{j=0}^{k} a_{kj}w_j+b_{kj}\bar{w}_j+c_k=0, \tag{3.1.29}$$

where $a_{k+s,s}=a_{k0}$ and $b_{k+s,s}=b_{k0}$ for $s=1,\ldots,\ n-1-k$, and $k=0,\ldots,n-1$.

For solutions of (3.1.24) and (3.1.28), (3.1.29) many properties of complex function theory hold as in the case of one complex unknown; e.g. a local similarity principle (Carleman's Theorem 1.1.3) holds, whereas for general elliptic systems (3.1.1) the similarity principle is no longer valid (see [13] and Section 5.3 in this book).

Now let us return to the more general case supposing (3.1.17). Then to each eigenvalue λ_+ belongs one homeomorphism ζ satisfying (3.1.20). If all eigenvalues are different then there are n corresponding different $\zeta_1, \ldots, \zeta_n$ belonging to $w_1, \ldots, w_n$.

Definition The system (3.1.27) is called the *normal form.* The normal form depends on the point (x, y) and becomes in general

$$w_{j\bar{\zeta}_j} + \sum_{k \neq j} (\alpha_{jk} w_{k\zeta_j} + \beta_{jk} w_{k\bar{\zeta}_j}) + \mathbf{a}_j^T \mathbf{w} + \mathbf{b}_j^T \bar{\mathbf{w}} + c_j = 0 \tag{3.1.30}$$

where the coefficients α_{jk}, β_{jk}, for each $z \in \bar{\mathrm{G}}$, form a *rearranged* triangular matrix consisting of submatrices corresponding to (3.1.22) and (3.1.23). The *rearrangement* can be written in every *point* such that the differential operator becomes

$$\begin{pmatrix} \frac{\partial}{\partial \bar{\zeta}_1} & & & & & & 0 \\ \varepsilon_1 \frac{\partial}{\partial y} & \frac{\partial}{\partial \bar{\zeta}_1} & & & & & \\ & \varepsilon_k \frac{\partial}{\partial y} & \frac{\partial}{\partial \bar{\zeta}_1} & & & & \\ & & 0 & \frac{\partial}{\partial \bar{\zeta}_2} & & & \\ & & & \varepsilon_l \frac{\partial}{\partial y} & \frac{\partial}{\partial \bar{\zeta}_2} & & \\ & & & & & \ddots & \\ & 0 & & & & \varepsilon_n \frac{\partial}{\partial y} & \frac{\partial}{\partial \bar{\zeta}_n} \end{pmatrix} \tag{3.1.31}$$

where the $\varepsilon_j \in C^\alpha(\bar{\mathrm{G}})$; *the rearrangement depends on z.* If (3.1.31) holds *without* rearrangements everywhere in $\bar{\mathrm{G}}$ then we get the system which was investigated by Bojarski [4, 5].

For the boundary value problems corresponding to (3.1.1) let us consider Riemann–Hilbert boundary conditions which generalize (2.6.2),

$$\mathbf{r}\mathbf{u} = \boldsymbol{\psi} \quad \text{on} \quad \dot{\mathrm{G}} \tag{3.1.32}$$

or, in components,

$$\sum_{k=1}^{2n} r_{jk}u_k = \psi_j, \qquad j=1, \ldots, n. \tag{3.1.33}$$

$\mathsf{r} \in C^{1+\alpha}(\dot{\mathsf{G}})$ is a given $n \times 2n$ real valued boundary matrix. Let us denote the row vectors of r by $\mathbf{r}_j$,

$$\mathbf{r}_j\mathbf{u} = \psi_j, \qquad j=1, \ldots, n, \tag{3.1.34}$$

then (3.1.14) leads to the transformed boundary conditions

$$\operatorname{Re} \sum_{k=1}^{n} 2\mathbf{r}_j\mathbf{q}_k w_k = \psi_j, \qquad j=1, \ldots, n \tag{3.1.35}$$

or

$$\operatorname{Re} \mathsf{P}\mathbf{w} = \boldsymbol{\psi} \quad \text{with} \quad P_{jk} = 2\mathbf{r}_j\mathbf{q}_k. \tag{3.1.36}$$

The solvability theory for the boundary value problems depends on the following *Lopatinski condition,*

$$\operatorname{Det} \mathsf{P} = \operatorname{Det} (2\mathbf{r}_j\mathbf{q}_k) \neq 0. \tag{3.1.37}$$

Remark 3.1.3 The Lopatinski condition is independent of the smoothness properties of $\mathbf{q}_k$. It classifies the possible boundary conditions. For the Pascali systems the formulation (3.1.37) corresponds to [3, p. 150 (6.15)]. The following example of Bitsadze shows that even some Dirichlet problems lead to problems with a violated Lopatinski condition.

$$\left.\begin{aligned} v^1_{xx} - v^1_{yy} - 2v^2_{xy} &= 0 \\ 2v^1_{xy} + v^2_{xx} - v^2_{yy} &= 0 \end{aligned}\right\} \text{in G and} \tag{3.1.38}$$

$$v^1 = \psi^1, \qquad v^2 = \psi^2 \quad \text{on} \quad \dot{\mathsf{G}}. \tag{3.1.39}$$

With the new unknowns

$$u^1 = v^1_y, \qquad u^2 = -v^1_x, \qquad u^3 = v^2_y, \qquad u^4 = -v^2_x, \tag{3.1.40}$$

the system (3.1.38) becomes

$$\mathbf{u}_x + \begin{pmatrix} 0 & 1 & 0 & 0 \\ 1 & 0 & 0 & -2 \\ 0 & 0 & 0 & 1 \\ 0 & 2 & 1 & 0 \end{pmatrix} \mathbf{u}_y = 0 \tag{3.1.41}$$

with eigenvalues and homeomorphism

$$\lambda_+ = \mathrm{i}, \qquad \zeta = z. \tag{3.1.42}$$

Here exists only *one* eigenvector and hence one generalized eigenvector

$$\mathbf{p}_0^T = (1, \mathrm{i}, \mathrm{i}, -1), \qquad \boldsymbol{p}_1^T = (0, 1, 0, \mathrm{i}), \qquad \mu = 1. \tag{3.1.43}$$

The complex valued unknowns are

$$\begin{aligned} w_0 &= v_y^1 - iv_x^1 + iv_y^2 + v_x^2, \\ w_1 &= -v_x^1 - iv_x^2. \end{aligned} \tag{3.1.44}$$

The transformed system (3.1.22) and (3.1.23) is

$$\begin{aligned} &w_{0\bar{z}} = 0 \\ &w_{1\bar{z}} + i(w_{0z} - w_{0\bar{z}}) = 0. \end{aligned} \tag{3.1.45}$$

The boundary conditions (3.1.39) become by differentiation

$$\mathbf{ru} = \begin{pmatrix} \dot{y} & -\dot{x} & 0 & 0 \\ 0 & 0 & \dot{y} & -\dot{x} \end{pmatrix} \mathbf{u} = \begin{pmatrix} \dot{\psi}^1 \\ \dot{\psi}^2 \end{pmatrix}, \tag{3.1.46}$$

and with

$$\mathbf{q}_1^T = \tfrac{1}{2}(1, 0, -i, 0), \qquad \mathbf{q}_2^T = \tfrac{1}{2}(-i, 1, -1, -i), \tag{3.1.47}$$

(3.1.46) is transformed into

$$\operatorname{Re} \begin{pmatrix} \dot{y} & -(\dot{x} + i\dot{y}) \\ -i\dot{y} & (i\dot{x} - \dot{y}) \end{pmatrix} \begin{pmatrix} w_1 \\ w_2 \end{pmatrix} = \begin{pmatrix} \dot{\psi}_1 \\ \dot{\psi}_2 \end{pmatrix} \tag{3.1.48}$$

where the new boundary matrix P violates the Lopatinski condition *on the whole boundary*

$$\operatorname{Det} \mathsf{P} = 0. \tag{3.1.49}$$

Boundary value problems with violated Lopatinski condition do not define Fredholm operators from, e.g. $C^{1+\alpha}(\bar{\mathrm{G}})$ into $C^{\alpha}(\bar{\mathrm{G}}) \times C^{1+\alpha}(\dot{\mathrm{G}})$. We shall come back to this problem in Section 3.5.

3.2 The Fredholm Property for Systems in Normal Form

For the special systems (3.1.30) where $\alpha_{jk} = \beta_{jk} = a_{jk} = b_{jk} = 0$ with boundary conditions having a diagonal matrix $P_{jk} = \delta_{jk} P_j$, the boundary value problem degenerates into n completely separated problems where each is of the form (1.0.1) and each is Fredholm according to Section 1.3. Hence these special systems are Fredholm. We shall show how to prove the Fredholm property for systems in normal form by the use of elementary linear algebra and Atkinson's theorems [2]. Let us recall these theorems without proofs in the formulation as in [23, chap. V]. Let us denote by X, Y, Z given Banach spaces and by $\phi(X, Y)$ the class of linear bounded Fredholm operators mapping X into Y. The Fredholm index of $A \in \phi$ is denoted by $\nu(A)$. By $B(X, Y)$ let us denote the bounded linear operators and by $K(X, Y)$ the ideal of completely continuous linear operators mapping X into Y.

Theorem 3.2.1 *If $A \in \phi(X, Y)$ and $B \in \phi(Y, Z)$ then $BA \in \phi(X, Z)$ and $\nu(BA) = \nu(B) + \nu(A)$.*

Theorem 3.2.2 *If $A \in \phi(X, Y)$ and $K \in K(X, Y)$ then $A + K \in \phi(X, Y)$ and $\nu(A + K) = \nu(A)$.*

Theorem 3.2.3 *$A \in \phi(X, Y)$ if and only if A has left and right regularizers L, $M \in B(Y, X)$ such that*

$$AM - I \in K(Y, Y), \qquad LA - I \in K(X, X)$$

where I denotes the identity in Y or X, respectively.
For $A \in \phi$ there even exist L and M with $L = M$.

Theorem 3.2.4 *The set of Fredholm operators $\phi(X, Y)$ is open with respect to the operator norm induced by the norms in X and Y. The index ν is continuous.*

Now we are in the position to prove the Fredholm properties for *Pascali's system* (3.1.24).

Theorem 3.2.5 *Let the coefficient matrices a, b consist of functions in $C^\alpha(\bar{\mathsf{G}})$ and let $\mathsf{P} \in C^{1+\alpha}(\dot{\mathsf{G}})$ satisfy the Lopatinski condition*

$$\det \mathsf{P} \neq 0 \text{ on } \dot{\mathsf{G}}.$$

Then the mapping

$$\mathsf{R}\mathbf{w} \equiv \begin{cases} \mathbf{w}_{\bar{z}} + \mathsf{a}\mathbf{w} + \mathsf{b}\bar{\mathbf{w}} & \text{in } \mathsf{G} \\ \operatorname{Re} \mathsf{P}\mathbf{w} & \text{on } \dot{\mathsf{G}} \end{cases} \tag{3.2.1}$$

from $X \equiv C^{1+\alpha}(\bar{\mathsf{G}})$ into $Y \equiv C^\alpha(\bar{\mathsf{G}}) \times C^{1+\alpha}(\dot{\mathsf{G}})$ is Fredholm.

The Theorem will be proved in several steps. First let us recall without proof the following Lemma of Rellich's type:

Lemma 3.2.6 *The embedding $C^\beta(\bar{\mathsf{G}}) \to C^\alpha(\bar{\mathsf{G}})$ is compact for $\beta > \alpha$* (see Wloka [28, p. 262 Section 8]).

Hence, according to Theorem 3.2.2 the mapping R is Fredholm if and only if the corresponding mapping with $\mathsf{a} = \mathsf{b} = 0$ is Fredholm.

Since Theorem 3.2.5 was already proved in Section 1.3 for the special case $n = 1$, it can be obtained by induction from the following:

Lemma 3.2.7 *Let us assume that Theorem 3.2.5 is true for systems with $n-1$ complex valued unknowns. Then it holds also true for systems with n complex valued unknowns.*

Proof We shall construct the operators L, M of Theorem 3.2.3 in a few single steps. We have to prove that the mapping

$$\mathbf{w} \mapsto \begin{cases} w_{1\bar{z}} = c_1 & \operatorname{Re} \sum_{k=1}^{n} P_{1k} w_k = \psi_1 \\ w_{2\bar{z}} = c_2 & \operatorname{Re} \sum_{k=1}^{n} P_{2k} w_k = \psi_2 \\ \quad \text{in } \mathsf{G}, & \qquad \text{on } \dot{\mathsf{G}} \\ \vdots & \vdots \qquad \vdots \\ w_{n\bar{z}} = c_n & \operatorname{Re} \sum_{k=1}^{n} P_{nk} w_k = \psi_n \end{cases} \tag{3.2.2}$$

is Fredholm if that is true for every reduced system for $w_2, \ldots, w_n$ satisfying the Lopatinski condition. Det $\mathsf{P} \neq 0$ implies for each point on $\dot{\mathsf{G}}$

$$P_{j1}(s) \neq 0 \quad \text{for at least one } j \tag{3.2.3}$$

depending on s. Since P_{j1} is continuous we can find a partition of unity, namely real valued C^∞ functions $\chi'_{11}(s), \ldots, \chi'_{1n}(s)$ such that with suitable $t_j = 0$ or 1 and

$$\begin{aligned} &\chi_{1j} = (-1)^{t_j} \chi'_{1j}, \\ &\tilde{P}_{11}(s) \equiv \sum_{j=1}^{n} \chi_{1j}(s) P_{j1}(s) \neq 0 \quad \text{for all } s \text{ on } \dot{\mathsf{G}}. \end{aligned} \tag{3.2.4}$$

The functions χ_{1j} can easily be completed to a system of real valued functions $\chi_{kj} \in C^\infty(\dot{\mathsf{G}})$ such that

$$\operatorname{Det}(\chi_{kj}) = 1 \quad \text{on } \dot{\mathsf{G}}. \tag{3.2.5}$$

Introducing an isomorphism $Y \to Y$ by

$$\chi(\mathbf{c}, \boldsymbol{\psi}) \equiv \left(\mathbf{c}, \left(\sum_{j=1}^{n} \chi_{kj} \psi_j\right)\right), \tag{3.2.6}$$

it follows from Theorem 3.2.1 that the mapping (3.2.2) is Fredholm if and only if the corresponding mapping with the *new boundary operator*

$$\tilde{\mathsf{P}} \equiv (\tilde{P}_{jk}) \equiv \left(\sum_{l=1}^{n} \chi_{jl} P_{lk} \right) \tag{3.2.7}$$

is Fredholm. The condition (3.2.5) ensures that $\tilde{\mathsf{P}}$ also satisfies the Lopatinski condition. Hence, it suffices to prove Lemma 3.2.7 under the additional assumption

$$\tilde{P}_{11}(s) \neq 0 \quad \text{on } \dot{\mathsf{G}}. \tag{3.2.8}$$

Now let us extend the boundary functions

$$p_{1k} = \frac{\tilde{P}_{1k}}{\tilde{P}_{11}}, \qquad k = 2, \ldots, n \tag{3.2.9}$$

to functions $p_{1k} \in C^{1+\alpha}(\bar{\mathsf{G}})$ and let us define an isomorphism $X \to X$ by

$$\tilde{w}_1 \equiv w_1 + \sum_{k=2}^{n} p_{1k} w_k, \qquad \tilde{w}_j \equiv w_j \quad \text{for} \quad j = 2, \ldots, n. \tag{3.2.10}$$

Inserting (3.2.10) into (3.2.2) we get the new problem,

$$\left.\begin{aligned} &\tilde{w}_{1\bar{z}} = c_1 + \sum_{k=2}^{n} p_{1k} c_k + \sum_{k=2}^{n} (p_{1k})_{\bar{z}} \tilde{w}_k \\ &\tilde{w}_{2\bar{z}} = c_2 \\ &\cdot \\ &\cdot \\ &\cdot \\ &\tilde{w}_{n\bar{z}} = c_n \end{aligned}\right\} \text{in } \mathsf{G},$$

$$\tag{3.2.11}$$

$$\left.\begin{aligned} &\operatorname{Re} \tilde{P}_{11} \tilde{w}_{11} = \tilde{\psi}_1 \\ &\operatorname{Re} \tilde{P}_{21} \tilde{w}_1 + \sum_{k=2}^{n} \left(\tilde{P}_{2k} - \tilde{P}_{21} \frac{\tilde{P}_{11}}{\tilde{P}_{11}} \right) \tilde{w}_k = \tilde{\tilde{\psi}}_2 \\ &\cdot \\ &\cdot \\ &\cdot \\ &\operatorname{Re} \tilde{P}_{n1} \tilde{w}_1 + \sum_{k=2}^{n} \left(\tilde{P}_{nk} - \tilde{P}_{n1} \frac{\tilde{P}_{1k}}{\tilde{P}_{11}} \right) \tilde{w}_k = \tilde{\psi}_n \end{aligned}\right\} \text{on } \dot{\mathsf{G}}.$$

Using Theorem 3.2.2 and Lemma 3.2.6 for the first equation in (3.2.11) and the corresponding transformation to (3.2.10) for the $c_1, \ldots, c_n$, we see that it suffices to prove Lemma 3.2.7 for the special case

$$\left.\begin{array}{l} w_{1\bar z}=c_1 \\ w_{2\bar z}=c_2 \\ \cdot \\ \cdot \\ \cdot \\ w_{n\bar z}=c_n \end{array}\right\} \text{in } \mathsf{G} \qquad \left.\begin{array}{l} \operatorname{Re} P_{11}w_1 = \psi_1 \\ \operatorname{Re}\left(P_{21}w_1+\sum_{k=2}^{n} P_{2k}w_k\right)=\psi_2 \\ \\ \\ \operatorname{Re}\left(P_{n1}w_1+\sum_{k=2}^{n} P_{nk}w_k\right)=\psi_n \end{array}\right\} \text{on } \dot{\mathsf{G}} \tag{3.2.12}$$

Here we skipped the notation of all the transformations: the functions in (3.2.12) are others than those in (3.2.2). Observe that the determinant of the new boundary coefficients in (3.2.11) and in (3.2.12) is still Det $\tilde{\mathsf{P}}$. Moreover, the reduced problem for $n-1$ unknowns

$$\begin{array}{ll} w_{2\bar z}=c_2, & \operatorname{Re}\sum_{k=2}^{n} P_{2k}w_k=\psi_2', \\ \cdot & \\ \cdot & \\ \cdot & \\ w_{n\bar z}=c_n, & \operatorname{Re}\sum_{k=2}^{n} P_{nk}w_k=\psi_n', \end{array} \tag{3.2.13}$$

satisfies the Lopatinski condition

$$\operatorname{Det}(P_{jk})_{j,k=2,\ldots,n}=\frac{1}{\tilde P_{11}}\operatorname{Det}\tilde{\mathsf{P}}\neq 0.$$

Hence, according to our assumption, (3.2.13) is a Fredholm problem and Theorem 3.2.3 assures the existence of regularizers satisfying

$$\sum_{j=2}^{n}\left(L_{kj}w_{j\bar z}+l_{kj}\operatorname{Re}\sum_{l=2}^{n}P_{jl}w_{l|\dot{\mathsf{G}}}\right)\sim w_k, \tag{3.2.14}$$

$$\left.\begin{array}{l} \dfrac{\partial}{\partial\bar z}\left[\sum_{j=2}^{n}(L_{kj}c_j+l_{kj}\psi_j)\right]\sim c_k, \\ \operatorname{Re}\sum_{l=2}^{n}P_{kl}\sum_{j=2}^{n}(L_{lj}c_j+l_{lj}\psi_j)\sim\psi_k, \qquad k=2,\ldots,n \end{array}\right. \tag{3.2.15}$$

where $\sim$ *is defined* here and in the following as to be *equal modulo compact mappings* according to the corresponding spaces. To the inhomogeneous Cauchy–Riemann system

$$w_{1\bar z}=c_1, \qquad \operatorname{Re}P_{11}w_1=\psi_1 \tag{3.2.16}$$

regularizers also exist according to Section 1.3 and Theorem 3.2.3

satisfying

$$L_{11}w_{1\bar{z}} + l_{11} \operatorname{Re} P_{11} w_{1|\dot{G}} \sim w_1, \tag{3.2.17}$$

$$\begin{aligned} &\frac{\partial}{\partial \bar{z}}(L_{11}c_1 + l_{11}\psi_1) \sim c_1, \\ &\operatorname{Re} P_{11}(L_{11}c_1 + l_{11}\psi_1) \sim \psi_1. \end{aligned} \tag{3.2.18}$$

As in the Gauss elimination method in linear algebra, it follows immediately that the operators defined by

$$\begin{aligned} &\tilde{L}_{11} \equiv L_{11},\ \tilde{L}_{1l} = 0, && \tilde{l}_{11} \equiv l_{11},\ \tilde{l}_{1l} = 0, \\ &\tilde{L}_{k1} \equiv -\sum_{j=2}^{n} l_{kj} \operatorname{Re} P_{j1} L_{11}, && \tilde{l}_{k1} \equiv -\sum_{j=2}^{n} l_{kj} \operatorname{Re} P_{j1} l_{11}, \\ &\tilde{L}_{kl} \equiv L_{kl}, && \tilde{l}_{kl} \equiv l_{kl} \text{ for } k, l = 2, \dots, n \end{aligned} \tag{3.2.19}$$

and by

$$\begin{aligned} &w_1 \equiv L_{11}c_1 + l_{11}\psi_1, \\ &w_k \equiv \sum_{j=2}^{n} \{L_{kj}c_j + l_{kj}\ [(\operatorname{Re} P_{j1}(L_{11}c_1 + l_{11}\psi_1)]\}, \qquad k = 2, \dots, n \end{aligned} \tag{3.2.20}$$

become left and right regularizers, respectively, to the system (3.2.12) with n complex valued unknowns. Thus, Theorem 3.2.3 yields the Fredholm property for (3.2.12) and, therefore, also for (3.2.2).

With the help of solutions to (1.1.1) we are able to show:

Theorem 3.2.8 *Let* a, $\mathsf{b} \in C^\alpha(\bar{\mathsf{G}})$ *and let* $\mathsf{P} \in C^{1+\alpha}(\bar{\mathsf{G}})$ *satisfy the Lopatinski condition. Then the mapping*

$$\mathsf{R}w \equiv \begin{cases} w_{j\bar{\zeta}_j} + \mathbf{a}_j^T \mathbf{w} + \mathbf{b}_j^T \bar{\mathbf{w}} & \text{in } \mathsf{G},\ j = 1, \dots, n \\ \operatorname{Re} \mathsf{P} w & \text{on } \dot{\mathsf{G}} \end{cases} \tag{3.2.21}$$

from X into Y is Fredholm.

Proof Again it suffices to prove the theorem only for the case $\mathbf{a}_j = \mathbf{b}_j = 0$. We shall construct left and right regularizers by the use of the regularizers to the Pascali system in Theorem 3.2.7 which satisfy

$$\begin{aligned} &\sum_{t=1}^{n} \left(L_{jt} w_{t\bar{z}} + l_{jt} \sum_{s=1}^{n} \operatorname{Re} P_{ts} w_{s|\dot{G}} \right) \sim w_j, \\ &\frac{\partial}{\partial \bar{z}} \left[\sum_{t=1}^{n} \left(L_{jt} c_t + l_{jt} \sum_{s=1}^{n} P_{ts} \psi_s \right) \right] \sim c_j, \\ &\operatorname{Re} \sum_{k=1}^{n} P_{jk} \left[\sum_{t=1}^{n} \left(L_{kt} c_t + l_{kt} \sum_{s=1}^{n} P_{ts} \psi_s \right) \right] \sim \psi_j,\ j = 1, \dots, n. \end{aligned} \tag{3.2.22}$$

Identifying z with ζ_l and choosing $w_t=0$ for $t\neq j$ and c_t, ψ_t correspondingly, we find for the corresponding regularizers $L_{jl}^{(l)}$, $l_{jl}^{(l)}$ the equations

$$L_{jl}^{(l)} w_{l\bar{\zeta}_l} + \sum_{t=1}^{n} l_{jt}^{(l)} \operatorname{Re} P_{tl} w_{l|\dot{G}} \sim \delta_{jl} w_j \tag{3.2.23}$$

$$\frac{\partial}{\partial \bar{\zeta}_j} L_{jl}^{(l)} c_l \qquad \sim \delta_{jl} c_j,\ l,\ j=1,\ldots,n \tag{3.2.24}$$

$$\frac{\partial}{\partial \bar{\zeta}_j} \sum_{t=1}^{n} l_{jt}^{(l)} \psi_t \qquad \sim 0, \tag{3.2.25}$$

$$\operatorname{Re} \sum_{k=1}^{n} P_{jk} L_{kl}^{(l)} c_l \qquad \sim 0, \tag{3.2.26}$$

$$\operatorname{Re} \sum_{k=1}^{n} P_{jk} l_{kt}^{(l)} \psi_t \qquad \sim \delta_{jt} \psi_t,\ t,\ j=1,\ldots,n. \tag{3.2.27}$$

In (3.2.23) we already found the desired left regularizer if there were the same $l_{jt}^{(l)}$ for different l. To get rid of this difficulty we observe (3.2.25): these functions are holomorphic with respect to ζ_l. Therefore let us define the following mapping: let ζ_j, ζ_k be two of the homeomorphisms occurring in the normal form. Let ϕ_j denote the conformal mapping of the image domain of G under ζ_j onto the unit circle in the ϕ_j plane with $\phi_j(0)=0$ and e.g. $\phi_j[\zeta_j(z_0)]=1$ for some fixed $z_0\in\dot{G}$. Then we define the mapping from $C^{1+\alpha}(\bar{G})$ into itself,

$$M_j w(z) \equiv -\frac{i}{\pi} \oint_{\hat{z}\in\dot{G}} \operatorname{Re} w(\hat{z}) \frac{d\phi_j[\zeta_j(\hat{z})]}{\phi_j[\zeta_j(\hat{z})] - \phi_j[\zeta_j(z)]} \tag{3.2.28}$$

for all $z\in G$ and by its boundary values for $z\to\dot{G}$.
Observe that

$$(M_j w)_{\bar{\zeta}_j} = 0. \tag{3.2.29}$$

Lemma 3.2.9 *Let w be holomorphic with respect to ζ_k,*

$$w_{\bar{\zeta}_k} = 0 \quad \text{in G}. \tag{3.2.30}$$

Then the mapping

$$V_{jk} w \equiv w - M_j w \tag{3.2.31}$$

from the closed subspace of $C^{1+\alpha}(\bar{G})$ consisting of all functions w with (3.2.30) into $C^{1+\alpha}(\bar{G})$ is completely continuous. The index k indicates that the space of definition depends on k according to (3.2.30).

Proof The proof follows from (1.1.30) since

$$w(z)=-\frac{\mathrm{i}}{\pi}\oint_{\hat{z}\in\dot{G}} \operatorname{Re} w(\hat{z})\frac{d\phi_k[\zeta_k(\hat{z})]}{\phi_k[\zeta_k(\hat{z})]-\phi_k[\zeta_k(z)]}$$
$$+\frac{\mathrm{i}}{\pi}\operatorname{Re}\oint_{\hat{z}\in\dot{G}}[\operatorname{Re} w(\hat{z})]\frac{d\phi_k[\zeta_k(\hat{z})]}{\phi_k[\zeta_k(\hat{z})]}+\mathrm{i}\operatorname{Im} w(0).$$

With

$$1=-\frac{\mathrm{i}}{\pi}\oint_{\hat{z}\in\dot{G}}\frac{d\phi_k[\zeta_k(\hat{z})]}{\phi_k[\zeta_k(\hat{z})]-\phi_k[\zeta_k(z)]}$$

we get

$$V_{jk}w(z)=-\frac{\mathrm{i}}{\pi}\oint_{\hat{z}\in\dot{G}}\operatorname{Re}[w(z)-w(\hat{z})]\cdot\left\{\frac{d\phi_j[\zeta_j(\hat{z})]}{\phi_j[\zeta_j(\hat{z})]-\phi_j[\zeta_j(z)]}\right.$$
$$\left.-\frac{d\phi_k[\zeta_k(\hat{z})]}{\phi_k[\zeta_k(\hat{z})]-\phi_k[\zeta_k(z)]}\right\}$$
$$+\frac{\mathrm{i}}{\pi}\operatorname{Re}\oint_{\hat{z}\in\dot{G}}[\operatorname{Re} w(\hat{z})]\frac{d\phi_k[\zeta_k(\hat{z})]}{\phi_k[\zeta_k(\hat{z})]}+\mathrm{i}\operatorname{Im} w(0). \tag{3.2.33}$$

The right-hand side of (3.2.33) is an integral of the Cauchy type and by Privalov's theorem [7, II, p. 401], the $C^{1+\alpha}$ norm of $V_{jk}w$ is bounded by the $C^{1+\alpha}$ norm of its boundary values times some constant depending on the ζ_j, ζ_k and the domain G only. The formula (3.2.33) holds even up to the boundary and there, for $z\in\dot{G}$, *the kernel in* (3.2.33) *becomes smooth* for smooth $\dot{G}$. Hence, this boundary integral defines a compact mapping [28]. The last two terms are compact mappings also since they map continuously onto the two-dimensional space of constant functions. The lemma is proved.

Repeating the whole proof we find the following:

Corollary 3.2.10 *Lemma* 3.2.9 *is also true for the mapping defined by*

$$\bar{M}_j w\equiv-\frac{\mathrm{i}}{\pi}\oint_{\hat{z}\in\dot{G}}\operatorname{Re} w(\hat{z})\frac{\overline{d\phi_j[\zeta_j(\hat{z})]}}{\overline{\phi_j[\zeta_j(\hat{z})]-\phi_j[\zeta_j(z)]}} \tag{3.2.34}$$

where

$$(\bar{M}_j w)_{\zeta_j}=0. \tag{3.2.35}$$

Now let us return to the proof of Theorem 3.2.8. By adding the completely continuous mapping

$$-V_{1l}\left(\sum_{t=1}^{n} l_{jt}^{(l)}\operatorname{Re} P_{tl}w_{l|\dot{G}}\right)$$

from (3.2.31) to (3.2.23) we find after summation over l the equations

$$\sum_{l=1}^{n} L_{jl}^{(l)} w_{l\bar{\zeta}_l} + \sum_{l,t=1}^{n} M_1 l_{jt}^{(l)} \text{ Re } P_{tl} w_{l|\dot{G}} \sim w_j \tag{3.2.36}$$

defining the desired left regularizers.

For the right regularizers we shall prove that the operators given by

$$L_{ll}^{(l)} c_l + M_l \sum_{j=1}^{n} l_{lj}^{(1)} \left(\psi_j - \sum_{k=1}^{n} \text{ Re } P_{jk} L_{kk}^{(k)} c_k \right) \equiv u_l \tag{3.2.37}$$

define the desired regularizers. For the differential equations we find with (3.2.29) and (3.2.24)

$$\frac{\partial}{\partial \bar{\zeta}_l} u_l \sim c_l \tag{3.2.38}$$

and for the boundary conditions with (3.2.31):

$$\begin{aligned} \text{Re} \sum_{l=1}^{n} P_{tl} \left[L_{ll}^{(l)} c_l + M_l \sum_{j=1}^{n} l_{lj}^{(1)} \left(\psi_j - \sum_{k=1}^{n} \text{ Re } P_{jk} L_{kk}^{(k)} c_k \right) \right] & \\ \sim \text{Re } \sum_{k=1}^{n} P_{tk} L_{kk}^{(k)} C_k + \text{Re } \sum_{k=1}^{n} P_{tk} l_{kj}^{(1)} (\cdots) & \\ \sim \psi_t. & \end{aligned} \tag{3.2.39}$$

Thus, Theorem 3.2.3 implies the proposed Theorem 3.2.8.

The most general case of systems in normal form was formulated in (3.1.30). Now let us assume that the normal form (3.1.30) admits coefficients α_{jk}, β_{jk} defining either a lower or an upper *triangular matrix all over the region* $\bar{G}$. Such systems were investigated by Bojarski [4, 5] and let us prove the Fredholm properties for these systems.

Theorem 3.2.11 *Let* $\mathbf{a}_j, \mathbf{b}_j \in C^{\alpha}(\bar{G})$; $\alpha_{jk}, \beta_{jk} \in C^{1+\alpha}(\bar{G})$, $1 > \alpha > 0$, $\alpha_{jk} = \beta_{jk} \equiv 0$ *for* $k \geq j$. *Let the boundary operator* $\mathsf{P} \in C^{1+\alpha}(\dot{G})$ *satisfy the Lopatinski condition* (3.1.37). *Then the mapping* R,

$$\mathsf{R}w \equiv \begin{cases} w_{j\bar{\zeta}_j} + \sum_{k<j} \alpha_{jk} w_{k\zeta_j} + \beta_{jk} w_{k\bar{\zeta}_j} + \mathbf{a}_j^T w + \mathbf{b}_j^T \bar{\mathbf{w}} & \text{in } \bar{G}, \quad j = 1, \ldots, n, \\ \text{Re } \mathsf{P}\mathbf{w} & \text{on } \dot{G}. \end{cases} \tag{3.2.40}$$

from $X = C^{1+\alpha}(\bar{G})$ *into* $Y = C^{\alpha}(\bar{G}) \times X^{1+\alpha}(\dot{G})$ *is Fredholm.*

The proof follows immediately by induction from

Lemma 3.2.12 *Let Theorem* 3.2.11 *be true for all systems* (3.2.40) *under the additional conditions*

$$\alpha_{2k} = \alpha_{3k} = \cdots = \alpha_{tk} = \beta_{2k} = \cdots = \beta_{tk} = 0, \qquad k = 1, \ldots, n \quad (3.2.41)$$

with some t, $2 \leq t \leq n$.
Then Theorem 3.2.11 *holds true for all systems with*

$$\alpha_{2k} = \cdots = \alpha_{t-1,k} = \beta_{2k} = \cdots = \beta_{t-1,k} = 0. \quad (3.2.42)$$

Moreover, if in case (3.2.41) *regularizers are known which are left and right regularizers, then regularizers can be constructed in case* (3.2.42) *which are also left and right at the same time.*

According to Theorem 3.2.2 we can skip all terms in R without derivatives. For $t = n$ in (3.2.41), the Theorem 3.2.11 coincides with Theorem 3.2.9. Hence, the initial proposition for the induction is satisfied.

Proof of Lemma 3.2.12 According to the assumption we know that the BVP,

$$\begin{aligned} &D_1 \mathbf{w} \equiv w_{1\bar{\zeta}_1}, \\ &\cdot \\ &\cdot \\ &\cdot \\ &D_t \mathbf{w} \equiv w_{t\bar{\zeta}_t}, \\ &D_{t+1} \mathbf{w} \equiv w_{t+1\overline{\zeta_{t+1}}} + \alpha_{t+1,t} w_{t\zeta_{t+1}} + \beta_{t+1,t} w_{t\overline{\zeta_{t+1}}}, \\ &\cdot \\ &\cdot \\ &\cdot \\ &D_n \mathbf{w} \equiv w_{n\bar{\zeta}_n} + \sum_{k=t}^{n-1} \alpha_{n,k} w_{k\zeta_n} + \beta_{n,k} w_{k\bar{\zeta}_n}, \\ &\operatorname{Re} \mathbf{Pw}_{|\dot{G}}, \end{aligned} \quad (3.2.43)$$

provides a regularizer satisfying

$$\sum_{j=1}^{n} \left(L_{sj} D_j \mathbf{w} + l_{sj} \operatorname{Re} \sum_{k=1}^{n} P_{jk} w_{k|\dot{G}} \right) \sim w_s, \quad (3.2.44)$$

$$D_s \left[\sum_{j=1}^{n} (L_{tj} c_j + l_{tj} \psi_j) \right]_{(t=1,\ldots,n)} \sim c_s, \quad (3.2.45)$$

$$\operatorname{Re} \sum_{k=1}^{n} P_{sk} \left[\sum_{j=1}^{n} (L_{kj} c_j + l_{kj} \psi_j)_{|\dot{G}} \right] \sim \psi_s, \qquad s = 1, \ldots, n. \quad (3.2.46)$$

For (3.2.44) we have

$$w_{t-1} \sim L_{t-1,t-1} w_{t-1\bar{\zeta}_{t-1}} + \sum_{j=1}^{n} l_{t-1,j} \text{ Re } P_{j,t-1} w_{t-1|\dot{G}} \tag{3.2.47}$$

From (3.2.44) we compute for the more general operator belonging to (3.2.41) with (3.2.47):

$$\begin{aligned}&\sum_{j=1}^{n} L_{sj}\left[D_j \mathbf{w} + \left(\alpha_{j,t-1} \frac{\partial}{\partial \zeta_j} + \beta_{j,t-1} \frac{\partial}{\partial \bar{\zeta}_j}\right) w_{t-1}\right] \\ &- \sum_{j=1}^{n} L_{sj}\left(\alpha_{j,t-1} \frac{\partial}{\partial \zeta_j} + \beta_{j,t-1} \frac{\partial}{\partial \bar{\zeta}_j}\right)\left(L_{t-1,t-1} D_{t-1} \mathbf{w} + \sum_{l=1}^{n} l_{t-1,l} \text{ Re } P_{l,t-1} w_{t-1|\dot{G}}\right) \\ &+ \sum_{j,k=1}^{n} l_{sj} \text{ Re } P_{jk} w_{k|\dot{G}} \sim w_s, \qquad s=1, \ldots, n.\end{aligned} \tag{3.2.48}$$

If (3.2.48) did not contain the term

$$u \equiv \sum_{l=1}^{n} l_{t-1,l} \text{ Re } P_{l,t-1} w_{t-1|\dot{G}} \tag{3.2.49}$$

it would already define a left regularizer. Corresponding to (3.2.25), the function u is ζ_{t-1}-holomorphic,

$$u_{\bar{\zeta}_{t-1}} = 0. \tag{3.2.50}$$

Using M_j and (3.2.31) this implies with (3.2.28) that

$$\frac{\partial u}{\partial \bar{\zeta}_j} \sim \frac{\partial}{\partial \bar{\zeta}_j} (M_j u) = 0 \tag{3.2.51}$$

and analogously

$$\frac{\partial u}{\partial \zeta_j} \sim \frac{\partial}{\partial \zeta_j} (\bar{M}_j u) = 0. \tag{3.2.52}$$

Therefore the corresponding terms in (3.2.48) can be skipped and the relations

$$\begin{aligned}&\sum_{j=1}^{n}\left[L_{sj} - \sum_{r=1}^{n} L_{sr}\left(\alpha_{r,t-1} \frac{\partial}{\partial \zeta_r} + \beta_{r,t-1} \frac{\partial}{\partial \bar{\zeta}_r}\right) L_{t-1,t-1} \delta_{t-1,j}\right] \\ &\qquad \times \left[D_j \mathbf{w} + \left(\alpha_{j,t-1} \frac{\partial}{\partial \zeta_j} + \beta_{j,t-1} \frac{\partial}{\partial \bar{\zeta}_j}\right) w_{t-1}\right] \\ &\qquad + \sum_{j,k=1}^{n} l_{sj} \text{ Re } P_{jk} w_{k|\dot{G}} \sim w_s, \qquad s=1, \ldots, n\end{aligned} \tag{3.2.53}$$

define a left regularizer.

On the other hand, (3.2.53) defines already a right regularizer by

$$\sum_{j=1}^{n}\left[L_{sj}-\sum_{r=1}^{n}L_{sr}\left(\alpha_{r,t-1}\frac{\partial}{\partial\zeta_r}+\beta_{r,t-1}\frac{\partial}{\partial\bar{\zeta}_r}\right)L_{t-1,t-1}\delta_{t-1,j}\right]c_j$$

$$+\sum_{j=1}^{n}l_{sj}\psi_j\equiv v_s+u_s,\qquad s=1,\ldots,n,\quad (3.2.54)$$

as we shall show in the following.
For the differential equations we have from (3.2.45)

$$D_s\mathbf{v}\sim c_s-\left(\alpha_{s,t-1}\frac{\partial}{\partial\zeta_s}+\beta_{s,t-1}\frac{\partial}{\partial\bar{\zeta}_s}\right)L_{t-1,t-1}c_{t-1}\qquad(3.2.55)$$

and from (3.2.42)

$$D_{t-1}\mathbf{v}\sim c_{t-1}.\qquad(3.2.56)$$

Using (3.2.44) for $w_j=\delta_{j,t-1}v_{t-1}$, we find

$$L_{t-1,t-1}c_{t-1}+\sum_{j=1}^{n}l_{t-1,j}\text{ Re }P_{j,t-1}v_{t-1|\dot{G}}\sim v_{t-1}.\qquad(3.2.57)$$

Therefore (3.2.55) takes the form

$$D_s\mathbf{v}+\left(\alpha_{s,t-1}\frac{\partial}{\partial\zeta_s}+\beta_{s,t-1}\frac{\partial}{\partial\bar{\zeta}_s}\right)v_{t-1}$$

$$\sim c_s+\left(\alpha_{s,t-1}\frac{\partial}{\partial\zeta_s}+\beta_{s,t-1}\frac{\partial}{\partial\bar{\zeta}_s}\right)\sum_{j=1}^{n}l_{t-1,j}\text{ Re }P_{jt-1}v_{t-1|\dot{G}}.\quad(3.2.58)$$

Since $l_{t-1,j}$ is ζ_{t-1}-holomorphic it follows similarly to (3.2.51) and (3.2.52) that the last term is a completely continuous operator acting on v_{t-1}. Hence we have with (3.2.45) and (3.2.57)

$$D_s(\mathbf{v}+\mathbf{u})+\left(\alpha_{s,t-1}\frac{\partial}{\partial\zeta_s}+\beta_{s,t-1}\frac{\partial}{\partial\bar{\zeta}_s}\right)(v_{t-1}+u_{t-1})$$

$$\sim c_s+\left(\alpha_{s,t-1}\frac{\partial}{\partial\zeta_s}+\beta_{s,t-1}\frac{\partial}{\partial\bar{\zeta}_s}\right)\sum_{j=1}^{n}l_{t-1,j}\psi_j$$

$$\sim c_s\qquad(3.2.59)$$

by using the ζ_{t-1}-holomorphy of $l_{t-1,j}\psi_j$ again.
The boundary conditions

$$\text{Re }\sum_{s=1}^{n}P_{ks}(v_s+u_s)_{|\dot{G}}=\text{Re }\sum_{s=1}^{n}P_{ks}\left\{\sum_{j=1}^{n}[L_{sj}(\cdot\cdot\cdot)+l_{sj}\psi_j]\right\}$$

$$\sim\psi_k\qquad(3.2.60)$$

follow immediately from (3.2.46).

Hence, the operators defined by (3.2.53) or (3.2.54) are both left and right regularizes and Lemma 3.2.11 is proved.

Lemma 3.2.13 *Let L_{sj} belong to the regularizer to* (3.2.40) *with* (3.2.41) i.e. *satisfying* (3.2.44–46). *Then the operator defined by*

$$L_{sj}(\alpha_1+i\alpha_2)v-\alpha_1 L_{sj}v-\alpha_2 L_{sj}iv\equiv W_{sj}v \tag{3.2.61}$$

where $\alpha_1, \alpha_2 \in C^{1+\alpha}(\bar{G})$ are fixed real valued functions, maps C^α completely continuously into $C^{1+\alpha}(\bar{G})$.

Proof For the differences

$$L_{sj}\alpha_1 v-\alpha_1 L_{sj}v\equiv W'_{sj}v \tag{3.2.62}$$

we find, according to (3.2.45), the differential equations

$$\begin{aligned} D_s W'_{sj}v &\sim \delta_{sj}\alpha_1 v-(D_s\alpha_1)L_{sj}v-\alpha_1\delta_{sj}v,\\ D_s W'_{sj}v &\sim -(D_s\alpha_1)L_{sj}v, \qquad s=1,\dots,n. \end{aligned} \tag{3.2.63}$$

Since L_{sj} maps C^α continuously into $C^{1+\alpha}$ and the latter is compactly embedded in C^α, the right-hand side of (3.2.63) defines a compact mapping in $C^\alpha(\bar{G})$. From (3.2.46) it follows that

$$\sum_{s=1}^{n} \text{Re } P_{ks}W'_{sj}v\sim 0 \tag{3.2.64}$$

holds. Inserting (3.2.63) and (3.2.64) into (3.2.44) we arrive at the proposed compactness for W'_{sj}. For the second term

$$W''_{sj}v\equiv L_{sj}\alpha_2 iv-\alpha_2 L_{sj}iv,$$

the compactness follows in the same way.

The Lemma 3.2.13 leads immediately to the following proposition:

Proposition 3.2.14 *The left regularizer of Lemma* 3.2.12 *in* (3.2.53), (3.2.54) *can be replaced by*

$$\begin{aligned} \tilde{L}_{sk} \equiv L_{sk}-\delta_{t-1,k}\sum_{j=1}^{n}\Big[&(\text{Re }\alpha_{j,t-1})L_{sj}\frac{\partial}{\partial\zeta_j}L_{t-1,t-1}\\ &+(\text{Im }\alpha_{j,t-1})L_{sj}\, i\frac{\partial}{\partial\zeta_j}L_{t-1,t-1}+(\text{Re }\beta_{j,t-1})L_{sj}\frac{\partial}{\partial\bar{\zeta}_j}L_{t-1,t-1}\\ &+(\text{Im }\beta_{j,t-1})L_{sj}\, i\frac{\partial}{\partial\bar{\zeta}_j}L_{t-1,t-1}\Big] \end{aligned}$$

and the former l_{sj}.

Repeating this interchanging procedure in connection with the induction proof of Theorem 3.2.11 by using Lemma 3.2.12 the following proposition holds owing to the triangular shape of the differential operator in (3.2.40):

Proposition 3.2.15 *For every triangular system, the regularizer* (3.2.65) *can be represented in the form*

$$\tilde{L}_{sk} \sim L_{sk} + \sum \gamma^{p_1}_{j_1,k} T^{p_1}_{j_1 sk} + \sum \gamma^{p_1}_{j_1,j_2} \gamma^{p_2}_{j_2,k} T^{p_1 p_2}_{j_1 j_2 sk}$$

$$+ \cdots + \sum \gamma^{p_1}_{j_1 j_2} \gamma^{p_2}_{j_2 j_3} \cdots \gamma^{p_{n-1}}_{j_{n-1} k} T^{p_1 \cdots p_{n-1}}_{j_1 \cdots j_{n-1} sk} \qquad (3.2.66)$$

together with the l_{jk} *belonging to* (3.2.21).
Here the summation indices $j_1, \ldots, j_{n-1} \neq k$ *are different mutually and running from* 1 *to* n, *where* $p_t = 1, \ldots, 4$ *and*

$$\gamma^1_{lt} = \text{Re } \alpha_{lt}, \qquad \gamma^2_{lt} = \text{Im } \alpha_{lt}, \qquad \gamma^3_{lt} = \text{Re } \beta_{lt}, \qquad \gamma^4_{lt} = \text{Im } \beta_{lt}$$

and where the $T^{p_1 \cdots p_t}_{j_1 \cdots j_t sk}$ *are operators defined by compositions of the* L_{sk} *belonging to the problem* (3.2.21) *and of derivatives. But they do not depend on the* α_{lt} *or the* β_{lt}.

Remark Since the α_{lt} and β_{lt} form a triangular matrix the representation (3.2.66) is of triangular shape, too.

Since the representation (3.2.66) holds also for a system (3.2.40) where the triangular differential operator is *reordered* we have the following:

Theorem 3.2.16 *Let* $\mathbf{a}_j, \mathbf{b}_j \in C^\alpha(\bar{\mathsf{G}})$, $\alpha_{jk}, \beta_{jk} \in C^{1+\alpha}(\bar{\mathsf{G}})$ $0 < \alpha < 1$. *Let the boundary operator* $\mathsf{P} \in C^{1+\alpha}(\dot{\mathsf{G}})$ *satisfy the Lopatinski condition* (3.1.37). *Then the mapping* R *belonging to the general normal form* (3.1.30),

$$\mathbf{Rw} \equiv \begin{cases} w_{j\bar{\zeta}_j} + \sum\limits_{k \neq j} \alpha_{jk} w_{k\zeta_j} + \beta_{jk} w_{k\bar{\zeta}_j} + \mathbf{a}_j^T \mathbf{w} + \mathbf{b}_j^T \bar{\mathbf{w}} & \text{in } \bar{\mathsf{G}}, \quad j = 1, \ldots, n \\ \text{Re}\, \mathbf{Pw} & \text{on } \dot{\mathsf{G}} \end{cases} \qquad (3.2.67)$$

from $X = C^{1+\alpha}(\bar{\mathsf{G}})$ *into* $Y = C^\alpha(\bar{\mathsf{G}}) \times C^{1+\alpha}(\dot{\mathsf{G}})$ *is Fredholm.*

Proof If we consider an auxiliary operator to (3.2.67) where the coefficients,

$$\alpha_{jk} = \alpha_{jk}(z_0), \qquad \beta_{jk} = \beta_{jk}(z_0), \qquad z_0 \in \bar{\mathsf{G}}, \qquad (3.2.68)$$

are considered to be constant with any z_0 then the regularizer (3.2.66) is well defined corresponding to z_0. Identifying now z_0 with z, we obtain operators (3.2.66), (3.2.53, 54). Multiplying the components of R with these operators from the left and the right, correspondingly, one obtains the equations (3.2.44, 45) at every point $z=z_0$ modulo compact perturbations produced by the derivatives of α_{jk} and β_{jk}. The relations (3.2.46) also hold since (3.2.60) hold now for every operator corresponding to any $z_0 \in \bar{\mathsf{G}}$.

3.3 The Index for Systems in Normal Form

Using the continuity of the Fredholm index and the Fredholm property of the boundary value problem we can obtain Bojarski's explicit index formula in a very easy way. The result is formulated in

Theorem 3.3.1 *Under the same assumptions as in Theorem* 3.2.16, *the Fredholm index ν, belonging to* R *in* (3.2.67) *can be computed by*

$$\nu = n - \frac{1}{\pi} \oint_{\dot{\mathsf{G}}} \mathrm{d} \arg \operatorname{Det} \mathsf{P}. \tag{3.3.1}$$

(Recall that the Lopatinski condition $\operatorname{Det} \mathsf{P} \neq 0$ is assumed).

Remarks For elliptic systems with the Laplacian as principal part, the corresponding formula was found by Bitsadze [3, p. 151 (6.16)]. This corresponds to the index formula by Pascali [20]. For systems with normal form with lower triangular matrix in the principal part as (3.2.40), the index formula was showed by Bojarski [5]. Later we shall show its connection to Vol'pert's formula for general systems [27].

Proof For the proof we construct a family of regular boundary value problems $\mathsf{R}(t)$ with operators depending continuously on a real parameter $t \in [0, 1]$ satisfying the side conditions:

$$\mathsf{R}(t) \in \phi(X, Y) \quad \text{for every} \quad t \in [0, 1], \tag{3.3.2}$$

$$\operatorname{Det} \mathsf{P}(t) = \operatorname{Det} \mathsf{P} \neq 0 \quad \text{for every} \quad t \in [0, 1] \tag{3.3.3}$$

$$\mathsf{R}(0) = \mathsf{R}, \tag{3.3.4}$$

$\mathsf{R}(1)$ is completely decomposed in the following sense: this boundary value problem is defined by (3.3.5)

$$w_{j\bar{\zeta}_j} = c_j \quad \text{in } \mathsf{G} \text{ and } \quad \operatorname{Re} w_j P_{jj}(t=1) = \psi_j \quad \text{on } \dot{\mathsf{G}} \qquad j = 1, \ldots, n. \tag{3.3.6}$$

If this family is constructed, then for its indexes the equation

$$\nu(R(t)) = \nu(R) = \nu(R(1)) \tag{3.3.7}$$

will hold because ν depends continuously on $t \in [0, 1]$ (Theorem 3.2.4) on one hand and ν takes only integer images on the other hand. For (3.3.6) the index is explicitly known because each single equation is Fredholm with the single index:

$$\nu_j = 1 - \frac{1}{\pi} \oint_{\dot{G}} d \arg P_{jj}(1) \tag{3.3.8}$$

(Theorem 2.5.1). Hence we get with (3.3.3), (3.3.7) and (3.3.8) Bojarski's formula:

$$\begin{aligned}\nu = \sum_{j=1}^{n} \nu_j &= n - \frac{1}{\pi} \oint_{\dot{G}} d \sum_{j=1}^{n} \arg P_{jj}(1) = n - \frac{1}{\pi} \oint_{\dot{G}} d \arg \prod_{j-1}^{n} P_{jj}(1) \\ &= n - \frac{1}{\pi} \oint d \arg \operatorname{Det} \mathsf{P}.\end{aligned} \tag{3.3.9}$$

It remains to construct the family $\mathsf{R}(t)$.

First of all we construct the boundary conditions by $n+1$ steps in a manner related to Gaussian elimination. To this end we can assume beforehand, by using transformations (3.2.7), that Gaussian elimination can be applied to P with pivots in the diagonal only.

1st step:

$$P_{jk}(\tau_1) \equiv P_{jk} - \tau_1(1 - \delta_{k1}) \frac{P_{j1}}{P_{11}} P_{1k}, \qquad \tau_1 \in [0, 1],$$

$$t = \frac{\tau_1}{n+1} \in \left[0, \frac{1}{n+1}\right]. \tag{3.3.10}$$

For $\tau_1 = 1$ we have $P_{1k}(\tau_1 = 1) = 0$ for all $k = 2, \ldots, n$; and, furthermore,

$$\operatorname{Det} \mathsf{P}(\tau_1) = \operatorname{Det} \mathsf{P} \quad \text{for all} \quad \tau_1 \in [0, 1]. \tag{3.3.11}$$

2nd step:

$$P_{jk}(\tau_2) \equiv \begin{cases} P_{jk} \quad \text{for} \quad k = 1, 2 \\ P_{jk} - \tau_2 \dfrac{P_{j2}}{P_{22}} P_{2k}, \quad \tau_2 \in [0, 1], \\ t = \dfrac{\tau_2 + 1}{n+1} \in \left[\dfrac{1}{n+1}, \dfrac{2}{n+1}\right] \end{cases} \tag{3.3.12}$$

where $P_{jk} = P_{jk}(\tau_1 = 1)$ on the right hand side.

Repeating this construction n steps, we get for

$$P_{jk}\left(t = \frac{n}{n+1}\right) = \tilde{P}_{jk}$$

a lower triangular matrix:

$$\tilde{P}_{jk}=0 \quad \text{for} \quad k>j. \tag{3.3.13}$$

The last step perturbs the triangular matrix into the diagonal form:

$$P_{jk}(\tau_{n+1})\equiv\begin{cases}\tilde{P}_{jk} & \text{for} \quad j=k,\\ (1-\tau_{n+1})\tilde{P}_{jk} & \text{for} \quad j>k, \quad \tau_{n+1}\in[0,1],\end{cases} \tag{3.3.14}$$

$$t=\frac{\tau_{n+1}+n}{n+1}$$

Hence, the desired family of boundary conditions is constructed. The differential operator family can be chosen e.g. as

$$w_{j\bar{\zeta}_j}+(1-t)\left(\sum_{k\neq j}\alpha_{jk}w_{k\zeta_j}+\beta_{jk}w_{k\bar{\zeta}_j}+\mathbf{a}_j^T\mathbf{w}+\mathbf{b}_j^T\bar{\mathbf{w}}\right). \tag{3.3.15}$$

3.4 One Adjoint System and Normal Solvability

If the boundary value problem is given having the differential equations in normal form (3.1.30),

$$\mathsf{R}\mathbf{w}\equiv\begin{cases}D_j\mathbf{w}\equiv w_{j\bar{\zeta}_j}+\sum_{k\neq j}(\alpha_{jk}w_{k\zeta_j}+\beta_{jk}w_{k\bar{\zeta}_j})+\sum_{k=1}^{n}(a_{jk}w_k+b_{jk}\bar{w}_k), \text{ in } \mathsf{G}\\ \operatorname{Re}\mathsf{P}\mathbf{w} \quad \text{on } \dot{\mathsf{G}}\end{cases} \tag{3.4.1}$$

then the formal adjoint system will essentially have the same principal part. As usual, the adjoint boundary value problem can be found by the use of Green's formula, e.g.

$$\begin{aligned}&\operatorname{Re}\sum_{j=1}^{n}\iint_{\mathsf{G}}\left(w_{j\bar{\zeta}_j}+\sum_{k\neq j}(\alpha_{jk}w_{k\zeta_j}+\beta_{jk}w_{k\bar{\zeta}_j})+\sum_{k=1}^{n}(a_{jk}w_k+b_{jk}\bar{w}_k)\right)v_j\mathrm{i}[\mathrm{d}\zeta_j,\mathrm{d}\bar{\zeta}_j].\\ &=\operatorname{Re}\iint_{\mathsf{G}}\sum_{j,k=1}^{n}\{a_{jk}w_kv_j+b_{jk}\bar{w}_kv_j\\ &\qquad -(v_{k\bar{\zeta}_k}\delta_{jk}+(\alpha_{jk}v_j)_{\zeta_j}+(\beta_{jk}v_j)_{\bar{\zeta}_j})w_k\}\mathrm{i}[\mathrm{d}\zeta_j,\mathrm{d}\bar{\zeta}_j]\\ &\quad-\operatorname{Re}\oint_{\dot{\mathsf{G}}}\sum_{j=1}^{n}\left\{\mathrm{i}w_jv_j\,\mathrm{d}\zeta_j-\mathrm{i}\sum_{k\neq j}w_k\alpha_{jk}v_j\,\mathrm{d}\bar{\zeta}_j+\mathrm{i}\sum_{k\neq j}w_k\beta_{jk}v_j\,\mathrm{d}\zeta_j\right\}.\end{aligned} \tag{3.4.2}$$

The surface integral on the right has to be arranged such that a differential operator for $\mathbf{v}$ in normal form occurs and the boundary integrals have to be reformulated by the use of the boundary matrix P. To this end we assume once again the *Lopatinski condition*,

$$\text{Det } \mathsf{P} \neq 0. \tag{3.4.3}$$

Consequently, there exists the inverse Q to P^{T},

$$\mathsf{P}^{\mathsf{T}}\mathsf{Q} = \mathsf{I},^{\dagger} \tag{3.4.4}$$

$\mathbf{1}$ denotes the unit matrix. With Q, (3.4.2) takes the form

$$\text{Re} \sum_{j=1}^{n} \iint_{\mathsf{G}} D_j \mathbf{w} v_j \, \mathrm{i}[\mathrm{d}\zeta_j, \mathrm{d}\bar{\zeta}_j]$$

$$= \text{Re} \sum_{k=1}^{n} - \iint_{\mathsf{G}} \Big\{ v_{k\bar{\zeta}_k} + \sum_{j=k} ((\alpha_{jk} v_j)_{\zeta_j} + (\beta_{jk} v_j)_{\bar{\zeta}_j}) \frac{\partial(\xi_j, \eta_j)}{\partial(\xi_k, \eta_k)}$$

$$- \sum_{j=1}^{n} (a_{jk} v_j + \bar{b}_{jk} \bar{v}_j) \frac{\partial(\xi_j, \eta_j)}{\partial(\xi_k, \eta_k)} \Big\} w_k \, \mathrm{i}[\mathrm{d}\zeta_k, \mathrm{d}\bar{\zeta}_k]$$

$$- \oint_{\dot{\mathsf{G}}} \text{Re } (\mathsf{P}\mathbf{w})^T \text{ Re } (\mathrm{i}\mathsf{Q}\mathsf{S}\mathbf{v}) \, \mathrm{d}s + \oint_{\dot{\mathsf{G}}} \text{Im } (\mathsf{P}\mathbf{w})^T \text{ Re } (\mathsf{Q}\mathsf{S}\mathbf{v}) \, \mathrm{d}s \tag{3.4.5}$$

where the matrix S is defined by

$$S_{jk} \equiv (\delta_{jk} + \beta_{kj})\dot{\zeta}_k - \alpha_{kj}\overline{\dot{\zeta}_k}. \tag{3.4.6}$$

Referring to Theorem 1.3.4 the relation (3.4.5) leads us to introduce the following bilinear forms:

$$\langle\langle |-|, \mathbf{v} \rangle\rangle \equiv \sum_{j=1}^{n} \iint_{\mathsf{G}} \text{Re } c_j v_j \, \mathrm{i}[\mathrm{d}\zeta_j, \mathrm{d}\bar{\zeta}_j] + \oint_{\dot{\mathsf{G}}} \boldsymbol{\psi}^T \text{ Re } \mathrm{i}\mathsf{Q}\mathsf{S}\mathbf{v} \, \mathrm{d}s$$

$$\text{for } (|-|, \mathbf{v}) = ((\mathbf{c}, \boldsymbol{\psi}), \mathbf{v}) \in Y \times X, \tag{3.4.7}$$

$$\langle \mathbf{w}, E \rangle \equiv -\sum_{j=1}^{n} \iint_{\mathsf{G}} \text{Re } f_j w_j \; \mathrm{i}[\mathrm{d}\zeta_j, \mathrm{d}\bar{\zeta}_j] + \oint_{\dot{\mathsf{G}}} \boldsymbol{\phi}^T \text{ Im } \mathsf{P}\mathbf{w} \, \mathrm{d}s$$

$$\text{for } (\mathbf{w}, E) = (\mathbf{w}, (\mathbf{f}, \boldsymbol{\phi})) \in X \times Y \tag{3.4.8}$$

where $X = C^{1+\alpha}(\bar{\mathsf{G}})$ and $Y = C^{\alpha}(\bar{\mathsf{G}}) \times C^{1+\alpha}(\dot{\mathsf{G}})$.
Since α_{jk} and β_{jk} in the normal form (3.4.1) are (rearranged) triangular matrices with vanishing diagonal elements we observe the identities

$$\det \mathsf{Q} = (\det \mathsf{P})^{-1} \quad \text{and} \quad \det \mathsf{S} = \prod_{k=1}^{n} \dot{\zeta}_k \neq 0. \tag{3.4.9}$$

† Here and in the following I denotes the identity matrix.

Therefore the boundary matrices occurring in the boundary integrals in (3.4.7) and (3.4.8) are regular. Now an elementary computation shows that both bilinear forms are *regular.* By the use of (3.4.7) and (3.4.8) the identity (3.4.5) can be written as

$$\langle\langle \mathsf{R}\mathbf{w}, \mathbf{v}\rangle\rangle = \langle \mathbf{w}, \mathsf{R}^*\mathbf{v}\rangle \quad \text{for all} \quad \mathbf{v}, \mathbf{w} \in C^{1+\alpha}(\bar{\mathsf{G}}) \tag{3.4.10}$$

if we define the *adjoint boundary value problem* by

$$\mathsf{R}^*\boldsymbol{v} \equiv \begin{cases} v_{j\bar{\zeta}_j} + \sum_{k=1}^{n} \{(\alpha_{kj} v_k)_{\zeta_k} + (\beta_{kj} v_k)_{\bar{\zeta}_k} \\ \qquad\qquad - a_{kj} v_k - \bar{b}_{kj} \bar{v}_k\} \dfrac{\partial(\xi_n, \eta_k)}{\partial(\xi_j, \eta_j)} \quad \text{in } \mathsf{G}, \\ \operatorname{Re} \mathsf{QS}\mathbf{v} \quad \text{on } \dot{\mathsf{G}} \end{cases} \tag{3.4.11}$$

Since (3.4.9) implies that for the boundary matrix QS of R^* the Lopatinski condition also holds, Theorem 3.2.16 yields $\mathsf{R}^* \in \phi(X, Y)$. In addition, we find from (3.3.1) and (3.4.9) the relation,

$$\begin{aligned} \nu(\mathsf{R}^*) &= n - \frac{1}{\pi} \oint_{\dot{\mathsf{G}}} d \arg \det (\mathsf{QS}) \\ &= n + \frac{1}{\pi} \oint_{\dot{\mathsf{G}}} d \arg \det \mathsf{P} - 2 \sum_{k=1}^{n} \frac{1}{2\pi} \oint_{\dot{\mathsf{G}}} d \arg \zeta_k \\ &= \frac{1}{\pi} \oint_{\dot{\mathsf{G}}} d \arg \det \mathsf{P} - n = -\nu(\mathsf{R}), \end{aligned} \tag{3.4.12}$$

between the Fredholm indexes of R and R^*.

Hence all the assumptions to Theorem 1.3.4 are fulfilled and we find

Theorem 3.4.1 R *and* R^* *are normally solvable*; i.e. *the boundary value problem*

$$\mathsf{R}\mathbf{w} = (\mathbf{c}, \boldsymbol{\psi})$$

has a solution if and only if the solvability conditions

$$\begin{aligned} \langle\langle(\mathbf{c}, \boldsymbol{\psi}), \mathbf{v}\rangle\rangle = \sum_{j=1}^{n} \operatorname{Re} \iint_{\mathsf{G}} (c_j v_j) i[d\zeta_j, d\bar{\zeta}_j] \\ + \oint_{\dot{\mathsf{G}}} \boldsymbol{\psi}^T \operatorname{Re}(i\mathsf{QS}\mathbf{v}) \, ds = 0 \end{aligned} \tag{3.4.13}$$

are satisfied for all $\mathbf{v}$ *with*

$$\mathsf{R}^*\mathbf{v} = 0,$$

i.e. for all eigensolutions of the homogeneous adjoint boundary value problem. Moreover, the eigenspaces of R *and* R* *are both finite dimensional.*

Conclusion 3.4.2 *For* R *given by* (3.4.1) *there exists a constant* γ *such that the* a priori *estimate*

$$\|\mathbf{w}\|_{1+\alpha} \leqslant \gamma\{\|\mathsf{R}\mathbf{w}\|_Y + \|\mathbf{w}\|_0\} \tag{3.4.14}$$

holds for all $\mathbf{w} \in C^{1+\alpha}(\bar{\mathsf{G}})$.

Remark (3.4.14) is one of the Agmon–Douglis–Nirenberg estimates corresponding to Theorem 2.5.2 with $n=1$, [1].

Proof If $(\mathbf{c}, \boldsymbol{\psi}) \in$ range (R), then the general solution is given by

$$\mathbf{w} = \mathbf{w}_p + \sum_{j=1}^{N} \alpha_j \mathbf{w}_{j0} \tag{3.4.15}$$

with $\alpha_j \in R$ and w_{j_0} a basis of the nullspace N(R). Because the $\mathbf{w}_{j_0}$ are linearly independent, there exist functions $\mathbf{g}_k \in C^\alpha$ such that

$$\operatorname{Re} \iint_{\mathsf{G}} \mathbf{w}_{j_0}^T \cdot \mathbf{g}_k [dx, dy] = \delta_{jk}. \tag{3.4.16}$$

Then the side conditions

$$\operatorname{Re} \iint_{\mathsf{G}} \mathbf{w}_{j_0}^T \cdot \mathbf{g}_j [dx, dy] = \gamma_j \in R, \; j = 1, \ldots, N \tag{3.4.17}$$

make the solution $\mathbf{w}$ unique. Thus the mapping defined by R and (3.4.17),

$$\tilde{\mathsf{R}} : X \to \text{range } (\mathsf{R}) \times R^N \tag{3.4.18}$$

has a closed range and is bijective, hence it has a continuous inverse and that means by the use of Banach's theorem on the bounded inverse (see e.g. Theorem 4.1, p. 63 in [23]):

$$\|\mathbf{w}\|_X \leqslant \tilde{\gamma}\{\|\mathsf{R}\mathbf{w}\|_Y + \sum_{j=1}^{N} |\gamma_j|\}. \tag{3.4.19}$$

The replacement of the γ_j by (3.4.17) implies (3.4.14).

3.5 Two Problems with Violated Lopatinski Condition

As we have seen, the Lopatinski condition (3.1.37) implied that the operator R to the boundary value problem (3.4.1) became Fredholm from

$X = C^{1+\alpha}(\bar{\mathrm{G}})$ into $Y = C^{\alpha}(\bar{\mathrm{G}}) \times C^{1+\alpha}(\dot{\mathrm{G}})$. For violated Lopatinski condition we can not expect $R \in \phi(X, Y)$ anymore. Since the violation of the Lopatinski condition may happen in many different ways there is no systematic investigation of these problems available yet. But some special problems have already been investigated and we shall consider two of them: One where the determinant vanishes in isolated points with a positive order and another problem where the determinant vanishes identically on the whole boundary. For brevity we shall not present all the details.

In the case of two real unknowns, $n = 1$, for pointwise violations of (3.1.37) such problems have been investigated by Bruhn [6] and Jaenicke [15] (*see also* [12] Sections 11.7–11.9). After removing the right hand side of the differential equation as in (1.2.3) such boundary value problems take the form

$$\begin{aligned} & w_{\bar{z}} - \bar{a}w - b\bar{w} = 0 && \text{in } \mathrm{G} \\ & \operatorname{Re} pw = \phi && \text{on } \dot{\mathrm{G}} \end{aligned} \tag{3.5.1}$$

where

$$p(z_j) = 0,\ j = 1, \ldots, r, \qquad z_j \in \dot{\mathrm{G}} \quad \text{and} \quad p \neq 0 \quad \text{for} \quad z \neq z_j. \tag{3.5.2}$$

The method for solving (3.5.1) is based on a factorization similar to (1.2.11) which leads to Fredholm mappings in different function spaces with weighted norms where the functions have zeros of prescribed orders in z_j. The crucial assumption on p is that the zeros in (3.5.2) are of orders τ_j; more precisely, the function of s defined by

$$\frac{p}{|p|}\,\sigma(s) = \mathrm{e}^{\mathrm{i}\tau(s)} \tag{3.5.3}$$

has an argument $\tau(s)$ which is piecewise C^{α} in $s_j \leq s \leq s_{j+1}$ where σ is suitably chosen,

$$\sigma(s) = 1 \text{ or } -1 \quad \text{in} \quad s_j \leq s < s_{j+1} \tag{3.5.4}$$

such that

$$0 \leq \tau(s_j - 0) - \tau(s_j + 0) = \tau_j \begin{cases} < \pi & \text{for } j = 2, \ldots, r \\ < 2\pi & \text{for } j = 1. \end{cases} \tag{3.5.5}$$

The function

$$\mathrm{e}^{\mathrm{i}\tau(s)} \frac{(z - z_j)^{\tau_j}}{|z - z_j|^{\tau_j}} \equiv \mathrm{e}^{\mathrm{i}\hat{\tau}(s)} \tag{3.5.6}$$

where the branchlines of the power functions are chosen in the exterior of

$\bar{\mathsf{G}}-\{z_j\}$ belongs to $C^\alpha(\dot{\mathsf{G}})$ and

$$\hat{n}\equiv\frac{1}{2\pi}\oint_{\dot{\mathsf{G}}} \mathrm{d}\hat{\tau}=0, \pm 1, \dots \tag{3.5.7}$$

is well defined. The proposed factorization is defined by

$$w=\hat{w}\prod_{j=1}^{r}(z-z_j)^{\nu_j} \quad \text{with} \quad \nu_j=\tau_j \tag{3.5.8}$$

which leads to the boundary value problem for $\hat{w}$,

$$\begin{aligned}\hat{w}_{\bar{z}}&=\bar{a}\hat{w}+b\prod_{j=1}^{r}\frac{(\bar{z}-\bar{z}_j)^{\nu_j}}{(z-z_j)^{\nu_j}}\bar{\hat{w}} \quad \text{in G and}\\ \operatorname{Re} e^{i\hat{\tau}}\hat{w}&=\hat{\phi}\equiv\phi\frac{\sigma}{|p|}\prod_{j=1}^{r}\frac{1}{|z-z_j|^{\nu_j}} \quad \text{on } \dot{\mathsf{G}}.\end{aligned} \tag{3.5.9}$$

Since this is a Fredholm problem for $\hat{w}$ we introduce the corresponding Banach spaces:

$$\begin{aligned}X&\equiv\{w \text{ having generalized derivatives satisfying}\\ &\qquad w_{\bar{z}}=\bar{a}w+b\bar{w} \text{ in G and } \|w\|_X\equiv\|\hat{w}\|_{C^\alpha(\bar{\mathsf{G}})}<\infty\}\\ Y&\equiv\{\phi \text{ with } \hat{\phi}=\phi\frac{\sigma}{|p|}\prod_{j=1}^{r}\frac{1}{|z-z_j|^{\nu_j}}\in C^\alpha(\dot{\mathsf{G}}) \quad \text{and}\\ &\qquad \|\phi\|_Y\equiv\|\hat{\phi}\|_{C^\alpha(\dot{\mathsf{G}})}\}.\end{aligned} \tag{3.5.10}$$

With these notations Jaenicke's results imply the following

Theorem 3.5.1 *The boundary value problem* (3.5.1) *defines a Fredholm mapping* $\operatorname{Re} pw_{|\dot{\mathsf{G}}}: X\to Y$ *with Fredholm index* $1-2\hat{n}$.

The whole approach can be extended to the case $\nu_j=2\kappa_j\pi+\tau_j$ with $\kappa_j=0, \pm 1, \pm 2, \dots$ leading to other spaces, corresponding mappings and indexes. Especially Bruhn's results in [6] belong to the choice $\kappa_j=-1$.

In Chapter 5 we shall see that the boundary value problems are connected with certain systems of singular integral equations of Cauchy type on the boundary $\dot{\mathsf{G}}$. Using factorization techniques for these integral equations which were obtained by Prößdorf [2], Kopp [17] generalized the above problems partially to the Pascali system (3.2.1):

$$\mathbf{R}\mathbf{w}\equiv\begin{cases}\mathbf{w}_{\bar{z}}+\mathbf{a}\mathbf{w}+\mathbf{b}\bar{\mathbf{w}}=0 & \text{in G},\\ \operatorname{Re}\mathbf{P}\mathbf{w}=\boldsymbol{\phi} & \text{on } \dot{\mathsf{G}}\end{cases} \tag{3.5.11}$$

where the determinant of $\mathbf{P}$ has zeros of *integer orders* m_j:

$$\det\mathbf{P}=\Delta_0(s)\prod_{j=1}^{r}(z-z_j)^{m_j}, \quad \Delta_0\neq 0. \tag{3.5.12}$$

Let us assume that the coefficients of $\mathsf{P} \in C^\alpha(\dot{\mathsf{G}})$ have the form

$$P_{lt} = \sum_{k=0}^{m_j-1} P_{ltk}(z-z_j)^k + \tilde{P}_{lt}(z)(z-z_j)^{m_j} \tag{3.5.13}$$

in the neighbourhood of z_j where the P_{ltk} are constants and where the $\tilde{P}_{lt}$ are functions in $C^\alpha(\dot{\mathsf{G}})$. For such matrices Prößdorf [21, Section 8.1] proved the following

Lemma 3.5.2 *To* P *there exist single integer indices* $\mu_{lj} \geq 0$ *and two matrix valued functions* T *and* D_0 *such that*

$$\mathsf{P} = \mathsf{TDD}_0 \quad \text{on } \dot{\mathsf{G}} \tag{3.5.14}$$

where

$$D_{lk} = \prod_{j=1}^{r} (z-z_j)^{\mu_{lj}} \delta_{lk}, \tag{3.5.15}$$

T *is a polynomial matrix with constant determinant,* $\det \mathsf{T} = \text{const} \neq 0$, D_0 *is regular.*

In addition, Kopp assumes that

$$\mathsf{T} \textit{ in } (3.5.14) \textit{ is real valued.} \tag{3.5.16}$$

This implies that T is *constant* [17, Chapter 4, Section 4].

In this case the complex conjugated to P has a factorization with

$$\bar{\mathsf{P}} = \mathsf{TDC}_0 \quad \text{on } \dot{\mathsf{G}} \tag{3.5.17}$$

where C_0 is also regular.

Introducing the Banach spaces X corresponding to (3.5.10) and

$$Y \equiv \{\boldsymbol{\phi} = \mathsf{TD}\mathbf{f} \text{ where } \boldsymbol{\phi} \text{ has real valued components and } \mathbf{f} \in C^\alpha(\dot{\mathsf{G}});\ \|\boldsymbol{\phi}\|_Y \equiv \|\mathbf{f}\|_{C^\alpha(\dot{\mathsf{G}})}\}, \tag{3.5.18}$$

Kopp proves the following

Theorem 3.5.3 *Under the assumption* (3.5.17) *the boundary value problem* (3.5.11) *defines an operator* $\mathsf{R} \in \phi(X, Y)$ *with Fredholm index*

$$\nu(\mathsf{R}) = n - \frac{1}{2\pi} \oint_{\dot{\mathsf{G}}} d\{\arg \operatorname{Det} \mathsf{D}_0 - \arg \operatorname{Det} \mathsf{C}_0\}. \tag{3.5.19}$$

In the above approach the spaces X and Y were chosen such that the operator R became a *bounded* Fredholm operator. Using another approach from Prößdorf [22] for singular integral operators as unbounded

Fredholm mappings leads to an interpretation of

$$\mathsf{R}\mathbf{w} = \begin{cases} \mathbf{w}_{\bar{z}} + \mathsf{a}\mathbf{w} + \mathsf{b}\bar{\mathbf{w}} & \text{in } \mathsf{G}, \\ \operatorname{Re} \mathsf{P}\mathbf{w} & \text{on } \dot{\mathsf{G}} \end{cases} \tag{3.5.20}$$

as an *unbounded* Fredholm operator $R: X \to Y$ in different spaces; i.e. R is a densely defined closed operator from X to Y with closed range in Y, finite nullity $\alpha(\mathsf{R}) < \infty$ and finite codimension $\beta(\mathsf{R}) < \infty$ of the range in Y (see e.g. [23, p. 161 ff.]). Here we choose

$$X \equiv \text{closure of } \{\mathbf{w} \in C^{m+\alpha}(\bar{\mathsf{G}})\} \text{ with respect to}$$
$$\|\mathbf{w}\|_X \equiv \|\mathbf{w}\|_{C^\alpha(\bar{\mathsf{G}})} + \|\mathbf{w}_{\bar{z}}\|_{C^{m+\alpha-1}(\bar{\mathsf{G}})},$$
$$Y \equiv C^{m+\alpha-1}(\bar{\mathsf{G}}) \times C^{m+\alpha}(\dot{\mathsf{G}}) \tag{3.5.21}$$

where

$$m \equiv \max\{m_j\}. \tag{3.5.22}$$

Instead of (3.5.16) we assume $\dot{\mathsf{G}} \in C^{2m+2+\alpha}$, $\mathsf{P} \in C^{2m+\alpha}(\dot{\mathsf{G}})$,

$$\mathsf{a}, \mathsf{b} \in C^{m+\alpha}(\bar{\mathsf{G}}) \quad \text{and} \quad \mathsf{a}, \mathsf{b} \text{ have compact supports in } \mathsf{G}. \tag{3.5.23}$$

It is easily shown that R with

$$\text{domain } \mathsf{R} \equiv \{\mathbf{w} \in X \text{ with } \operatorname{Re} \mathsf{P}\mathbf{w}_{|\dot{\mathsf{G}}} \in C^{m+\alpha}(\dot{\mathsf{G}})\} \tag{3.5.24}$$

is a *densely defined closed operator* from X into Y. Since for any sequence $\mathbf{w}_j \in (\text{domain } \mathsf{R})$ with $\mathbf{w}_j \to \mathbf{w}$ in Y and $\mathsf{R}\mathbf{w}_j \to \binom{\mathbf{c}}{\psi}$ in X the latter also converges in C^α implying $\mathsf{R}\mathbf{w} = \binom{\mathbf{c}}{\psi}$ in $C^\alpha(\bar{\mathsf{G}}) \times C^\alpha(\dot{\mathsf{G}})$. Rewriting the boundary value problem into a system of singular integral equations on $\dot{\mathsf{G}}$, Prößdorf's results [22] imply

Theorem 3.5.4 *The above defined operator* R *is an unbounded Fredholm operator.*

We shall come back to the proof in Section 5.1.

In the case

$$\det \mathsf{P} \equiv 0 \quad \text{on } \dot{\mathsf{G}} \tag{3.5.25}$$

the properties of R change completely.

The homogeneous boundary value problem (3.1.45), (3.1.48) for example, with

$$\mathsf{R}\mathbf{w} \equiv \begin{cases} w_{0\bar{z}} & = 0 \\ w_{1\bar{z}} + \mathrm{i} w_{0z} & = 0 \\ \operatorname{Re} \begin{bmatrix} \dot{y}, & -(\dot{x} + \mathrm{i}\dot{y}) \\ -\mathrm{i}\dot{y}, & (\mathrm{i}\dot{x} - \dot{y}) \end{bmatrix} \begin{bmatrix} w_0 \\ w_1 \end{bmatrix} & = 0 \end{cases} \tag{3.5.26}$$

corresponds to the homogeneous Dirichlet problem with Bitsadze's operator B,

$$\mathsf{B}v \equiv v_{xx} - v_{yy} + 2iv_{xy} = 0 \tag{3.5.27}$$

which belongs to the class of problems investigated by Szilagyi [24]. One of his results can be formulated as

Theorem 3.5.5 *For the unit disk* $\mathsf{G} = \{z \| z| < 1\}$, R (3.5.26) *has the nullity* $\alpha = \infty$ *as mapping from the Sobolev space* $W^{1,2}(\mathsf{G})$ *into* $L^2(\mathsf{G})$.

Hence R is no longer a Fredholm operator.

Using the fundamental solution and representation formula, Tovmasjan [25, 26] investigated similar problems for elliptic systems with constant coefficients, especially 'Dirichlet problems' where the boundary values of one of the complex unknowns are given. There the degeneration (3.5.25) holds on $\dot{\mathsf{G}}$ and Tovmasjan points out in [26] that these boundary value problems can have completely different properties. For the problem

$$\begin{aligned} w_{1\bar{z}} - w_2 &= 0, \\ w_{2\bar{z}} \qquad &= h \quad \text{in } \mathsf{G} \end{aligned} \tag{3.5.28}$$

with boundary conditions

$$\operatorname{Re} \begin{pmatrix} 1 & 0 \\ i & 0 \end{pmatrix} \begin{pmatrix} w_1 \\ w_2 \end{pmatrix} = 0 \quad \text{on } \dot{\mathsf{G}} \tag{3.5.29}$$

Tovmasjan shows the following properties extending results from Bitsadze which we collect in

Theorem 3.5.6 (i) *Let* $\dot{\mathsf{G}}$ *be the unit circle. Then* (3.5.28), (3.5.29) *has infinite nullity and is solvable if and only if h satisfies infinitely many linear solvability conditions, i.e. the problem is still normally solvable* ([16, p. 234]).
(ii) *If* $\dot{\mathsf{G}}$ *is an analytic Jordan curve then* (3.5.28), (3.5.29) *is normally solvable if and only if* G *is the image of a rational conformal mapping from the unit disk onto* G.

Hence, in case (i) the above problem is neither Fredholm nor semi-Fredholm but it is normally solvable. In [26] Tovmasjan presents another example of a boundary value problem which has nullity zero but it is not normally solvable. Tovmasjan also investigates second-order systems and characterizes two classes of boundary value problems, one is uniquely solvable for every right hand side and the second is normally solvable but has infinitely many eigensolutions and infinitely many solvability conditions.

References

1 Agmon, S., Douglis, A., and Nirenberg, L. Estimates near the boundary for solutions of elliptic partial differential equations satisfying general boundary conditions I, II. *Comm. Pure Appl. Math.* **12,** 623–727, 1959; **17,** 35–92, 1964.

2 Atkinson, F. V. On relatively regular operators. *Acta. Sci. Math. Szeged.* **15,** 38–56, 1953.

3 Bitsadze, A. *Boundary value problems.* North Holland, Amsterdam, 1968.

4 Bojarski, B. Some boundary value problems for systems of $2m$ equations of elliptic type in the plane. *Dokl. Akad. Nauk. SSSR* **124,** I, 15–18, 1958 (Russian).

5 Bojarski, B. Theory of generalized analytic vectors. *Ann. Pol. Math.* **17,** 281–320, 1966 (Russian).

6 Bruhn, G. Stetig lösbare Randwertprobleme für elliptische Systeme bei Randvorgaben mit endlich vielen Unstetigkeiten 1. *Art. Math. Nachr.* **31,** 363–378, 1966.

7 Courant, R. and Hilbert, D. *Methods of mathematical physics* I, II. Interscience, New York, 1962.

8 Douglis, A. A function-theoretic approach to elliptic systems of equations in two variables. *Comm. Pure Appl. Math.* **VI,** 259–289, 1953.

9 Gantmacher, F. R. *Matrizenrechnung.* Dt. Verl. d. Wiss., Berlin, 1958.

10 Gilbert, R. *Constructive methods for elliptic equations. Lecture Notes* No. 365, Springer, Berlin, 1974.

11 Gilbert, R. and Hile, G. Generalized hyperanalytic function theory. *Trans. Am. Math. Soc.* **195,** 1–29, 1974.

12 Haack, W. and Wendland, W. *Lectures on pfaffian and partial differential equations.* Pergamon Press, Oxford, 1972.

13 Habetha, K. On zeros of elliptic systems of first order in the plane. *In* R. P. Gilbert and R. J. Weinacht (Eds.), *Function theoretic methods in differential equations,* pp. 45–62. Pitman Publ. Research Notes Math. Nr. 8, London, 1976.

14 Hörmander, L. *Linear partial differential operators.* Springer, Berlin, New York, Heidelberg, 1964.

15 Jaenicke, J. Über lineare Randwertprobleme der Systeme von zwei linearen partiellen Differentialgleichungen erster Ordnung vom elliptischen Typus. *Dissertation.* T.U., Berlin D 83, 1957.

16 Kato, T. *Perturbation theory for linear operators.* Springer, Berlin, 1966.

17 Kopp, P. Über eine klasse von Randwertproblemen mit verletzter Lopatinski-Bedingung für Elliptische Systeme erster Ordnung in der Ebene. *Dissertation.* T. H., Darmstadt D 17, 1977.

18 Labrousse, J. Homotopy invariants associated with elliptic partial differential operators. *Lecture Notes.* University of California at Berkeley, 1963.

19 Miranda, C. *Partial differential equations of elliptic type.* Springer, Berlin, Heidelberg, New York, 1970.

20 Pascali, D. Vecteurs analytiques généralisés. *Revue Roumaine Máth. Pure. App.* **10**, 779–808, 1965.
21 Prößdorf, S. *Einige Klassen singulärer Gleichungen.* Birkhäuser, Basel, Stuttgart, 1974.
22 Prößdorf, S. Über eine Klasse singulärer Integralgleichungen nicht normalen Typs. *Math. Annalen* **183,** 130–150, 1969.
23 Schechter, M. *Principles of functional analysis.* Academic Press, New York, 1971.
24 Szilagyi, P. Die Eigenwerte des Operators von A. V. Bitsadze. Vortrag, Math. Forschungsinst. Oberwolfach, *Tagung Partielle Differentialgleichungen* 15. 2–3. 3, 1973.
25 Tovmasjan, N. E. General boundary value problems for elliptic systems of second order with constant coefficients I. *Differentialnye Uravnenija* **2,** 3–23, 1966.
26 Tovmasjan, N. E. General boundary value problems for elliptic systems of second order with constant coefficients II. *Differentialnye Uravnenija* **2,** 163–171, 1966.
27 Vol'pert, A. On the index and normal solvability of boundary value problems for elliptic systems of differential equations in the plane. *Trudy Moskovskogo Mat. Obštšestva* **10,** 41–87, 1961 (Russian).
28 Wloka, J. *Funktionalanalysis und Anwendungen.* de Gruyter, 1971.

Additional References

29 Bitsadze, A. V. The theory of non-Fredholm elliptic boundary value problems. *Am. Math. Soc. Transl.* (2) **105,** 95–103, 1976.
30 Tovmasjan, N. E. Some boundary value problems for a system of second-order elliptic equations which do not satisfy Lopatinski's condition. *Dokl. Akad. Nauk. SSSR* **160,** 1028–1031, 1965.

4

The General Regular Elliptic Boundary Value Problem

In this chapter we consider general elliptic problems and show that they are also Fredholm if the Lopatinski Shapiro condition for the boundary conditions is satisfied. Results of this type were first proved by Lopatinski [21] and for the general first order systems by Vol'pert [36] and Avantaggiati [2]. Palamodov showed in [27] the connections between the boundary value problems and the Atiyah–Singer theory.

In Section 4.1 we prove the Agmon–Douglis–Nirenberg *a priori* estimates [1] for the general linear elliptic boundary value problems by using the usual technique with partition of unity and the corresponding *a priori* estimates for problems in normal form. Here we follow the ideas by Hörmander [14] although for the first-order plane systems the analysis is more primitive. The *a priori* estimates imply the semi-Fredholm properties of the corresponding operators immediately. For second-order Dirichlet problems many aspects of the presented technique can also be found in Wloka's book [45].

In Section 4.2 the general problem is connected homotopically with a boundary value problem with constant principal part within the semi-Fredholm operators. Now Atkinson's stability theorem for semi-Fredholm operators due to Kato [17] implies that the general problem is Fredholm and the explicit index formula holds. It is pointed out through an example by Hile and Protter [13] that the general system does not always have a normal form. However, in every case the system can be transformed into a quasinormal form due to Vinogradov, who used the quasinormal form for solving boundary value problems for general systems having only Hölder continuous coefficients in the principal part [33, 34, 35]. Hence, in general the question arises whether every elliptic problem can be approximated by systems having normal forms. Due to ideas of Waid it is pointed out by the use of the implicit function theorem that in general an elliptic system *cannot* be approximated by problems having normal forms with *separated eigenvalues*. But on the boundary as well as for systems with constant coefficients such an approximation is possible.

In Section 4.3, by using the index formula and dualities, the normal solvability with respect to the formal adjoint problem is shown.

In Section 4.4 we consider semilinear and nonlinear problems. We start with a review on results by Tutschke [30, 31] which are based on the Banach fixed point theorem. Then we sketch the approach for semilinear and nonlinear problems in connection with Schauder's continuation method. This approach goes back to the classical work by Bers and Nirenberg [6] and some of the ideas coincide with those in the book of Bers, John and Schechter [5]. First we consider semilinear equations by using a fixed point theorem of Schäfer [28] and we review results by Begehr and Gilbert [3], Gilbert [9] and Wendland [44]. Then a version of Caccioppoli's fixed point principle [23] is formulated which shows the necessity of *a priori* estimates. As an example, the nonlinear boundary value problems of Section 2.6 are presented also in this framework. Then a short review of interior *a priori* estimates is given which were obtained by v. Wahl [39, 40, 41], Hildebrandt and Widman [12] and Oskolkov [26]. The Schauder continuation method can even be used for the construction of numerical solutions (cf. Kellog, Li and Yorke [18]) but for the elliptic first-order problems this is yet to be done in detail. For the last part of this section and for many of the additional references I have to thank Dr. B. Rüpprich for his complete collection and a good survey of references.

4.1 *A Priori* Estimate and Semi-Fredholm Property

In this section we shall prove an *a priori* estimate for the general regular elliptic boundary value problem (3.1.3), (3.1.32) with

$$\mathsf{R}\mathbf{u} \equiv \begin{cases} \mathsf{D}\mathbf{u} \equiv \mathbf{u}_x + \mathsf{B}\mathbf{u}_y + \mathsf{C}\mathbf{u} = \mathbf{f} & \text{in } \mathsf{G}, \\ \mathsf{r}\mathbf{u} = \boldsymbol{\psi} & \text{on } \dot{\mathsf{G}}. \end{cases} \tag{4.1.1}$$

which is the *a priori* estimate by Agmon, Douglis, Nirenberg [1] in the case of Hölder-spaces. As before, we assume that D is *elliptic* in $\bar{\mathsf{G}}$ and that $\mathsf{B} \in C^{1+\alpha}(\bar{\mathsf{G}})$, $\mathsf{C} \in C^{\alpha}(\bar{\mathsf{G}})$ and $\mathsf{r} \in C^{1+\alpha}(\dot{\mathsf{G}})$. Moreover, we assume again that the *Lopatinski condition* (3.1.37) is fulfilled:

$$\operatorname{Det}(2\mathbf{r}_j\mathbf{q}_k) = \operatorname{Det}\mathsf{P} \neq 0. \tag{4.1.2}$$

Systems with these properties will be called *regular elliptic boundary value problems*. Since the *a priori* estimates are already available for systems with normal form, especially when B has constant coefficients, the proof can be obtained from *local arguments*. To this end we shall extend the boundary operator r to a neighbourhood of $\dot{\mathsf{G}}$ such that the Lopatinski condition remains valid. The latter, (4.1.2), is formulated *pointwise* where the generalized eigenvectors $\mathbf{q}_k$ do not need to be continuous. But for the use of continuity arguments let us reformulate (4.1.2) with respect to another basis of the $\mathbf{q}_k$ similarly to Vol'pert's approach [36, p. 46]. Let us denote by

$$L = \{\mathbf{p}_1, \ldots, \mathbf{p}_n\} \tag{4.1.3}$$

the n-dimensional subspace of $\mathbb{C}^{2n}$ spanned by $\mathbf{p}_1, \ldots, \mathbf{p}_n$. Then the space

$$L^* = \{\mathbf{q}_1, \ldots, \mathbf{q}_n\} \tag{4.1.4}$$

with the basis $\mathbf{q}_1, \ldots, \mathbf{q}_n$ is just it's dual. Hence, the Lopatinski condition

$$\text{Det}\,(\mathbf{r}_j \mathbf{q}_k) \neq 0$$

is a condition on r and L^* and it is fulfilled if and only if

$$\text{Det}\,(\mathbf{r}_j \mathbf{q}'_k) \neq 0 \tag{4.1.5}$$

holds for *any* basis $\mathbf{q}'_1, \ldots, \mathbf{q}'_n$ of L^*. Since unfortunately, the original basis may not be continuously dependent on (x, y) in the neighbourhood of $\dot{\mathsf{G}}$, let us replace it by another basis.

For that purpose, the spectral projection, the matrix valued function,

$$\mathsf{M} \equiv -\frac{1}{2\pi i} \oint_{\mathsf{c}} (\mathsf{B}(x, y) - \lambda \mathsf{I})^{-1} \, d\lambda, \tag{4.1.6}$$

will be used where c denotes a suitably chosen fixed smooth Jordan curve in the upper λ-half plane circumscribing the union of all eigenvalues $\lambda_+(x, y)$ as (x, y) traces $\bar{\mathsf{G}}$. For fixed (x, y) M maps $\mathbb{C}^{2n}$ onto the space L^* spanned by the corresponding algebraic eigenspaces to $\mathsf{B}(x, y)$, [17 p. 67]. As will be shown in Theorem 4.1.4, M yields the existence of a basis to L^*,

$$\mathbf{q}'_1(x, y), \ldots, \mathbf{q}'_n(x, y), \tag{4.1.7}$$

having *the same smoothness properties* as $\mathsf{B}(x, y)$, i.e., in our case $C^{1+\alpha}(\bar{\mathsf{G}})$. Now the Lopatinski condition in the formulation (4.1.5) contains only *smooth* functions. This enables us to prove a local version of the desired *a priori* estimate.

Lemma 4.1.1 *Let* D *in* (4.1.1) *be elliptic in* $\bar{\mathsf{G}}$ *and let* r *satisfy the Lopatinski condition* (4.1.5). *Then for every* $z_0 \in \bar{\mathsf{G}}$ *there exist constants* δ, $\gamma > 0$ *such that the* a priori *estimate,*

$$\|\chi \mathbf{u}\|_{1+\alpha} \leq \gamma \{\|\mathsf{D}\chi \mathbf{u}\|_\alpha + \|\chi\|_{1+\alpha} \|\mathbf{u}\|_\alpha + |\chi \mathbf{r} \mathbf{u}|_{1+\alpha}\}, \tag{4.1.8}$$

holds for all $\mathbf{u} \in \mathrm{C}^{1+\alpha}(\bar{\mathsf{G}})$ *and all* $\chi \in C^{1+\alpha}(R^2)$ *with*

$$\text{supp}\,(\chi) \subset \{z \mid |z - z_0| < \delta\} \tag{4.1.9}$$

Proof First of all let us extend the vectors $\mathbf{r}_j$ to a neighbourhood of $\dot{\mathsf{G}}$. Then we can suppose that there exists a $\delta_0>0$ such that the extension of the boundary functions $\mathbf{r}_j$ satisfies

$$\operatorname{Det}(\mathbf{r}_j\mathbf{q}_k')(x, y)\neq 0 \quad \text{for all} \quad (x, y)\in \mathsf{G}_0 \tag{4.1.10}$$

where G_0 is defined by

$$\mathsf{G}_0\equiv\{z\in\bar{\mathsf{G}}\mid \operatorname{dist}(z,\dot{\mathsf{G}})\leqslant\delta_0\}. \tag{4.1.11}$$

All δ, which will be chosen later on, shall satisfy $\delta\leqslant\delta_0$. Let $z_0\in\bar{\mathsf{G}}$ be any point, fixed in the following. To z_0 we assign the following boundary value problem with *constant coefficients*:

$$\mathsf{R}_0\mathbf{u}\equiv\begin{cases}\mathsf{D}_0\mathbf{u}\equiv\mathbf{u}_x+\mathsf{B}(z_0)\mathbf{u}_y+\mathsf{C}(z_0)\mathbf{u} & \text{in} \quad \bar{\mathsf{G}},\\ \mathsf{r}_0\mathbf{u}\equiv(\mathbf{r}_{j0}\mathbf{u}) & \text{on} \quad \dot{\mathsf{G}},\end{cases} \tag{4.1.12}$$

where the boundary conditions r_0 are defined in two different ways:

(i) If $z_0\in\mathsf{G}_0$ then we choose

$$\mathbf{r}_{j0}\equiv\mathbf{r}_j(z_0) \quad \text{on the whole boundary } \dot{\mathsf{G}}. \tag{4.1.13}$$

(ii) If $z_0\in\mathsf{G}\backslash\mathsf{G}_0$ then we choose arbitrary vectors $\mathbf{r}_{j0}$ satisfying

$$\operatorname{Det}(\mathbf{r}_{j0}\mathbf{q}_k'(z_0))\neq 0. \tag{4.1.14}$$

Then R_0 can be transformed into the normal form (3.1.30) and for that holds an *a priori* estimate (3.4.14). Hence, for the expression

$$\mathsf{R}_0\chi\mathbf{u}=\begin{cases}\mathsf{D}_0\chi\mathbf{u}=\mathsf{D}\chi\mathbf{u}+(\mathsf{D}_0\chi\mathbf{u}-\mathsf{D}\chi\mathbf{u}),\\ \mathsf{r}_0\chi\mathbf{u}=\mathsf{r}\chi\mathbf{u}+(\mathbf{R}_0\chi\mathbf{u}-\mathbf{R}\chi\mathbf{u})\end{cases} \tag{4.1.15}$$

we get the estimate

$$\|\chi\mathbf{u}\|_{1+\alpha}\leqslant\gamma_0\{\|\mathsf{D}\chi\mathbf{u}\|_\alpha+\|\mathsf{D}_0\chi\mathbf{u}-\mathsf{D}\chi\mathbf{u}\|_\alpha + |\mathsf{r}\chi\mathbf{u}|_{1+\alpha}+|\mathsf{r}_0\chi\mathbf{u}-\mathsf{r}\chi\mathbf{u}|_{1+\alpha}+\|\chi\mathbf{u}\|_0\} \tag{4.1.16}$$

where γ_0 does not depend on χ nor $\mathbf{u}$.

The two remainder terms in (4.1.16) can be estimated by

$$\|\mathsf{D}_0\chi\mathbf{u}-\mathsf{D}\chi\mathbf{u}\|_\alpha\leqslant\varepsilon(z_0,\delta)\,\|\chi\mathbf{u}\|_{1+\alpha}+k\,\|\chi\mathbf{u}\|_1 \tag{4.1.17}$$

and

$$|\mathsf{r}_0\chi\mathbf{u}-\mathsf{r}\chi\mathbf{u}|_{1+\alpha}\leqslant\varepsilon(z_0,\delta)\,\|\chi\mathbf{u}\|_{1+\alpha}+k\,\|\chi\mathbf{u}\|_1, \tag{4.1.18}$$

where $\varepsilon(z_0,\delta)\to 0$ with $\delta\to 0$ and where the constant k depends only on the coefficients in R.

In G_0 the inequality (4.1.18) follows from $\mathsf{r}\in C^{1+\alpha}(\mathsf{G}_0)$; for $z_0\in\mathsf{G}\backslash\mathsf{G}_0$

we have $\text{supp}(\chi)\cap\dot{\mathsf{G}}=\varnothing$ and, hence, both sides in (4.1.18) vanish. Inserting (4.1.17) and (4.1.18) into (4.1.16) we find

$$\|\chi\mathbf{u}\|_{1+\alpha}\leqslant\gamma_0\cdot 2\varepsilon(x_0,\delta)\,\|\chi\mathbf{u}\|_{1+\alpha} + \gamma_0\{\|\mathsf{D}\chi\mathbf{u}\|_\alpha+|\mathsf{r}\chi\mathbf{u}|_{1+\alpha}+2k\,\|\chi\mathbf{u}\|_1+\|\chi\mathbf{u}\|_0\}, \quad (4.1.19)$$

which becomes

$$\|\chi\mathbf{u}\|_{1+\alpha}\leqslant\gamma_1\{\|\mathsf{D}\chi\mathbf{u}\|_\alpha+|\mathsf{r}\chi\mathbf{u}|_{1+\alpha}+2k\,\|\chi\mathbf{u}\|_1+\|\chi\mathbf{u}\|_0\} \quad (4.1.20)$$

for $\delta>0$ suitably small. According to the *compact embeddings*

$$C^{1+\alpha}\hookrightarrow C^1\hookrightarrow C^0(\bar{G})$$

(Wloka [45, p. 260, Section 7]) we can use the abstract version of Ehrling's inequality [45, p. 264, Section 9] which yields in our case: To every $\varepsilon>0$ there exists a $k(\varepsilon)$ such that

$$\|\chi\mathbf{u}\|_1\leqslant\varepsilon\,\|\chi\mathbf{u}\|_{1+\alpha}+k(\varepsilon)\,\|\chi\mathbf{u}\|_0 \quad (4.1.21)$$

holds. With $\varepsilon=1/2\gamma_1$ we get from (4.1.20) the estimate

$$\|\chi\mathbf{u}\|_{1+\alpha}\leqslant\tfrac{1}{2}\,\|\chi\mathbf{u}\|_{1+\alpha}+\gamma_1\{\|\mathsf{D}\chi\mathbf{u}\|_\alpha+|\mathsf{r}\chi\mathbf{u}|_{1+\alpha} +(1+k(\varepsilon))\,\|\chi\mathbf{u}\|_0\}, \quad (4.1.22)$$

which implies the desired inequality (4.1.8).

From the local *a priori* estimate it follows from the Heine–Borel theorem and a suitable partition of unity that the *a priori* estimate is also valid globally.

Theorem 4.1.2 *For* R *there exists a constant* γ *such that the* a priori *estimate*

$$\|\mathbf{u}\|_{1+\alpha}\leqslant\gamma\{\|\mathsf{D}\mathbf{u}\|_\alpha+|\mathsf{r}\mathbf{u}|_{1+\alpha}+\|\mathbf{u}\|_0\} \quad (4.1.23)$$

holds for all $\mathbf{u}\in C^{1+\alpha}\,(\bar{\mathsf{G}})$.

Proof From Lemma 4.1.1 we know that for each $z_0\in\bar{\mathsf{G}}$ there exists a $\delta>0$ where the local estimate (4.1.8) is valid. Let us designate to z_0 the open neighbourhood $U_{(\delta/2)}(z_0)=\{z\mid|z-z_0|<(\delta/2)\}$. Then the open covering of $\bar{\mathsf{G}}$,

$$\bigcup_{z_0\in\bar{\mathsf{G}}} U_{(\delta/2)}(u_0)\supset\bar{\mathsf{G}} \quad (4.1.24)$$

contains a finite covering by Heine–Borel's theorem:

$$\bigcup_{j=1}^{m} U_{(\delta_j/2)}(z_j)\supset\bar{\mathsf{G}}. \quad (4.1.25)$$

To this covering let us choose a partition of unity:

$$\chi_j \in C^{1+\alpha}(R^2) \quad \text{with} \quad \sum_{j=1}^{m} \chi_j(z) = 1, 0 \leq \chi_j \leq 1 \quad \text{for all} \quad z \in \bar{\mathsf{G}} \quad \text{and}$$
$$\operatorname{supp}(\chi_j) \subset U_{\delta_j}(z_j), j = 1, \ldots, m. \tag{4.1.26}$$

Then we have $\mathbf{u} = \sum_{j=1}^{m} \chi_j \mathbf{u}$ and for each term we can use (4.1.8):

$$\begin{aligned}\|\mathbf{u}\|_{1+\alpha} &= \left\| \sum_{j=1} \chi_j \mathbf{u} \right\|_{1+\alpha} \leq \sum_{j=1} \|\chi_j \mathbf{u}\|_{1+\alpha} \\ &\leq \gamma_{\max} \left\{ \sum_{j=1}^{m} \|\mathsf{D}\chi_j \mathbf{u}\|_\alpha + \left(\sum_{j=1}^{m} \|\chi_j\|_{1+\alpha} \right) \|\mathbf{u}\|_\alpha \right. \\ &\quad \left. + \sum_{j=1}^{m} |\chi_j \mathbf{r}\mathbf{u}|_{1+\alpha} \right\} \\ &\leq \gamma \{ \|\mathsf{D}\mathbf{u}\|_\alpha + |\mathbf{r}\mathbf{u}|_{1+\alpha} + \|\mathbf{u}\|_\alpha \}\end{aligned} \tag{4.1.27}$$

For $\|\mathbf{u}\|_\alpha$ we use again the abstract Ehrling's inequality regarding $C^{1+\alpha} \hookrightarrow C^\alpha \hookrightarrow C^0$, [45, p. 264 Section 9]:

$$\|\mathbf{u}\|_\alpha \leq \frac{1}{2\gamma} \|\mathbf{u}\|_{1+\alpha} + k \|\mathbf{u}\|_0 \tag{4.1.28}$$

Using (4.1.28) in (4.1.27) we get the desired estimate (4.1.23).

Theorem 4.1.3 *The operator* $\mathsf{R} : C^{1+\alpha}(\bar{\mathsf{G}}) = X \to Y = C^\alpha(\bar{\mathsf{G}}) \times C^{1+\alpha}(\dot{\mathsf{G}})$ *is semi-Fredholm; that means that*

$$\alpha = \operatorname{Dim} N(\mathsf{R}) < \infty \textit{ and } \operatorname{range}(\mathsf{R}) = \overline{\operatorname{range}(\mathsf{R})}^{Y}. \tag{4.1.29}$$

Proof (see [29, p. 127, Theorem 6.2])

(i) Let $N(\mathsf{R})$ be the nullspace of R furnished with the maximum norm $\|\cdot\|_0$. Then for the unit ball in N there holds

$$\|u_0\|_{1+\alpha} \leq \gamma \quad \text{for all} \quad u_0 \in N(\mathsf{R}) \quad \text{with} \quad \|u\|_0 \leq 1. \tag{4.1.30}$$

Hence, the unit ball in N is compact and, therefore, finite dimensional [45, p. 197, Section 14].

(ii) Let $\mathbf{u}_k \in X$ be a sequence with $\mathsf{R}\mathbf{u}_k \to (\mathbf{c}, \boldsymbol{\psi}) \in Y$.
Then we may use the elements

$$\mathbf{u}'_k = \mathbf{u}_k + \mathbf{u}_{0k} \quad \text{with} \quad \|\mathbf{u}'_k\|_0 = \min_{\mathbf{u}_0 \in N(\mathsf{R})} \|\mathbf{u}_k - \mathbf{u}_0\|_0 \tag{4.1.31}$$

instead of the $\mathbf{u}_k$, and since $N(\mathsf{R})$ is finite dimensional,

$$\mathsf{R}\mathbf{u}'_k \to (\mathbf{c}, \boldsymbol{\psi}). \tag{4.1.32}$$

Now let us distinguish two cases (a), (b), depending on whether or not $\|\mathbf{u}_k\|_0$ is bounded.

(a) $\|\mathbf{u}_k\|_0 = c_k \leq K.$ (4.1.33)

Then the *a priori* estimate (4.1.23) implies with (4.1.33)

$$\|\mathbf{u}'_k\|_{1+\alpha} \leq K'. \tag{4.1.34}$$

Hence, we may choose a convergent subsequence $\mathbf{u}''_k$

$$\|\mathbf{u}''_k - \mathbf{u}^*\|_0 \to 0. \tag{4.1.35}$$

For this subsequence there holds again from (4.1.19):

$$\|\mathbf{u}''_k - \mathbf{u}''_j\|_{1+\alpha} \leq \gamma\{\|\mathbf{c}_k - \mathbf{c}_j\|_\alpha + |\boldsymbol{\psi}_k - \boldsymbol{\psi}_j|_{1+\alpha} + \|\mathbf{u}''_k - \mathbf{u}''_j\|_0\} \tag{4.1.36}$$

Thus $\mathbf{u}''_k$ converges in $C^{1+\alpha}$ also and (4.1.35), (4.1.36) yield for the limit function $\mathbf{u}^*$ the properties

$$\mathbf{u}^* \in C^{1+\alpha}, \quad \mathsf{R}\mathbf{u}^* = \lim_{k\to\infty} \mathsf{R}\mathbf{u}''_k = (\mathbf{c}, \boldsymbol{\psi}). \tag{4.1.37}$$

Hence, as proposed the accumulation element $(\boldsymbol{c}, \boldsymbol{\psi})$ of the range of R also belongs to the range.

(b) Now we consider the case that $\mathbf{u}'_k$ is unbounded,

$$\beta_k = \|\mathbf{u}'_k\|_0 \to \infty. \tag{4.1.38}$$

Here let us consider the sequence $\mathbf{v}'_k := \mathbf{u}'_k/\beta_k$, for which the properties

$$\|\mathbf{v}'_k\|_0 = 1, \|\mathsf{R}\mathbf{v}'_k\|_Y = \|(\mathbf{c}_k, \boldsymbol{\psi}_k)\|_Y/\beta_k \to 0 \tag{4.1.39}$$

hold. The *a priori* estimate (4.1.23) yields the boundedness of $\|\mathbf{v}'_k\|_{1+\alpha}$ and, hence, according to the compact embedding $C^{1+\alpha} \hookrightarrow C^0$ we may use a convergent subsequence $\mathbf{v}''_k$,

$$\|\mathbf{v}''_k - \mathbf{v}_0\|_0 \to 0. \tag{4.1.40}$$

As in the first case one finds $\mathbf{v}_0 \in N(\mathsf{R})$. On the other hand, from (4.1.31) we have with (4.1.39) and (4.1.40) the contradiction

$$1 = \|\mathbf{v}''_k\|_0 \leq \|\mathbf{v}''_k - \mathbf{v}_0\|_0 \to 0.$$

Thus, this case (b) cannot appear and in case (a) it was already shown that range (R) is closed.

Appendix The construction of the smooth basis to L^*

Another construction of the proposed basis was made by Vinogradov [34]. Since $\bar{\mathsf{G}}$ is bounded with smooth boundary $\dot{\mathsf{G}} \in C^{1+\alpha}$, all given functions in B and C can be extended to $C^{1+\alpha}$ functions in the whole plane. This extension can even be chosen such that the ellipticity of D is preserved. Hence let us assume, without loss of generality, that

$$\mathsf{G} \subset \mathsf{K}_1 \equiv \{z \mid |z| < 1\} \tag{4.1.41}$$

and that $\mathsf{B} \in C^{1+\alpha}(\bar{\mathsf{K}}_1)$.

In the following we shall use the family of matrices defined by

$$\mathsf{B}(x, y, t) \equiv \mathsf{B}(tx, ty) \quad \text{for} \quad 0 \leqslant t \leqslant 1 \tag{4.1.42}$$

which is in $C^{1+\alpha}$ for $(x, y, t) \in \bar{\mathsf{K}}_1 \times [0, 1]$. For any fixed (x, y, t) the eigenvalues $\lambda^j_\pm(x, y, t)$ of B are nonreal and to the $\lambda^j_+(x, y, t)$ there belongs $L^*(x, y, t)$, the span of the corresponding algebraic eigenspaces. Letting (x, y, t) trace over $\mathsf{K}_1 \times [0, 1]$, $L^*(x, y, t)$ becomes a family of spaces for which the following theorem holds:

Theorem 4.1.4 *To the family* $\mathsf{B}(x, y, t)$ *there exists a family of orthonormalized basis vectors*

$$\mathbf{q}_j(x, y, t) \in C^{1+\alpha}(\mathsf{K}_1 \times [0, 1]), \quad j = 1, \ldots, n,$$

spanning $L^*(x, y, t)$.

Proof For $t = 0$ we have $\mathsf{B} = \mathsf{B}(0, 0)$, i.e. constant coefficients. To this matrix there exists an orthonormalized basis,

$$\mathbf{q}_j(x, y, 0) \equiv \mathbf{q}_j(0), \quad j = 1, \ldots, n, \tag{4.1.43}$$

defined by the generalized eigenvectors (3.1.9), (3.1.13) to $\lambda^j_+(0, 0, 0)$. The $\mathbf{q}_j(0)$ can be understood as constant valued vector functions in $\bar{\mathsf{K}}_1$.

For $t > 0$ we shall construct the basis successively, beginning with $\mathbf{q}_1$. For the spectral projection, i.e. the matrix

$$\mathsf{M}(x, y, t) \equiv -\frac{1}{2\pi i} \oint_{\mathsf{c}} (\mathsf{B}(x, y, t) - \lambda \mathsf{I})^{-1} \, d\lambda, \tag{4.1.44}$$

we have

$$\mathbf{q}_1(0) = \mathsf{M}(x, y, 0)\mathbf{q}_1(0) \tag{4.1.45}$$

and the vector

$$\mathbf{s}_1(x, y, t) \equiv \mathsf{M}(x, y, t)\mathbf{q}_1(0) \tag{4.1.46}$$

is a continuous continuation for all $(x, y, t) \in \mathsf{K}_1 \times [0, 1]$.

The path of integration c is chosen in the upper half plane circumscribing the union of all $\lambda^j_+(x, y, t)$ which is compact since $\bar{\mathsf{K}}_1 \times [0, 1]$ is compact and the λ^j_+ are continuous.

For $t=0$ the *Euclidean length* of $\mathbf{s}_1$ is one,

$$|\mathbf{s}_1(x, y, 0)|=1. \tag{4.1.47}$$

Hence, by continuity there exists a $t_1>0$ such that

$$|\mathbf{s}_1(x, y, t)|\geq\tfrac{1}{2} \quad \text{for all } t \text{ with} \quad 0\leq t\leq t_1. \tag{4.1.48}$$

Therefore the vector

$$\mathbf{q}_1(x, y, t)\equiv\mathbf{s}_1(x, y, t)\,|\mathbf{s}_1(x, y, t)|^{-1}, \quad 0\leq t\leq t_1, \tag{4.1.49}$$

belongs to $L^*(x, y, t)$ and

$$\mathbf{q}_1(x, y, t)\equiv \mathsf{M}(x, y, t)\mathbf{q}_1(x, y, t) \quad \text{for} \quad 0\leq t\leq t_1.$$

Again we can continue by defining

$$\mathbf{s}_1(x, y, t)\equiv \mathsf{M}(x, y, t)\mathbf{q}_1(x, y, t) \quad \text{for} \quad t_1\leq t \tag{4.1.50}$$

and find $t_2>t_1$ with

$$|\mathbf{s}_1(x, y, t)|\geq\tfrac{1}{2} \quad \text{for} \quad t_1\leq t\leq t_2. \tag{4.1.51}$$

Hence, $\mathbf{q}_1$ can be continued by

$$\mathbf{q}_1(x, y, t)\equiv\mathbf{s}_1(x, y, t)\,|\mathbf{s}_1(x, y, t)|^{-1} \quad \text{for} \quad t_1\leq t\leq t_2. \tag{4.1.52}$$

Repeating this procedure we find $\mathbf{q}_1(x, y, t)$ for $t_\nu<t\leq t_{\nu+1}$, $t_0=0$, $\nu=0, 1, \ldots$. The construction provides $\mathbf{q}_1\in C^{1+\alpha}$ for the corresponding x, y, t. What remains to show is, that this procedure leads after finitely many steps to the definition of $\mathbf{q}_1$ in the whole region $\bar{\mathsf{K}}_1\times[0, 1]$. This is assured by Lemma 4.1.5 since $t_{\nu+1}-t_\nu\geq\delta>0$ with δ independent of x, y and t. Hence the above construction provides the smooth basis vector

$$\mathbf{q}_1(x, y)\equiv\mathbf{q}_1(x, y, 1). \tag{4.1.53}$$

In the same way we may construct $\mathbf{q}_2(x, y, t)$ obeying the additional condition of orthogonality to $\mathbf{q}_1$:

$$\mathbf{s}_2(x, y, t)\equiv \mathsf{M}(x, y, t)\mathbf{q}_2(x, y, t), \tag{4.1.54}$$

$$|\mathbf{s}_2(x, y, t)|\geq\tfrac{1}{2} \quad \text{for} \quad t_\nu\leq t\leq t_{\nu+1}, \quad t_\nu<t_{\nu+1}, \quad \nu=0, 1, \ldots, \tag{4.1.55}$$

$$\mathbf{q}_2(x, y, t)\equiv(\mathbf{s}_2(x, y, t)-\mathbf{q}_1^T(x, y, t)\mathbf{s}_2(x, y, t)\mathbf{q}_1(x, y, t))\cdot|\,(\mathbf{s}_2-\mathbf{q}_1^T\mathbf{s}_2\mathbf{q}_1)|^{-1} \tag{4.1.56}$$

Again, Lemma 4.1.5 holds for $\mathbf{s}_2$ in (4.1.55): $t_{\nu+1}-t_\nu\geq\delta>0$. After finitely many steps, the second basis vector,

$$\mathbf{q}_2(x, y)\equiv\mathbf{q}_2(x, y, 1)$$

is constructed.

Repeating the above construction in connection with the Schmidt orthonormalization procedure the proposed basis is finally found.

Lemma 4.1.5 *To* B *there exists a constant* $\delta>0$ *independent of* (x, y, t) $(x, y, t)\in\bar{\mathsf{K}}_1\times[0, 1]$ *and* t_ν *such that*

$$|\mathbf{s}_j(x, y, t)|\geq\tfrac{1}{2} \tag{4.1.57}$$

holds for all t *with* $t_\nu\leq t\leq t_\nu+\delta$, $j=1, \ldots, n$.

Proof The inequality (4.1.57) is a consequence of the equicontinuity of

$$(\mathsf{B}(x, y, t)-\lambda\mathsf{I})^{-1} \quad \text{in} \quad \bar{\mathsf{K}}_1\times[0, 1]\times\mathsf{c}.$$

Hence, there exists a constant $\delta>0$ such that

$$|\mathsf{M}(x, y, t)-\mathsf{M}(x, y, t_\nu)|\leq\frac{1}{2\pi}\oint_{\mathsf{c}} |(\mathsf{B}(x, y, t)-\lambda\mathsf{I})^{-1} - (\mathsf{B}(x, y, t_\nu)-\lambda\mathsf{I})^{-1}|\,|\mathrm{d}\lambda|\leq\tfrac{1}{2} \tag{4.1.58}$$

holds for all $(x, y)\in\bar{\mathsf{K}}_1$, t, $t_\nu\in[0, 1]$ with $|t-t_\nu|\leq\delta$ where $|\cdot|$ denotes the matrix norm corresponding to the Euclidean vector norm in $\mathbb{C}^{2n}$. Since

$$\mathbf{s}_j(x, y, t_\nu)=\mathbf{q}_j(x, y, t_\nu)=\mathsf{M}(x, y, t_\nu)\mathbf{q}_j(x, y, t_\nu)$$

the estimate (4.1.58) implies the proposed inequality:

$$\begin{aligned}|\mathbf{s}_j(x, y, t)| &= |\mathbf{s}_j(x, y, t_\nu)+\mathbf{s}_j(x, y, t)-\mathbf{s}_j(x, y, t_\nu)| \\ &\geq 1-|(\mathsf{M}(x, y, t)-\mathsf{M}(c, y, t_\nu))\mathbf{q}_j(x, y, t_\nu)| \\ &\geq 1-\tfrac{1}{2}=\tfrac{1}{2}.\end{aligned}$$

4.2 The Fredholm Property of the General System and its Approximation

First let us show that the general system (4.1.1) is not only semi-Fredholm but even Fredholm provided the Lopatinski–Shapiro condition (4.1.5) is satisfied. The following theorem coincides with the result by Vol'pert [36] who used singular integral equations and by Vinogradov [33, 34, 35] who modified the method for $\mathsf{B}\in C^\alpha$.

Theorem 4.2.1 *The elliptic boundary value problem* (4.1.1) *is Fredholm with the index*

$$\nu(\mathsf{R})=n-\frac{1}{\pi}\oint_{\dot{\mathsf{G}}} \mathrm{d}\arg\det(\mathbf{r}_j\mathbf{q}_k') \tag{4.2.1}$$

provided the Lopatinski–Shapiro condition (4.1.5) *is satisfied.*

For the proof we shall need the following stability theorem due to Kato which we formulate only for bounded operators:

Theorem 4.2.2 ([17] p. 235 Theorem 5.17) *Let* $\mathsf{R} \in B(X, Y)$ *be any semi-Fredholm operator. Then there exists a* $\gamma > 0$ *such that all* $\mathsf{T} \in B(X, Y)$ *with*

$$\|\mathsf{R} - \mathsf{T}\| < \gamma$$

are also semi-Fredholm and

$$\nu(\mathsf{R}) = \nu(\mathsf{T}) \tag{4.2.2}$$

even if ν *is infinite.*

Proof of Theorem 4.2.1 Kato's Theorem shows that the set of bounded semi-Fredholm operators is open in $B(X, Y)$ and that the index is stable. Therefore let us join R with a Fredholm operator R_0 having the index (4.2.1) by a path within the semi-Fredholm operators. For this purpose we may use the family of coefficient matrices $\mathsf{B}(x, y, t)$ of (4.1.42) defining the family of elliptic differential operators

$$\mathsf{D}_t \mathbf{u} \equiv \mathbf{u}_x + \mathsf{B}(x, y, t)\mathbf{u}_y + t\,\mathsf{C}(x, y)\mathbf{u}, \quad 0 \leq t \leq 1. \tag{4.2.3}$$

For a corresponding family of boundary operators $\mathsf{r}(s, t)$ the Lopatinski–Shapiro condition has to be required for all t. Such a family can be defined by the following $2n^2$ linear equations for the $2n^2$ coefficients of $\mathsf{r}(s, t)$:

$$\begin{aligned} &\mathbf{r}_j(s, t)\mathbf{q}_k(x(s), y(s), t) = \mathbf{r}_j(s)\mathbf{q}_k(x(s), y(s), 1), \\ &\mathbf{r}_j(s, t)\bar{\mathbf{q}}_k(x(s), y(s), t) = \mathbf{r}_j(s)\bar{\mathbf{q}}_k(x(s), y(s), 1) \end{aligned} \tag{4.2.4}$$

where the right hand sides are given functions. If we use the basis $\mathbf{p}_l(x, y, t)$, belonging to the inverse of the $\mathbf{q}_j, \bar{\mathbf{q}}_j$ and satisfying (3.1.9), (3.1.10), then the equations (4.2.4) yield the explicit equations

$$r_{jl}(s, t) = 2\,\mathrm{Re} \sum_{k=1}^{n} \mathbf{r}_j(s)\mathbf{q}_k(x(s), y(s), 1)p_{k(l)}(x(s), y(s), t), \quad 0 \leq t \leq 1. \tag{4.2.5}$$

This special choice yields

$$\mathrm{Det}\,(\mathbf{r}_j(s, t)\mathbf{q}_k(x(s), y(s), t)) = \mathrm{Det}\,(\mathbf{r}_j(s)\mathbf{q}_k(x(x), y(s), 1)) \neq 0 \quad \text{on } \dot{\mathsf{G}} \text{ for } 0 \leq t \leq 1. \tag{4.2.6}$$

Thus the family of boundary value problems

$$\mathsf{R}_t \mathbf{u} \equiv \begin{cases} \mathsf{D}_t \mathbf{u} = \mathbf{f} & \text{in } \mathsf{G} \\ \mathsf{r}_t \mathbf{u} = \boldsymbol{\psi} & \text{on } \dot{\mathsf{G}} \end{cases} \quad \text{for } 0 \leq t \leq 1 \tag{4.2.7}$$

defines a family of semi-Fredholm operators $\mathsf{R}_t : C^{1+\alpha}(\bar{G}) \to C^{\alpha}(\bar{G}) \times C^{1+\alpha}(\dot{G}))$ according to Theorem 4.1.3. R_t depends on t continuously. Hence, Kato's Theorem 4.2.2 implies

$$\nu(\mathsf{R}_1) = \nu(\mathsf{R}_t) = \nu(\mathsf{R}_0) \quad \text{for} \quad 0 \leq t \leq 1. \tag{4.2.8}$$

The differential operator for $t = 0$, D_0, has constant coefficients and possesses the Bojarski normal form. Therefore for D_0 Theorems 3.2.16 and 3.3.1 are valid. The index formula (3.3.1) yields with (4.2.6) and (4.2.8) the desired formula (4.2.1):

$$\begin{aligned} \nu(\mathsf{R}_1) = \nu(\mathsf{R}_0) &= n - \frac{1}{\pi} \oint_{\dot{G}} \mathrm{d} \arg \operatorname{Det} (\mathbf{r}_j(s, 0) \mathbf{q}_k(0)) \\ &= n - \frac{1}{\pi} \oint_{\dot{G}} \mathrm{d} \arg \operatorname{Det} (\mathbf{r}_j(s) \mathbf{q}_k(x(s), y(s), 1)). \end{aligned} \tag{4.2.9}$$

Since $\nu(\mathsf{R})$ is finite the operator R is not only semi-Fredholm but even Fredholm.

Although the general system could have been perturbed homotopically into a system with constant coefficients, in general it is not clear whether R can be *approximated* by problems R_ε with normal forms whose differential operators converge to that of R. In other words: does there exist to every $\varepsilon > 0$ a matrix $B(x, y, \varepsilon) = ((b_{jk\varepsilon}))$ possessing normal form such that

$$\|b_{jk}(.\,,.) - b_{jk}(.\,,.\,, \varepsilon)\|_{C^{1+\alpha}(\bar{G})} < \varepsilon? \tag{4.2.10}$$

By introducing new unknowns with the bases of $L(x, y, 1)$ and $L^*(x, y, 1)$ constructed in Theorem 4.1.4 it is possible to transform the general system (4.1.1) algebraically into a system which allows an approximation and the use of Vekua's approach [32] as for the Beltrami system in our Sections 2.4 and 2.6. This approach was investigated by Vinogradov [33, 34, 35].

Introducing the new unknowns by using the *smooth* basis,

$$w_j \equiv \mathbf{p}_j^T \mathbf{u}, \qquad j = 1, \ldots, n, \qquad \mathbf{u} = \sum_{k=1}^{n} \{\mathbf{q}_k w_k + \bar{\mathbf{q}}_k \bar{w}_k\}, \tag{4.2.11}$$

the general differential operator in (4.1.1) takes the form

$$w_{jx} + \sum_{k=1}^{n} \mathbf{p}_j^T \mathsf{B}(\mathbf{q}_k w_{ky} + \bar{\mathbf{q}}_k \bar{w}_{ky}) + \text{lower order terms.} \tag{4.2.12}$$

For this system we have the following

Proposition 4.2.3 (*See also* [36, Section 3.1])

$$\mathbf{p}_j^T \mathsf{B}(x, y) \bar{\mathbf{q}}_k = 0 \quad \text{for} \quad j, k = 1, \ldots, n, (x, y) \in \bar{G}. \tag{4.2.13}$$

Proof To $\mathsf{B}(x, y)$ at any fixed point (x, y) there belongs a system of generalized eigenvectors $\boldsymbol{\pi}_j$ corresponding to the eigenvalues λ_+^j, such that

$$\mathbf{p}_j^T = \sum_{l=1}^{n} \alpha_{jl} \boldsymbol{\pi}_l^T, \qquad j = 1, \ldots, n \tag{4.2.14}$$

with an orthogonal transformation α_{jl}. Inserting (4.2.14) into (4.2.13) we get with

$$\mathbf{p}_j^T \mathbf{q}_k = \delta_{jk}, \ \mathbf{p}_j^T \bar{\mathbf{q}}_k = 0$$

$$\mathbf{p}_j^T \mathsf{B} \bar{\mathbf{q}}_k = \sum_{l=1}^{n} \alpha_{jl} \lambda_+^l \boldsymbol{\pi}_l^T \bar{\mathbf{q}}_k = \sum_{l,j=1}^{n} \alpha_{jl} \lambda_+^l \bar{\alpha}_{jk} \mathbf{p}_j^T \bar{\mathbf{q}}_k = 0, \tag{4.2.15}$$

the proposed relations (4.2.13).

Hence, (4.2.12) reduces to the n complex equations

$$\mathbf{w}_x + \Lambda \mathbf{w}_y + \cdots \quad \text{where} \quad \Lambda = ((\mathbf{p}_j^T \mathsf{B} \mathbf{q}_k)). \tag{4.2.16}$$

Since Λ has the eigenvalues λ_+^j, $j = 1, \ldots, n$ in the upper halfplane, the inverse of $(\mathsf{I} - \mathrm{i}\Lambda)$ exists and (4.2.16) can be transformed with (1.1.4) into

$$\mathbf{w}_{\bar{z}} - \mathsf{Q}(x, y) \mathbf{w}_z + \mathsf{a}\mathbf{w} + \mathsf{b}\bar{\mathbf{w}} = \mathbf{c} \tag{4.2.17}$$

where the $n \times n$ matrix valued function Q is defined by

$$\mathsf{Q}(x, y) = -(\mathsf{I} - \mathrm{i}\Lambda)^{-1}(\mathsf{I} + \mathrm{i}\Lambda) \tag{4.2.18}$$

and has the eigenvalues

$$q_j(x, y) = \frac{\lambda_+^j - \mathrm{i}}{\lambda_+^j + \mathrm{i}} \quad \text{with} \quad |q_j| < 1. \tag{4.2.19}$$

Let us call (4.2.17) the *quasinormal form* of the general elliptic system (4.1.1).

Remark 4.2.4 As Hile and Protter [13] show, the system

$$\mathbf{w}_{\bar{z}} + \left(1 - \frac{x^4}{4}\right)^{-1} \begin{pmatrix} \dfrac{x^4}{4} & \mathrm{i}x\dfrac{|x|}{2} \\ -\mathrm{i}x\dfrac{|x|}{2} & \dfrac{x^4}{4} \end{pmatrix} \mathbf{w}_z + \cdots \quad \text{with} \quad \mathbf{w} = \begin{pmatrix} w_1 \\ w_2 \end{pmatrix}$$

does *not admit* a normal form in any neighbourhood of $x = 0$. Here Λ is given by

$$\Lambda = \begin{pmatrix} \mathrm{i} & x|x| \\ -x|x| & \mathrm{i} \end{pmatrix}.$$

Another system without normal form can be found in Vinogradov's work [35].

Remark 4.2.5 The above matrices Λ and Q are in the same class of smoothness as B according to the construction in Theorem 4.1.4. For $\mathsf{B} \in \overline{C^{1+\alpha}(\bar{\mathsf{G}})^{C^{\alpha}(\bar{\mathsf{G}})}}$ we also have $\mathsf{Q} \in \overline{C^{1+\alpha}(\bar{\mathsf{G}})^{C^{\alpha}(\bar{\mathsf{G}})}}$. Approximating the bases $\mathbf{p}_j$, $\mathbf{q}_k$ by $C^{1+\alpha}$ functions $\mathbf{q}_{j\varepsilon}$, $\mathbf{p}_{k\varepsilon}$ with respect to the C^{α} norm we find an approximate system to (4.2.17),

$$\mathbf{w}_{\bar{z}}^{(\varepsilon)} - \mathsf{Q}^{(\varepsilon)}\mathbf{w}_z^{(\varepsilon)} + \tilde{\mathsf{Q}}^{(\varepsilon)}\bar{\mathbf{w}}_z^{(\varepsilon)} + \mathsf{a}^{(\varepsilon)}\mathbf{w}^{(\varepsilon)} + \mathsf{b}^{(\varepsilon)}\bar{\mathbf{w}}^{(\varepsilon)} = \mathbf{c}^{(\varepsilon)} \tag{4.2.20}$$

for

$$w_j^{(\varepsilon)} \equiv \mathbf{p}_{j\varepsilon}^T\mathbf{u}$$

where

$$\|\mathsf{Q} - \mathsf{Q}^{(\varepsilon)}\|_{C^{\alpha}(\bar{\mathsf{G}})} + \|\tilde{\mathsf{Q}}^{(\varepsilon)}\|_{C^{\alpha}(\bar{\mathsf{G}})} \to 0(\varepsilon \to 0) \tag{4.2.21}$$

whereas $\mathsf{a}^{(\varepsilon)}$, $\mathsf{b}^{(\varepsilon)}$ and $\mathbf{c}^{(\varepsilon)}$ might become rather big for $\varepsilon \to 0$. The system in the form (4.2.20) is the basis for Vinogradov's investigations of boundary value problems [33, 34, 35] since it allows Vekua's approach even for C^{α} coefficients B.

Clearly, the question of approximation (4.2.10) is equivalent to the corresponding approximation of (4.2.17) with matrices Q having normal forms—or even simple eigenvalues. For constant coefficients in the whole domain on one hand and for variable coefficients but only on the boundary on the other hand an approximation is possible. The following considerations are due to Waid [42].

Theorem 4.2.6 *Let $\mathsf{Q}(x, y) = \mathsf{Q}_0$ have constant coefficients. Then to every $\varepsilon > 0$ there exists a constant matrix Q_ε with simple characteristic roots such that*

$$|\mathsf{Q}_o - \mathsf{Q}_\varepsilon| < \varepsilon. \tag{4.2.22}$$

Proof Since the theorem is trivial for $n = 1$, let us suppose $n \geqslant 2$. With the characteristic polynomial,

$$f_Q(\lambda) \equiv \mathrm{Det}\,(\lambda\mathsf{I} - \mathsf{Q}) \tag{4.2.23}$$

let us consider the resultant of f_Q and its derivative,

$$D(\mathsf{Q}) \equiv \mathrm{Resultant}\,(f_Q, f'_Q) \tag{4.2.24}$$

(van der Waerden [38, p. 93]). Since $D(\mathsf{Q})$ is a determinant of the coefficients in f_Q and f'_Q, $D(\mathsf{Q})$ becomes a polynomial in the coefficients

q_{jk} of Q. It defines an algebraic mapping

$$D:\mathbb{C}^{n^2}\to\mathbb{C} \tag{4.2.25}$$

whose zeros are those matrices having repeated eigenvalues. The surface S defined by $D(S)=0$ is of complex codimension at least 1 and hence of real codimension at least 2. According to a theorem in dimension theory [15, Theorem IV 4 p. 48], S cannot separate any open ball in $\mathbb{C}^{n^2}$. Moreover, X has there Lebesgue measure zero as an analytic set [22, 15 p. 105]. Therefore any 'point' Q on S can be approximated by 'points', i.e. matrices $\mathsf{Q}_\varepsilon\notin S$.

Theorem 4.2.7 *For every $\varepsilon>0$ there exists to $\mathsf{Q}|_{\dot{\mathsf{G}}}\in C^0(\dot{\mathsf{G}})$ a matrix $\mathsf{Q}_\varepsilon(s)$ such that*

$$\sup_{\dot{\mathsf{G}}}\|\mathsf{Q}_\varepsilon(s)-\mathsf{Q}(s)\|<\varepsilon \tag{4.2.26}$$

and such that Q_ε has only simple eigenvalues at each s.

Proof Since $\dot{\mathsf{G}}$ is compact and Q is continuous we can find a finite number of points s_ν, $\nu=1,\ldots,N$ such that for each pair of points $s_\nu, s_{\nu+1}$ the matrices $\mathsf{Q}(s_\nu)$, $\mathsf{Q}(s_{\nu+1})$ lie in the intersection of ε-balls in $\mathbb{C}^{n^2}$ around $\mathsf{Q}(s)$ for $s_\nu\leq s\leq s_{\nu+1}$. As we have seen in Theorem 4.2.6, we can find matrices $\mathsf{Q}_\varepsilon(s_\nu)$ in these balls with $\mathsf{Q}_\varepsilon(s_\nu)\in\mathbb{C}^{n^2}\backslash S$. Moreover, $\mathbb{C}^{n^2}\backslash S$ is polygonally arcwise connected for s_ν fixed, and the union $\bigcup_{s_\nu\leq s\leq s_{\nu+1}}(\mathbb{C}^{n^2}\backslash S)$ has the same property for N big and $|s_\nu-s_{\nu+1}|$ small enough. Hence, $\mathsf{Q}_\varepsilon(s_\nu)$ and $\mathsf{Q}_\varepsilon(s_{\nu+1})$ can be connected by a path of continuous matrices in the intersection of both ε-balls within $\mathbb{C}^{n^2}\backslash S$. Hence inequality (4.2.26) is valid for all s.

The following example by Waid shows that there are systems with matrix Q which *cannot* be approximated in the C^1 norm by systems with *simple* characteristic roots:

$$\mathsf{Q}\equiv\begin{pmatrix}0 & x+iy\\ 1 & 0\end{pmatrix}. \tag{4.2.27}$$

For $z\neq0$, Q has the two different eigenvalues

$$q_1=+\sqrt{z},\qquad q_2=-\sqrt{z}$$

collapsing at $z=0$. For any matrix

$$\mathsf{Q}_\varepsilon=\mathsf{Q}+((\varepsilon_{jk}))$$

where $\|\varepsilon_{jk}\|_{C^1(\bar{G})}\leq\varepsilon$ we find the characteristic equation

$$q^2-(\varepsilon_{11}+\varepsilon_{22})q+\varepsilon_{11}\varepsilon_{22}-(\varepsilon_{12}\varepsilon_{21}+z+\varepsilon_{21}z+\varepsilon_{12})=0.$$

It has a double zero where

$$x+\mathrm{i}y=-\frac{(\varepsilon_{11}-\varepsilon_{22})^2+4\varepsilon_{12}}{4(1+\varepsilon_{21})}\equiv\phi_\varepsilon(x,y) \tag{4.2.28}$$

holds. But since ϕ_ε together with it's first derivatives is supposed to become small, the two real equations to (4.2.27) have always a fixed point (x, y) near the origin according to the implicit function theorem. Hence it is impossible to approximate Q in (4.2.26) in C^1 by matrices Q_ε having *simple* eigenvalues.

4.3 The Adjoint Problem

For the general system (4.1.1) the adjoint boundary value problem and the solvability conditions can be formulated as in Section 3.4. Green's identity yields

$$\iint_{\mathsf{G}} \mathbf{v}^T(\mathbf{u}_x+\mathsf{B}\mathbf{u}_y+C\mathbf{u})[\mathrm{d}x,\mathrm{d}y]$$

$$=-\iint_{\mathsf{G}} \mathbf{u}^T(\mathbf{v}_x+(\mathsf{B}^T\mathbf{v})_y-\mathsf{C}^T\mathbf{v})[\mathrm{d}x,\mathrm{d}y]+\oint_{\dot{\mathsf{G}}} \mathbf{v}^T\mathbf{u}\,\mathrm{d}y-\mathbf{v}^T\mathsf{B}\mathbf{u}\,\mathrm{d}x. \tag{4.3.1}$$

For the reformulation of the boundary integral let us assume again that the Lopatinski condition (4.1.5) is fulfilled where $\mathbf{q}_1(x, y), \ldots, \mathbf{q}_n(x, y)$ and $\mathbf{p}_1(x, y), \ldots, \mathbf{p}_n(x, y)$ denote the *smooth* bases to L^* and L, respectively, satisfying (3.1.9) and (3.1.10). With (4.2.13), the boundary integral takes the form

$$\oint_{\dot{\mathsf{G}}} (\mathbf{v}^T\mathbf{u}\dot{y}-\mathbf{v}^T\mathsf{B}\mathbf{u}\dot{x})\,\mathrm{d}s$$

$$=\oint_{\dot{\mathsf{G}}}\left\{\sum_{j,k=1}^{n}\mathbf{q}_j^T\mathbf{v}(\dot{y}\delta_{jk}-\dot{x}\mathbf{p}_j^T\mathsf{B}\mathbf{q}_k)\mathbf{p}_k^T\mathbf{u}+\bar{\mathbf{q}}_j^T\mathbf{v}(\dot{y}\delta_{jk}-\dot{x}\bar{\mathbf{p}}_j^T\mathsf{B}\bar{\mathbf{q}}_k)\bar{\mathbf{p}}_k^T\mathbf{u}\right\}\mathrm{d}s$$

$$=2\,\mathrm{Re}\oint_{\dot{\mathsf{G}}}\sum_{j,k=1}^{n}\mathbf{q}_j^T\mathbf{v}(\dot{y}\delta_{jk}-\dot{x}\Lambda_{jk})\mathbf{p}_k^T\mathbf{u}\,\mathrm{d}s \tag{4.3.2}$$

where Λ is defined by (4.2.16). The Lopatinski condition (4.1.5) implies that the matrix

$$\mathsf{P}\equiv((\mathbf{r}_j\mathbf{q}_k)),\qquad j,k=1,\ldots,n$$

is invertible. Let us denote the inverse by

$$\mathsf{Q}\equiv((Q_{jk})),\quad \mathsf{PQ}=\mathsf{I}. \tag{4.3.3}$$

Then (4.3.2) becomes

$$\oint_{\dot{G}} (\mathbf{v}^T\mathbf{u}\dot{y} - \dot{x}\mathbf{v}^T\mathsf{B}\mathbf{u})\,ds$$

$$= \oint_{\dot{G}} \sum_{l=1}^{n} \operatorname{Re} \sum_{j,k=1}^{n} \mathbf{q}_j^T\mathbf{v}(\dot{y}\delta_{jk} - \dot{x}\Lambda_{jk})Q_{kl}\mathbf{r}_l\mathbf{u}\,ds$$

$$- \oint_{\dot{G}} \sum_{l=1}^{n} 2\operatorname{Im} \sum_{j,k=1}^{n} \mathbf{q}_j^T\mathbf{v}(\dot{y}\delta_{jk} - \dot{x}\Lambda_{jk})Q_{kl} \operatorname{Im} \sum_{t=1}^{n} \mathbf{r}_l\mathbf{q}_t\mathbf{p}_t^T\mathbf{u}\,ds \tag{4.3.4}$$

For the definition of the adjoint problem the last term must be split up into $\mathbf{ru}$ and a remaining part. To this end let us complete the n given *row vectors* $\mathbf{r}_l$ to a system of $2n$ real linearly independent vectors

$$\begin{matrix} \mathbf{r}_1(s) \\ \cdot \\ \cdot \\ \mathbf{r}_n(s) \quad \text{with } \operatorname{Det}((\mathbf{r}_j)) \neq 0 \quad \text{on } \Gamma. \\ \mathbf{r}_{n+1}(s) \\ \cdot \\ \cdot \\ \mathbf{r}_{2n}(s) \end{matrix} \tag{4.3.5}$$

Of course, this completion is not unique. To the matrix $((\mathbf{r}_j))$ there exists the inverse given by $2n$ column vectors $\boldsymbol{\rho}_1, \ldots, \boldsymbol{\rho}_{2n}$ satisfying

$$\mathbf{r}_j\boldsymbol{\rho}_k = \delta_{jk} \quad \text{and} \quad \sum_{k=1}^{2n} \boldsymbol{\rho}_k\mathbf{r}_k = \mathsf{I}. \tag{4.3.6}$$

Inserting the latter identity into the last term of (4.3.4) we have

$$\oint_{\dot{G}} (\mathbf{v}^T\mathbf{u}\dot{y} - \dot{x}\mathbf{v}^T\mathsf{B}\mathbf{u})\,ds$$

$$= \oint \sum_{l=1}^{n} \left\{ \operatorname{Re}\left(\sum_{j,k=1}^{n} \mathbf{q}_j^T\mathbf{v}(\dot{y}\delta_{jk} - \dot{x}\Lambda_{jk})Q_{kl}\right) \right.$$

$$\left. -2\left(\operatorname{Im} \sum_{j,k,m=1}^{n} \mathbf{q}_j^T\mathbf{v}(\dot{y}\delta_{jk} - \dot{x}\Lambda_{jk})Q_{km}\right) \operatorname{Im} \sum_{t=1}^{n} \mathbf{r}_m\mathbf{q}_t\mathbf{p}_t^T\boldsymbol{\rho}_l \right\} \mathbf{r}_l\mathbf{u}\,ds$$

$$- \oint_{\dot{G}} \sum_{h=n+1}^{2n} \left\{ 2\left(\operatorname{Im} \sum_{j,k,m=1}^{n} \mathbf{q}_j^T\mathbf{v}(\dot{y}\delta_{jk} - \dot{x}\Lambda_{jk})Q_{km}\right)\right.$$

$$\left. \times \left(\operatorname{Im} \sum_{t=1}^{n} \mathbf{r}_m\mathbf{q}_t\mathbf{p}_t^T\boldsymbol{\rho}_h\right) \right\} \mathbf{r}_h\mathbf{u}\,ds$$

$$= \oint_{\dot{G}} \sum_{l=1}^{n} \mathbf{v}^T(\dot{y}\mathsf{I} - \dot{x}\mathsf{B})\boldsymbol{\rho}_l\mathbf{r}_l\mathbf{u}\,ds + \oint_{G} \sum_{h=n+1}^{2n} \mathbf{v}^T(\dot{y}\mathsf{I} - \dot{x}\mathsf{B})\boldsymbol{\rho}_h\mathbf{r}_h\mathbf{u}\,ds. \tag{4.3.7}$$

The equations (4.3.1) and (4.3.7) imply the following definition of an *adjoint boundary value problem*:

$$\mathsf{R}^*\mathbf{v} \equiv \begin{cases} \mathsf{D}^*\mathbf{v} \equiv \mathbf{v}_x - (\mathsf{B}^T\mathbf{v})_y + \mathsf{C}^T\mathbf{v}, \\ \mathbf{r}_h^*\mathbf{v} \equiv -2 \sum_{j,k,m,t=1}^{n} \operatorname{Im}(\mathbf{r}_m \mathbf{q}_t \mathbf{p}_t^T \boldsymbol{\rho}_h) \operatorname{Im}((\dot{y}\delta_{jk} - \dot{x}\Lambda_{jk}) Q_{km} \mathbf{q}_j^T)\mathbf{v} \\ \quad = \boldsymbol{\rho}_h^T(\dot{y}\mathsf{I} - \dot{x}\mathsf{B}^T)\mathbf{v}, \\ \qquad h = n+1, \ldots, 2n. \end{cases} \tag{4.3.8}$$

The equality of the two versions of $\mathbf{r}_h^*$ is a simple conclusion from the regularity of the bilinear form (4.3.7) which results from $\operatorname{Det}(\dot{y}\mathsf{I} - \dot{x}\mathsf{B}) \neq 0$ due to the ellipticity.

Theorem 4.3.1 R^* *satisfies the Shapiro–Lopatinski condition if* R *does. For the corresponding Fredholm indices there holds*

$$\nu(\mathsf{R}^*) = -\nu(\mathsf{R}). \tag{4.3.9}$$

Proof Since D^* contains B^T, the transpose of B, the determinant in the Lopatinski condition for R^* has the form

$$\operatorname{Det}((\mathbf{r}_h^*\mathbf{p}_l)) = \operatorname{Det}\left(\left(\operatorname{Im} \sum_{t=1}^{n} \mathbf{r}_m \mathbf{q}_t \mathbf{p}_t^T \boldsymbol{\rho}_h\right)\right)_{\substack{m=1,\ldots,n,\\ h=n+1,\ldots,2n}}$$

$$\times \operatorname{Det}\left(\left(\mathrm{i} \sum_{j,k=1}^{n} \{(\dot{y}\delta_{jk} - \dot{x}\Lambda_{jk}) Q_{km} \mathbf{q}_j^T - (\dot{y}\delta_{jk} - \dot{x}\bar{\Lambda}_{jk}) \bar{Q}_{km} \bar{\mathbf{q}}_j^T\} \mathbf{p}_l\right)\right)_{m,l=1,\ldots,n}$$

$$= \operatorname{Det}\left(\left(\operatorname{Im} \sum_{t=1}^{n} \mathbf{r}_m \mathbf{q}_t \mathbf{p}_t^T \boldsymbol{\rho}_h\right)\right)_{\substack{m=1,\ldots,n\\ h=n+1,\ldots,2n}} \mathrm{i}^n \operatorname{Det}(\dot{y}\mathsf{I} - \dot{x}\Lambda) \operatorname{Det} \mathsf{Q} \tag{4.3.10}$$

The first determinant is always *real*. It never vanishes if we can show that the system of linear equations,

$$\sum_{m=1}^{n} \alpha_m \operatorname{Im} \mathbf{r}_m \sum_{t=1}^{n} \mathbf{q}_t \mathbf{p}_t^T \boldsymbol{\rho}_h = 0, \qquad h = n+1, \ldots, 2n \tag{4.3.11}$$

admits only the trivial solution $\alpha_m = 0$. In addition to (4.3.11) the equations (3.1.10) and (4.3.6) imply

$$\sum_{m=1}^{n} \alpha_m \operatorname{Re} \mathbf{r}_m \sum_{t=1}^{n} \mathbf{q}_t \mathbf{p}_t^T \boldsymbol{\rho}_h = \frac{1}{2} \sum_{m=1}^{n} \alpha_m \mathbf{r}_m \boldsymbol{\rho}_h = 0 \tag{4.3.12}$$

since $h \neq m$ for the above indices. Hence (4.3.11) implies

$$\sum_{t,m=1}^{n} \alpha_m (\mathbf{r}_m \mathbf{q}_t)(\mathbf{p}_t^T \boldsymbol{\rho}_h) = 0, \qquad h = n+1, \ldots, 2n. \tag{4.3.13}$$

The Lopatinski condition implies that the first matrix in (4.3.13) does not vanish. But then the second matrix in (4.3.13) is also nonsingular since the equations

$$\sum_{m=1}^{2n} (\mathbf{p}_t^T \boldsymbol{\rho}_m)(\mathbf{r}_m \mathbf{q}_\tau) = \delta_{t\tau}; \; \sum_{m=1}^{2n} (\mathbf{p}_t^T \boldsymbol{\rho}_m)(\mathbf{r}_m \bar{\mathbf{q}}_j) = 0 \tag{4.3.14}$$

yield

$$\sum_{h=n+1}^{2n} (\mathbf{p}_t^T \boldsymbol{\rho}_h)\left\{(\mathbf{r}_h \mathbf{q}_\tau) - \sum_{j,l=1}^{n} (\mathbf{r}_h \bar{\mathbf{q}}_j) Q_{jl} (\mathbf{r}_l \mathbf{q}_\tau)\right\} = \delta_{t\tau} \tag{4.3.15}$$

after multiplication of the second group of equations in (4.3.14) by $\bar{Q}_{jl}$ and inserting $(\mathbf{p}_t^T \boldsymbol{\rho}_l)$ for $l = 1, \ldots, n$ into the first equations. Hence, (4.3.13) admits only the trivial solution $\alpha_1 = \cdots = \alpha_n = 0$ and

$$\operatorname{Det}\left(\left(\operatorname{Im} \sum_{t=1}^{n} \mathbf{r}_m \mathbf{q}_t \mathbf{p}_t^T \boldsymbol{\rho}_h\right)\right)_{\substack{m=1,\ldots,n \\ h=n+1,\ldots,2n}} \neq 0. \tag{4.3.16}$$

For the second determinant in (4.3.10) we also find

$$\operatorname{Det}(\dot{y}\mathsf{I} - \dot{x}\Lambda) \neq 0 \tag{4.3.17}$$

since Λ has exactly the n eigenvalues $\lambda_+^j = \lambda_1^j + i\lambda_2^j$ with $\lambda_2^j \neq 0$. The determinant of Q is not zero thanks to (4.3.3).

Consequently, we have already proved that R^* satisfies the Lopatinski condition and defines a Fredholm mapping according to Theorem 4.2.1. The index of R^* can be computed by formula (4.2.1) with $\mathbf{r}_j^*$ instead of $\mathbf{r}_j$ and $\mathbf{p}_k$ instead of $\mathbf{q}_k$. The sum of indices becomes

$$\nu(\mathsf{R}) + \nu(\mathsf{R}^*) = 2n - \frac{1}{\pi} \oint_{\dot{G}} d \arg \operatorname{Det}((\mathbf{r}_j \mathbf{q}_k)) - \frac{1}{\pi} \oint_{\dot{G}} d \arg \operatorname{Det}((\mathbf{r}_l^* \mathbf{p}_t)). \tag{4.3.18}$$

Inserting (4.3.10) into (4.3.18) we have with $\mathsf{PQ} = \mathsf{I}$

$$\nu(\mathsf{R}) + \nu(\mathsf{R}^*) = 2 - 2\frac{1}{2\pi} \oint_{\dot{G}} d \arg \operatorname{Det}(\dot{y}\mathsf{I} - \dot{x}\Lambda) - \frac{2}{2\pi} \oint_{\dot{G}} d \arg \operatorname{Det}\left(\left(\operatorname{Im} \sum_{t=1}^{n} \mathbf{r}_m \mathbf{q}_t \mathbf{p}_t^T \boldsymbol{\rho}_h\right)\right). \tag{4.3.19}$$

Since (4.3.16) holds, the determinant in the last integral is real and nonvanishing. It can be considered to be a vector field in $\mathbb{C}$ given on $\dot{\mathsf{G}}$ which never leaves the (positive or negative) direction of the real axis and, hence, never turns to the direction of i. Therefore it has the revolution zero (*see* e.g. Krasnoselski *et al.* [19, p. 25, Section 4.7]). For the remaining integral in (4.3.19) let us consider the family of matrices,

$$\dot{y}\mathsf{I}-\dot{x}(\mathrm{i}\tau\mathsf{I}+(1-\tau)\Lambda), \quad 0\leq\tau\leq 1. \tag{4.3.20}$$

The determinants of this family are never zero since

$$\mathrm{Im}\,(\dot{y}-\mathrm{i}\tau\dot{x}):(1-\tau)\dot{x}\leq 0$$

holds and the eigenvalues of Λ are in the upper halfplane. Therefore the winding number in (4.3.10) can be calculated explicitly.

$$\frac{1}{2\pi}\oint \mathrm{d}\arg\mathrm{Det}\,(\dot{y}\mathsf{I}-\dot{x}\Lambda)=\frac{1}{2\pi}\oint_{\dot{\mathsf{G}}} \mathrm{d}\arg\mathrm{Det}\,\{(\dot{y}-\mathrm{i}\dot{x})\mathsf{I}\}=n, \tag{4.3.21}$$

(*see* e.g. Krasnoselski *et al.* [19, p. 13 Section 2.3]) and this implies (4.3.9).

Now we are in the position to formulate the normal solvability for (4.1.1) and (4.3.8) similarly to Theorem 3.4.1. Let us introduce the following two regular bilinear forms:

$$\langle\langle(\mathbf{c},\boldsymbol{\psi}),\mathbf{v}\rangle\rangle\equiv\iint_{\mathsf{G}}\mathbf{v}^T\mathbf{c}[\mathrm{d}x,\mathrm{d}y]-\oint_{\dot{\mathsf{G}}}\sum_{l=1}^{n}\mathbf{v}^T(\dot{y}\mathsf{I}-\dot{x}\mathsf{B})\boldsymbol{\rho}_l\psi_l\,\mathrm{d}s \tag{4.3.22}$$

$$\langle\mathbf{u},(\mathbf{c},\boldsymbol{\psi})\rangle\equiv-\iint_{\mathsf{G}}\mathbf{u}^T\mathbf{c}[\mathrm{d}x,\mathrm{d}y]+\oint_{\dot{\mathsf{G}}}\sum_{h=n+1}^{2n}\mathbf{r}_h\mathbf{u}\psi_{h-n}\,\mathrm{d}s. \tag{4.3.23}$$

Formula (4.3.7) shows that

$$\langle\langle\mathsf{R}\mathbf{u},\mathbf{v}\rangle\rangle=\langle\mathbf{u},\mathsf{R}^*\mathbf{v}\rangle \quad \text{holds for all} \quad \mathbf{u},\mathbf{v}\in C^{1+\alpha}(\bar{\mathsf{G}}). \tag{4.3.24}$$

Therefore Theorem 1.3.4 implies

Theorem 4.3.2 R (4.1.1) *and* R* (4.3.8) *are normally solvable*; i.e. the boundary value problem (4.1.1),

$$\mathsf{R}\mathbf{u}=(\mathbf{f},\boldsymbol{\psi})$$

has a solution if and only if the solvability conditions

$$\langle\langle(\mathbf{f},\boldsymbol{\psi}),\mathbf{v}\rangle\rangle=\iint_{\mathsf{G}}\mathbf{v}^T\mathbf{f}[\mathrm{d}x,\mathrm{d}y]-\oint_{\dot{\mathsf{G}}}\sum_{l=1}^{n}\mathbf{v}^T(\dot{y}\mathsf{I}-\dot{x}\mathsf{B})\boldsymbol{\rho}_l\psi_l\,\mathrm{d}s=0 \tag{4.3.25}$$

are satisfied for all $\mathbf{v}$ *with*

$$\mathsf{R}^*\mathbf{v}=0,$$

i.e. *for all eigensolutions of the homogeneous adjoint boundary value problem. Moreover, the eigenspaces of* R *and* R* *are both finite dimensional.*

4.4 Some Remarks on Nonlinear Systems

The most general first order nonlinear system can be written as

$$\mathbf{\Phi}(x, y, \mathbf{u}, \mathbf{u}_x, \mathbf{u}_y) = 0 \quad \text{in} \quad \mathsf{G} \tag{4.4.1}$$

where $\mathbf{\Phi}$ denotes a vector valued function with $2n$ components. Here we shall consider only linear boundary conditions of the form

$$\mathbf{r}\mathbf{u} = \boldsymbol{\psi} \quad \text{on} \quad \dot{\mathsf{G}}. \tag{4.4.2}$$

For the treatment of such general problems the main tools are the Banach fixed point theorem on one hand and the Schauder–Leray technique based on the Brouwer fixed point theorem on the other hand. Whereas Banach's fixed point theorem provides the convergent successive approximation, the Schauder–Leray method is not constructive in general. But in connection with the continuation method and Brouwer's theorem constructive methods are also available (e.g., Wacker [37] and Kellog, Li and Yorke [18]). For general nonlinear elliptic boundary value problems in connection with the Leray–Schauder theory there has been done already a vast amount of work. Let us mention only a few authors, such as Browder [8], Nečas [24], Berger and Schechter [4] and Nirenberg [25].

Let us begin the treatment of (4.4.1) by the use of the Banach fixed point theorem following Tutschke [30, 31]. As before we assume that G is a bounded simply connected plane domain with a $C^{1+\alpha}$-boundary curve $\dot{\mathsf{G}}$. Tutschke introduces complex valued unknown functions by

$$w_j \equiv u_j + \mathrm{i} u_{n+j}, \quad j = 1, \ldots, n \tag{4.4.3}$$

and assumes that the equations (4.4.1) can be resolved such that they become

$$\mathbf{w}_{\bar{z}} = \mathbf{H}(z, \mathbf{w}, \mathbf{w}_z) \quad \text{in} \quad \mathsf{G}. \tag{4.4.4}$$

As in Section 2.6, the H_j are complex-valued functions of all the arguments but they are *not* assumed to be *holomorphic*. The boundary conditions become

$$\operatorname{Re} \mathsf{P}\mathbf{w} = \boldsymbol{\psi} \quad \text{on} \quad \dot{\mathsf{G}}. \tag{4.4.5}$$

Similarly to (2.6.4), it is assumed that the right hand side in (4.4.4) satisfies a Lipschitz condition:

$$\|\mathbf{H}(z, \mathbf{w}(z), \mathbf{h}(z)) - \mathbf{H}(z, \mathbf{v}(z), \mathbf{g}(z))\|_\alpha \leq L\{\|\mathbf{w} - \mathbf{v}\|_\alpha + \|\mathbf{h} - \mathbf{g}\|_\alpha\} \tag{4.4.6}$$

and, further $\mathbf{H}(\cdot, \mathbf{w}, \mathbf{h}) \in C^\alpha(\bar{\mathsf{G}})$ for fixed constants $(\mathbf{w}, \mathbf{h}) \in \mathbb{C}^{2n}$. If L is small enough and $\mathsf{P} = \mathsf{I}$, the following theorem holds:

Theorem 4.4.1 [31] *If $L \leq L_0$ with L_0 sufficiently small then the Dirichlet problem,* (4.4.4), (4.4.5) *with* $\mathsf{P} = \mathsf{I}$ *and*

$$\operatorname{Im} \mathbf{w}(z_1) = \boldsymbol{\kappa}, \quad z_1 \in \mathsf{G}, \quad \boldsymbol{\kappa} \in \mathbb{R}^n \tag{4.4.7}$$

has exactly one solution.

Proof For convenience let us transform the domain G into the unit disk and $z_1 = 0$, e.g. by using a conformal mapping. Then the above Dirichlet problem can be solved as in Section 2.6 using the ansatz

$$\mathbf{w} = P'\boldsymbol{\rho} + \boldsymbol{\theta} \tag{4.4.8}$$

where P' is the operator defined by (2.6.32) and

$$\boldsymbol{\theta} \equiv \frac{1}{2\pi \mathrm{i}} \oint_{|t|=1} \frac{t+z}{t-z} \boldsymbol{\psi}\, \mathrm{d}t + \mathrm{i}\boldsymbol{\kappa} + \frac{1}{2\pi \mathrm{i}} \oint_{|t|=1} \boldsymbol{\psi}\, \mathrm{d}t. \tag{4.4.9}$$

The Dirichlet problem becomes now equivalent to the nonlinear functional equation for the density $\boldsymbol{\rho}$,

$$\boldsymbol{\rho}(z) = \mathbf{H}(z, P'\boldsymbol{\rho} + \boldsymbol{\theta}, T\boldsymbol{\rho} + \boldsymbol{\theta}_z). \tag{4.4.10}$$

The proposition of Theorem 4.4.1 can be proved in two different ways:
(1) Use the same argumentation as in Section 2.6.
(2) (2.4.9) shows that T maps $C^\alpha(\bar{\mathsf{G}})$ into itself continuously as well as P' does. Hence (4.4.6) with $L \leq L_0$ and L_0 small enough implies the contraction of the mapping of $\boldsymbol{\rho}$ given by the right hand side of (4.4.10). The Banach fixed point theorem supplies a unique density $\boldsymbol{\rho} \in C^\alpha$ satisfying (4.4.10). The desired solution $\mathbf{w}$ is given by (4.4.8).

Remark 4.4.2 The above method was introduced by Vekua [32]. In [10] Gilbert and Hsiao extended this approach to second order Dirichlet problems in higher dimensions.

Tutschke also solved in [30] the more general problem (4.4.4) and (4.4.5). To this end he needs an extension of P to the whole of $\bar{\mathsf{G}}$. For this extension let us assume that the boundary conditions (4.4.5) have already been transformed as in (3.2.7) such that P has the properties

$$U_r \equiv \operatorname{Det}((P_{jk}))_{1 \leq j, k \leq r} = \begin{vmatrix} P_{11} & \cdots & P_{1r} \\ \cdot & & \\ \cdot & & \\ \cdot & & \\ P_{r1} & & P_{rr} \end{vmatrix} \neq 0$$

for all points on $\dot{\mathsf{G}}$ and $r = 1, \ldots, n$.

Lemma 4.4.3 *Let* $\mathsf{P} \in C^{1+\alpha}(\dot{\mathsf{G}})$ *satisfy the Lopatinski condition and let all* U_r *have index zero;* i.e.

$$U_r \neq 0 \quad \text{and} \quad \frac{1}{2\pi} \oint_{\dot{\mathsf{G}}} \mathrm{d} \arg U_r = 0 \quad \text{for} \quad r = 1, \ldots, n. \tag{4.4.11}$$

Then the field P *can be extended to* $\bar{\mathsf{G}}$ *with* $\mathsf{P} \in C^{1+\alpha}(\bar{\mathsf{G}})$ *and*

$$\operatorname{Det} \mathsf{P} \neq 0 \quad \text{in} \quad \bar{\mathsf{G}}.$$

It seems that this Lemma is even true if (4.4.11) is satisfied only for $r = n$ according to investigations by Bojarski (loc. cit. (j) in [32]). But this is yet to be proved. The following proof is due to Tutschke (private communication).

Proof Let us first extend the functions

$$U_r = |U_r| \, \mathrm{e}^{\mathrm{i}\tau_r}$$

from $\dot{\mathsf{G}}$ to $\bar{\mathsf{G}}$ which can be explicitly done by

$$U_r \equiv -\oint_{\dot{\mathsf{G}}} U_r| \, \mathrm{d}_n G^I \exp\left\{-\mathrm{i} \oint_{\dot{\mathsf{G}}} \tau_r \, \mathrm{d}_n G^I\right\}, \quad r = 1, \ldots, n.$$

Since τ_r is periodic on $\dot{\mathsf{G}}$ according to (4.4.11) we have $U_r \in C^{1+\alpha}(\bar{\mathsf{G}})$. Moreover $U_r \neq 0$ in $\bar{\mathsf{G}}$ due to the maximum principle which holds for the first harmonic factor. With $P_{11} = U_1$ we have already extended P_{11}.

For the remaining P_{jk} the extension can be found by induction. Let P_{jk} be already extended to $\bar{\mathsf{G}}$ for $1 \leq j, k \leq r$. Then extend $P_{1,r+1}, \ldots, P_{r,r+1}$ and $P_{r+1,1}, \ldots, P_{r,r+1}$ *arbitrarily* to functions in $C^{1+\alpha}(\bar{\mathsf{G}})$. Using the Lagrange development of the determinant U_{r+1} with respect to the $r+1$-th column,

$$U_{r+1} = P_{r+1,r+1} U_r + \sum_{j=1}^{r} P_{j,r+1} \Delta_j,$$

let us define $P_{r+1,r+1}$ in $\bar{\mathsf{G}}$ by

$$P_{r+1,r+1} \equiv U_r^{-1} \left\{ U_{r+1} - \sum_{j=1}^{r} P_{j,r+1} \Delta_j \right\}$$

where all functions on the right hand side are already defined in $\bar{\mathsf{G}}$ since the minors Δ_j contain only P_{lk} for $1 \leq l \leq r+1$ and $1 \leq k \leq r$.

Introducing new unknown functions $\mathbf{W}$ by

$$\mathbf{w} = \mathsf{P}\mathbf{W}, \ \mathbf{W} = \mathsf{P}^{-1}\mathbf{w} \tag{4.4.12}$$

one finds for $\mathbf{W}$ the Dirichlet problem

$$\mathbf{W}_{\bar{z}} = (\mathsf{P}_{\bar{z}}^{-1})\mathsf{P}\mathbf{W} + \mathsf{P}^{-1}\mathbf{H}(z, \mathsf{P}\mathbf{W}, \mathsf{P}_z \mathbf{W} + \mathsf{P}\mathbf{W}_z) \tag{4.4.13}$$

$$\operatorname{Re} \mathbf{W} = \boldsymbol{\psi} \quad \text{on} \quad \dot{\mathsf{G}}. \tag{4.4.14}$$

Theorem 4.4.4 (Tutschke [30]) *Let the assumptions of Lemma 4.4.3 be satisfied and let the Lipschitz constant* (4.4.6) *of the new system* (4.4.13) *be small enough. Then the problem*

$$\mathbf{w}_{\bar{z}} = \mathbf{H}(z, \mathbf{w}, \mathbf{w}_z) \quad \text{in} \quad \mathsf{G}, \quad \operatorname{Re} \mathsf{P} w = \boldsymbol{\psi} \quad \text{on} \quad \dot{\mathsf{G}},$$
$$\operatorname{Im} \mathsf{P}\mathbf{w}(z_1) = \boldsymbol{\kappa} \qquad (4.4.15)$$

with (4.4.11) *has exactly one solution.*

Here we demanded by (4.4.6) that $\mathbf{H}$ or the transformed right hand sides have a contraction property with respect to $\mathbf{w}_z$ *and* to $\mathbf{w}$. The first contraction corresponds in some way to the ellipticity in view of the quasinormal form (4.2.17) for linear problems. In the case $n = 1$ we have seen in Section 2.6 that a contraction with respect to $\mathbf{w}$ was not necessary. There the solution of the problem hinged on the *a priori* estimates for the corresponding linearized problems. The latter were based on uniqueness theorems for corresponding Fredholm equations of index zero and the uniqueness was based on the similarity principle.

Therefore we may expect other results by the Schauder–Leray theory if we can find suitable *a priori* estimates for the solution. For one connection between *a priori* estimates and solvability let us quote the following theorem by Schäfer:

Theorem 4.4.5 [28] *Let E be a linear, locally convex, complete Hausdorff space and let* T *denote a completely continuous mapping from E into itself. Then either there exists a fixed point*

$$X = t\mathsf{T}(X) \qquad (4.4.16)$$

to each $t \in [0, 1]$ *or the set* $\{X : X = t\mathsf{T}(X), 0 < t < 1\}$ *is bounded in E.*

This theorem is very useful especially for semilinear problems. For $n = 1$ the desired *a priori* estimates can be found relatively simply for (1.4.1), (1.4.2), (1.4.5). With $w(t)$, the solution of (1.4.6), (1.4.7), we may introduce $h(t)$ by

$$w(t) = w(0) + h(t). \qquad (4.4.17)$$

Then h satisfies

$$h_{\bar{z}} = tH(z, w(0) + h), \qquad (4.4.18)$$

$$\operatorname{Re} e^{i\tau} h \,|_{\dot{\mathsf{G}}} = 0, \qquad \oint_{\dot{\mathsf{G}}} \operatorname{Im} e^{i\tau} h\sigma \, ds = 0, \qquad h(z_j) = 0. \qquad (4.4.19)$$

and for E we can choose the linear subspace of C^α which consists of all functions satisfying (4.4.19). The equation (4.4.16) corresponds to the solutions of (4.4.18), (4.4.19). If H satisfies the boundedness assumptions

(1.4.3) then the *a priori* estimate (1.2.33) implies that all solutions h of (4.4.18), (4.4.19) are uniformly bounded in E. Hence, Theorem 4.4.5 yields Theorem 1.4.1 except the uniqueness.

In case of an unbounded right-hand side in (4.4.18) we can use the cut off technique as in Corollary 1.4.2, i.e. define for the differential equation

$$w_{\bar{z}} = f(z, w) \tag{4.4.20}$$

an auxiliary equation with

$$v_{\bar{z}} = H(z, v) \equiv f(z, v)\chi(v \cdot \bar{v}) \tag{4.4.21}$$

where $\chi \in C^\infty$ with

$$\chi(\rho) = \begin{cases} 1 & \text{for} \quad \rho \leq K^2 \\ 0 & \text{for} \quad \rho \geq 2K^2 \end{cases} \tag{4.4.22}$$

with a suitable constant K. If one can show by *a priori* estimates that every solution of the boundary value problem corresponding to the auxiliary equation has the property

$$|v| \leq K \tag{4.4.23}$$

then v is also the desired solution of (4.4.20). Hence it turns out that the crucial property for the existence is the *a priori* estimate.

Now let us consider some examples where *a priori* estimates are available. Using the Green functions G^I and G^{II} of Section 1.1, Begehr and Gilbert [3] solved the first boundary value problem

$$w_{\bar{z}} = f(z, w) \quad \text{in } \mathsf{G}, \quad \operatorname{Re} w \quad \text{on } \dot{\mathsf{G}}, \quad \oint_{\dot{\mathsf{G}}} \operatorname{Im} w \, d\phi = 2\pi\kappa \tag{4.4.24}$$

where ϕ denotes the conformal mapping used in (1.1.10).
With

$$k \equiv \sup_{z,\zeta \in \mathsf{G}} \left| \frac{\phi'(z)(\zeta - z)}{\phi(\zeta) - \phi(z)} \right|$$

the assumptions on f are the following:

$$f(z, w(z)) \in L_p(\bar{\mathsf{G}}) \quad \text{for some} \quad p > 2 \quad \text{and every} \quad w \in C^0(\bar{\mathsf{G}}),$$

$$|f(z, w) - f(z, w')| \leq \frac{1}{8k}\, \omega(z, |w - w'|) \quad \text{for all} \quad z \in \bar{\mathsf{G}};\ w, w' \in \mathbb{C}$$

where $\omega(z, \rho)$ is a nonnegative function monotonically increasing with $\rho \geq 0$ and

$$\omega(z, 0) = 0, \ \omega(z, \gamma(z)) \in L_p(\bar{\mathsf{G}}) \quad \text{for every} \quad \gamma \in C^0(\bar{\mathsf{G}});$$

there exists a constant K such that

$$\int_{G} \omega(\zeta, K) \frac{d\xi\, d\eta}{|\zeta - z|} \leq K$$

and the integral equation

$$\gamma(z) = \int_{G} \omega(\zeta, \gamma(\zeta)) \frac{d\xi\, d\eta}{|\zeta - z|} \quad \text{in} \quad \bar{G}$$

admits for $\gamma \leq K$ only the trivial solution. (4.4.25)

In this case the constant K corresponds to K in (4.4.22) and (4.4.23) and E is the subspace of $C^0(\bar{G})$ with (4.4.19).

Another case was considered by Wendland [44] where

$$w_{\bar{z}} = aw + b\bar{w} + c + g(z, w) \quad \text{in} \quad G,$$

$$\operatorname{Re} we^{i\tau} = \psi \quad \text{on } \dot{G}, \quad \oint_{\dot{G}} \operatorname{Im} we^{i\tau} \sigma\, ds = \kappa \tag{4.4.26}$$

$$w(z_j) = \gamma_j \quad \text{for} \quad j = 1, \ldots, |n| \quad \text{where} \quad n = \frac{1}{2\pi} \oint_{\dot{G}} d\tau \leq 0.$$

(4.4.26)

Here the corresponding *a priori* estimate is based on an estimate

$$\|h\|_{1+\alpha} \leq \|h_{\bar{z}} - ah - b\bar{h}\|_{\alpha} \cdot \gamma \cdot \exp(\gamma'(\|a\|_{\alpha} + \|b\|_{\alpha})) \tag{4.4.27}$$

for h with (4.4.19) which corresponds to (1.2.55) and a Gronwall inequality (see Walter [43] p. 16). The assumptions in [44] are:

The mappings $w \to g(\cdot, w(\cdot))$, $w \to (\partial g/\partial w)(\cdot, w(\cdot))$ and $w \to (\partial g/\partial \bar{w})(\cdot, w(\cdot))$ are bounded and continuous from $C^{\alpha}(\bar{G})$ into itself where $(\partial/\partial w)$ *and* $(\partial/\partial \bar{w})$ are defined as in (1.1.4).
$g(z, 0) = 0$ and g satisfies the growth conditions

$$\|g(\cdot, w(\cdot))\|_{\alpha} \leq \omega(\|w\|_{\alpha}),$$

$$\left\|\frac{\partial g}{\partial w}(\cdot, w(\cdot))\right\|_{\alpha} + \left\|\frac{\partial g}{\partial \bar{w}}(\cdot, w(\cdot))\right\|_{\alpha} \leq \omega'(\|w\|_{\alpha}) \tag{4.4.28}$$

with monotonic increasing smooth functions ω, ω'; $\omega(0) = \omega'(0) = 0$.
The Volterra equation

$$\rho(t) = \gamma\omega(\|w(0)\|_{\alpha} + \int_0^t \rho(\tau)\, d\tau) \exp\{\gamma'(\|a\|_{\alpha} + \|b\|_{\alpha} +$$

$$+ t\omega'(\|w(0)\|_{\alpha} + \int_0^t \rho(\tau)\, d\tau))\}$$

admits a solution for $0 \leq t \leq 1$.

In this case the constant K is given by

$$K = \|w(0)\|_\alpha + \int_0^1 \rho(\tau)\,dt.$$

Both approaches can be carried over to systems with $n > 1$ with the Douglis normal form (3.1.26), (3.1.27).

For $\mathsf{P} = \mathsf{I}$ (the first boundary value problem for Douglis systems) this was done by Gilbert [9] with assumptions corresponding to (4.4.25). For P with the special form

$$P_{jk} = \begin{cases} 0 & \text{for} \quad j < k \\ P_{j-k+1,1} & \text{for} \quad j \geq k \end{cases} \tag{4.4.29}$$

the corresponding assumptions to (4.4.28) were made in [44] by Wendland.

For more complicated equations the continuity method can no longer be formulated in a simple way like (4.4.16). Here let us embed the problem into a family of equations

$$\mathsf{T}_\tau(\mathbf{u}) = 0 \quad \text{for} \quad 0 \leq \tau \leq 1 \tag{4.4.30}$$

where (4.4.30) is solvable for $\tau = 0$ and for which the principle of Caccioppoli and Schauder can be used (see Miranda [23, Section 36] and for more special situations Bers–John–Schechter [5, II 7]).

To this end let us assume that the above family has the following properties:

$\mathsf{T}_\tau(\cdot)$ maps $E \times [0, 1]$ continuously into Y where E denotes some open set in X and where X and Y are Banach spaces (4.4.31)

A. The Gateaux derivative with respect to $\mathbf{u}$,

$\mathsf{T}_\tau(\mathbf{u}) \cdot$ is continuous in $E \times [0, 1]$ and *has a bounded linear inverse from Y to X.* (4.4.32)

B. Every solution $\mathbf{u} \in E$ at $\tau \in [0, 1]$ is bounded; i.e.

$\mathsf{T}_\tau(\mathbf{u}) = 0$ *with* $\mathbf{u} \in E$ *implies* $\mathbf{u} \in B \subset E$ *where* B *is bounded.*

(4.4.33)

C. Every sequence of solutions $\mathbf{u}_j$ to τ_j, i.e.

$$\mathsf{T}_{\tau_j}(\mathbf{u}_j) = 0 \quad \text{in } E, \qquad 0 \leq \tau_j \leq 1 \tag{4.4.34}$$

is a *relatively compact subset* of B.

Then there holds

Theorem 4.4.6 *Under the assumptions* A, B and C *the equation* (4.4.30) *has a solution for every* $\tau \in [0, 1]$.

Proof Let us just sketch the proof. If $\sum \subset [0, 1]$ denotes the set of τ values for which (4.4.30) has a solution then A assures by the implicit function theorem that $\sum$ is open (e.g. Kantorowitsch–Akilow [16, XVII Section 4]). From C and B it follows immediately that $\sum$ is also closed. Hence $\sum = [0, 1]$.

For B and C the *a priori* estimates play again the essential role. Therefore let us review some of them connected with equations (4.4.1) or (4.4.4). In case $n = 1$ the nonlinear system (2.6.31) provides an *a priori* estimate by the use of (2.6.38) and (2.6.39):

$$\begin{aligned}\|\omega\|_p &\leqslant \|h(z, w, T\omega + \theta_z)\|_p \\ &\leqslant q_0' \|\omega\|_p + \|h(z, w, \theta_z)\|_p.\end{aligned} \tag{4.4.35}$$

Hence, for $w = P'\omega + \theta$ we find the *a priori* estimates

$$\begin{aligned}\|w\|_\beta &\leqslant \gamma\{\|\omega\|_p + \|\theta\|_\beta\} \\ &\leqslant \frac{\gamma}{1 - q_0'} \|h(z, w, \theta_z)\|_p + \gamma \|\theta\|_\beta \\ &\leqslant k(\|w\|_0, |\kappa|, |\phi|_{1+\beta})\end{aligned} \tag{4.4.36}$$

and

$$\|\omega\|_p \leqslant k(\|w\|_0, |\kappa|, |\phi|_{1+\beta}), \qquad \beta = 1 - 2/p \tag{4.4.37}$$

where k is a suitable continuous function of its three arguments. Let us define:

$$\begin{aligned}X \equiv \{&w \in C^{\beta'}(\bar{\mathrm{G}}) \text{ having derivatives } w_{\bar{z}} \in L_{p'}(G) \\ &\text{equipped with } \|w\|_X \equiv \|w\|_{C^{\beta'}} + \|w_{\bar{z}}\|_{L_{p'}}\},\end{aligned} \tag{4.4.38}$$

$$Y \equiv L_{p'}(\bar{\mathrm{G}}) \times C^{\beta'}(\dot{\mathrm{G}}) \times \mathbb{R} \tag{4.4.39}$$

and

$$\mathsf{T}_\tau(w) \equiv \begin{cases} w_{\bar{z}} - \tau h(z, w, w_z) = 0 & \text{in } \mathsf{G}, \\ \operatorname{Re} w - \psi = 0 & \text{on } \dot{\mathsf{G}}, \\ \operatorname{Im} w(0) - \kappa = 0, \end{cases} \tag{4.4.40}$$

where $2 < p' < p$ and $\beta' = (p' - 2)/p' < \beta$. Then it can be shown that (2.6.35) implies

$$\|h(z, P'\chi + \theta, \theta_z)\|_{p'} \leqslant K(1 - q_0') \tag{4.4.41}$$

for all χ with $\|\chi\|_{p'} \leqslant K + \delta$ where $\delta > 0$. Choosing

$$E \equiv \{w = \theta + P'\chi \quad \text{where} \quad \|\chi\|_{p'} < K + \delta\} \tag{4.4.42}$$

and

$$B \equiv \{w = \theta + P'\chi \quad \text{where} \quad \|\chi\|_{p'} \leq K\} \tag{4.4.43}$$

we find from (4.4.41) the property B and from (4.4.36) and (4.4.37) the property C since B is relatively compact in $C^0(\bar{\mathrm{G}})$.

The above example shows that the compactness property C follows from *a priori* estimates like (4.4.36), (4.4.37) which provide bounds for 'higher order norms' in terms of 'lower order norms' as e.g. $\|w\|_0$. The boundedness of $\|w\|_0$ must be assured by an additional assumption which corresponds in the above case to (4.4.41).

For exterior problems Beyer [7] proved *a priori* estimates with weighted Hölder norms for a subclass of the equations (2.6.31).

Of course, Tutschke's results can also be obtained with the above technique since his assumptions in [30, 31] yield *a priori* estimates for the family of problems

$$\begin{aligned} &\mathbf{w}_{\bar{z}} - \tau \mathbf{H}(z, \mathbf{w}, \mathbf{w}_z) = 0 \quad \text{in} \quad \mathrm{G}, \\ &\operatorname{Re} \mathbf{P}\mathbf{w} - \boldsymbol{\psi} \qquad\quad = 0 \quad \text{on} \quad \dot{\mathrm{G}}, \\ &\operatorname{Im} \mathbf{P}\mathbf{w}(z_1) - \kappa \qquad = 0. \end{aligned} \tag{4.4.44}$$

In the general case there is much work done for *second* order elliptic systems based on the work by Ladyshenskaja and Uraltseva [20]. Here it turns out that the growth of $\mathbf{H}$ with respect to $\mathbf{w}$ becomes the crucial point. Heinz presented an example in [11] that shows that quadratic growth in $\mathbf{w}$ already spoils estimates for $\|\mathbf{w}\|_0$ unless $\mathbf{H}$ satisfies additional quantitative conditions as in (4.4.41). For general systems so called interior estimates have been proved in the last years. These are useful in the case of elliptic systems on compact manifolds without boundary conditions. In the following let us review a few of these results for semilinear and nonlinear systems.

W. v. Wahl considered in [39, 41] the following *second order systems*

$$\sum_{i,j=1}^{2} \frac{\partial}{\partial x_j}\left(a_{ij}(x)\frac{\partial u^l}{\partial x_i}\right) = f^l(x, \mathbf{u}, \mathbf{u}_{x_k}), \quad l = 1, \ldots, m \tag{4.4.45}$$

with measurable coefficients a_{ij} and f^l.

Assuming that the differential operator is uniformly elliptic i.e.

$$\nu|\xi|^2 \leq \sum_{i,j=1}^{2} a_{ij}\xi_i\xi_j \leq \mu|\xi|^2, \quad 0 < \nu, \mu \tag{4.4.46}$$

where $a_{ij} = a_{ji}$ and a growth condition for f^l,

$$|f^l(x, \mathbf{u}, \mathbf{p}_k)| \leq \tilde{\mu}(|\mathbf{u}|)\left(1 + \sum_{k=1}^{2} |\mathbf{p}_k|^2\right) \tag{4.4.47}$$

where the monotonic increasing function $\tilde{\mu}$ satisfies some additional growth conditions, v. Wahl proves

Theorem 4.4.7 [39] *Let* $\mathbf{u} \in L_\infty(\mathrm{G}) \cap H^1(\mathrm{G})$ *be a generalized solution of* (2.4.45) *in* G. *Let* Ω *be a subdomain of* G *with* $\bar{\Omega} \subset \mathrm{G}$. *Then* $\mathbf{u} \in C^\alpha(\bar{\Omega})$ *where* α *depends on* ν, μ, $\tilde{\mu}$ *and* $\|\mathbf{u}\|_{L_\infty(\mathrm{G})}$.

An estimate similar to (4.4.36) is only possible if $\|\mathbf{u}\|_{L_\infty(\mathrm{G})}$ is small enough. Hildebrandt and Widman [12] also obtained *a priori* estimates and for systems in the plane they showed that the smallness assumption

$$\|\mathbf{u}\|_{L_\infty(\bar{\mathrm{G}})}\tilde{\mu}(\|\mathbf{u}\|_{L_\infty(\bar{\mathrm{G}})}) < \nu \tag{4.4.48}$$

with ν from (4.4.46) is sufficient and even necessary in order to find an interior *a priori* estimate of the kind

$$\|\mathbf{u}\|_{C^\alpha(\bar{\Omega})} \leqslant \kappa(\nu, \mu, \tilde{\mu}, \Omega, \|\mathbf{u}\|_{L_\infty(\bar{\mathrm{G}})}). \tag{4.4.49}$$

This result contains a corresponding estimate by v. Wahl in [39, 41].

For nonlinear elliptic second-order systems v. Wahl extended in [40] results by Oskolkov [26]. These systems have the form

$$\Phi^l(x, u^k, u^k_{x_i}, u^l_{x_ix_j}) = 0,\ l, k, i, j = 1, 2, \tag{4.4.50}$$

and v. Wahl proves an interior *a priori* estimate for the second derivatives. He assumes for Φ^l:

$$a_{ij} \equiv \frac{\partial \Phi^1}{\partial q^1_{ij}}(x, \mathbf{u}, \mathbf{p}_k, q^l_{ij}) = \frac{\partial \Phi^2}{\partial q^2_{ij}},$$

$$\nu\,|\xi|^2 \leqslant \sum_{i,j=1}^{2} a_{ij}\xi_i\xi_j \leqslant \mu\,|\xi|^2$$

where the a_{ij} satisfy (4.4.46) uniformly,

$$\frac{\partial \Phi^l}{\partial x_m}, \frac{\partial \Phi^l}{\partial u^k}, \frac{\partial \Phi^l}{\partial p^k_m}, \frac{\partial a_{ij}}{\partial x_m}, \quad \text{and} \quad \frac{\partial a_{ij}}{\partial u^k} \tag{4.4.51}$$

are uniformly bounded and

$$\left|\frac{\partial a_{ij}}{\partial q_{ij}}\right| \leqslant \frac{\mu^2}{\tilde{\rho}(\sum_{i,j=1}^2 q^2_{ij})\{\sum_{i,j=1}^2 q^2_{ij}\}}, \quad \left|\frac{\partial a_{ij}}{\partial p^k_m}\right| \leqslant \frac{\mu_1}{\tilde{\rho}(\sum_{i,j=1}^2 q^2_{ij})}$$

where $\tilde{\rho}(\rho) \to \infty$ for $\rho \to \infty$.

Theorem 4.4.8 [40] *For every solution of* (4.4.50) *with* $\mathbf{u} \in C^1(G) \cap H^3_{loc}(G)$ *there holds the interior estimate*

$$\|\mathbf{u}_{x_i x_j}\|_{L_\infty(\Omega)} \leq k(\tilde{\rho}, \|\mathbf{u}\|_0, \|\mathbf{u}_{x_k}\|_0)$$

where k also depends on the functions ϕ^l *and their derivatives.*

4.5 Remarks on the Additional References

The Fredholm properties of linear Dirichlet problems for second order plane systems with continuous and also discontinuous coefficients and their classification in connection with the connected components within the Fredholm mappings have been investigated in [50] and [51].

The rest of the additional references refers to nonlinear problems (see also Section 1.5). Most of the more recent investigations are based on [87] and [97] where even problems with nonlinear boundary conditions and semilinear equations have been solved.

In [54] Dirichlet problems and in [86] and [110] more general boundary value problems for semilinear equations and systems are solved by the use of function theoretic and constructive methods.

The functional analytic methods for the nonlinear problems are mostly based on monotonicity, Schauder's method, Caccioppoli's principle or the degree of mappings. General presentations and surveys can be found in [4], [8], [20], [24], [25], [53], [65], [66] and [99] and for bifurcation problems in [85].

The monotonicity methods were used in [46–48], [62], [79] and [115], in [62] with Orlicz spaces. Schauder's method in connection with *a priori* estimates are used in [59], [64–66], [68], [73], [78], [93], [98], [105], [106] and [112]. *A priori* estimates are the main topics of [52], [63], [69], [72], [74], [77], [80], [83], [88], [89], [100], [101] and [103]; L^∞-estimates for generalized solutions are obtained in [60], [75], [114] and [115] and estimates for the first derivatives are proved in [61], [68], [70], [71], [84], [90] and [113]. Regularity properties have been investigated in [55–57], [74], [95], [96], [102] and [107], especially by the use of variational methods. Whereas all these estimates are of a qualitative nature, the constants of the estimates in [58] are explicitly constructed and the solution is obtained by an iterative procedure. The *a priori* estimates depend significantly on the growth as can be seen in [81–84] and [92]; its connection with *a priori* estimates have been studied in [12], [104] and [116]. The influence of boundary properties on the estimates was investigated in [108] and [109]. By the use of weighted norms, in [67], [91] and [94] the Schauder method was extended to unbounded domains. In [111] this method was investigated for degenerating quasilinear problems.

The Caccioppoli principle was used in [52] for solving nonlinear transition problems. In [47] the degree of mappings plays the basic role.

Nonlinear eigenvalue problems for higher order equations are solved in [49]. In [76] bifurcation problems for semilinear problems have been investigated.

References

1 Agmon, A., Douglis, A. and Nirenberg, L., Estimates near the boundary for solutions of elliptic partial differential equations satisfying general boundary conditions, II. *Comm. Pure Appl. Math.* **XVII,** 35–92, 1964.

2 Avantaggiati, A., Nuovi contributi allo studio dei problemi al contorno per i sistemi ellittici del primo ordine. *Ann. Mat. Pura Appl.* **69,** 107–170, 1965.

3 Begehr, H. and Gilbert, R. P., Über das Randwert-Normproblem für ein nichtlineares elliptisches System. *In Proc. Function Theoretic Methods for Partial Diff. Eqns.* Lecture Notes in Math. **561,** Springer, Berlin, 112–122, 1976.

4 Berger, M. S. and Schechter, M., On the solvability of semilinear operator equations and elliptic boundary value problems. *Bull. Am. Math. Soc.* **78,** No. 5, 1972.

5 Bers, L., John, F. and Schechter, M., *Partial differential equations.* Interscience, New York, London, Sydney, 1964.

6 Bers, L. and Nirenberg, L., On linear and nonlinear boundary value problems in the plane. *Atti del Conv. internazionale sulle Equazioni alle derivate parziali,* pp. 111–140, Trieste, 1954.

7 Beyer, K., Nichtlineare Randwertprobleme für elliptische Systeme 1. Ordnung für zwei Funktionen von zwei Variablen. *Beiträge zur Analysis* **4,** 31–34, 1972.

8 Browder, F. E., Nonlinear elliptic boundary value problems. *Bull. Am. Math. Soc.* **69,** 862–874, 1963.

9 Gilbert, R. P., *Nonlinear boundary value problems for elliptic systems in the plane.* University of Delaware, Technical Report G 3, 1976.

10 Gilbert, R. P. and Hsiao, G. C., On Dirichlet's problem for quasilinear elliptic equations. Proc. *In Lecture Notes* No. 430, pp. 184–236. Springer, Berlin, 1974.

11 Heinz, E., On certain nonlinear elliptic differential equations and universal mappings. *J. Analysis Math.* **5,** 197–272, 1956.

12 Hildebrandt, S. and Widman, K. O., Some regularity results for quasilinear systems of second order. *Math. Zeitschr.* **142,** 67–86, 1975.

13 Hile, G. N. and Protter, M. H., Unique continuation and the Cauchy problem for first order systems of partial differential equations, *Comm. Partial Diff. Eqns.* **1**, 437–465, 1976.

14 Hörmander, L., *Linear partial differential operators.* Springer, Berlin, Heidelberg, New York, 1969.

15 Hurewitz, W. and Wallman, W., *Dimension theory.* Princeton University Press, 1941.

16 Kantorowitsch, L. W. and Akilow, G. P., *Funktionalanalysis in normierten Räumen*. Akademie-Verlag, Berlin, 1964.

17 Kato, T., *Perturbation theory for linear operators*. Springer, Berlin, Heidelberg, New York, 1966.

18 Kellog, R., Li, T. and Yorke, J., A constructive proof of the Brouwer fixed-point theorem and computational results. *SIAM J. Numer. Anal.* **13,** 473–483, 1976.

19 Krasnoselski, M. A., Derow, A. I., Powolowzki, A. I. and Sabrejko, P. P., *Vektorfelder in der Ebene*. Akademie Verlag, Berlin, 1966.

20 Ladyshenskaja, O. A. and Uraltseva, N. N., *Linear and quasilinear elliptic equations*. Academic Press, New York, 1968.

21 Lopatinski, Y. B., On a method of reducing boundary problems for a system of differential equations of elliptic type to regular equations. *Ukrain. Mat. Z.* **5,** 123–151, 1953.

22 Malgrange, B., *Ideals of differentiable functions*. Oxford University Press, 1966.

23 Miranda. C., *Partial differential equations of elliptic type*. Springer-Verlag, Berlin, Heidelberg, New York, 1970.

24 Nečas, J., On the existence and regularity of solutions of nonlinear elliptic equations. *In Diff. equations and their applications, equadiff.* II, (Proc. Conf., Bratislava, 1966), Slov. Ped. Nakladetel., Bratislava, pp. 101–119, 1967.

25 Nirenberg, L., An application of generalized degree to a class of nonlinear problems. *Coll. Analyse Fonctionelle*, pp. 57–74. Liège, 1972.

26 Oskolkov, A. P., Interior estimates for the first order derivatives for a class of quasilinear elliptic systems. *Proc. Steklov Inst. Math.* **110,** 116–121, 1970.

27 Palamodov V. P., Systems of linear differential equations. *Itogi Nauki: Mat. Analiza* **1968,** 5–37, 1969.

28 Schäfer, H., Über die Methode der *a priori*-Schranken. *Math. Ann.* **129,** 415–416, 1955.

29 Schechter, M., *Principles of functional analysis*. Academic Press, New York, 1971.

30 Tutschke, W., The Riemann Hilbert problem for nonlinear systems of differential equations in the plane, 1977, to appear.

31 Tutschke, W., Lösung nichtlinearer partieller Differentialgleichungssysteme erster Ordnung in der Ebene durch Verwendung einer komplexen Normalform. *Math. Nachr.* **75,** 283–298, 1976.

32 Vekua, I. N., *Generalized analytic functions*. Pergamon Press, Oxford, 1962.

33 Vinogradov, V. S., A certain boundary value problem for an elliptic system of special form. *Differential'nye Uravnenija* **77,** No. 7, 1226–1234; 1341, 1971.

34 Vinogradov, V. S., A boundary value problem for a first order elliptic equation on the plane. *Differential'nye Uravnenija* **77,** No. 8, 1440–1448; 1541, 1971.

35 Vinogradov, V. S., On a method of solution of a boundary value problem

for a first-order elliptic system on the plane. *Dokl. Akad. Nauk SSSR*, Tom 201, 1971; *Soviet Math. Dokl.* **12,** 1699–1703, 1971.
36 Vol'pert, A. I., On the index and the normal solvability of boundary value problems for elliptic systems of differential equations in the plane. *Trudy Mosovskogo Matem. Obstsestva,* **10,** 41–87, 1961.
37 Wacker, H. J., Eine Lösungsmethode zur Behandlung nichtlinearer Randwertprobleme. *In* 'Iterationsverfahren, Numerische Mathematik, Approximationstheorie' *ISNM* **15,** 245–257, Birkhäuser, 1970.
38 van der Waerden, B. L., *Algebra I.* Springer, Berlin, Göttingen, Heidelberg, 1960.
39 v. Wahl, W., Über die Hölderstetigkeit der schwachen Lösungen gewisser semilinearer elliptischer Systeme. *Math. Zeitschr.* **130,** 149–57, 1973.
40 v. Wahl, W., Einige neue innere Abschätzungen für nichtlineare elliptische Gleichungen und Systeme. *Math. Zeitschr.* **134,** 119–128, 1973.
41 v. Wahl, W., Berichtigung zur Arbeit 'Über die Hölderstetigkeit der schwachen Lösungen gewisser semilinearer Systeme'. *Math. Zeitschr.* **136,** 93–94, 1974.
42 Waid, C., On the approximation of matrices. *Private communication.* Delaware, 1974.
43 Walter, W., *Differential and integral inequalities.* Springer, Berlin, Heidelberg, New York, 1970.
44 Wendland, W., On a class of semilinear boundary value problems for certain elliptic systems in the plane, Acad. Sc. USSR, in memory of I. N. Vekua. N. Bogolubov and N. Bogolubov (Eds.), to appear.
45 Wloka, J., Funktionalanalysis und Anwendungen. de Gruyter, 1971.

Additional references

46 Amann, H., On the existence of positive solutions of nonlinear elliptic boundary value problems. *Indiana University Math. J.* **21,** No. 2, 1971.
47 Amann, H., Existence of multiple solutions for nonlinear elliptic boundary value problems. *Indiana University Math. J.* **21,** No. 10, 1972.
48 Amann, H., A uniqueness theorem for nonlinear elliptic boundary value problems. *Arch. Rat. Mech. Anal.* **44,** 178–181, 1972.
49 Beckert, H., Über nichtlineare Eigenwertprobleme von Differentialgleichungssystemen höherer Ordnung. Ell. Differentialgleichungen II. *Schriftenreihe Inst. Math. D. Akad. Wiss. Berlin* A **8,** 79–90, 1971.
50 Bojarski, B., On the first boundary value problem for elliptic systems of second order in the plane. *Bull. Polskoi Akad. Nauk Ser. Mat. Astr. Phys. Nauk* **7,** 565–570, 1959.
51 Bojarski, B. and Iwaniec, T., On systems of two second order elliptic equations with nonregular coefficients, to appear.
52 Borsuk, M. V., A-priori estimates of quasilinear elliptic equations with nonlinear boundary conditions. *Proc. Steklov Inst.* **103,** 13–52, 1968.
53 Browder, F. E., Existence theory for boundary value problems for

quasilinear elliptic systems with strongly nonlinear lower order terms. *Proc. Symp. Pure Math.* **13,** Am. Math. Soc., 1973.

54 Colton, D. and Gilbert, R. P., Rapidly convergent approximations to Dirichlet's problem for semilinear elliptic equations. *Appl. Analysis* **2,** 229–240, 1972.

55 Dshabrailov, A. D., A study of some classes of quasilinear elliptic equations of second order, I. *Differential Equations* **5,** 1683–1693, 1969.

56 Dshabrailov, A. D., An investigation of certain classes of second order quasilinear elliptic equations, II, III. *Differential Equations* **6,** 684–693 and 977–982, 1970.

57 Dshabrailov, A. D., Estimates, close to the boundary, of generalized solutions of quasilinear elliptic equations. *Differential Equations* **7,** 1108–1117, 1971.

58 Elcrat, A. R., Constructive existence for semilinear elliptic equations with discountinuous coefficients. *SIAM J. Math. Anal.*, to appear.

59 Fiorenza, R., Sui problemi di derivate oblique per le equationes elliptiques quasi lineari. *Ricerche di Mat. Napoli* **15,** 74–186, 1964.

60 Frehse, J., On the boundedness of weak solutions of higher order nonlinear elliptic partial differential equations. *Boll. Un. Mat. Italiana* **4,** 607–627, 1970.

61 Gariepy, R. and Ziemer, W. P., A gradient estimate at the boundary for solutions of quasilinear elliptic problems. *Bull. Am. Math. Soc.* **82,** 629–631, 1976.

62 Gossez, J. P., Boundary value problems for quasilinear elliptic equations with rapidly increasing coefficients. *Bull. Am. Math. Soc.* **78,** 753–758, 1972.

63 Ivanov, A. V., Interior estimates of the first derivatives of the solutions of quasilinear elliptic systems. *Sem. Math.* **14,** 9–21, 1971.

64 Ivanov, A. V., Structure of quasilinear second order elliptic equations. *J. Sov. Math.* **1,** 393–415, 1973.

65 Ivanov, A. V., On the Dirichlet problem for quasilinear nonuniformly elliptic equations of second order. *Proc. Steklov Inst.* **116,** 29–51, 1971.

66 Ivanov, A. V., On the question of admissible limiting growth for quasilinear elliptic equations. *Proc. Steklov Inst.* **125,** 80–86, 1973.

67 Ivančov, N. I., Über die Lösbarkeit des Dirichlet-Problems für quasilineare elliptische Gleichungen in Funktionenklassen mit vorgegebenem Verhalten im Unendlichen. *Izvestija vysš učebn. Zaved., Mat.* **11** (114), 72–77, 1971.

68 Ivočkina, N. M., The Dirichlet problem for two dimensional quasilinear second order elliptic equations. *Probl. Math. Analysis* **2,** 115–126, 1971.

69 Ivočkina, N. M., A-priori estimates for solutions of the Dirichlet problem for multidimensional quasilinear elliptic equations. *Proc. Steklov Inst.* **125,** 87–97, 1973.

70 Ivočkina N. M. and Oskolkov, A. P., Nonlocal estimates of the first derivatives of solutions to the Dirichlet problem for nonuniformly elliptic equations. *Sem. Math.* **5,** 12–34, 1969.

71 Ivočkina, N. M. and Oskolkov, A. P., Nonlocal estimates of the first

derivatives of the solutions of nonuniformly elliptic equations. *Sem. Math.* **11,** 1–25, 1970.

72 Ivočkina, N. M. and Oskolkov, A. P., Nonlocal estimates for certain classes of nonuniformly elliptic systems. *Proc. Steklov Inst.* **110,** 72–115, 1970.

73 Jakovlev, G. N., On the solution of a class of second order quasi-linear elliptic equations. *Sov. Math. Dokl.* **13,** 255–258, 1972.

74 Jakovlev, G. N., Properties of the solution of a certain class of quasilinear second order elliptic equations in divergence form. *Trudy Mat. Inst. Steklov* **131,** 232–242, 48, 1974.

75 Kačur, J., On boundedness of the weak solution for some quasilinear partial differential equations. *Časopis Pešt. Math.* **98,** 43–55, 1973.

76 Kirchgässner, K. and Scheurle, J., Verzweigung und Stabilität von Lösungen semilinearer elliptischer Randwertprobleme, to appear.

77 Komkov, V., Certain estimates for solutions of nonlinear elliptic differential equations applicable to the theory of thin plates. *SIAM J. Appl. Math.* **28,** 24–34, 1975.

78 Kostenko, V. G. and Sarii, T. A., The investigation of boundary value problems for a certain quasilinear equation. *Visnik L'viv. Derz. Univ. Ser. Meh.-Mat.* **7,** 22–28, 132, 1972.

79 Kuiper, H. J., Eigenvalue problems for noncontinuous operators associated with quasilinear elliptic equations. *Arch. Rat. Mech. Anal.* **53,** 178–186, 1974.

80 Ladyshenskaja, O. A., Quasilinear nonuniformly elliptic equations. Elliptische Differentialgleichungen II. *Schriftenreihe Inst. Math. Dt. Akad. Wiss. Berlin,* Reihe A **8,** 167–175, 1971.

81 Ladyshenskaja, O. A. and Ural'tseva, N. N., Certain classes of nonuniformly elliptic equations. *Sem. Math.* **5,** 67–69, 1969.

82 Ladyshenskaja, O. A. and Ural'tseva, N. N., On some classes of nonuniformly elliptic equations. *Sem. Math.* **11,** 47–53, 1970.

83 Ladyshenskaja, O. A. and Ural'tseva, N. N., Local estimates for gradients of solutions of nonuniformly elliptic equations. *Comm. Pure Appl. Math.* **23,** 687–703, 1970.

84 Ladyshenskaja, O. A. and Ural'tseva, N. N., Global estimates of the solutions of quasilinear elliptic equations. *Sem. Math.* **14,** 63–77, 1971.

85 Marsden, J. E. and McCracken, N., *The Hopf bifurcation and its applications.* Springer, New York, Heidelberg, Berlin, 1976.

86 Naas, J. and Tutschke, W., Some probabilistic aspects in partial complex differential equations, 1977, to appear.

87 Nirenberg, L., On nonlinear elliptic partial differential equations and Hölder continuity. *Comm. Pure Appl. Math.* **6,** 103–156, 1953.

88 Oskolkov, A. P., A-priori estimates for two dimensional quasilinear strongly elliptic systems. *Proc. Steklov Inst.* **92,** 219–232, 1966.

89 Oskolkov, A. P. Some estimates for nonuniformly elliptic equations and systems. *Proc. Steklov Inst.* **92,** 233–267, 1966.

90 Oskolkov, A. P., A-priori estimates of the first derivatives of solutions of the Dirichlet problem for nonuniformly elliptic equations. *Proc. Steklov Inst.* **102,** 119–144, 1967.

91 Oskolkov, A. P., Solvability of the Dirichlet problem for quasilinear elliptic equations in unbounded domains, I, II. *Proc. Steklov Inst.* **102,** 145–155,1967; and *Sem. Math.* **14,** 88–96, 1971.
92 Oskolkov, A. P., Some classes of quasilinear nonuniformly elliptic equations, I, II. *Sem. Math.* **14,** 78–87, 1971; and *Proc. Steklov Inst.* **116,** 140–155, 1971.
93 Oskolkov, A. P. Solvability of the Dirichlet problem for quasilinear elliptic systems. *J. Soviet Math.* **1,** 471–476, 1973.
94 Oskolkov, A. P. and Tarasov, V. A., A-priori estimates for weighted first derivatives for certain classes of quasilinear nonuniformly elliptic equations in an unbounded domain. *Proc. Steklov Inst.* **116,** 156–166, 1971.
95 Parenti, C., Tracce delle soluzioni di certe equazioni ellitiche. *Boll. Un. Mat. Ital.* **4,** 968–984, 1971.
96 Piccinini, L. C. and Spagnolo, S., On the Hölder continuity of solutions of second order elliptic equations in two variables. *Ann. Scuola Norm. Sup. Pisa* (3) **26,** 391–402, 1972.
97 Pogorzelski, W., Problème aux dérivées tangentielles discontinues pour une équation elliptique. *Ann. Polon. Math.* **13,** 3–56, 1963.
98 Sadowska, D., Problème aux limites aux dérivées tangentielles pour l'equation elliptique dont les coefficients dépendent d'une fonction inconnue. *Ann. Polon. Math.* **10,** 7–33, 1961.
99 Serrin, J., The problem of Dirichlet for quasilinear elliptic equations with many independent variables. *Roy. Soc. Phil. Trans.* **264,** 413–496, 1969.
100 Simon, L., Global estimates of Hölder-continuity for a class of divergence from elliptic equations. *Arch. Rat. Mech. Anal.* **56,** 253–272, 1975.
101 Siukae, S. N., A-priori estimates for the solutions of quasilinear differential equations of elliptic type. *Soviet Math. Dokl.* **14,** 1262–1265, 1973.
102 Skrypnik, I. V., A regularity condition for generalized solutions of higher-order quasilinear elliptic equations. *Math. USSR, Izvestija* **7,** 1371–1421, 1973; *Orig. Izv. Akad. Nauk SSSR* **35,** 1973.
103 Solomyak, T. B., Dirichlet problem for nonuniformly elliptic quasilinear equations. *Probl. Math. Analysis* **1,** 81–101, 1968.
104 Sperner, E., Zur Regularitätstheorie gewisser Systeme nichtlinearer elliptischer Differentialgleichungen. *Dissertation,* Bonn, 1975.
105 Szeptycki, P., Existence theorem for the first boundary problem for a quasilinear elliptic system. *Bull. Acad. Polon. Sér. Sci. math. astron. phys.* **7,** 419–424, 1959.
106 Tomi, F., Über semilineare elliptische Differentialgleichungen zweiter Ordnung. *Math. Z.* **111,** 350–366, 1969.
107 Tomi, F., Ein einfacher Beweis eines Regularitätssatzes für Lösungen gewisser elliptischer Systeme. *Math. Z.* **112,** 214–218, 1969.
108 Trudinger, N. S., Some existence theorems for quasilinear nonuniformly elliptic equations. *J. Math. Mech.* **18,** 909–920, 1969.
109 Trudinger, N. S., The boundary gradient estimates for quasilinear elliptic differential equations. *Indiana Univ. Math. J.* **21,** 657–670, 1972.
110 Tutschke, W., Die neuen Methoden der komplexen Analysis und ihre

Anwendung auf nichtlineare Differentialgleichungssysteme. *Sitz. Ber. Akad. Wiss. DDR* 1976, to appear.

111 Ural'tseva, N. N., Degenerate elliptic systems. *Math. Z.* **130,** 149–157, 1973.

112 v. Wahl, W., Über quasilineare elliptische Systeme in der Ebene. *Manuscripta Math.* **8,** 59–67, 1973.

113 v. Wahl, W., Innere Abschätzungen für nichtlineare elliptische Gleichungen und Systeme in der Ebene. *Manuscripta Math.* **9,** 375–382, 1973.

114 v. Wahl, W., Über die Hölderstetigkeit der zweiten Ableitungen der Lösungen nichtlinearer elliptischer Gleichungen. *Math. Z.* **136,** 151–162, 1974.

115 v. Wahl, W., Über semilineare elliptische Gleichungen mit monotoner Nichtlinearität. *Nachr. Akad. Wiss. Göttingen*, II. Math.-Phys. Klasse 1975 Nr. 3, 27–33.

116 Wiegner, M., Ein optimaler Regularitätssatz für schwache Lösungen gewisser elliptischer Systeme. *Math. Z.* **147,** 21–28, 1976.

5

Singular Integral Equations, Representation Formulas and some Function Theoretic Properties

This chapter is devoted to singular integral equation methods, to some problems in connection with zeros of the solution, Carleman's theorem and unique continuation and to a representation formula providing a Bergman–Vekua operator which maps analytic vectors onto solutions of elliptic systems with analytic coefficients.

We begin, in Section 5.1 with singular integral equations for systems in normal form. This is one of the explicit methods for the practical treatment of boundary value problems. It is based on Cauchy integrals for holomorphic functions and leads to systems of integral equations consisting of weakly singular Fredholm integral equations over the domain G and singular integral equations with Cauchy kernel over the boundary curve $\dot{\mathsf{G}}$. In an Appendix we apply these equations to boundary value problems with violated Lopatinski condition which have already been presented in Section 3.5.

In Section 5.2 the above method is refined by the use of Bojarski's fundamental solution [8], [9] and its specialized version for generalized hyperanalytic systems of Gilbert [22] and Gilbert and Hile [24]. Another specialization shows the connection with an approach introduced by Fichera [16] and investigated by Fichera and Ricci [17], [59]. Related results have been obtained independently by Tovmasjan [62], [63].

In all these cases we find generalized Cauchy formulas and representations of the solution which can be used for its construction. Following Fichera and Ricci, we find for a certain class of boundary conditions first kind integral equations having a logarithmic principal part. These integral equations have already found some interest for practical treatment (see e.g., [39], [40], [18], [60] and references given there).

It should be pointed out that these methods can be applied to many other problems which we do not present here but which are of much interest. For example, Mme. Wolska-Bochenek [70], [71] investigated boundary value problems for generalized analytic functions and nonlinear boundary conditions which can surely be extended to systems with $n > 1$.

All our investigations assume smoothness properties for $\dot{\mathsf{G}}$. For Pascali systems and the related singular integral equations with the classical Cauchy kernel Daniljuk [13] has already developed the complete theory in $C^\alpha(\dot{\mathsf{G}})$ and $L^p(\dot{\mathsf{G}})$ for a wide class of irregular boundary curves and it should be not too difficult to generalize the application of his theory to integral equations for more general systems with $n > 1$. Another kind of problem arises if the boundary conditions change discontinuously on parts of $\dot{\mathsf{G}}$, the so called 'mixed boundary conditions'. For the Laplacian these problems with many applications have already been rather thoroughly investigated (see e.g., the survey [30] and the important work by Kondratiev [42]). For the more general equations and boundary conditions of our presentation one can find equivalent singular integral equations having discontinuous coefficients in the principal part. For such singular integral equations the solvability theory is already available in the development of Gakhov, Gohberg and Krupnik (see [27]). Hence it should be possible to investigate mixed boundary value problems in the plane with these methods; but this is yet to be done. In connection with these results there are recent results available on such singular integral equations of the non-normal type obtained by Schüppel [61]. Hence mixed boundary value problems with violated Lopatinskii condition could be investigated. For an introduction to singular integral equations and a survey of some recent developments for this purpose see Meister's forthcoming work [48] and for the singular integral equations with discontinuous coefficients the forthcoming book by Gohberg and Krupnik [26].

In Section 5.3 we present results on the similarity principle, the zeros of the solution vector, and unique continuation for elliptic systems with $n > 1$. The similarity principle only holds in a rather special generalization of the case $n = 1$ for a certain class of generalized hyperanalytic systems. For these a treatment of boundary value problems is possible as in Chapter 1 due to Begehr and Gilbert [3] and in the case of some semilinear equations treated by Gilbert [23] and Wendland [69]. For Carleman's theorem on the zeros we present results by Habetha [34] and Kühn [44] without proof which show its validity only for generalized hyperanalytic systems and Pascali systems. As shown by Hile and Protter [37], unique continuation holds in addition for elliptic systems with normal coefficient matrix B. We conclude this section with the Runge approximation property which has been proved yet only for purely hypercomplex generalized hyperanalytic systems by Goldschmidt [28]; there also the Cousin problem is solved which serves as a generalization of analytic continuation to these systems.

In Section 5.4 we collect some special problems for the analytic theory for elliptic systems. We begin with Carleman's theorem following

Habetha [34]. Then we present the generalization of Vekua's representation formula [65, Section 15] to generalized hyperanalytic systems with analytic coefficients following Gilbert and Wendland [25]. The key idea is to extend the equations analytically into some domain of the $2n$ hypercomplex valued space where the system is treated as a hyperbolic system in two independent variables. We arrive at a representation formula mapping holomorphic vectors onto the solutions. This Bergman–Vekua operator can be used for the constructive generation of solutions of systems in order to approximate, e.g., solutions of boundary value problems. These representations can also be used for proving Lewy's reflection principle generalizing some of the results of Chung Ling Yu [72] from $n=1$ to generalized hyperanalytic systems with $n>1$. As a corollary we also present a proof for Carleman's theorem of Section 5.3. These are only a few topics connected with recent results in the general analytic theory which is rather far developed, see for instance the books by Vekua [65], Garabedian [20], Gilbert [21], [22] and Colton [11], [12]. The analytic theory has many applications but it would go beyond the limits of this chapter if we would try to present this theory completely.

5.1 Singular Integral Equations for Systems in Normal Form

The normal form of the first order system allows a direct application of Vekua's representation of holomorphic functions by means of Cauchy integrals with real layers to these rather general systems. This leads to singular integral equations for the layers on the boundary and coupled equations over the domain which can be used for finding the solution of the original boundary value problem. These equations were used by Kopp [43] and Wendland for the boundary value problems with violated Lopatinski condition (see Section 3.5).

Vekua's representation theorem can be formulated as follows:

Lemma 5.1.1 [52, p. 172, p. 179]: *Every holomorphic function* Ψ *in* G *with* $\Psi \in C^{\alpha}(\bar{\mathrm{G}})$ *can be represented by*

$$\Psi(\zeta)=\frac{1}{2\pi\mathrm{i}}\oint_{\dot{\mathrm{G}}}\frac{\nu(t)\,\mathrm{d}t}{t-\zeta}+\mathrm{i}k,\ \zeta\in\mathrm{G}, \tag{5.1.1}$$

where the boundary layer $\nu\in C^{\alpha}(\dot{\mathrm{G}})$ *is real valued and* k *is a real constant.*

Proof To any real layer ν the right hand side in (5.1.1) defines a holomorphic function in G belonging to $C^{\alpha}(\bar{\mathrm{G}})$. Its harmonic real part U

has the boundary values

$$U|_{\dot{G}} = \tfrac{1}{2}\nu(\zeta) + \frac{1}{4\pi i}\oint_{\dot{G}} \nu(t)\left\{\frac{dt}{t-\zeta} - \frac{d\bar{t}}{\bar{t}-\bar{\zeta}}\right\} \tag{5.1.2}$$

according to the Plemelj formulae ([52, p. 42]). The integral in (5.1.2) is a Cauchy principal value. But the special combination appearing there is just

$$\frac{dt}{t-\zeta} - \frac{d\bar{t}}{\bar{t}-\bar{\zeta}}\bigg|_{\dot{G}} = 2i\, d_{n_t} \log|t-\zeta|\bigg|_{\dot{G}}, \tag{5.1.3}$$

which is a multiple of the kernel belonging to the classical double layer potential and which is weakly singular for $\dot{G} \in C^{1+\alpha}$ and Hölder continuous for $\dot{G} \in C^{2+\alpha}$.

Hence, for the representation (5.1.1) we demand $U = \text{Re}\ \Psi$ on the boundary and the desired density ν must satisfy the integral equation

$$2\text{Re}\ \Psi = \nu + \frac{1}{\pi}\oint_{\dot{G}} \nu\, d_{n_t} \log|t-\zeta| \quad \text{for } \zeta \in \dot{G}. \tag{5.1.4}$$

This is the well known integral equation determining the double layer ν which generates the harmonic solution of the Dirichlet problem. Equation (5.1.4) has exactly one solution ν to any given $\text{Re}\ \Psi|_{\dot{G}}$ [33, Section 4.11]. Since the kernel is weakly singular ν becomes $C^\alpha(\dot{G})$ if $\text{Re}\ \Psi \in C^\alpha(\dot{G})$. With this ν the real part of the Cauchy integral in (5.1.1) has the same boundary values as Re Ψ. Hence, the integral differs from Ψ at most by an imaginary constant.

Corollary 5.1.2 Every function $\Phi \in C^\alpha(\bar{G})$ with $\Phi_{\bar{z}} \in C^\alpha(\bar{G})$ can be represented by

$$\Phi(\zeta) = \frac{1}{2\pi i}\iint_G \Phi_{\bar{t}} \frac{1}{t-\zeta}[dt, d\bar{t}] + \frac{1}{2\pi i}\oint_{\dot{G}} \frac{\nu(t)}{t-\zeta}\, dt + ik, \quad \zeta \in G, \tag{5.1.5}$$

where $\nu \in C^\alpha$ is a suitable real valued boundary layer and k is a suitable real constant.

This Corollary follows immediately from Lemma 5.1.1 by using (2.4.3) for the surface integral in (5.1.5) (see also Theorem 1.1.1 (ii) and its proof).

The Corollary enables us to represent the complex solution $\mathbf{w}$ of the system (3.1.30) in normal form,

$$w_{j\bar{\zeta}_j} + \sum_{k\neq j} \beta_{jk} w_{k\bar{\zeta}_j}$$

$$= -\sum_{k\neq j} \alpha_{jk} w_{k\zeta_j} - \mathbf{a}_j^T \mathbf{w} - \mathbf{b}_j^T \bar{\mathbf{w}} - c_j. \tag{5.1.6}$$

From the construction (3.1.23) it is clear that $\alpha_{jk}(z) \neq 0$ only where $\zeta_j = \zeta_k$. Moreover, the matrices

$$\underset{\sim}{\beta} \equiv ((\beta_{jk})), \qquad \underset{\sim}{\alpha} \equiv ((\alpha_{jk})) \tag{5.1.7}$$

at every point are both the same rearrangements of triangular matrices with *zero diagonals*. Hence we have

$$(\mathsf{I} + \underset{\sim}{\beta})^{-1} = \mathsf{I} + \sum_{l=1}^{n-1} (-1)^l \underset{\sim}{\beta}^l, \qquad \beta^n = 0 \tag{5.1.8}$$

and the $\underset{\sim}{\beta}^l$ are reordered triangular matrices of the same shape as $\underset{\sim}{\alpha}$. Therefore multiplication of (5.1.6) with (5.1.8) yields

$$w_{j\bar{\zeta}_j} = -\sum_{k \neq j} (\gamma_{jk} w_{k\zeta_k} + \mathbf{a}_j^T \mathbf{w} + \mathbf{b}_j^T \bar{\mathbf{w}}) - c_j, \qquad j = 1, \ldots, n \tag{5.1.9}$$

where the matrix valued function

$$\gamma \equiv ((\gamma_{jk}(x, y))) \tag{5.1.10}$$

is of the same shape as $\underset{\sim}{\alpha}$ and $\underset{\sim}{\beta}$ with

$$\gamma^n = 0. \tag{5.1.11}$$

Now we use the representation formula (5.1.5) for w_j in the transformed ζ_j-domain and insert (5.1.9) obtaining

$$w_j(z) = \frac{-1}{2\pi i} \iint_{\mathsf{G}} \frac{1}{\zeta_j(t) - \zeta_j(z)} \left\{ \sum_{k \neq j} \gamma_{jk}(t) w_{k\zeta_k} + \underset{\sim}{\mathbf{a}}_j \mathbf{w} + \underset{\sim}{\mathbf{b}}_j \bar{\mathbf{w}} + c_j \right\} [d\zeta_j(t), d\bar{\zeta}_j(t)]$$

$$+ \frac{1}{2\pi i} \oint_{\dot{\mathsf{G}}} \frac{\nu_j \, d\zeta_j(t)}{\zeta_j(t) - \zeta_j(z)} + ik_j, \quad j = 1, \ldots, n. \tag{5.1.12}$$

These equations can be used for converting the boundary value problems of Chapter 3 into a system of integral equations for ν_j, k_j and also the desired w_j. Let us consider first:

Case $\gamma = 0$ The boundary conditions (3.1.36) are

$$\sum_{k=1}^{n} (P_{jk} w_k + \bar{P}_{jk} \bar{w}_k) = 2\psi_j \quad \text{on} \quad \dot{\mathsf{G}}, \tag{5.1.13}$$

to which we may apply the boundary values of (5.1.12). They follow from the Plemelj formula [52, p. 42]:

$$w_k(z) = -\frac{1}{2\pi i}\iint_G \frac{1}{\zeta_k(t)-\zeta_k(z)} \sum_{l=1}^{n} (a_{kl}w_l + b_{kl}\bar{w}_l)[d\zeta_k(t), d\bar{\zeta}_k(t)]$$

$$+\tfrac{1}{2}\nu_k(z) + \frac{1}{2\pi i}\oint_{\dot{G}} \frac{\nu_k \, d\zeta_k(t)}{\zeta_k(t)-\zeta_k(z)} + ik_k \quad \text{for} \quad z \in \dot{G}. \tag{5.1.14}$$

Hence, on the boundary we have the equations

$$\sum_{k=1}^{n} \tfrac{1}{2}(P_{jk}+\bar{P}_{jk})\nu_k(z)$$

$$+\frac{1}{2\pi i}\sum_{k=1}^{n}\left\{P_{jk}(z)\oint_{\dot{G}} \frac{\nu_k(t)\, d\zeta_k(t)}{\zeta_k(t)-\zeta_k(z)} - \bar{P}_{jk}(z)\oint_{\dot{G}} \frac{\nu_k(t)\, d\bar{\zeta}_k(t)}{\overline{\zeta_k(t)}-\overline{\zeta_k(z)}}\right\}$$

$$+i\sum_{k=1}^{n}(P_{jk}(z)-\bar{P}_{jk}(z))k_k$$

$$-\frac{1}{2\pi i}\sum_{k=1}^{n}\left\{P_{jk}(z)\iint_G \frac{1}{\zeta_k(t)-\zeta_k(z)}\sum_{l=1}^{n}(a_{kl}w_l + b_{kl}\bar{w}_l)[d\zeta_k(t), d\bar{\zeta}_k(t)]\right.$$

$$\left.+\bar{P}_{jk}(z)\iint_G \frac{1}{\overline{\zeta_k(t)}-\overline{\zeta_k(z)}}\sum_{l=1}^{n}(\overline{a_{kl}}\overline{w_l} + \overline{b_{kl}}w_l)[d\zeta_k(t), d\bar{\zeta}_k(t)]\right\}$$

$$= 2\psi_j(z) \quad \text{for} \quad z \in \dot{G}, \quad j = 1, \ldots, n. \tag{5.1.15}$$

If we replace $d\bar{\zeta}(t)(\bar{\zeta}(t)-\bar{\zeta}(z))^{-1}$ by the corresponding formula to (5.1.3) then the first part becomes an operator belonging to an 'elliptic' system of singular integral equations and we find the system of equations

$$\sum_{k=1}^{n}\left\{\tfrac{1}{2}(P_{jk}(z)+\bar{P}_{jk}(z))\nu_k(z)\right.$$

$$+\tfrac{1}{2}(P_{jk}(z)-\bar{P}_{jk}(z))\frac{1}{\pi i}\oint_{\dot{G}}\frac{\nu_k(t)\, d\zeta_k(t)}{\zeta_k(t)-\zeta_k(z)}$$

$$+\bar{P}_{jk}(z)\frac{1}{\pi}\oint_{\dot{G}}\nu_k(t)d_{n_{\zeta_k(t)}}\log|\zeta_k(t)-\zeta_k(z)|$$

$$+i(P_{jk}(z)-\bar{P}_{jk}(z))k_k$$

$$\left.-P_{jk}(z)T_k(\mathbf{w})-\bar{P}_{jk}(z)\bar{T}_k(\mathbf{w})\right\} = 2\psi_j(z) \quad \text{for } z \in \dot{G},\ j = 1, \ldots, n \tag{5.1.16}$$

where $d_{n_{\zeta_k}}$ denotes the operator (1.1.8) defined on the image of $\dot{G}$ in the ζ_k plane and where the T_k are defined from (5.1.15). The equations (5.1.12) with $\gamma=0$,

$$w_j(z)+\frac{1}{2\pi i}\sum_{k=1}^{n}\iint_G \frac{1}{\zeta_j(t)-\zeta_j(z)}(a_{jk}w_k+b_{jk}\bar{w}_k)[d\zeta_j(t),\overline{d\zeta_j(t)}]$$

$$-\frac{1}{2\pi i}\oint_{\dot{G}}\frac{\nu_j\, d\zeta_j(t)}{\zeta_j(t)-\zeta_j(z)}-ik_j$$

$$=\frac{-1}{2\pi i}\iint_G \frac{c_j(t)[d\zeta_j, d\bar{\zeta}_j]}{\zeta_j(t)-\zeta_j(z)}\quad \text{for}\quad z\in G \tag{5.1.17}$$

and (5.1.16) together form a system of integral equations for $w_1,\ldots,w_n$; $\nu_1,\ldots,\nu_n$; $k_1,\ldots,k_n$. Considering $\mathbf{w}\in C^\alpha(\bar{G})$ and $\boldsymbol{\nu}=(\nu_1,\ldots,\nu_n)\in C^\alpha(\dot{G})$, the operators in (5.1.16) and (5.1.17) have the following properties: the operators T_k in (5.1.16) defined by the surface integrals in (5.1.15) have weakly singular kernels mapping $C^\alpha(\bar{G})$ continuously into $C^{1+\alpha}(\bar{G})$ [64, (8.25)] and also into $C^{1+\alpha}(\dot{G})$ which is compactly embedded into $C^\alpha(\dot{G})$. Therefore the coupling of $\boldsymbol{\nu}$ and $\mathbf{w}$ in (5.1.16) is *compact* and the principal part in these equations is the system of singular integral equations on $\dot{G}$ *having the fundamental matrices* P *and* $\bar{P}$ [52, (119, 10)].

Hence, the Lopatinski–Shapiro condition (3.4.3) is equivalent to the normality of the singular integral equations (5.1.16). The equations (5.1.17) are Fredholm integral equations of the second kind for the desired solution $\mathbf{w}$. Since the surface integrals define completely continuous operators $C^\alpha(\bar{G})\to C^\alpha(\bar{G})$ these equations do not influence the Fredholm properties of the whole system (5.1.16) and (5.1.17)

Let us collect these results in the following:

Proposition 5.1.3 *If the Lopatinski–Shapiro condition* $\det P\neq 0$ *is fullfilled then the boundary value problem* (5.1.9) *with* $\gamma=0$ *and* (5.1.13) *can be found by solving the system of normal type integral equations* (5.1.16), (5.1.17) *for* $\boldsymbol{\nu}$, $\mathbf{k}$ *and* $\mathbf{w}$.

For the solvability of such systems see Vekua [66].

In the case of violated Lopatinski condition the integral equations (5.1.16) and (5.1.17) form the basic tool of the investigations by Kopp [43]. Let us return first to the more general:

Case $\gamma \neq 0$ The boundary conditions (5.1.13) yield now the equations

$$\sum_{k=1}^{n} \Big\{ \tfrac{1}{2}(P_{jk}(z)+\bar{P}_{jk}(z))\nu_k(z) + \tfrac{1}{2}(P_{jk}(z)-\bar{P}_{jk}(z)) \frac{1}{\pi i} \oint_{\dot{G}} \frac{\nu_k(t)\, d\zeta_k(t)}{\zeta_k(t)-\zeta_k(z)}$$

$$+\bar{P}_{jk}(z)\frac{1}{\pi}\oint_{\dot{G}} \nu_k(t)\, d_{n_{\zeta k(t)}} \log|\zeta_k(t)-\zeta(z)|$$

$$+i(P_{jk}(z)-\bar{P}_{jk}(z))k_k$$

$$-2\mathrm{Re}\left(P_{jk}(z)\frac{1}{2\pi i}\iint_{G} \frac{1}{\zeta_k(t)-\zeta_k(z)}\Big(\sum_l(\gamma_{kl}(t)\rho_l(t)\right.$$

$$\left.+\underset{\sim}{a}_{kl}w_l+b_{kl}\bar{w})+c_k\Big)[d\zeta_k(t), d\bar{\zeta}_k(t)]\Big)\Big\}$$

$$=2\psi_j(z) \quad \text{for} \quad z\in\dot{G}, \quad j=1,\ldots,n \tag{5.1.18}$$

where $\rho_l \equiv w_{l_{\bar{\zeta}_l}}$. To this system of singular integral equations on $\dot{G}$ let us add the equations (5.1.12),

$$w_j(z) = \frac{-1}{2\pi i}\iint_{G} \frac{1}{\zeta_j(t)-\zeta_j(z)}\Big\{\sum_{k=1}^{n}(\gamma_{jk}(t)\rho_k(t)$$

$$+\underset{\sim}{a}_{jk}w_k+\underset{\sim}{b}_{jk}\overline{w_k})+c_j\Big\}[d\zeta_j(t), \overline{d\zeta_j(t)}]$$

$$+\frac{1}{2\pi i}\oint_{\dot{G}} \frac{\nu_j\, d\zeta_j(t)}{\zeta_j(t)-\zeta_j(z)}+ik_j, \quad j=1,\ldots,n. \tag{5.1.19}$$

For the ρ_l we also add the derivatives of (5.1.19),

$$\rho_j(z) = \frac{-1}{2\pi i}\iint_{G} \frac{1}{(\zeta_j(t)-\zeta_j(z))^2}\Big\{\sum_{k=1}^{n}(\gamma_{jk}(t)\rho_k(t)$$

$$+\underset{\sim}{a}_{jk}w_k+\underset{\sim}{b}_{jk}\bar{w}_k)+c_j\Big\}[d\zeta_j(t), \overline{d\zeta_j(t)}]$$

$$+\frac{1}{2\pi i}\oint_{\dot{G}} \frac{\nu_j(t)\, d\zeta_j(t)}{(\zeta_j(t)-\zeta_j(z))^2}, \quad z\in G. \tag{5.1.20}$$

Proposition 5.1.4 *The system of integral equations* (5.1.18), (5.1.19), (5.1.20) *for* $(\boldsymbol{\nu}, \mathbf{k}, \mathbf{w}, \boldsymbol{\rho}) \in C^{1+\alpha}(\dot{G})\times R^n \times C^{\beta}(\bar{G})\times C^{\alpha}(\bar{G})$ *is of normal type if the Lopatinski–Shapiro condition is satisfied and provided that* $\dot{G}\in C^{2+\alpha}$, $\underset{\sim}{\alpha}$, $\underset{\sim}{\beta}\in C^{1+\alpha}(\bar{G})$, a, b, $\mathbf{c}\in C^{\beta}(\bar{G})$, P, $\boldsymbol{\psi}\in C^{1+\alpha}(\dot{G})$, $0<\alpha<\beta<1$.

Proof The proof hinges on the special triangular shape of γ in (5.1.20) and (5.1.18).

First we observe that the commutators

$$\gamma_{jk}\prod_j - \prod_j \gamma_{jk} \quad \text{where} \quad \prod_j \equiv \frac{1}{2\pi i}\iint_{\mathsf{G}} \frac{[d\zeta_j, d\bar{\zeta}_j]}{(\zeta_j(t)-\zeta_j(z))^2} \tag{5.1.21}$$

have weakly singular kernels defining completely continuous operators in $C^\alpha(\bar{\mathsf{G}})$. Therefore the finite Neumann series

$$\mathsf{I} + \sum_{\mu=1}^{n-1} \left(\left(\gamma_{jk}\prod_j\right)\right)^\mu \tag{5.1.22}$$

defines left and right regularizers of the equations (5.1.20) for $\boldsymbol{\rho}$. This allows one to solve (5.1.20) formally for $\boldsymbol{\rho}$. The result can be inserted into (5.1.18) eliminating $\boldsymbol{\rho}$. Then (5.1.18) becomes a system of singular integral equations for $\boldsymbol{\nu}$ alone with some completely continuous additional terms. After a long computation which also takes care of the composition of the boundary integral in (5.1.20) with the weakly singular integral operator in (5.1.18) and by using Plemelj's formula, one finally finds singular integral equations on $\dot{\mathsf{G}}$ possessing the characteristic matrices

$$\mathsf{P}(\mathsf{I}+\underset{\sim}{\gamma}) \quad \text{and} \quad \bar{\mathsf{P}}(\mathsf{I}+\underset{\sim}{\bar{\gamma}})$$

which are regular if and only if $\det \mathsf{P} \neq 0$.
Let us omit the detailed computations here.

Appendix *The singular integral equations assigned to problems with violated Lopatinski-condition in Section* 3.5.
For the system in the Pascali normal form,

$$\mathsf{R}\mathbf{w} = \begin{cases} \mathbf{w}_{\bar{z}} + \mathsf{a}\mathbf{w} + \mathsf{b}\bar{\mathbf{w}} = \mathbf{c} & \text{in } \mathsf{G}, \\ \operatorname{Re} \mathsf{P}\mathbf{w} = \boldsymbol{\psi} & \text{on } \dot{\mathsf{G}}, \end{cases} \tag{5.1.23}$$

the equations (5.1.12) reduce to

$$\mathbf{w}(z) + \frac{1}{2\pi \mathrm{i}} \iint_{\mathsf{G}} \frac{1}{t-z} \{\mathsf{a}\mathbf{w} + \mathsf{b}\bar{\mathbf{w}}\}(t)[dt, d\bar{t}]$$

$$= \mathbf{ik} + \frac{1}{2\pi \mathrm{i}} \oint_{\dot{\mathsf{G}}} \frac{\boldsymbol{\nu}(t)\, dt}{t-z} + \frac{1}{2\pi \mathrm{i}} \iint_{\mathsf{G}} \frac{1}{t-z} \mathbf{c}(t)[dt, d\bar{t}]. \tag{5.1.24}$$

If we consider equation (5.1.24) as an integral equation for $\mathbf{w}$ with given right hand side this equation might have *eigensolutions*. Therefore let us add to the above operator a finite dimensional operator with holomorphic image in order to find an invertible sum. To this end let us use the scalar product

$$\langle \mathbf{w}, \mathbf{u} \rangle \equiv -\frac{1}{2\mathrm{i}} \iint_{\mathsf{G}} (\mathbf{w}(z)^T \mathbf{h}(z) + \bar{\mathbf{w}}(z)^T \bar{\mathbf{h}}(z))[dz, d\bar{z}] \tag{5.1.25}$$

over the complex valued functions establishing a linear space over the *reals* and let us prove the following:

Lemma 5.1.5 *Let* $\mathbf{h}$ *be a solution of the homogeneous system of integral equations adjoint to* (5.1.24) *and let* $\mathbf{h}$ *satisfy*

$$\langle \mathbf{h}, \boldsymbol{\phi} \rangle = 0 \tag{5.1.26}$$

for all vectors $\boldsymbol{\phi} \in C^{m+\alpha}(\bar{\mathsf{G}})$ *holomorphic in* G *where* $m \geq 0$ *is any fixed integer. Then* $\mathbf{h}$ *vanishes identically.*

Proof From (5.1.24) and (5.1.25) after changing order of integrations and reordering the terms we find for $\mathbf{h}$ the integral equation adjoint to (5.1.24) with respect to (5.1.25):

$$\mathbf{h}(t) = \frac{1}{2\pi \mathrm{i}} \mathsf{a}^T(t) \iint_{\mathsf{G}} \frac{\mathbf{h}(z)}{z-t}[dz, d\bar{z}] + \frac{1}{2\pi \mathrm{i}} \bar{\mathsf{b}}^T(t) \iint_{\mathsf{G}} \frac{\bar{\mathbf{h}}(z)}{\bar{z}-\bar{t}}[dz, d\bar{z}] \tag{5.1.27}$$

Since both integral equations (5.1.24) and (5.1.27) have weakly singular kernels, Fredholm's classical alternative holds [49, Chap. 8] and it suffices to consider only continuous $\mathbf{h}$ which become automatically $C^\alpha(\bar{\mathsf{G}})$ due to (5.1.27). To $\mathbf{h}$ let us assign the potential.

$$\mathbf{v}(t) \equiv \frac{1}{2\pi \mathrm{i}} \iint_{\mathsf{G}} \frac{\mathbf{h}(z)}{z-t}[dz, d\bar{z}] \tag{5.1.28}$$

which is well defined in the whole plane and satisfies

$$\mathbf{v}_{\bar{t}} = \begin{cases} \mathbf{h}(t) = \mathsf{a}^T \mathbf{v} + \bar{\mathsf{b}}^T \bar{\mathbf{v}} & \text{for} \quad t \in \mathsf{G}, \\ 0 & \text{for} \quad t \notin \bar{\mathsf{G}}. \end{cases} \tag{5.1.29}$$

Moreover from (5.1.26) we conclude for $\boldsymbol{\phi}$ and $\mathrm{i}\boldsymbol{\phi}$, both holomorphic, that

$$\iint_{\mathsf{G}} \{\mathbf{h}(z)^T \boldsymbol{\phi}(z) + \bar{\mathbf{h}}(z)^T \bar{\boldsymbol{\phi}}(z)\} \frac{1}{2\mathrm{i}}[dz, d\bar{z}] = 0 \quad \text{and}$$

$$\mathrm{i} \iint_{\mathsf{G}} \{\mathbf{h}(z)^T \boldsymbol{\phi}(z) - \bar{\mathbf{h}}(z)^T \bar{\boldsymbol{\phi}}(z)\} \frac{1}{2\mathrm{i}}[dz, d\bar{z}] = 0$$

hold, implying

$$\iint_G \boldsymbol{\phi}^T(z)\mathbf{h}(z)[dz, d\bar{z}] = 0 \tag{5.1.30}$$

for all $\boldsymbol{\phi}$ holomorphic in G. Choosing $\boldsymbol{\phi}_j(z) \equiv 1/(z-t)((\delta_{jk}))$ with $t \notin \bar{G}$ we find from (5.1.28)

$$\mathbf{v}(t) \equiv 0 \quad \text{for all} \quad t \notin \bar{G}. \tag{5.1.31}$$

From $\mathbf{v} \in C^\alpha$ it follows by continuity that $\mathbf{v} = 0$ on $\dot{G}$ implying by continuity $\mathbf{h}(t) = 0$ on $\dot{G}$ and $\mathbf{v} \in C^{1+\alpha}$ in the plane.

Consequently, the pointwise inequality,

$$|\mathbf{v}_{\bar{t}}| \leq \alpha\, |\mathbf{v}| \quad , \quad \alpha \equiv \text{Max}\, \{\|a_{jk}\|_0, \|b_{jk}\|_0\}, \tag{5.1.32}$$

holds and

$$\mathbf{v} \equiv 0 \quad \text{for} \quad t \notin \bar{G}. \tag{5.1.33}$$

Thus the unique continuation property can be applied according to Hile and Protter [37] (see Corollary 5.3.8) yielding

$$\mathbf{v} \equiv 0 \quad \text{in the whole plane.} \tag{5.1.34}$$

Hence $\mathbf{h} \equiv 0$ from (5.1.29).

Now we are in the position to prove the following.

Theorem 5.1.6† *To* a, b $\in C^{m+\alpha-1}(\bar{G})$, $m \geq 1$ *and* G *there exist N vector functions* $\mathbf{g}_j \in C^\alpha(\bar{G})$ *and N holomorphic vectors* $\boldsymbol{\phi}_j \in C^{m+\alpha}(\bar{G})$ *such that the operator given by*

$$\mathsf{U}\mathbf{w} \equiv \mathbf{w}(z) + \frac{1}{2\pi i}\iint_G \frac{1}{t-z}\{\mathsf{a}\mathbf{w} + \mathsf{b}\bar{\mathbf{w}}\}(t)[dt, d\bar{t}] + \sum_{j=1}^{N} \boldsymbol{\phi}_j(z)\langle \mathbf{w}, \mathbf{g}_j\rangle \tag{5.1.35}$$

is continuously invertible in $C^{m'+\alpha}(\bar{G})$ *where* $0 \leq m' \leq m$. *Here N equals the dimension of the eigenspace for* (5.1.24).

Proof In the case that (5.1.24) has no eigensolutions the Theorem is trivial with $\mathbf{g}_j = \boldsymbol{\phi}_j = 0$ according to the classical Fredholm alternative.

In the case that (5.1.24) has N linearly independent eigensolutions $\mathbf{w}_1, \ldots, \mathbf{w}_N$ we choose a biorthogonal system $\mathbf{g}_j$ with

$$\langle \mathbf{w}_k, \mathbf{g}_j\rangle = \delta_{kj}, \qquad j, k = 1, \ldots, N. \tag{5.1.36}$$

† The idea for this theorem is due to Goldschmidt, oral communication.

Let $\mathbf{h}_1, \ldots, \mathbf{h}_N$ be a basis of the eigenspace for (5.1.27) which has same dimension N due to the classical Fredholm Theorem. To the first vector $\mathbf{h}_1$ there exists a holomorphic $\boldsymbol{\phi}_1$ with

$$\langle \boldsymbol{\phi}_1, \mathbf{h}_1 \rangle \neq 0$$

according to Lemma 5.1.5 since $\mathbf{h}_1 \neq 0$. Now let $\boldsymbol{\phi}_1, \ldots, \boldsymbol{\phi}_{n-1}$ be holomorphic satisfying

$$g \equiv \text{Det}\,(((\langle \boldsymbol{\phi}_j, \mathbf{h}_k \rangle))_{j,k=1,\ldots,n-1} \neq 0 \tag{5.1.37}$$

and let us consider for n linearly independent $\mathbf{h}_1, \ldots, \mathbf{h}_n$ the bigger determinant

$$\text{Det}\,(((\langle \boldsymbol{\phi}_j, \mathbf{h}_k \rangle))_{j,k=1,\ldots,n} = \langle \boldsymbol{\phi}_n, \mathbf{h} \rangle \tag{5.1.38}$$

evaluated by means of the Laplace development with respect to the nth column where

$$\mathbf{h} = (-1)^n \gamma \mathbf{h}_n + \sum_{l=1}^{n-1} (-1)^l \,\text{Det}\,(((\langle \boldsymbol{\phi}_j, \mathbf{h}_k \rangle))_{\substack{k \neq l \\ j \neq n}} \mathbf{h}_l \tag{5.1.39}$$

is an eigensolution of (5.1.27) and $\gamma \neq 0$. If (5.1.38) were identically zero for all holomorphic $\boldsymbol{\phi}_n$ then we would have $\mathbf{h} \equiv 0$ due to Lemma 5.1.5, contradicting the linear independence. Hence, there exists a holomorphic $\boldsymbol{\phi}_n \in C^{m+\alpha}(\bar{\mathsf{G}})$ such that

$$\text{Det}\,(((\langle \boldsymbol{\phi}_j, \mathbf{h}_k \rangle))_{j,k=1,\ldots,n} \neq 0. \tag{5.1.40}$$

Induction with $n = 1, \ldots, N$ yields the existence of N holomorphic $\boldsymbol{\phi}_1, \ldots, \boldsymbol{\phi}_N$ such that

$$\text{Det}\,(((\langle \boldsymbol{\phi}_j, \mathbf{h}_k \rangle))_{j,k=1,\ldots,N} \neq 0. \tag{5.1.41}$$

Using these functions in (5.1.35) we see that U has no eigensolutions since

$$\mathsf{U}\mathbf{v} = 0 \quad \text{with} \quad \mathbf{v} = \sum_{k=1}^{N} \kappa_k \mathbf{w}_k$$

would imply with (5.1.27)

$$\langle \mathsf{U}\mathbf{v}, \mathbf{h}_l \rangle = 0 + \sum_{j=1}^{N} \langle \boldsymbol{\phi}_j, \mathbf{h}_l \rangle \langle \mathbf{v}, \mathbf{g}_j \rangle = 0$$

for $l = 1, \ldots, N$, and with (5.1.41) and (5.1.36)

$$\langle \mathbf{v}, \mathbf{g}_j \rangle = \sum_{k=1}^{N} \kappa_k \langle \mathbf{w}_k, \mathbf{g}_j \rangle = \kappa_j = 0 \quad \text{for} \quad j = 1, \ldots, N$$

i.e. $\mathbf{v} \equiv 0$. Hence U is an injective Fredholm mapping of index zero from

$C^{m+\alpha}(\bar{\mathsf{G}})$ into itself possessing a continuous inverse due to the classical Fredholm alternative [49, Chap. 8].

Taking into account that we alter the holomorphic functions on the right hand side of (5.1.24) in order to change the left hand side into U we find the new representation for solutions of the differential equations,

$$\mathbf{w}(z) = -\frac{1}{2\pi i}\iint\limits_{\mathsf{G}} \frac{1}{t-z}\{\mathsf{a}\mathbf{w}+\mathsf{b}\bar{\mathbf{w}}\}(t)[dt, d\bar{t}] - \sum_{j=1}^{N} \boldsymbol{\phi}_j(z)\langle \mathbf{w}, \mathbf{g}_j\rangle$$

$$+\mathrm{i}\mathbf{k} + \frac{1}{2\pi i}\oint\limits_{\dot{\mathsf{G}}} \frac{\boldsymbol{\nu}(t)\,dt}{t-z} + \frac{1}{2\pi i}\iint\limits_{\mathsf{G}} \frac{\mathbf{c}(t)}{t-z}[dt, d\bar{t}]. \quad (5.1.42)$$

Note that the densities $\boldsymbol{\nu}$ and constants $\mathbf{k}$ are different from those in (5.1.24). Of course, (5.1.42) is equivalent to

$$\mathbf{w} = \mathsf{U}^{-1}\left\{\mathrm{i}\mathbf{k} + \frac{1}{2\pi i}\oint\limits_{\dot{\mathsf{G}}} \frac{\boldsymbol{\nu}(t)\,dt}{t-z} + \frac{1}{2\pi i}\iint\limits_{\mathsf{G}} \frac{\mathbf{c}(t)}{t-z}[dt, d\bar{t}]\right\}$$

$$\equiv \mathsf{S}_1\mathbf{k} + \mathsf{S}_2\boldsymbol{\nu} + \mathsf{S}_3\mathbf{c}. \quad (5.1.43)$$

Inserting this representation into (5.1.16) which has to be altered corresponding to (5.1.42) we find for $\mathbf{k}$ and $\boldsymbol{\nu}$ the equations

$$\tfrac{1}{2}(\mathsf{P}(z)+\bar{\mathsf{P}}(z))\boldsymbol{\nu}(z) + \tfrac{1}{2}(\mathsf{P}(z)-\bar{\mathsf{P}}(z))\frac{1}{\pi i}\oint_{\dot{\mathsf{G}}} \frac{\boldsymbol{\nu}(t)\,dt}{t-z} + \mathrm{i}(\mathsf{P}(z)-\bar{\mathsf{P}}(z))\mathbf{k}$$

$$+\sum_{j=1}^{N} 2\{\mathrm{Re}\,\mathsf{P}\boldsymbol{\phi}_j\}\langle \mathsf{S}_1\mathbf{k}+\mathsf{S}_2\boldsymbol{\nu}, \mathbf{g}_j\rangle + \mathsf{P}(z)\frac{1}{\pi}\oint_{\dot{\mathsf{G}}} \boldsymbol{\nu}(t)\, d_{n_t} \log|t-z|$$

$$-\frac{2}{\pi}\,\mathrm{Re}\,\mathsf{P}(z)\iint\limits_{\mathsf{G}} \frac{1}{t-z}\{\mathsf{a}(\mathsf{S}_1\mathbf{k}+\mathsf{S}_2\boldsymbol{\nu})$$

$$+\mathsf{b}(\bar{\mathsf{S}}_1\mathbf{k}+\bar{\mathsf{S}}_2\boldsymbol{\nu})\}\frac{1}{2i}[dt, d\bar{t}] + \frac{1}{2\pi i}\iint\limits_{\mathsf{G}} \frac{\mathbf{c}(t)}{t-z}[dt, d\bar{t}]$$

$$= 2\boldsymbol{\psi}(z) - \sum_{j=1}^{N} 2\{\mathrm{Re}\,\mathsf{P}\boldsymbol{\phi}_j\}\langle \mathsf{S}_3\mathbf{c}, \mathbf{g}_j\rangle + \frac{1}{2\pi i}\iint\limits_{\mathsf{G}} \frac{\mathbf{c}(t)[dt, d\bar{t}]}{t-z}$$

$$+\frac{2}{\pi}\,\mathrm{Re}\left(\mathsf{P}(z)\iint\limits_{\mathsf{G}} \frac{1}{t-z}\{\mathsf{a}\mathsf{S}_3\mathbf{c}+\mathsf{b}\overline{\mathsf{S}_3\mathbf{c}}\}[dt, d\bar{t}]\right) \quad \text{for} \quad z \in \dot{\mathsf{G}}. \quad (5.1.44)$$

These equations enable us to replace the boundary value problem (5.1.23) in the spaces

$$X_m^\alpha \equiv \{\mathbf{w} \in C^\alpha(\bar{\mathrm{G}}) \quad \text{with} \quad \mathbf{w}_{\bar{z}} \in C^{m+\alpha-1}(\bar{\mathrm{G}}) \quad \text{and}$$
$$\|\mathbf{w}\|_{X_m^\alpha} \equiv \|\mathbf{w}\|_{C^\alpha(\bar{\mathrm{G}})} + \|\mathbf{w}_{\bar{z}}\|_{C^{m+\alpha-1}(\bar{\mathrm{G}})}\},$$
$$m \geq 1 \text{ an integer,} \tag{5.1.45}$$

by the system of singular integral equations (5.1.44).

Theorem 5.1.7

(i) *Let* a, b $\in C^{m+\alpha-1}(\bar{\mathrm{G}})$. *Then to every* $\mathbf{w} \in X_m^\alpha$ *the equation in* (5.1.23) *defines* $\mathbf{c} \in C^{m+\alpha-1}(\mathrm{G}) \cap C^\alpha(\bar{\mathrm{G}})$ *and to* $\mathbf{w}$ *and this* $\mathbf{c}$ *there exist exactly one* $\boldsymbol{\nu} \in C^\alpha(\dot{\mathrm{G}})$ *and* $\mathbf{k} \in R^n$ *satisfying the integral equations* (5.1.44) *with* $\boldsymbol{\psi} = \mathrm{Re}\, \mathsf{P}\mathbf{w}|_{\dot{\mathrm{G}}}$. *If* a, b *have compact supports in* G *then* $\mathbf{c} \in C^{m+\alpha-1}(\bar{\mathrm{G}})$.

(ii) *Let* $\mathbf{c} \in C^\alpha(\bar{\mathrm{G}})$ *and* $\boldsymbol{\psi} \in C^\alpha(\dot{\mathrm{G}})$ *be given and let* $\boldsymbol{\nu} \in C^\alpha(\dot{\mathrm{G}})$, $\mathbf{k} \in R^n$ *be a solution of* (5.1.44). *Then* (5.1.43) *defines a solution* $\mathbf{w} \in X_1^\alpha$ *of* (5.1.23). *For* a, b, $\mathbf{c} \in C^{m+\alpha-1}(\bar{\mathrm{G}})$ *and* $\boldsymbol{\nu} \in C^{m+\alpha}(\dot{\mathrm{G}})$ *we have* $\mathbf{w} \in C^{m+\alpha}(\bar{\mathrm{G}})$. *For* a, b $\in C^{m+\alpha-1}(\bar{\mathrm{G}})$ *having compact supports in* G *and* $\mathbf{c} \in C^{m+\alpha-1}(\bar{\mathrm{G}})$, $\boldsymbol{\nu} \in C^\alpha(\dot{\mathrm{G}})$, *we have* $\mathbf{w} \in X_m^\alpha$.

Proof The first part (i) is clear from the preceding considerations in connection with the definition (5.1.45); let us omit the details. The second part (ii) is also clear: Any $\mathbf{c} \in C^\alpha(\bar{\mathrm{G}})$, $\mathbf{k} \in R^n$, $\boldsymbol{\nu} \in C^\alpha(\dot{\mathrm{G}})$, define $\mathbf{w} \in C^\alpha(\bar{\mathrm{G}})$ by (5.1.43) satisfying (5.1.42). Hence, this $\mathbf{w}$ also satisfies the differential equation in (5.1.23) implying $\mathbf{w} \in X_1^\alpha$. Since $\mathbf{c}$, $\boldsymbol{\nu}$, $\mathbf{k}$ solve (5.1.44) the corresponding $\mathbf{w}$ (5.1.43) satisfies the boundary condition in (5.1.23). In the case a, b, $\mathbf{c} \in C^{m+\alpha-1}(\bar{\mathrm{G}})$ and $\boldsymbol{\nu} \in C^{m+\alpha}(\dot{\mathrm{G}})$ the expression $\{\cdots\}$ in (5.1.43) defines a function in $C^{m+\alpha}(\bar{\mathrm{G}})$. Thus, $\mathbf{w} \in C^{m+\alpha}(\bar{\mathrm{G}})$ according to Theorem 5.1.6. If a, b have compact supports in G and $\mathbf{w} \in X_m^\alpha$ then the representation

$$\mathbf{w}(z) = \frac{1}{2\pi \mathrm{i}} \iint_{\mathrm{G}} \frac{1}{t-z} \mathbf{w}_{\bar{t}}[\mathrm{d}t, \mathrm{d}\bar{t}] + \frac{1}{2\pi \mathrm{i}} \oint_{\dot{\mathrm{G}}} \frac{\mathbf{w}}{t-z} \mathrm{d}t \quad \text{for} \quad z \in \mathrm{G} \tag{5.1.46}$$

shows that the operator

$$-\frac{1}{2\pi \mathrm{i}} \iint_{\mathrm{G}} \frac{1}{t-z} \{\mathrm{a}\mathbf{w} + \mathrm{b}\bar{\mathbf{w}}\}[\mathrm{d}t, \mathrm{d}\bar{t}] - \sum_{j=1}^{N} \boldsymbol{\phi}_j(z) \langle \mathbf{w}, \mathbf{g}_j \rangle$$

maps X_m^α completely continuously into $C^{m+\alpha}(\bar{\mathrm{G}}) \subset X_m^\alpha$.

Hence now the Fredholm alternative for (5.1.42) also holds in X_m^α yielding that U^{-1} maps $\{\cdots\} \in X_m^\alpha$ in (5.1.43) onto $\mathbf{w} \in X_m^\alpha$.

In the following let us distinguish two different cases of problems with violated Lopatinski condition; in both cases further restrictions on the boundary matrix P or on the coefficients a, b allow the application of Prößdorf's results [58] to (5.1.44).

The case with further restrictions on the boundary conditions.

This is the approach investigated by Kopp [43].

The integral operators in (5.1.44) and the right hand side except ψ possess all the factors P or $\bar{\mathsf{P}}$. Therefore the assumptions (3.5.14) and (3.5.17),

$$\mathsf{P} = \mathsf{TDD}_0, \bar{\mathsf{P}} = \mathsf{TDC}_0, \tag{5.1.47}$$

allow the multiplication of (5.1.44) by $\mathsf{D}^{-1}\mathsf{T}^{-1}$ converting (5.1.44) into a *new system of integral equations.* This is equivalent to (5.1.44) considered as a mapping from the real valued $C^\alpha(\dot{\mathsf{G}}) \times R^n$ into the space Y defined in (3.5.18).

Kopp proves:

Theorem 5.1.8 *The operator defined by the left hand side of* (5.1.44) *from* $C^\alpha(\dot{\mathsf{G}}) \times R^n$ *into* Y (3.5.18) *is a bounded Fredholm operator of index*

$$\nu = n - \frac{1}{2\pi} \oint_{\dot{\mathsf{G}}} \mathrm{d} \arg \mathrm{Det}\, \mathsf{D}_0 \mathsf{C}_0^{-1}.$$

For the proof see [43]. There Kopp also investigates (5.1.44) and the corresponding boundary value problems for other spaces Y.

With Theorem 5.1.7 the proposed Theorem 3.5.3 on the solvability of the boundary value problem in X_1^α follows from Theorem 5.1.8.

The case with further restrictions on the coefficients a, b†.

In this case let us assume that the coefficients a_{jk}, b_{jk} *in* (5.1.23) *have compact supports in* G.

Now let us consider the singular integral equations (5.1.44) assuming (3.5.17), and (3.5.13) for P,

$$P_{lt} = \sum_{k=0}^{m_j - 1} P_{ltk}(z - z_j)^k + \tilde{P}_{lt}(z)(z - z_j)^{m_j}$$

$$\text{in the neighbourhood of } z_j \text{ and} \tag{5.1.48}$$

$$\det \mathsf{P} = \Delta_0(s) \prod_{j=1}^{r} (z - z_j)^{m_j} \quad \text{with} \quad \Delta_0(s) \neq 0.$$

In order to apply Prößdorf's results in [57] we assume either

$$(A) \quad \dot{\mathsf{G}} \in C^{2m+\beta}, \quad \mathsf{a}, \mathsf{b} \in C^{m+\beta}(\bar{\mathsf{G}}),$$
$$\tilde{\mathsf{P}} \in C^{m+\beta}(\dot{\mathsf{G}}), \quad 0 < \alpha < \beta$$

† These results have been submitted to the 'Colloquium on Mathematical Analysis' in Jyväskylä, Finland 1973 where the author has given a lecture on these problems.

where

$$m \equiv \max_{j=1,\ldots,r}\{m_j\} \quad \text{and} \quad \tilde{\mathsf{P}} = ((\tilde{p}_{lt})) \tag{5.1.49}$$

or

$$(B) \quad \dot{\mathsf{G}} \in C^{2m+2+\beta}, \quad \mathsf{a}, \mathsf{b} \in C^{2m+\beta}(\bar{\mathsf{G}}),$$
$$\tilde{\mathsf{P}} \in C^{2m+\beta}(\dot{\mathsf{G}}), \quad 0 < \alpha < \beta.$$

Under these assumptions the operators in (5.1.44) have the following properties:

Lemma 5.1.9 *The following operators are completely continuous:*

(i) $\mathrm{i}(\mathsf{P}-\bar{\mathsf{P}})\mathbf{k}$ *from* R^n *into* $C^{m+\alpha}(\dot{\mathsf{G}})$ *in case* (A) *and into* $C^{2m+\alpha}(\dot{\mathsf{G}})$ *in case* (B).

(ii) $\mathsf{P}(z)\dfrac{1}{\pi}\oint_{\dot{\mathsf{G}}} \boldsymbol{\nu}(t)\,\mathrm{d}_{n_t} \log|t-z|$ *from* $C^{\alpha}(\dot{\mathsf{G}})$ *into* $C^{m+\alpha}(\dot{\mathsf{G}})$ *in case* (A) *and into* $C^{2m+\alpha}(\dot{\mathsf{G}})$ *in case* (B).

(iii) $\displaystyle\sum_{j=1}^{N} 2\{\mathrm{Re}\,\mathsf{P}\boldsymbol{\phi}_j\}\langle \mathsf{S}_1\mathbf{k}+\mathsf{S}_2\boldsymbol{\nu}, \mathbf{g}_j\rangle$

$$-\frac{2}{\pi}\,\mathrm{Re}\left[\mathsf{P}(z)\iint \frac{1}{t-z}\left\{\mathsf{a}(\mathsf{S}_1\mathbf{k}+\mathsf{S}_2\boldsymbol{\nu})+\mathsf{b}\Big((\bar{\mathsf{S}}_1\mathbf{k}+\bar{\mathsf{S}}_2\boldsymbol{\nu})\right\}\frac{1}{2\mathrm{i}}[\mathrm{d}t,\mathrm{d}\bar{t}]\Big)\right]$$

from $C^{\alpha}(\dot{\mathsf{G}})$ *into* $C^{m+\alpha}(\dot{\mathsf{G}})$ *in case* (A) *and into* $C^{2m+\alpha}(\dot{\mathsf{G}})$ *in case* (B).

Proof

(i) is trivial since the image is finite dimensional.

(ii) follows from the fact that the kernel of this integral operator,

$$\mathsf{P}(z)\frac{1}{\pi}\,\mathrm{d}_{n_t}\log|t(s)-z|\,\mathrm{d}s^{-1} = \mathsf{P}(z)\frac{1}{\pi}\left((\eta(s)-y)\frac{\mathrm{d}\xi}{\mathrm{d}s}-(\xi(s)-x)\frac{\mathrm{d}\eta}{\mathrm{d}s}\right)$$
$$\{(\xi(s)-x)^2+(\eta(s)-y)^2\}^{-1} \quad \text{for} \quad z \in \dot{\mathsf{G}} \tag{5.1.50}$$

can be extended to a function in $C^{m+\beta}([0,L]\times\dot{\mathsf{G}})$ in case (A) and in $C^{2m+\beta}([0,L]\times\dot{\mathsf{G}})$ in case (B). Let us omit the details here. (A similar analysis can be found in [40, Appendix].)

(iii) Since (5.1.42) is a Fredholm equation of the second kind, the definition of $\mathsf{S}_2\boldsymbol{\nu}$ in (5.1.43) provides the representation

$$\mathsf{S}_2\boldsymbol{\nu} = \frac{1}{2\pi\mathrm{i}}\oint_{\dot{\mathsf{G}}}\frac{\boldsymbol{\nu}(t)\,\mathrm{d}t}{t-z} + U^{-1}\left[\frac{1}{2\pi\mathrm{i}}\iint_{\mathsf{G}}\frac{1}{t-z}\left\{\mathsf{a}(t)\frac{1}{2\pi\mathrm{i}}\oint_{\dot{\mathsf{G}}}\frac{\boldsymbol{\nu}(s_\tau)}{\tau-t}\,\mathrm{d}\tau\right.\right.$$
$$\left.\left. -\mathsf{b}(t)\frac{1}{2\pi\mathrm{i}}\oint_{\dot{\mathsf{G}}}\frac{\boldsymbol{\nu}(s_\tau)}{\bar{\tau}-\bar{t}}\,\mathrm{d}\bar{\tau}\right\}[\mathrm{d}t,\mathrm{d}\bar{t}]\right]. \tag{5.1.51}$$

With $\underset{\sim}{\bar{G}} \equiv \bigcup_{j,k=1}^{n}$ supp $(a_{jk}) \cup$ supp $(b_{jk}) \subset G$ and dist $(\underset{\sim}{\bar{G}}, \dot{G}) > 0$ it is clear that the Cauchy integral $\oint_{\dot{G}} \boldsymbol{\nu}(t)\,dt/(t-z)$ defines a mapping $C^{\alpha}(\dot{G}) \to C^{2m+\beta}(\underset{\sim}{\bar{G}})$ which is completely continuous. Since in (5.1.50) and (5.1.51) the surface integrals have only $\underset{\sim}{G}$ as the domain of integration they define continuous mappings from $C^{2m+\beta}(\underset{\sim}{\bar{G}})$ into $C^{m+\alpha}(\bar{G})$ or $C^{2m+\alpha}(\bar{G})$ corresponding to (A) or (B). Therefore the compositions appearing in (5.1.50) and (5.1.51) with the above compact mapping become completely continuous from $C^{\alpha}(\dot{G})$ into $C^{m+\alpha}(\dot{G})$ or $C^{2m+\alpha}(\dot{G})$. The remaining operators in (5.1.50) have finite dimensional images in $C^{m+\alpha}(\dot{G})$ or $C^{2m+\alpha}(\dot{G})$ hence they are also completely continuous.

The propositions of Lemma 5.1.9 assure that all perturbations of the principal part in (5.1.44) define completely continuous operators from $C^{\alpha}(\dot{G})$ into $C^{m+\alpha}(\dot{G})$ or $C^{2m+\alpha}(\dot{G})$, respectively. Thus we may apply Prößdorf's results [57, Sections 1 and 3] finding

Theorem 5.1.10 *The mapping from* $D \times R^n$ *defined by the left hand side of* (5.1.44) *with*

$$D \equiv \left\{ \boldsymbol{\phi} \in C^{\alpha}(\dot{G}) \mid (P+\bar{P})\boldsymbol{\phi} + (P-\bar{P}) \frac{1}{\pi i} \oint_{\dot{G}} \frac{\boldsymbol{\phi}(t)\,dt}{t-z} \in C^{m+\alpha}(\dot{G}) \right\} \tag{5.1.52}$$

into $C^{m+\alpha}(\dot{G})$ *is a semi-Fredholm operator in case* (A) *with finite dimensional null space and is a Fredholm operator in case* (B) *which is a densely defined closed operator.*

In case (B) *the Fredholm index is given by*

$$\nu = n - 2\sum_{j=1}^{r} m_j - \frac{1}{\pi} \oint_{\dot{G}} d \arg \Delta_0(s). \tag{5.1.53}$$

Remark 5.1.11 n appears in (5.1.53) since the domain of definition contains R^n as one of the factors. For the application of [57, (21)] one has to choose

$$\Delta_1^0 \cdot \Delta_2^{0^{-1}} = \bar{\Delta}_0 \Delta_0^{-1} \prod_{j=1}^{r} \left(\frac{\overline{(z-z_j)}}{z-z_j} \right)^{m_j}.$$

Now it is clear that Theorem 5.1.10 can be carried over to the boundary value problem (5.1.23) by the use of Theorem 5.1.6. Since $\underset{\sim}{\bar{G}} \subset G$ all the mappings in Theorem 5.1.6 become continuous in the corresponding spaces. For R let us define the domain by

$$D_R \equiv \{ \mathbf{w} \in X_m^{\alpha} \quad \text{with} \quad \operatorname{Re} \mathbf{Pw}\,|_{\dot{G}} \in C^{m+\alpha}(\dot{G}) \}. \tag{5.1.54}$$

Then (5.1.44) provides with $\mathbf{c}$ and $\boldsymbol{\psi}$ defined by (5.1.23) and with Lemma 5.1.9 that to any $\mathbf{w} \in D_R$ the corresponding $\boldsymbol{\nu} \in D$ (5.1.52). Conversely, every $\boldsymbol{\nu} \in D$ and $\mathbf{k} \in R^n$ generates by (5.1.43) a $\mathbf{w} \in D_R$. Correspondingly, the null space of R in D_R is mapped bijectively onto the null space of (5.1.44) in $D \times R^n$ and the nullities of R and (5.1.44) are the same.

For the range of R in $C^{m+\alpha-1}(\bar{\mathsf{G}})\times C^{m+\alpha}(\dot{\mathsf{G}})$ we observe that it is mapped bijectively and continuously onto $C^{m+\alpha-1}(\bar{\mathsf{G}})\times\{$range of (5.1.44) in $C^{m+\alpha}(\dot{\mathsf{G}})\}$. Since the latter is closed and $C^{m+\alpha-1}(\bar{\mathsf{G}})$ is the whole space, the range of R is also closed and R is a semi-Fredholm mapping in both cases (A) and (B). In case (B) we find moreover that a solution $\mathbf{w}$ of (5.1.43) belongs to D_R if and only if the right hand side in (5.1.44) consisting of $\boldsymbol{\psi}$ and $\mathbf{c}$ satisfies the finitely many solvability conditions corresponding to the Fredholm property proposed in Theorem 5.1.10. Since $\boldsymbol{\psi}$ appears in (5.1.44) explicitly the dimension of solvability conditions remains unchanged. Let us collect all these properties on the following Theorem which corresponds to Theorem 3.5.3:

Theorem 5.1.12 *In the cases* (A) *and* (B) R *in* (5.1.23) *is a densely defined closed semi Fredholm operator from* D_R *into* $C^{m+\alpha-1}(\bar{\mathsf{G}})\times C^{m+\alpha}(\dot{\mathsf{G}})$. *In case* (B), R *is then even a Fredholm operator with Fredholm index* (5.1.53).

Remark 5.1.13 It remains still open how the additional properties on P or on a, b can be weakened or whether such additional conditions are really necessary for the solvability of (5.1.23) under the assumptions (5.1.48).

In [63] Tovmasjan introduced integral representations similar to the Cauchy formula for second order elliptic systems with constant coefficients. Using these representations in [62] he investigated exterior and interior Dirichlet, Neumann and special Poincaré problems. There he also showed that the Dirichlet problem in the case of violated Lopatinski condition is still normally solvable but not semi-Fredholm anymore since both the nullity of R and the codimension of the range of R, are infinite.

5.2 Fundamental Solutions and Singular Integral Equations

The integral equations (5.1.18)–(5.1.20) have been constructed by the use of $(\zeta_j(z)-\zeta_j(t))^{-1}$, the fundamental solution of the differential operator $\partial/\partial\bar{\zeta}_j$. Hence these integral equations can be simplified and the method can be extended by using fundamental solutions of the complete principal part of the differential equations. For first-order systems there exist three such methods which have been worked out in detail. Bojarski [8], [9] developed an iteration technique for the fundamental solution of systems in the quasinormal form (4.2.17) which admit a normal form (3.1.30) where the differential operator takes the form of a *lower triangular matrix* throughout G. This approach is similar to the 'generating function' introduced by Douglis in [15] for the more special systems (3.1.29). With the fundamental solution, generalized Cauchy kernels are obtained which allow representations of the solution by singular integrals of boundary layers. Inserting these representations into the boundary conditions we arrive at a system of singular integral equations which is simpler than in 5.1 but whose Cauchy kernels are not explicitly known as in Section 5.1.

The whole approach simplifies for the Douglis system (3.1.29) for which we shall introduce the Douglis hypercomplex algebra. Here Gilbert [22] and Gilbert and Hile [24] developed the corresponding representation formulas, Cauchy integrals, and L^p-theory. A different approach is based on the fundamental solution for systems with constant coefficients; in this case the fundamental solution can be constructed by the use of suitable integrals in the complex plane as Agmon [1] did or with Fourier transforms developed extensively by Hörmander [38]. For higher order equations Fichera [16] used this fundamental solution as a paramatrix [50, p. 92 ff.] to find the complete fundamental solution. With the latter he reformulated the corresponding Dirichlet problem into a system of singular integral equations of the first kind. For a single even order equation with constant coefficients Ricci [59] characterized the boundary conditions which lead to first kind integral equations and Fichera and Ricci [17] extended this approach to even order systems with constant coefficients. Here we shall show that our approach in Section 5.1 together with the Dunford Taylor integrals (4.1.44) for the generalized eigenvectors leads in case of constant coefficients to Fichera's and Ricci's integral equations. Independently, Tovmasjan developed in [62], [63] a similar method for such systems finding equivalent Fredholm integral equations of the first and also of the second kind with logarithmic and with smooth kernels, respectively. In [63] for $\dot{G}$ being an ellipse he also showed how to solve these equations by an explicit Fourier expansion method which leads to an infinite system of linear equations for the coefficients. The solvability and nullity of these equations are explicitly connected with those of the boundary value problem. In [62] Tovmasjan extended this approach to the Neumann problem and to certain Poincaré problems including problems with violated Lopatinski condition.

For the general system (4.1.1), Vol'pert [67] used the complete fundamental solution and proved the solvability properties of the boundary value problems by using the corresponding singular integral equations. This approach is based on Lopatinskiĭ's classical investigations of elliptic equations [46]. Here we shall present only a few constructive aspects of this theory.

Let us consider first the Bojarski fundamental solution for the general system in quasinormal form

$$\mathbf{w}_{\bar{z}} - \mathsf{Q}\mathbf{w}_z + \mathsf{a}\mathbf{w} + \mathsf{b}\bar{\mathbf{w}} = \mathbf{c}, \tag{5.2.1}$$

and let us assume that this system admits the Bojarski normal form, i.e. there exists a system of generalized eigenvectors $\mathbf{p}_j$ satisfying (3.1.17). Then Q becomes a triangular matrix for which we further assume that Q *is a lower triangular matrix throughout* G. More precisely, from (3.1.12) and (4.2.18) it follows by (4.2.19) that

$$\mathsf{Q}=\left\{((q_{jk}\delta_{jk}))-\mathrm{i}\left(\left(\frac{1}{1-\mathrm{i}\lambda_+^j}\delta_{jk}\right)\right)((\mu_j\delta_{j-1,k}))\right\}$$
$$\times\left\{\mathsf{I}-\mathrm{i}\left(\left(\frac{1}{1-\mathrm{i}\lambda_+^j}\delta_{jk}\right)\right)((\mu_j\delta_{j-1,k}))\right\}^{-1} \tag{5.2.2}$$

holds. Since the functions μ_k in (3.1.5) can be multiplied by an arbitrary nonzero number changing only the magnitude of $\mathbf{p}_k^T$ (and of $\mathbf{q}_j$) we can assume without loss of generality that the μ_k are so small that

$$\sum_{j=1}^{n}|\kappa_{lj}(x,y)|\leq q_0<1 \quad \text{in the whole plane where}$$
$$\mathsf{Q}=((\kappa_{lj})) \quad \text{and} \quad \mathsf{Q}\equiv 0 \quad \text{for} \quad |z|\geq R \tag{5.2.3}$$

with R sufficiently large.

Now Bojarski constructs the fundamental solution $V(t,z)$ of the principal part of (5.2.1) as follows.

Theorem 5.2.1 *For given* Q *there exists* $V(t,z)\in C^{1+\alpha}$ *in the whole plane but* $z\neq t$ *with the following properties:*
V is a lower triangular matrix and satisfies

$$V_{\bar z}-\mathsf{Q}V_z=0 \quad \text{for all} \quad z\neq t \tag{5.2.4}$$

and the adjoint system

$$V_{\bar t}^T-(\mathsf{Q}^TV^T)_t=0 \quad \text{for} \quad t\neq z. \tag{5.2.5}$$

At infinity V has the properties

$$\lim_{|t|\to\infty} tV(t,z)=\mathsf{I} \quad \text{for each } z, \tag{5.2.6}$$

$$\lim_{|z|\to\infty} zV(t,z)=\tilde V(t) \quad \text{for each } t \text{ where} \tag{5.2.7}$$

$$\operatorname{Det}\tilde V(t)\neq 0 \quad \text{and} \quad \operatorname{Det}\lim_{|t|\to\infty}\tilde V(t))\neq 0.$$

Remark Bojarski assumes only that $\mathsf{Q}\in L^\infty$ and that (5.2.3) holds with L^∞-norms. Then V is a generalized solution of (5.2.4), (5.2.5) and $V\in C^\alpha$

for $z \neq t$. Following Bojarski, functions $\mathbf{W}$ satisfying

$$\mathbf{W}_{\bar{z}} - \mathsf{Q}\mathbf{W}_z = 0$$

will be called Q-*holomorphic.*

Proof The solution will be constructed in two steps. First we introduce a generalization Ξ of the homeomorphisms ζ_j,

$$\Xi(z) = z\mathsf{I} + \frac{1}{2\pi \mathrm{i}} \iint\limits_{\mathsf{G}} \frac{\omega(t)}{t-z} [\mathrm{d}t, \mathrm{d}\bar{t}] \tag{5.2.8}$$

with $\omega \in L^p(\bar{\mathsf{G}})$ for some $p > 2$, satisfying

$$\Xi_{\bar{z}} - \mathsf{Q}\Xi_z = 0. \tag{5.2.9}$$

Consequently, the density matrix ω has to satisfy the singular integral equation

$$\omega - \mathsf{Q}T\omega \equiv \omega - \mathsf{Q}(z) \frac{1}{2\pi \mathrm{i}} \iint\limits_{\mathbb{C}} \frac{\omega(t)}{(t-z)^2} [\mathrm{d}t, \mathrm{d}\bar{t}] = \mathsf{Q}. \tag{5.2.10}$$

The assumption (5.2.3) guarantees by (2.4.8) the estimate

$$\|\mathsf{Q}T\omega\|_2 \leqslant q_0 \|T\omega\|_2 \leqslant q_0 \|\omega\|_2. \tag{5.2.11}$$

Now we may proceed word for word as in Section 2.4 finding the solution $\Xi \in C^{1+\alpha}(R^2)$ by the successive approximation for ω,

$$\omega = \sum_{j=0}^{\infty} (\mathsf{Q}T)^j \mathsf{Q}, \tag{5.2.12}$$

converging in $L^p(R^2)$ with some $p > 2$.

All the compositions in (5.2.12) are defined by compositions of *lower triangular matrices*, therefore, ω and Ξ are lower triangular matrices as well. The diagonal elements of Ξ, ζ_{jj} satisfy the equations

$$\zeta_{jj\bar{z}} - q_j \zeta_{jjz} = 0 \tag{5.2.13}$$

and

$$\omega_{jj} - q_j \frac{1}{2\pi \mathrm{i}} \iint\limits_{\mathbb{C}} \frac{\omega_{jj}}{(t-z)^2} [\mathrm{d}t, \mathrm{d}\bar{t}] = q_j \tag{5.2.14}$$

with

$$\zeta_{jj} = z + \frac{1}{2\pi \mathrm{i}} \iint\limits_{\mathbb{C}} \frac{\omega_{jj}}{t-z} [\mathrm{d}t, \mathrm{d}\bar{t}], \tag{5.2.15}$$

so that we observe: The diagonal elements of Ξ are just the homeomorphisms of the normal form (3.1.30),

$$\zeta_{jj} = \zeta_j. \tag{5.2.16}$$

This implies the existence of

$$(\Xi(t) - \Xi(z))^{-1} \text{ for all } t \neq z \text{ and of } (\Xi_z)^{-1}. \tag{5.2.17}$$

With this 'generating solution' Ξ Bojarski defines the desired fundamental solution by

$$V(t, z) \equiv (\Xi(t))_t \{\Xi(t) - \Xi(z)\}^{-1} = \{\Xi(t) - \Xi(z)\}^{-1} (\Xi(t))_t. \tag{5.2.18}$$

The order of the product in (5.2.18) does not matter because lower triangular matrices commute. The properties of Ξ yield

$$\begin{aligned} (V[\Xi(t) - \Xi(z)])_{\bar{z}} &= V_{\bar{z}}[\Xi(t) - \Xi(z)] - V\Xi_{\bar{z}} = 0, \\ (V[\Xi(t) - \Xi(z)])_z &= V_z[\Xi(t) - \Xi(z)] - V\Xi_z = 0, \end{aligned} \tag{5.2.19}$$

which implies with (5.2.9)

$$V_{\bar{z}} - \mathsf{Q} V_z = (V\mathsf{Q} - \mathsf{Q} V)\Xi_z[\Xi(t) - \Xi(z)]^{-1} = 0 \tag{5.2.20}$$

since V *and* Q *commute.* In a similar way we get

$$\begin{aligned} V_{\bar{t}} - (V\mathsf{Q})_t &= \mathsf{Q}_t V - V\mathsf{Q}_t + \mathsf{Q}\Xi_{tt}[\Xi(t) - \Xi(z)]^{-1} - \Xi_{tt}[\Xi(t) - \Xi(z)]^{-1}\mathsf{Q} \\ &\times - \Xi_t[\Xi(t) - \Xi(z)]^{-2}\mathsf{Q}\Xi_t + \Xi_t[\Xi(t) - \Xi(z)]^{-2}\Xi_t\mathsf{Q} = 0, \end{aligned} \tag{5.2.21}$$

which also vanishes because of the commuting properties.

The property (5.2.6) follows from (5.2.8) and (5.2.10) corresponding to $\mathsf{Q} \equiv 0$ and $\omega \equiv 0$ for $|z| > R$:

$$tV(t, z) = \left\{ \mathsf{I} + \frac{1}{2\pi i} \iint_{|\tau| \leq R} \frac{\omega(\tau)[d\tau, d\bar{\tau}]}{(\tau - t)^2} \right\}$$

$$\cdot \frac{t}{t - z} \left[\mathsf{I} + \frac{1}{2\pi i} \iint_{|\tau| \leq R} \omega(\tau) \frac{1}{(t - z)} \left\{ \frac{1}{\tau - z} - \frac{1}{\tau - t} \right\} [d\tau, d\bar{\tau}] \right]^{-1}. \tag{5.2.22}$$

If $|t|$ tends to infinity for any fixed value of z then the right hand side tends to I. Property (5.2.7) follows similarly from

$$zV(t, z) = \Xi_t(t) \frac{z}{t - z} \left[\mathsf{I} + \frac{1}{2\pi i} \iint_{|\tau| \leq R} \omega(t) \right.$$

$$\left. \times \left\{ \frac{1}{(t - z)} \left\{ \frac{1}{\tau - z} - \frac{1}{\tau - t} \right\} [d\tau, d\bar{\tau}] \right]^{-1} \right\} \tag{5.2.23}$$

implying

$$\tilde{V}(t) = -\Xi_t(t) = \left\{\mathsf{I} + \frac{1}{2\pi i} \iint\limits_{|\tau|\le R} \frac{\omega(\tau)[d\tau, d\bar{\tau}]}{(\tau - t)^2}\right\}. \tag{5.2.24}$$

This equation shows also that $\tilde{V}(\infty) = \mathsf{I}$.

Theorem 5.2.2 *The fundamental solution provides the relationship*

$$\lim_{\rho \to 0} \frac{1}{2\pi i} \oint_{|t-z|=\rho} V(t, z)\, d_Q t \mathbf{w}(t) = \mathbf{w}(z) \tag{5.2.25}$$

for any continuous vector $\mathbf{w}(z)$ *and hence, the representation formula*

$$\begin{aligned}\mathbf{w}(z) &= \frac{1}{2\pi i} \oint\limits_{\dot{\mathsf{G}}} V(t, z)\, d_Q t \mathbf{w}(t) + \frac{1}{2\pi i} \iint\limits_{\mathsf{G}} V(t, z)(\mathbf{w}_{\bar{t}} - \mathsf{Q}\mathbf{w}_t)[dt, d\bar{t}] \\ &= \frac{1}{2\pi i} \oint_{\dot{\mathsf{G}}} \{\Xi(t) - \Xi(z)\}^{-1} d\Xi(t) \mathbf{w}(t) \\ &\quad + \frac{1}{2\pi i} \iint\limits_{\mathsf{G}} \{\Xi(t) - \Xi(z)\}^{-1} \Xi_t(t)(\mathbf{w}_{\bar{t}} - \mathsf{Q}\mathbf{w}_t)[dt, d\bar{t}]\end{aligned} \tag{5.2.26}$$

for every $\mathbf{w} \in C^1(\bar{\mathsf{G}})$ *where* $d_Q t$ *is defined by*

$$d_Q t = (dt\, \mathsf{I} + \mathsf{Q}(t)\, d\bar{t}). \tag{5.2.27}$$

Proof From the definition (5.2.18) it follows by the mean value theorem that

$$\begin{aligned}V(t, z) &= (t - z)^{-1} \Xi_t(t) \left\{\Xi_t + \Xi_{\bar{t}} \frac{\bar{t} - \bar{z}}{t - z} + \varepsilon_1(z, t)\right\}^{-1} \\ &= (t - z)^{-1} \left[\mathsf{I} + \mathsf{Q}(t) \frac{\bar{t} - \bar{z}}{t - z} + \varepsilon_2(z, t)\right]^{-1}\end{aligned} \tag{5.2.28}$$

holds where $|\varepsilon_2| \le k\, |t - z|^\alpha$ with some $\alpha > 0$.

Proceeding as in (1.1.18), (1.1.19) we find with (5.2.28) the limit (5.2.25).

Formula (5.2.26) follows directly from Green's Theorem:

$$\frac{1}{2\pi i}\oint_{\dot{G}} V(t,z)\,d_Q t\mathbf{w}(t)-\lim_{\rho\to 0}\frac{1}{2\pi i}\oint_{|z-t|=\rho} V(t,z)\,d_Q t\mathbf{w}(t)$$

$$=\frac{1}{2\pi i}\iint_G [d_{(t)}, V(t,z)(dt\,\mathsf{I}+\mathsf{Q}(t)\,d\bar{t})\mathbf{w}(t)]$$

$$=\frac{1}{2\pi i}\iint_G \{((V(t,z)\mathsf{Q}(t))_t - V_{\bar{t}})\mathbf{w}(t)$$

$$-V(t,z)(\mathbf{w}_{\bar{t}}-\mathsf{Q}\mathbf{w}_t)\}[dt, d\bar{t}]. \tag{5.2.29}$$

Property (5.2.5) and (5.2.25) yield the desired representation (5.2.26).

Now let us consider a generalized Cauchy formula.

Lemma 5.2.3 *Let $\boldsymbol{\nu}\in C^\alpha(\dot{G})$ be a given boundary layer. Then the Cauchy integral*

$$W(z)\equiv\frac{1}{2\pi i}\oint_{\dot{G}} V(t,z)\,d_Q t\boldsymbol{\nu}(t)=\frac{1}{2\pi i}\oint_{\dot{G}}\{\Xi(t)-\Xi(z)\}^{-1}d\Xi(t)\boldsymbol{\nu}(t) \tag{5.2.30}$$

has the following properties:
W is Q-holomorphic,

$$W_{\bar{z}}-\mathsf{Q}W_z=0 \text{ in } G \text{ and in } \mathbb{C}\backslash\bar{G}. \tag{5.2.31}$$

For W holds Plemelj's formula,

$$\lim_{z\to z_0\in\dot{G}} W^{\pm}(z)=\pm\tfrac{1}{2}\boldsymbol{\nu}(z_0)+\frac{1}{2\pi i}\oint_{\dot{G}} V(t,z_0)\,d_Q t\boldsymbol{\nu}(t)$$

$$=\pm\tfrac{1}{2}\boldsymbol{\nu}(z_0)+\frac{1}{2\pi i}\oint_{\dot{G}}\{\Xi(t)-\Xi(z_0)\}^{-1}d\,\Xi(t)\boldsymbol{\nu}(t) \tag{5.2.32}$$

where $^+$ denotes the interior and $^-$ the exterior boundary values. Together with its appropriate boundary values, W becomes $C^\alpha(\bar{G})$ or $C^\alpha(\mathbb{C}\backslash G)$, respectively.

Proof The Q-holomorphy (5.2.31) follows immediately from (5.2.4). The other properties can be deduced from the representation

$$W(z)=\frac{1}{2\pi i}\oint_{\dot{G}}\left[\mathsf{I}+\frac{\bar{t}-\bar{z}}{t-z}\mathsf{Q}+\varepsilon_2(z,t)\right]^{-1}\left[\mathsf{I}+\frac{\dot{\bar{t}}}{\dot{t}}\mathsf{Q}\right]\frac{1}{t-z}\boldsymbol{\nu}(t)\,dt \tag{5.2.33}$$

with the properties of the ordinary Cauchy integral (see [52, p. 55 ff.]).

Theorem 5.2.4 *Every Q-holomorphic vector* $\mathbf{w} \in C^\alpha(\bar{\mathrm{G}})$ *can be represented by*

$$\mathbf{w}(z) = \frac{1}{2\pi i} \oint_{\dot{G}} V(t, z)\, d_Q t \boldsymbol{\nu}(t) + i\mathbf{k}$$

$$= \frac{1}{2\pi i} \oint_{\dot{G}} \{\Xi(t) - \Xi(z)\}^{-1}\, d\Xi(t) \boldsymbol{\nu}(t) + i\mathbf{k} \quad (5.2.34)$$

with a real valued layer $\boldsymbol{\nu} \in C^\alpha(\dot{\mathrm{G}})$ *and real constants* $\mathbf{k} \in R^n$.

Proof Using the homeomorphism ζ_1 we find for the first component w_1 of $\mathbf{w}$ that it is holomorphic with respect to ζ_1,

$$w_{1\bar{\zeta}_1} = 0 \quad \text{in } \mathrm{G}_1 = \zeta_1(\mathrm{G}). \quad (5.2.35)$$

Therefore Lemma 5.1.1 implies for w_1 the desired representation

$$w_1(z) = \frac{1}{2\pi i} \oint_{\dot{G}} \frac{d\zeta_1(t)}{\zeta_1(t) - \zeta_1(z)} \nu_1(t) + ik_1$$

$$= \frac{1}{2\pi i} \oint_{\dot{G}} \frac{\zeta_{1t}}{\zeta_1(t) - \zeta_1(z)} [dt + q_1 d\bar{t}] \nu_1(t) + ik_1. \quad (5.2.36)$$

Then the special vector

$$\tilde{\mathbf{w}} \equiv \frac{1}{2\pi i} \oint_{\dot{G}} V(t, z)\, d_Q t\{(\nu_1, 0, \ldots, 0)^T\} \quad (5.2.37)$$

is Q-holomorphic and its second component satisfies the equation

$$\tilde{w}_{2\bar{z}} - \kappa_{21} w_{1z} - q_2 \tilde{w}_{2z} = 0. \quad (5.2.38)$$

Introducing the difference

$$v_2 \equiv w_2 - \tilde{w}_2 \quad (5.2.39)$$

and using that $\mathbf{w}$ is Q-holomorphic we find

$$v_{2\bar{z}} - q_2 v_{2z} = w_{2\bar{z}} - q_2 w_{2z} - \tilde{w}_{2\bar{z}} + q_2 \tilde{w}_{2\bar{z}}$$

$$= w_{2\bar{z}} - q_2 w_{2z} - \kappa_{21} w_{1z} = 0. \quad (5.2.40)$$

That means with ζ_2 that:

$$v_{2\bar{\zeta}_2} = 0. \quad (5.2.41)$$

Consequently, v_2 has the desired representation:

$$v_2 = \frac{1}{2\pi i} \oint_{\dot{G}} \frac{d\zeta_2(t)}{\zeta_2(t) - \zeta_2(z)} \nu_2(t) + ik_2. \quad (5.2.42)$$

Repeating this procedure n times we get (5.2.34).

From (5.2.26), (5.2.32) and (5.2.34) we have immediately the

Lemma 5.2.5 *The boundary value problem in Bojarski's quasinormal form is equivalent to the following system of singular integral equations for the layers* $\boldsymbol{\nu}$ *and the constants* $\mathbf{k}$:

$$\mathbf{w}(z)=\frac{1}{2\pi i}\oint_{\dot{G}} V(t,z)\,d_Q t\boldsymbol{\nu}(t)+i k$$

$$+\frac{1}{2\pi i}\iint_G V(t,z)\{\mathsf{a}\mathbf{w}+\mathsf{b}\bar{\mathbf{w}}-\mathbf{c}(t)\}[dt,d\bar{t}] \quad \text{for} \quad z\in \mathsf{G}, \qquad (5.2.43)$$

$$2\boldsymbol{\psi}=2\operatorname{Re}\mathsf{P}\mathbf{w}|_{\dot{G}}$$

$$=\tfrac{1}{2}(\mathsf{P}+\bar{\mathsf{P}})\boldsymbol{\nu}+\frac{1}{2\pi i}(\mathsf{P}-\bar{\mathsf{P}})\oint_{\dot{G}} V(t,z)\,d_Q t\boldsymbol{\nu}(t)$$

$$+i(\mathsf{P}-\bar{\mathsf{P}})\mathbf{k}+\frac{1}{2\pi i}\bar{\mathsf{P}}\oint_{\dot{G}}(V\,d_Q t-\bar{V}\,d_Q\bar{t})\boldsymbol{\nu}(t)$$

$$=\frac{1}{2\pi i}\left\{\mathsf{P}\iint_G V(t,z)\{\mathsf{a}\bar{\mathbf{w}}+\mathsf{b}\bar{\mathbf{w}}-\mathbf{c}(t)\}[dt,d\bar{t}]\right.$$

$$\left.+\bar{\mathsf{P}}\iint_G \bar{V}\{\bar{\mathsf{a}}\bar{\mathbf{w}}+\bar{\mathsf{b}}\mathbf{w}-\bar{\mathbf{c}}\}[dt,d\bar{t}]\right\} \text{ on } \dot{\mathsf{G}}. \qquad (5.2.44)$$

Remarks From (5.2.28) it follows that on $\dot{\mathsf{G}}$ the equation

$$V(t,z)\,d_Q t=\frac{1}{(t-z)}\left[\mathsf{I}+\mathsf{Q}\frac{\bar{t}}{t}+\varepsilon_3(z,t)\right]^{-1}\left[\mathsf{I}+\mathsf{Q}\frac{\bar{t}}{t}\right]dt$$

$$=\frac{1}{(t-z)}[\mathsf{I}+\varepsilon_4(z,t)]\,dt \qquad (5.2.45)$$

holds. Hence the difference $V\,d_Q t-\bar{V}\overline{d_Q t}$ in (5.2.44) has the form

$$V(t,z)\,d_Q t-\bar{V}(t,z)\,\overline{d_Q t}=2i\mathsf{I}\,d_{n_t}\log|t-z|+\frac{\varepsilon_5(z,t)}{(t-z)}dt, \qquad (5.2.46)$$

where $|\varepsilon_5|\leq K|z-t|^{\alpha}$ with $\alpha>0$. The system (5.2.43), (5.2.44) has the same properties as (5.1.14) and (5.1.15).

Another possibility for the construction of suitable integral equations arising from (5.2.43) can also be found in Bojarski's work. If we consider the difference

$$\mathbf{w}-\frac{1}{2\pi i}\iint_{G} V(t, z)\mathbf{c}(t)[dt, d\bar{t}],$$

instead of $\mathbf{w}$ then we can restrict ourselves to homogeneous equations (with $\mathbf{c}=0$) and (5.2.43) becomes

$$\mathbf{w}(z)=\boldsymbol{\phi}(z)+\frac{1}{2\pi i}\iint_{G} V(t, z)\{\mathsf{a}\mathbf{w}+\mathsf{b}\overline{\mathbf{w}}\}[dt, d\bar{t}] \tag{5.2.47}$$

with the Q*-holomorphic vector* $\boldsymbol{\phi}$. Hence, (5.2.47) can be considered as an integral equation for $\mathbf{w}$, where $\boldsymbol{\phi}$ is suitably given. In general this integral equation may have eigensolutions. Because $V(t, z)$ is weakly singular (5.2.45), the equation (5.2.47) is a Fredholm integral equation of the second kind with a completely continuous integral operator. On its range the inverse exists defined by the resolvent, a sum of the identity and weakly singular integral operators again [49, III]. Consequently, (5.2.47) provides the *representation formula*

$$\mathbf{w}(z)=\boldsymbol{\phi}(z)+\iint_{G}\Gamma_1(z, t)\boldsymbol{\phi}(t)[dt, d\bar{t}]$$

$$+\iint_{G}\Gamma_2(z, t)\overline{\boldsymbol{\phi}(t)}[dt, d\bar{t}]+\sum_{k=1}^{N} c_k(\boldsymbol{\phi})\mathbf{w}_k(z), \tag{5.2.48}$$

where the $\mathbf{w}_1, \ldots, \mathbf{w}_N$ *form a basis of the eigensolutions* satisfying

$$\mathbf{w}_j(z)=-\frac{1}{2\pi i}\iint_{G} V(t, z)\{\mathsf{a}\mathbf{w}_j+\mathsf{b}\bar{\mathbf{w}}_j\}[dt, d\bar{t}]. \tag{5.2.49}$$

Remarks (i) Setting $\mathsf{a}=\mathsf{b}=0$ for $z\notin\bar{G}$ we find that the $\mathbf{w}_j$ are well defined in the whole plane if they are known in G.

(ii) The resolvent kernels Γ_1, Γ_2 can be found by a suitable integral equation system, which can be perturbed with a finite dimensional kernel such that the new equations become *uniquely solvable* (see e.g. Pettineo [55]).

If we use the relation between $\boldsymbol{\phi}$ and $\mathbf{w}$,

$$\boldsymbol{\phi}(z)=\frac{1}{2\pi i}\oint_{\dot{G}} V(t, z)\, d_Q t\mathbf{w}(t) \tag{5.2.50}$$

and replace $\boldsymbol{\phi}$ in (5.2.48) by (5.2.50) we find a representation formula for $\mathbf{w}$ in G by means of its *values on the boundary*:

$$\mathbf{w}(z)=\frac{1}{2\pi i}\oint_{\dot{\mathsf{G}}}\left(V(t,z)+\iint_{\mathsf{G}}\Gamma_1(z,\tau)V(t,\tau)[d\tau,d\bar{\tau}]\right)d_Qt\mathbf{w}(t)$$
$$-\frac{1}{2\pi i}\oint_{\dot{\mathsf{G}}}\iint_{\mathsf{G}}\Gamma_2(z,\tau)\overline{V(t,\tau)}[d\tau,d\bar{\tau}])\,\overline{d_Qt}\,\overline{\mathbf{w}(t)}$$
$$+\sum_{k=1}^{N}c_k(\boldsymbol{\phi}(\mathbf{w}))\mathbf{w}_k(z). \qquad (5.2.51)$$

The $c_k=c_k(\boldsymbol{\phi})$ are functionals of $\boldsymbol{\phi}$. Thus they also depend only on the boundary values of $\mathbf{w}$ according to (5.2.50).

Let us denote the kernels in (5.2.51) by Ω_1 and Ω_2:

$$\Omega_1(t,z)=V(t,z)+\iint_{\mathsf{G}}\Gamma_1(z,\tau)V(t,\tau)[d\tau,d\bar{\tau}],$$
$$\Omega_2(t,z)=\iint_{\mathsf{G}}\Gamma_2(z,\tau)\overline{V(t,\tau)}[d\tau,d\bar{\tau}]. \qquad (5.2.52)$$

Then (5.2.51) takes the form of a *generalized Cauchy formula*:

$$\mathbf{w}(z)=\frac{1}{2\pi i}\oint_{\dot{\mathsf{G}}}(\Omega_1\,d\Xi(t)\mathbf{w}(t)-\Omega_2\,d\overline{\Xi}(t)\bar{\mathbf{w}}(t))+\sum_{k=1}^{N}c_k(\boldsymbol{\phi}(\mathbf{w}))\mathbf{w}_k(z). \qquad (5.2.53)$$

This is Bojarski's generalization of Vekua's Cauchy formula [64, p. 155 (14.1)]. It can also be used for the construction of singular integral equations on the boundary. Bojarski used this method for solving Hilbert problems and the boundary value problems (5.2.1), (5.1.13).

Furthermore, *if* Γ_1, Γ_2 *are known* then (5.2.48) allows the approximation of $\mathbf{w}$ by systems of Q-holomorphic functions $\boldsymbol{\phi}$ satisfying

$$\boldsymbol{\phi}_{\bar{z}}-\mathsf{Q}\boldsymbol{\phi}_z=0. \qquad (5.2.54)$$

They could be generated e.g. by (5.2.34) where the generating function Ξ is needed explicitly. Hence this method becomes applicable only if we have a *simple method for generating* $\boldsymbol{\phi}$ and computing Γ_1, Γ_2, Ω_1 and Ω_2. For special cases and for analytic coefficients such methods are available.

The above constructions simplify if the Bojarski normal form is specialized to the hyperanalytic system (3.1.26), (3.1.27). These equations in the new coordinates take the form (5.2.1) with a special Q namely†

$$\begin{aligned} &w_{0\bar z} && + \sum_{j=0}^{n-1} (a_{0j} w_j + b_{0j} \bar w_j) + c_0 = 0 \\ &w_{k\bar z} + \sum_{l=1}^{k} (\mathrm{i}\mu)^l w_{k-lz} && + \sum_{j=0}^{n-1} (a_{kj} w_j + b_{kj} \bar w_j) + c_k = 0, \\ & && k = 1, \ldots, n-1. \end{aligned} \tag{5.2.55}$$

According to Douglis [15] let us introduce the matrix algebra generated by

$$\mathrm{e}^0 \equiv \mathsf{I},\ \mathrm{e} \equiv ((\delta_{j-1,k})) = \begin{pmatrix} 0 & & \\ 1 \diagdown & 0 & \\ 0 & 1 & 0 \end{pmatrix}, \tag{5.2.56}$$

and the complex numbers, the so called *algebra of hypercomplex numbers* or *Douglis algebra* where we identify the elements $v\mathsf{I}$ with v for $v \in \mathbb{C}$. This algebra can also be generated by the two elements i and e, subject to the multiplication rules

$$\mathrm{i}^2 = -1,\ \mathrm{ie} = \mathrm{ei} \text{ and } \mathrm{e}^n = 0. \tag{5.2.57}$$

The elements e^k, ie^k with $\mathrm{e}^0 = \mathsf{I}$, $k = 0, \ldots, n-1$ are linearly independent over the reals. Hypercomplex numbers c are written as

$$c = \sum_{k=0}^{n-1} c_k \mathrm{e}^k \tag{5.2.58}$$

where $c_0, \ldots, c_{n-1}$ are complex numbers. When $c_0 = 0$, c is said to be *nilpotent* and, according to (5.2.56), c has an inverse c^{-1} with $cc^{-1} = c^{-1}c = 1$ if and only if c is not nilpotent. For c one can introduce the module by

$$|c| \equiv \sum_{k=0}^{n-1} |c_k|. \tag{5.2.59}$$

This implies for hypercomplex valued functions the concept of continuity and norms corresponding to (5.2.59).

Multiplying the kth equation of (5.2.55) with e^k, we may write this system in the form

$$\left\{\frac{\partial}{\partial \bar z} + \mathrm{e}\alpha \frac{\partial}{\partial z}\right\}\left\{\sum_{k=0}^{n-1} w_k \mathrm{e}^k\right\} + \sum_{k=0}^{n-1} \mathrm{e}^k \left\{\sum_{j=0}^{n-1} (a_{kj} w_j + b_{kj} \bar w_j) + c_k\right\} = 0 \tag{5.2.60}$$

where the hypercomplex valued function α is defined by

$$\alpha(x, y) \equiv \mathrm{i}\mu \sum_{l=0}^{n-2} (\mathrm{i}\mu \mathrm{e})^l. \tag{5.2.61}$$

† Use (1.1.4) in (3.1.27) and replace $w_{k-l\bar z}$ by the preceding equations.

Since $\kappa_{j,j-l}=(\mathrm{i}\mu)^l$ the equations (5.2.9) and (5.2.10) for the generating solution Ξ take also a special form implying

$$t_k \equiv \zeta_{k+1,1} = \zeta_{k+2,2} = \cdots = \zeta_{n,n-k} \quad \text{for} \quad k=1,\ldots,n-1. \tag{5.2.62}$$

Thus, *the generating solution* itself can also be written in the Douglis algebra,

$$\Xi = z\mathsf{I} + \sum_{k=1}^{n-1} t_k \mathrm{e}^k. \tag{5.2.63}$$

Now the Cauchy formula (5.2.26) simplifies once more to

$$W(z) = \frac{1}{2\pi\mathrm{i}} \oint_{\dot{\mathsf{G}}} \frac{W(t)\,\mathrm{d}\Xi(t)}{\{\Xi(t)-\Xi(z)\}} + \frac{1}{2\pi\mathrm{i}} \iint_{\mathsf{G}} \frac{1}{\{\Xi(t)-\Xi(z)\}} \mathsf{D}_t W(t)[\mathrm{d}\Xi(t), \mathrm{d}\bar{t}] \tag{5.2.64}$$

where

$$W \equiv \sum_{k=0}^{n-1} w_k \mathrm{e}^k \tag{5.2.65}$$

and

$$\mathsf{D}_z W \equiv \left(\frac{\partial}{\partial \bar{z}} + \alpha \mathrm{e} \frac{\partial}{\partial z}\right) W. \tag{5.2.66}$$

This version can be found in the work of Gilbert [22] and Gilbert and Hile [24] where for systems (5.2.60) with coefficients a and b of lower triangular shape the generalized Cauchy formula (5.2.64) can be obtained more explicitly. This approach hinges on the fact that for

$$a_{jk} = b_{jk} = 0 \quad \text{for} \quad j<k \tag{5.2.67}$$

the following part of the similarity principle is still valid:

Theorem 5.2.6 (Gilbert and Hile [24]). *Let W be a continuous and bounded generalized solution of the generalized hyperanalytic system in the whole plane,*

$$\mathsf{D}_z W + \sum_{k=0}^{n-1} \mathrm{e}^k \left\{ \sum_{l=0}^{k} a_{kl} w_l + b_{kl} \bar{w}_l \right\} = 0 \tag{5.2.68}$$

with coefficients

$$\mathsf{a}, \mathsf{b} \in L^{p,2}(\mathbb{C}) \equiv \left\{ f \,\Big|\, \iint_{\mathbb{C}} (1+|z|)^{2p-4} |f|^p \,[\mathrm{d}z, \mathrm{d}\bar{z}] < \infty \right\},$$

$2<p<\infty$. *Then W has the form*

$$W(z)=C\exp[\omega(z)] \tag{5.2.69}$$

where C is a hypercomplex constant and ω *is a hypercomplex function in* $C^{\alpha}(\mathbb{C})$ *with* $\alpha=(p-2)/p$ *and* $|\omega(z)|\leq K|z|^{-\alpha}$ *for* $|z|\to\infty$.

Here the exponential function is defined by the power series

$$\exp[\omega(z)]\equiv\sum_{l=0}^{\infty}\frac{1}{l!}(\omega(z))^{l} \tag{5.2.70}$$

converging for every $\omega(z)$ thanks to

$$|\omega^2|\leq|\omega|^2 \text{ and } |\exp[\omega]|\leq\exp|\omega|$$

and (5.2.59).

The proof essentially generalizes the steps of the generalized analytic case in Theorem 1.1.3 due to Bers, Nirenberg and Vekua and can be found in [24]. We shall come back to this in Section 5.3.

As a consequence we have:

Proposition 5.2.7 *For systems* (5.2.55) *with triangular matrices* (5.2.67) *the homogeneous integral equation corresponding to* (5.2.47) *does not have eigensolutions* $\mathbf{w}_k$ *and* (5.2.48), (5.2.51), (5.2.53) *reduce to the case* $c_k\equiv 0$.

Proof From (5.2.7) it follows that

$$\lim_{|z|\to\infty}|\mathbf{w}_k(z)|=\lim_{|z|\to\infty}\frac{1}{2\pi}\left|\iint_G V(t,z)\{\mathbf{a}\mathbf{w}_k+\mathbf{b}\bar{\mathbf{w}}_k\}[dt,d\bar{t}]\right|=0. \tag{5.2.71}$$

On the other hand, the hypercomplex valued function W_k corresponding to $\mathbf{w}_k$ satisfies equation (5.2.68) in the whole plane and, according to Theorem 5.2.6, must have the form

$$W_k(z)=C\exp[\omega(z)]$$

with $W_k\to C$. Thus $C=0$ from (5.2.71) yields $W_k\equiv 0$.

In the more special case of purely hypercomplex equations,

$$\mathsf{D}_z W+AW+B\bar{W}=0 \tag{5.2.72}$$

corresponding to (3.1.29), the case of generalized hyperanalytic functions, the kernels Ω_1, Ω_2 in the generalized Cauchy formula (5.2.53) can be constructed by suitable generalized hyperanalytic functions. For this purpose let us assume

$$A=B=0 \quad\text{for}\quad |z|\geq R \quad\text{and}\quad A, B\in L^p \tag{5.2.73}$$

where $R>0$ is chosen big enough. (Gilbert and Hile [24] don't need (5.2.73) for some of the following results.) Now Gilbert and Hile construct the fundamental kernels Ω_1, Ω_2 and Γ_1, Γ_2 using Vekua's approach [64, III, Sections 8–14]:

Theorem 5.2.8 *There exist hypercomplex functions* $X^{(1)}(z,t)$, $X^{(2)}(z,t)$ *with the properties*:

$$\mathsf{D}_z X^{(j)}(z,t)+A(z)X^{(j)}(z,t)+B(z)\overline{X^{(j)}}(z,t)=0 \quad \text{in} \quad \mathbb{C}\backslash\{t\}, \tag{5.2.74}$$

$$X^{(j)}(z,t)=\frac{\exp\left[\omega^{(j)}(z,t)-\omega^{(j)}(t,t)\right]}{(\mathrm{i})^{j+1}2(\Xi(z)-\Xi(t))}, \quad j=1,2, \tag{5.2.75}$$

where $\omega^{(j)}\in C^{\alpha}(\mathbb{C}^2)$, $\alpha=(p-2)/p$, $\omega^{(j)}(z,\tau)\leqslant K\,|z|^{-\alpha}$.

Sketch of proof For $X^{(1)}$ the proof in [24] is based on the solution of the integral equation

$$W(z,\tau)=1-\frac{1}{2\pi\mathrm{i}}\iint_{\mathbb{C}}\left\{\frac{1}{\{\Xi(t)-\Xi(z)\}}-\frac{1}{\{\Xi(t)-\Xi(\tau)\}}\right\}$$

$$\times\left\{A(t)W(t,\tau)+B(t)\frac{\Xi(t)-\Xi(\tau)}{\overline{\Xi(t)}-\overline{\Xi(\tau)}}\overline{W(t,\tau)}\right\}[\mathrm{d}\Xi(t),\mathrm{d}\bar{t}] \tag{5.2.76}$$

which corresponds to the equations (5.2.47) and (5.2.64). It has a unique solution W according to a slight modification of Proposition 5.2.7. W takes the form (5.2.69),

$$W(z,\tau)=C(\tau)\exp\left[\omega(z,\tau)\right] \tag{5.2.77}$$

From $W(\tau,\tau)=1$ (5.2.77) implies

$$C(\tau)=\exp\left[-\omega(\tau,\tau)\right] \quad \text{and} \quad W(z,\tau)=\exp\left[\omega(z,\tau)-\omega(\tau,\tau)\right]. \tag{5.2.78}$$

Now for

$$X^{(1)}(z,\tau)\equiv\frac{W(z,\tau)}{-2(\Xi(z)-\Xi(\tau))}$$

one finds the desired properties. Similarly $X^{(2)}$ is constructed.

Theorem 5.2.9 *With* $X^{(1)}$, $X^{(2)}$, *the fundamental kernels* (5.2.52) *are defined by*

$$\begin{aligned}\Omega_1(t,z)&\equiv X^{(1)}(z,t)+\mathrm{i}X^{(2)}(z,t),\\ \Omega_2(t,z)&\equiv X^{(1)}(z,t)-\mathrm{i}X^{(2)}(z,t).\end{aligned} \tag{5.2.79}$$

They satisfy the equations

$$\begin{aligned}\mathsf{D}_z\Omega_1+A(z)\Omega_1+B(z)\bar{\Omega}_2&=0,\\ \mathsf{D}_z\Omega_2+A(z)\Omega_2+B(z)\bar{\Omega}_1&=0\end{aligned} \tag{5.2.80}$$

and

$$\begin{aligned}\mathsf{D}_t\Omega_1-A(t)\Omega_1+B(t)\overline{\Xi_t}(t)(\Xi_t(t))^{-1}\bar{\Omega}_2&=0,\\ \mathsf{D}_t\Omega_2-A(t)\Omega_2+B(t)\overline{\Xi_t}(t)(\Xi_t(t))^{-1}\bar{\Omega}_1&=0.\end{aligned} \tag{5.2.81}$$

and have the behaviour

$$\left|\Omega_1(t,z)-\frac{1}{\{\Xi(t)-\Xi(z)\}}\right|+|\Omega_2(t,z)|\leq K\,|z-t|^{-2/p} \tag{5.2.82}$$

for $z\to t$. *For every generalized continuous solution* $W\in C^0(\bar{\mathsf{G}})$ *of the homogeneous equation*

$$\mathsf{D}_zW+AW+B\bar{W}=0 \tag{5.2.83}$$

there holds the generalized Cauchy formula

$$\frac{1}{2\pi\mathrm{i}}\oint_{\dot{\mathsf{G}}}\{\Omega_1(t,z)W(t)\mathrm{d}\Xi(t)-\Omega_2(t,z)\overline{W(t)}\overline{\mathrm{d}\Xi(t)}\}=\begin{cases}W(z) & \text{for } z\in\mathsf{G},\\ 0 & \text{for } z\in\mathbb{C}\backslash\bar{\mathsf{G}}\end{cases} \tag{5.2.84}$$

With the hyperanalytic (Q*-holomorphic*) *function*

$$\Phi(z)=\frac{1}{2\pi\mathrm{i}}\oint_{\dot{\mathsf{G}}}\frac{W(t)}{\{\Xi(t)-\Xi(z)\}}\,\mathrm{d}\Xi(t), \tag{5.2.85}$$

corresponding to (5.2.50), *we also have the representation formulae*

$$W(z)=\frac{1}{2\pi\mathrm{i}}\oint_{\dot{\mathsf{G}}}\{\Omega_1(t,z)\Phi(t)\,\mathrm{d}\Xi(t)-\Omega_2(t,z)\overline{\Phi(t)}\,\overline{\mathrm{d}\Xi(t)}\} \tag{5.2.86}$$

and

$$W(z)=\Phi(z)+\iint_{\mathsf{G}}\{\Gamma_1(z,t)\Phi(t)+\Gamma_2(z,t)\overline{\Phi(t)}\}[\mathrm{d}t,\mathrm{d}\bar{t}] \tag{5.2.87}$$

corresponding to (5.2.48). *The kernels* Γ_1, Γ_2 *satisfy*

$$\begin{aligned}\Gamma_1(z,t)&=-\frac{1}{2\pi\mathrm{i}}\{\Xi_tA(t)\Omega_1(t,z)-\bar{\Xi}_t\overline{B(t)}\Omega_2(t,z)\},\\ \Gamma_2(z,t)&=\frac{1}{2\pi\mathrm{i}}\{\bar{\Xi}_t\overline{A(t)}\Omega_2(t,z)-\Xi_tB(t)\Omega_1(t,z)\}.\end{aligned} \tag{5.2.88}$$

Let us omit the proof, referring instead to Gilbert's and Hile's work [24].

Remark For coefficients A, B belonging only to $L^p_{loc}(G)$, $p>2$, Goldschmidt [28], [29] also constructed the fundamental kernels.

Both formulae (5.2.86) and (5.2.87) provide the generation of solutions to the system (5.2.83) by hyperanalytic functions. Hence, the set of solutions of the special system (5.2.54) is mapped into the set of solutions of the more general system (5.2.83). (5.2.86) and (5.2.87) define *generating operators* in the sense of Bergman [4] and Vekua [65]. For the general study of these operators see also the work of Gilbert [21], [22] and Colton [11], [12] and the references given there.

In our case, let us note that the hyperanalytic functions have local representation as power series,

$$\Phi(z)=\sum_{j=0}^{\infty} A_j(\Xi(z)-\Xi(z_0))^j, \tag{5.2.89}$$

about every point z_0 in the interior of the region where (5.2.31) holds. This was shown by Douglis [15]. This is an important property since e.g. *polynomials in* $\Xi(z)$ *can now be used for generating solutions* with (5.2.86) or (5.2.87).

Now let us consider a system

$$\mathbf{u}_x+\mathbf{B}\mathbf{u}_y=\mathbf{f} \tag{5.2.90}$$

where $\mathbf{B}$ has *constant coefficients.* To begin with let us assume that $\mathbf{B}$ admits a *complete system of eigenvectors* but no generalized eigenvectors of higher grade, i.e. $\mathbf{B}$ is diagonalizable and $\mathbf{Q}$ in the quasinormal form is a diagonal matrix. Here we have from (5.1.5) the representation

$$w_j(z)=\frac{1}{2\pi i}\iint_G \frac{1}{\zeta_j(t)-\zeta_j(z)}\, w_{j\bar\zeta_j}(t)[d\zeta_j(t), d\bar\zeta_j(t)]$$
$$+\frac{1}{2\pi i}\oint_{\dot G} \frac{\nu_j(t)}{\zeta_j(t)-\zeta_j(z)}\, d\zeta_j(t)+ik_j, \quad j=1,\dots,n \tag{5.2.91}$$

for $z\in G$. Since $\mathbf{B}$ is constant and ζ_j is an affine transformation this specializes with

$$\frac{d\zeta_j(t)}{\zeta_j(t)-\zeta_j(z)}=\frac{dt+(\lambda_+^j-i)(\lambda_+^j+i)^{-1}d\bar t}{t-z+(\lambda_+^j-i)(\lambda_+^j+i)^{-1}(\bar t-\bar z)}=\frac{\lambda_+^j\,d\xi-d\eta}{\lambda_+^j(\xi-x)-(\eta-y)} \tag{5.2.92}$$

to

$$w_j(z)=\frac{1}{2\pi i}\iint_G \frac{[\lambda_+^j d\xi-d\eta, dw_j]}{\lambda_+^j(\xi-x)-(\eta-y)}+\frac{1}{2\pi i}\oint_{\dot G}\frac{\nu_j(t)\{\lambda_+^j\, d\xi-d\eta\}}{\lambda_+^j(\xi-x)-(\eta-y)}+ik_j. \tag{5.2.93}$$

With (3.1.6), (3.1.11) and (3.1.14) this takes the form

$$\mathbf{u}(z) = \text{Re}\,\frac{1}{\pi \mathrm{i}} \iint_{G} \sum_{j=1}^{n} \frac{\mathbf{q}_j \mathbf{p}_j^T \{\mathbf{u}_\xi + \mathbf{B}\mathbf{u}_\eta\}[\mathrm{d}\xi, \mathrm{d}\eta]}{\lambda_+^j(\xi - x) - (\eta - y)}$$

$$+ \text{Re}\,\frac{1}{\pi \mathrm{i}} \oint_{\dot{G}} \sum_{j=1}^{n} \frac{\mathbf{q}_j \{\lambda_+^j \mathrm{d}\xi - \mathrm{d}\eta\} \nu_j(t)}{\lambda_+^j(\xi - x) - (\eta - y)} + 2\,\text{Re} \sum_{j=1}^{n} k_j \mathrm{i} \mathbf{q}_j. \tag{5.2.94}$$

According to the above assumption the resolvent $(\lambda \mathsf{I} - \mathsf{B})^{-1}$ has at $\lambda = \lambda_+^j$ a pole of order 1 and admits the Laurent expansion

$$(\lambda \mathsf{I} - \mathsf{B})^{-1} = \frac{1}{\lambda - \lambda_+^j} \sum_{l} \mathbf{q}_l \mathbf{p}_l^T \rho_l + \cdots \tag{5.2.95}$$

about λ_+^j where the ρ_l denote suitable scalar weights and where the sum is taken over all indices with $\lambda_+^j = \lambda_+^l$ (see e.g., [68, Satz 4.3]). Applying (5.2.95) to $(\lambda \mathsf{I} - \mathsf{B})\mathbf{q}$ we find with (3.1.9) that $\rho_l = 1$. Hence, the integrand of the first term in (5.2.94) takes the form

$$\sum_{j=1}^{n} \frac{\mathbf{q}_j \mathbf{p}_j^T}{\lambda_+^j(\xi - x) - (\eta - y)} = \sum_{j=1}^{n} \left\{ \text{Residue}\, \frac{(\lambda \mathsf{I} - \mathsf{B})^{-1}}{\lambda(\xi - x) - (\eta - y)} \bigg|_{\lambda = \lambda_+^j} \right\}$$

$$= \frac{1}{2\pi \mathrm{i}} \oint_{\mathsf{c}} \frac{(\lambda \mathsf{I} - \mathsf{B})^{-1}}{\lambda(\xi - x) - (\eta - y)} \mathrm{d}\lambda \tag{5.2.96}$$

where c denotes the same complex path in the upper λ-halfplane as in (4.1.6). Here we used the identity

$$\lambda(\xi - x) - (\eta - y) = \frac{(\lambda + \mathrm{i})}{2} \left\{ 1 + \frac{\lambda - \mathrm{i}}{\lambda + \mathrm{i}} \cdot \frac{(\bar{\zeta} - \bar{z})}{(\zeta - z)} \right\} (\zeta - z)$$

implying

$$|\lambda(\xi - x) - (\eta - y)| \geq \left| \frac{\lambda + \mathrm{i}}{2} \right| \{1 - |q|\} |\zeta - z| \quad \text{with} \quad |q| < 1$$

$$\text{for} \quad \text{Im}\ \lambda > 0. \tag{5.2.97}$$

In order to express the eigenvectors $\mathbf{q}_j$ in a way similar to (5.2.96) let us use the column vectors of

$$\text{cofactors of } (\lambda \mathsf{I} - \mathsf{B}) \text{ denoted by } \mathbf{E}_l(\lambda), \quad l = 1, \ldots, 2n. \tag{5.2.98}$$

Then we have that the expressions

$$\frac{1}{2\pi \mathrm{i}} \oint_{\mathsf{c}} \mathbf{E}_{l_j}(\lambda)\, (\det(\lambda \mathsf{I} - \mathsf{B}))^{-1} \mathrm{d}\lambda, \quad j = 1, \ldots, n \tag{5.2.99}$$

form a basis to the space spanned by the $\mathbf{q}_j$ if the indices l_j are suitably chosen. Using other densities $\mu_j(t)$, which can be obtained from the $\nu_j(t)$ by a regular affine transformation at every point t, and using also transformed constants $\tilde{k}_j$, the representation (5.2.94) can be written as

$$\mathbf{u}(z)=-\frac{1}{2\pi^2}\iint\limits_{G}\operatorname{Re}\left\{\oint\limits_{c}\frac{(\lambda\mathsf{I}-\mathsf{B})^{-1}}{\lambda(\xi-x)-(\eta-y)}\,d\lambda\right\}\{\mathbf{u}_\xi+\mathsf{B}\mathbf{u}_\eta\}[d\xi,d\eta]$$
$$-\frac{1}{2\pi^2}\oint_{\dot{G}}\sum_{j=1}^{n}\mu_j(t)\operatorname{Re}\left\{\frac{d}{ds_t}\oint_{c}\frac{\mathbf{E}_{l_j}(\lambda)}{\det(\lambda\mathsf{I}-\mathsf{B})}\log[\lambda(\xi(s)-x)\right.$$
$$\left.-(\eta(s)-y)]\,d\lambda\right\}ds_t+\frac{1}{\pi}\sum_{j=1}^{n}\operatorname{Re}\left\{\oint_{c}\frac{\mathbf{E}_{l_j}(\lambda)}{\det(\lambda\mathsf{I}-\mathsf{B})}\,d\lambda\right\}\tilde{k}_j. \tag{5.2.100}$$

This is essentially the same representation of $\mathbf{u}$ as with the Agmon fundamental solution [1]. Although this formula was only obtained for B with a resolvent having only simple poles it remains valid for the general case since it can be approximated by constant coefficient systems with the above property according to Theorem 4.2.6 and since on c all the expressions in (5.2.100) depend continuously on the coefficients of B. The above kernels correspond to the fundamental solution of the system and have been used by Fichera [16] for higher order plane equations in order to construct the fundamental solution of equations with variable coefficients with the so-called parametrix method of Levy (see Miranda [50, p. 92ff]). Then Fichera used the jump relations for (5.2.100) in connection with the boundary conditions in order to get singular integral equations for the densities μ_j. In the case of constant coefficients these integral equations can also be obtained from (5.1.16). Since we consider equations (5.2.90) the equation (5.2.100) provides the solution $\mathbf{u}$ if μ_j and $\tilde{k}_j$ are known. Particular solutions to (5.2.90) are given by

$$\mathbf{u}_p(z)=-\frac{1}{2\pi^2}\iint\limits_{G}\operatorname{Re}\left\{\oint\limits_{c}\frac{(\lambda\mathsf{I}-\mathsf{B})^{-1}}{\lambda(\xi-x)-(\eta-y)}\,d\lambda\right\}\mathbf{f}(\zeta)\,[d\xi,d\eta]. \tag{5.2.101}$$

Thus we may restrict ourselves in the following to the case $f=0$. With (3.1.36), (3.1.9) and with (5.2.99) instead of $\mathbf{q}_j$, (5.1.15) takes the form

$$\sum_{k=1}^{n}\mathbf{r}_j\left\{\operatorname{Re}\frac{1}{2\pi i}\oint_{c}\frac{\mathbf{E}_{l_k}(\lambda)\,d\lambda}{\det(\lambda\mathsf{I}-\mathsf{B})}\right\}\mu_k(z)$$
$$+\sum_{k=1}^{n}\mathbf{r}_j\left\{\frac{1}{\pi}\operatorname{Im}\frac{1}{2\pi i}\oint_{\dot{G}}\oint_{c}\frac{\mathbf{E}_{l_k}(\lambda)(\lambda\dot{\xi}-\dot{\eta})\,d\lambda}{(\det(\lambda\mathsf{I}-\mathsf{B}))(\lambda(\xi-x)-(\eta-y))}\right\}\mu_k(t)\,ds_t$$
$$-\sum_{k=1}^{n}\mathbf{r}_j\left\{\operatorname{Im}\frac{1}{2\pi i}\oint_{c}\frac{\mathbf{E}_{l_k}(\lambda)\,d\lambda}{\det(\lambda\mathsf{I}-\mathsf{B})}\right\}\tilde{k}_k=2\psi_j,\quad j=1,\ldots,n. \tag{5.2.102}$$

Now let us follow ideas of Ricci [59]. It is easy to see that from our smoothness assumptions for $\dot{\mathsf{G}}$ it follows that

$$\frac{\lambda\dot{\xi}-\dot{\eta}}{\lambda(\xi-x)-(\eta-y)}\,ds=\frac{dt}{t-z}+H(\lambda,t,z)\,ds \tag{5.2.103}$$

holds with a smooth remainder $H\in C^{\omega}(\mathsf{c})\times C^{\alpha}(\dot{\mathsf{G}})\times C^{\omega}(\bar{\mathsf{G}})$ (C^{ω} means analytic). Moreover, introducing the matrix

$$\prod\equiv((\delta_{j_{l_k}}))_{\substack{j=1,\dots,2n\\k=1,\dots,n}} \tag{5.2.104}$$

corresponding to the indices in (5.2.99) *we can write for* (5.2.102):

$$\begin{aligned}&\left\{\mathbf{r}_j\mathrm{Re}\,\frac{1}{2\pi i}\oint_{\mathsf{c}}(\lambda\mathsf{I}-\mathsf{B})^{-1}\,d\lambda\prod\right\}\boldsymbol{\mu}(z)\\&+\left\{\mathbf{r}_j\mathrm{Im}\,\frac{1}{2\pi i}\oint_{\mathsf{c}}(\lambda\mathsf{I}-\mathsf{B})^{-1}d\lambda\prod\right\}\frac{1}{\pi}\oint_{\dot{\mathsf{G}}}\frac{\boldsymbol{\mu}(t)dt}{t-z}\\&-\mathbf{r}_j\ \mathrm{Im}\,\frac{1}{2\pi i}\oint_{\mathsf{c}}(\lambda\mathsf{I}-\mathsf{B})^{-1}\,d\lambda\prod\tilde{\mathbf{k}}\\&+\mathbf{r}_j\mathrm{Im}\,\frac{1}{2\pi i}\oint_{\dot{\mathsf{G}}}\oint_{\mathsf{c}}(\lambda\mathsf{I}-\mathsf{B})^{-1}H(\lambda,t,z)\prod\frac{1}{\pi}\,d\lambda\,\boldsymbol{\mu}(t)\,ds_t=2\psi_j.\end{aligned} \tag{5.2.105}$$

Due to (5.1.3), the principal part of this system of singular integral equations for $\boldsymbol{\mu}$ and $\tilde{\mathbf{k}}$ and the corresponding coefficient matrices are

$$\begin{aligned}A&\equiv\mathsf{r}\frac{1}{2}\frac{1}{2\pi i}\left(\oint_{\mathsf{c}}(\lambda\mathsf{I}-\mathsf{B})^{-1}\,d\lambda-\overline{\oint_{\mathsf{c}}(\lambda\mathsf{I}-\mathsf{B})^{-1}\,d\lambda}\right)\prod\\&=\mathsf{r}\frac{1}{4\pi i}\oint_{\mathsf{c}'}(\lambda\mathsf{I}-\mathsf{B})^{-1}\,d\lambda\prod\quad\text{and}\\C&\equiv\mathsf{r}\frac{-1}{2\pi}\frac{1}{2\pi}\left(\oint_{\mathsf{c}}(\lambda\mathsf{I}-\mathsf{B})^{-1}\,d\lambda+\overline{\oint_{\mathsf{c}}(\lambda\mathsf{I}-\mathsf{B})^{-1}\,d\lambda}\right)\prod\\&=-\mathsf{r}\frac{1}{4\pi^2}\oint_{\mathsf{c}''}(\lambda\mathsf{I}-\mathsf{B})^{-1}\,d\lambda\prod\end{aligned} \tag{5.2.106}$$

where c' and c'' are paths of integration in the λ-plane as indicated in Fig. 3. The equations (5.2.105) are singular integral equations in standard form

$$A\boldsymbol{\mu}(z)+C\frac{1}{\pi}\oint_{\dot{\mathsf{G}}}\frac{\boldsymbol{\mu}(t)\,dt}{t-z}+C\tilde{\mathbf{k}}+\oint_{\dot{\mathsf{G}}}\kappa(z,t)\boldsymbol{\mu}(t)\,ds_t=2\boldsymbol{\psi} \tag{5.2.107}$$

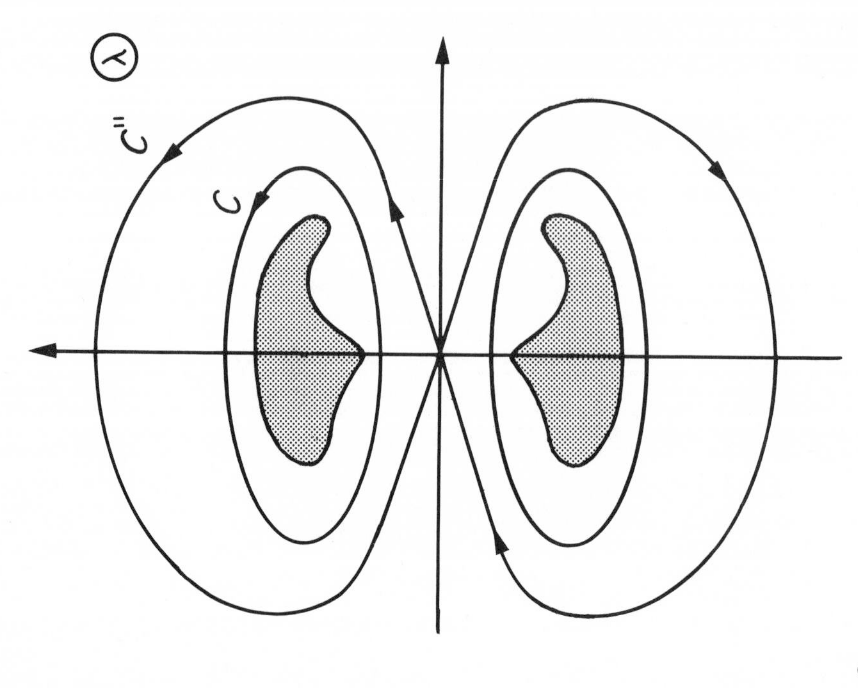

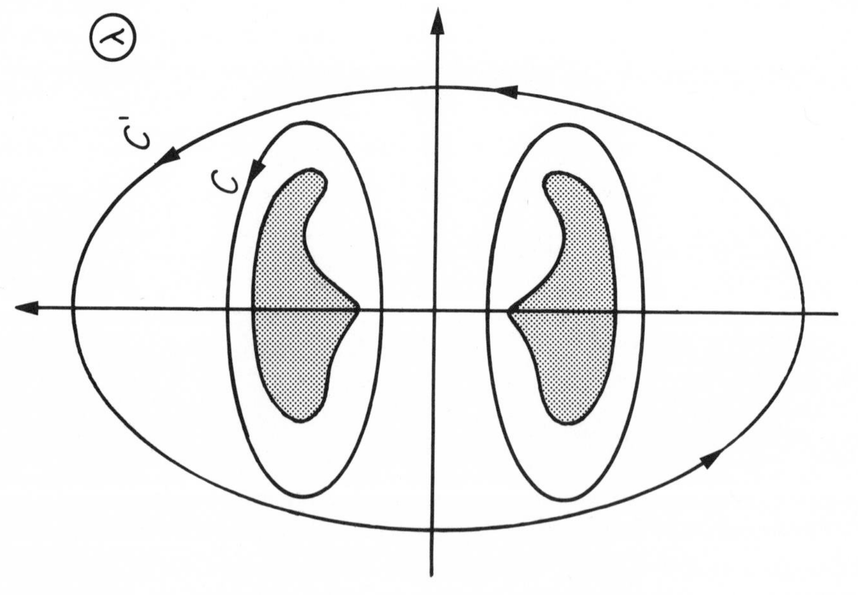

Fig. 3

where the kernel κ is given by

$$\kappa(z,t)\,\mathrm{d}s_t = \mathsf{r}\ \mathrm{Im}\frac{1}{2\pi^2\mathrm{i}}\oint_{\mathrm{c}}(\lambda\mathsf{I}-\mathsf{B})^{-1}H(\lambda,t,z)\,\mathrm{d}\lambda\prod\mathrm{d}s_t$$

$$+\frac{1}{\pi}(A-\mathrm{i}C)\,\mathrm{d}_{n_t}\log|t-z|, \quad (5.2.108)$$

defining a completely continuous integral operator with continuous kernel (or weakly continuous kernel if $\dot{\mathsf{G}}\in C^{1+\alpha}$).

For applications there are two cases where the system (5.2.107) becomes particularly handy for computations, namely if $A\equiv 0$ or $C\equiv 0$.

In case $C\equiv 0$ *the system* (5.2.107) *degenerates to a Fredholm integral equation of the second kind* for which many computational methods are available (see e.g. K. Atkinson [2] and Ben Noble [53]).

In case $A\equiv 0$ *the system* (5.2.107) *degenerates to singular integral equations of the first kind.* Since we have

$$\left.\frac{\mathrm{d}t}{z-t}\right|_{\dot{\mathsf{G}}} = -\frac{\mathrm{d}}{\mathrm{d}s}\log|z-t(s)|\,\mathrm{d}s+\mathrm{i}\,\mathrm{d}_{n_t}\log|z-t|, \quad (5.2.109)$$

for $A\equiv 0$ the system (5.2.107) becomes after multiplication with πC^{-1} and integration by parts

$$\oint_{\dot{\mathsf{G}}}\log|z-t(s)|\,\boldsymbol{\sigma}(s)\,\mathrm{d}s-\pi\tilde{\mathbf{k}}+\oint_{\dot{\mathsf{G}}}K(z,t(s'))\,\mathrm{d}s'\oint_{\dot{\mathsf{G}}}\boldsymbol{\sigma}\,\mathrm{d}s$$

$$-\int_{s=0}^{L}\int_{s'=0}^{s}K(z,t(s'))\,\mathrm{d}s'\boldsymbol{\sigma}(s)\,\mathrm{d}s = 2\pi C^{-1}\boldsymbol{\psi}\equiv\boldsymbol{\phi} \quad (5.2.110)$$

where

$$K(z,t)\equiv\pi C^{-1}\left\{\kappa(z,t)+\frac{\mathrm{i}}{\pi}C\frac{\partial}{\partial n_t}\log|t-z|\right\} \quad \text{and}$$

$$\boldsymbol{\sigma}(s)=\boldsymbol{\mu}'(s).$$

This is a system of Fredholm equations of the first kind with the logarithmic kernel as the principal part. Prescribing the values of $\oint_{\dot{\mathsf{G}}}\boldsymbol{\sigma}\,\mathrm{d}s = 0$ in (5.2.110), these equations can be used for finding $\boldsymbol{\sigma}(s)$ and the constants $\tilde{\mathbf{k}}$.

Although first kind equations have not so many benefits for computations as the Fredholm integral equations of the second kind, they still provide many advantages as e.g. good approximation properties for the so called Galerkin methods, and they find more and more interest in the field of numerical treatment and applications (see e.g. Hsiao and Wendland [40] and the references given there). It turns out that (5.2.110) defines a *Fredholm mapping of index zero* from $C^{\alpha}(\dot{\mathsf{G}})\times R^n$ into $C^{1+\alpha}(\dot{\mathsf{G}})$ if $\oint_{\dot{\mathsf{G}}}\boldsymbol{\sigma}\,\mathrm{d}s$ is a given constant vector.

Thus it is of practical interest to know whether one of the cases $A \equiv 0$ or $C \equiv 0$ is available. In the case of higher order strongly elliptic equations Fichera showed already in 1960 [16] that the corresponding Dirichlet problems lead to the case $A \equiv 0$. There he also investigated the corresponding equations for elliptic equations with variable coefficients using the fundamental solution. For these cases the role of $\tilde{\mathbf{k}}$ is also clarified in terms of suitable eigenfunctions of the elliptic differential equations. Hsiao and Mac Camy [39] investigated this approach for $(\Delta - \sigma')U = 0$ and for $(\Delta^2 - \sigma'\Delta)U = 0$ thoroughly and applied it to singular perturbations in connection with viscous flows. Their equations look like (5.2.110). Ricci [59] investigated the integral equations (5.2.105) for higher order elliptic differential equations with constant coefficients and formulated necessary and sufficient conditions for the boundary operators such that the matrix corresponding to A vanishes. Fichera and Ricci [17] extended this approach to even order complex elliptic systems in the plane with constant coefficients.

Let us conclude this section with a crude characterization of the cases $A = 0$ or $C = 0$:

Theorem 5.2.10 *The coefficient matrices A and C are given by*

$$A = \tfrac{1}{2}\mathsf{r}\prod = ((A_{jk})), \text{ i.e. } A_{jk} = \tfrac{1}{2}r_{jl_k} \quad \text{for} \quad j,k = 1, \ldots, n \tag{5.2.111}$$

and

$$C = -\mathsf{r}\frac{1}{2\pi}\int_{-\infty}^{+\infty} (\rho\mathsf{I} - \mathsf{B})^{-1}\, \mathrm{d}\rho \prod, \ \rho \in R. \tag{5.2.112}$$

Hence $A \equiv 0$ *if and only if the boundary conditions in* (4.1.1) *involve only the unknowns* u_m *with* $m \neq l_k$,

$$\mathbf{r}_j\mathbf{u} = \sum_{m \neq l_k} r_{jm}u_m = \psi_j, \qquad j = 1, \ldots, n. \tag{5.2.113}$$

Proof The proof is a simple consequence of the Cauchy Theorem for holomorphic functions.
Choosing for c' in (5.2.106) the circle

$$\lambda = R\mathrm{e}^{\mathrm{i}\alpha}, \qquad 0 \leq \alpha \leq 2\pi \quad \text{we have}$$

$$\begin{aligned} A &= \mathsf{r}\lim_{R\to\infty} \frac{1}{4\pi\mathrm{i}} \int_{\alpha=0}^{2\pi} (R\mathrm{e}^{\mathrm{i}\alpha}\mathsf{I} - \mathsf{B})^{-1}\,\mathrm{i}\,R\mathrm{e}^{\mathrm{i}\alpha}\,\mathrm{d}\alpha \prod, \\ &= \tfrac{1}{2}\mathsf{r}\lim_{R\to\infty} \frac{1}{2\pi} \int_{\alpha=0}^{2\pi} (\mathsf{I} - R^{-1}\,\mathrm{e}^{-\mathrm{i}\alpha}\mathsf{B})^{-1}\,\mathrm{d}\alpha \prod, \\ &= \tfrac{1}{2}\mathsf{r}\prod. \end{aligned}$$

In order to express C we choose for c the path $\lambda=\rho$ with $-R\leqslant\rho\leqslant R$ and $\lambda=Re^{i\alpha}$ with $0\leqslant\alpha\leqslant\pi$ finding

$$C=-\mathsf{r}\frac{1}{2\pi}\left\{\lim_{R\to\infty}\int_{-R}^{R}(\rho\mathsf{I}-\mathsf{B})^{-1}\,d\rho\right.$$

$$\left.+\lim_{R\to\infty}\operatorname{Re} i\int_{\alpha=0}^{\pi}(\mathsf{I}-R^{-1}e^{-i\alpha}\mathsf{B})^{-1}\,d\alpha\right\}=-\mathsf{r}\frac{1}{2\pi}\int_{-\infty}^{+\infty}(\rho\mathsf{I}-\mathsf{B})^{-1}\,d\rho,$$

the desired representation.

Remark 5.2.11 It is also possible to carry over Tovmasjan's approach [62], [63] to the first-order systems with constant coefficients. Then this method leads to Fredholm integral equations of the second kind having smooth kernels. But the details are yet to be done.

5.3 Generalized Hyperanalytic Functions, Zeros, Unique Continuation and the Runge Property

As we have already seen in Section 5.2, the generalized hyperanalytic systems

$$\mathsf{D}W+\sum_{k=0}^{n-1}e^{k}\sum_{l=0}^{k}\{a_{kl}w_l+b_{kl}\bar{w}_l\}=C \tag{5.3.1}$$

allow a treatment similar to that known from the Bers–Haack–Vekua Theory in the case $n=1$. It is even possible to use the approach of Chapter 1 for boundary value problems with special boundary conditions

$$\operatorname{Re}\left\{W\exp\left[i\sum_{k=0}^{n-1}e^{k}\tau_k(s)\right]\right\}=\sum_{k=0}^{n-1}e^{k}\psi_k \quad \text{on} \quad \dot{\mathsf{G}} \tag{5.3.2}$$

where the τ_k and ψ_k are real valued. For $n=1$, the solution of the corresponding boundary value problems in Sections 1.2 and 1.3 was based on the factorization of the solution with holomorphic functions. Here also a similar factorization with suitable hyperanalytic functions yields a stepwise construction of the solution. This was done by Begehr and Gilbert in [3]. Since the problems lead with suitable side conditions to well posed linear problems these can be used as linearizations of certain corresponding semilinear systems considered by Gilbert [23] and Wendland [69].

The factorization underlying the above treatment of (5.3.1), (5.3.2) in the corresponding case $n=1$, presented in Chapter 1, was connected with the *similarity principle* of Bers and Vekua for generalized analytic functions. But for $n>1$ the similarity principle is no longer valid for general systems. Kühn [44] showed for the generalized hyperanalytic systems (5.3.1) (with $C=0$) a certain similarity principle but there the factor carrying the zeros is no longer analytic or hyperanalytic but only $\tilde{\mathsf{Q}}$-holomorphic where $\tilde{\mathsf{Q}}$ depends on the solution. Except for Theorem 5.2.6 this is the only global result and according to Habetha's work in [34] Kühn's result seems to be the most general. Although the global similarity principle has to be given up, for homogeneous systems in the Pascali normal form (5.1.23) (with $\mathbf{c}=0$) a local similarity principle is valid, thus providing *Carleman's theorem*, namely that zeros of nontrivial solutions are isolated and of finite order. In contrast to the analytic case (see Section 5.4) Carleman's theorem holds in the case of nonanalytic coefficients only for the Pascali systems, as Habetha [34] and Heinz and Hildebrandt [36] remark, and for the generalized hyperanalytic systems (5.3.1) according to Kühn [44]. (For Habetha's theorem [34, Theorem 3] see the further remark following Theorem 5.3.5.) Hence it is still an open question for which elliptic first-order systems with or without normal form Carleman's theorem remains valid.

The still more general property of the *uniqueness of the Cauchy problem* or *unique continuation property* cannot hold for all elliptic systems according to the example by Pliš [56] but here Hile and Protter [37] showed this property for systems (4.1.1) with B being *normal.* This case contains systems having normal form (3.1.30) with $\alpha_{jk}=\beta_{jk}\equiv 0$ to which belong in particular the systems for which all the eigenvalues λ_+^j, $j=1,\dots,n$ are distinct and for which Carleman [10] already proved uniqueness of the Cauchy problem. These results will be presented here without proof. Hence unique continuation holds for the generalized analytic case (5.3.1) on the one hand and for the above mentioned systems on the other hand. Besides the above results there are many presented by Hörmander [38, VIII] but for the conditions given there it is not clear how they can be satisfied by simple conditions for the systems with $n>1$. Hence it is still open how to characterize the class of elliptic first-order systems for which unique continuation holds. Finally we shall present (without proof) a result by Goldschmidt [28] who showed the *Runge approximation property* for the purely hyperanalytic systems (5.2.72) with coefficients A, B belonging only to L^p_{loc} with $p>2$. Thus far this is the only generalization of the case $n=1$ to $n>1$.

Now let us begin with Begehr's and Gilbert's construction of solutions for (5.3.1), (5.3.2) which is based on the *first boundary value problem* with $\tau_k\equiv 0$. This special problem can be solved with an iterative procedure thanks to the triangular shape of (5.3.1) using the integral equations of Section 1.1.

Begin with w_0 by solving

$$w_{0\bar{z}} + a_{00}w_0 + b_{00}\bar{w}_0 = c_0 \quad \text{in} \quad \mathsf{G}, \qquad \operatorname{Re} w_0 = \psi_0 \quad \text{on} \quad \dot{\mathsf{G}}$$

and then evaluate w_1 from

$$w_{1\bar{z}} + a_{11}w_1 + b_{11}\bar{w}_1 = c_1 - \alpha_0 w_{0z} - a_{10}w_0 - b_{10}w_0$$
$$\text{in} \quad \mathsf{G}, \qquad \operatorname{Re} w_1 = \psi_1 \quad \text{on} \quad \dot{\mathsf{G}},$$

then evaluate w_2 and so on. The solution contains exactly n real constants which can be chosen arbitrarily.

If the characteristic of (5.3.2) vanishes,

$$n' \equiv \frac{1}{2\pi} \oint_{\dot{\mathsf{G}}} d\tau_0 = 0 \tag{5.3.3}$$

then the hypercomplex valued function $\exp\left[i \sum_{n=0}^{n-1} e^k \tau_k\right]$ admits a nonvanishing hypercomplex valued extension to a function $\gamma \in C^{1+\alpha}(\bar{\mathsf{G}})$ if $\tau_k \in C^{1+\alpha}(\dot{\mathsf{G}})$. Hence, the ansatz

$$w = \gamma^{-1}\omega \tag{5.3.4}$$

yields a first boundary value problem for ω.

If

$$n' = \frac{1}{2\pi} \oint_{\dot{\mathsf{G}}} d\tau_0 < 0 \tag{5.3.5}$$

then the ansatz similar to (1.2.11),

$$w = \prod_{j=1}^{|n'|} (\Xi(z) - \Xi(z_j))\omega, \qquad z_j \in \mathsf{G} \tag{5.3.6}$$

yields for ω a new boundary value problem of characteristic zero. Using this method Begehr and Gilbert [3] prove among other results the following.

Theorem 5.3.1 *The homogeneous boundary value problem* (5.3.1), (5.3.2) (*with* $\psi = 0$, $C = 0$) *with* $n' \leq 0$ *has exactly* $n(1\text{-}2n')$ *linearly independent solutions over R. The complex parts* w_0 *of these eigensolutions do not vanish identically.*

Using a basis of these eigensolutions constructed in [3] it can also be shown that the additional conditions

$$W(z_j) = \sum_{k=0}^{n-1} e^k \gamma_{kj} \equiv \gamma_j \quad \text{for} \quad j = 1, \ldots, |n'|, \tag{5.3.7}$$

$$\oint_{\dot{\mathsf{G}}} \operatorname{Im} \sigma W \exp\left[i\tau(s)\right] ds = \sum_{k=0}^{n-1} e^k \kappa_k \equiv \kappa \tag{5.3.8}$$

complete the boundary value problem (5.3.1), (5.3.2) for $n' \leq 0$ to a well posed problem which is uniquely solvable where the $z_j \in \mathsf{G}$ in (5.3.6) and (5.3.7) are distinct points which, as well as the real weight function $\sigma > 0$, can be chosen arbitrarily.

Theorem 5.3.2 *The boundary value problem* (5.3.1), (5.3.2), (5.3.7), (5.3.8) *with* $n' \leq 0$ *is well posed and defines a continuous mapping from the right-hand sides* $(C, \psi, \gamma_1, \ldots, \gamma_{|n'|}, \kappa) \in C^\alpha(\bar{\mathsf{G}}) \times C^{1+\alpha}(\dot{\mathsf{G}}) \times C^{n \cdot |n'|} \times R^n$ onto the solutions $W \in C^{1+\alpha}(\bar{\mathsf{G}})$.

Since this theorem corresponds to Theorem 1.2.4, the case $n = 1$, it is clear that the approach in Section 4.4 can be carried over to semilinear systems where the differential equations (5.3.1) are replaced by

$$\mathsf{D}W \equiv H(z, \bar{z}, W, \bar{W}). \tag{5.3.9}$$

For $\tau \equiv 0$, the Dirichlet problem, this has been done by Gilbert [23] using a suitable generalization of (4.4.25). For τ with $n' \leq 0$ Wendland [69] treated these problems under conditions corresponding to (4.4.28).

The above approach suggests that the similarity principle, Theorem 1.1.3, might also hold for the system (5.3.1) where the analytic functions have to be replaced by hyperanalytic functions. But unfortunately this is *not* true globally without making a change of the operator D depending on the solution except for systems of special type. For the generalized hyperanalytic functions the problems arise since the quotient $W\bar{W}^{-1}$ *does not remain bounded* if w_0 becomes zero with a higher order than $w_1, \ldots, w_{n-1}$. The only global results for (5.3.1) with $C = 0$ are Theorem 5.2.6 by Gilbert and Hile [24] and a similarity principle in case $b_{kl} = 0$ and $C = 0$ by Gilbert (*loc. cit.* [Gi 5] in [22]). Here we find for the first component $w_p \not\equiv 0$ of W that it satisfies

$$w_{p\bar{z}} + a_{pp} w_p + b_{pp} \bar{w}_p = 0$$

in the whole plane. Vekua's global result [64, 125 ff.] for generalized analytic functions provides the representation

$$w_p(z) = c_p \exp[\omega_p(z)] \quad \text{with } c_p \neq 0, \quad \omega_p \in C^{(2-q)/q}(\mathbb{C}) \quad \text{and} \tag{5.3.10}$$

$$\omega_p = 0(|z|^{(2-q)/q}, \quad \text{for } |z| \to \infty, \quad q > 2. \tag{5.3.11}$$

Introducing the *new hypercomplex* unknown function

$$\tilde{W} \equiv \sum_{j=0}^{n-p-1} \mathrm{e}^j w_{p+j} \tag{5.3.12}$$

we find with $w_0 = .. = w_{p-1} \equiv 0$ from (5.3.1) a new system for $\tilde{W}$,

$$\mathsf{D}\tilde{W} + \sum_{k=0}^{n-p-1} \mathrm{e}^k \sum_{l=0}^{n-p-1} \{a_{p+k,\, p+l} w_{p+l} + b_{p+k,\, p+l} \bar{w}_{p+l}\} = 0. \tag{5.3.13}$$

Now $\tilde{W}^{-1}$ is bounded due to (5.3.10) and $\omega \equiv \sum_{j=0}^{n-p-1} e^j \omega_{p+j}$ can be found from

$$\mathsf{D}\omega = -\tilde{W}^{-1} \sum_{k=0}^{n-p-1} e^k \sum_{l=0}^{n-p-1} \{a_{p+k,\,p+l} w_{p+l} + b_{p+k,\,p+l} \bar{w}_{p+l}\} \equiv \tilde{\mathsf{C}} \quad (5.3.14)$$

as

$$\omega(z) \equiv \frac{1}{2\pi i} \iint \frac{1}{\{\Xi(t) - \Xi(z)\}} \tilde{\mathsf{C}}(t)[d\Xi(t), d\bar{t}] \quad (5.3.15)$$

and the desired representation is

$$W = c_p\, e^p \exp[\omega]. \quad (5.3.16)$$

This is the proof for Theorem 5.2.6 given in [24] where also the continuity properties of ω are investigated in a more detailed way. For more general representations of the solution let us follow the investigations by Habetha [34] and Kühn [44]. From the latter it seems more appropriate to consider systems in the quasinormal form corresponding to Bojarski's normal form (5.2.1) *with* Q *a lower triangular matrix throughout* G *satisfying* (5.2.3). Then we have the following theorem as a similarity principle:

Theorem 5.3.3 *Let* $\mathbf{w} \in C^1(\mathsf{G})$ *be a solution of*

$$\mathbf{w}_{\bar{z}} - \mathsf{Q}\mathbf{w}_z + \mathsf{a}\mathbf{w} + \mathsf{b}\bar{\mathbf{w}} = 0 \quad (5.3.17)$$

with coefficients $\mathsf{a}, \mathsf{b} \in L^\infty(\bar{\mathsf{G}})$. *Then to every* $z_0 \in \mathsf{G}$ *there exists a neighbourhood* $U(z_0)$ *in which*

$$\mathbf{w} = \mathsf{E}\boldsymbol{\phi} \quad \text{for } z \in U(z_0) \quad (5.3.18)$$

where E *is a* C^α *matrix valued function with* $\det \mathsf{E} \neq 0$ *in* U *and* $\boldsymbol{\phi}$ *satisfies*

$$\boldsymbol{\phi}_{\bar{z}} - \mathsf{E}^{-1}\mathsf{Q}\mathsf{E}\boldsymbol{\phi}_z = 0 \quad \text{in } U \quad (5.3.19)$$

i.e. $\boldsymbol{\phi}$ *is* $(\mathsf{E}^{-1}\mathsf{Q}\mathsf{E})$*-holomorphic.* E *depends on* $\mathbf{w}$ *but* $U(z_0)$ *can be found independent of* $\mathbf{w}$.

Proof As in [44] let us define the measurable matrix

$$c_{jk} \equiv a_{jk} + \begin{cases} b_{jk} w_k / \bar{w}_k & \text{for } w_k \neq 0, \\ 0 & \text{for } w_k = 0. \end{cases} \quad (5.3.20)$$

Then the integral equation for E,

$$\mathsf{E}(z) = \frac{-1}{2\pi i} \iint_{|z - z_0| \leq \delta} \{\Xi(t) - \Xi(z)\}^{-1} \Xi_t(t) \mathsf{C}\mathsf{E}(t)[dt, d\bar{t}] + \mathsf{I} \quad (5.3.21)$$

admits a contractive weakly singular integral operator for a sufficiently small constant $\delta>0$ from $C^\alpha(\bar{\mathsf{G}})$ into itself. The choice of $\delta>0$ depends only on Ξ and $\|a_{jk}\|_{L\infty}+\|b_{jk}\|_{L\infty}$ independently of $\mathbf{w}$ and for such $\delta>0$ (5.3.21) possesses exactly one solution $\mathsf{E}\in C^\alpha(\bar{\mathsf{G}})$ with nonvanishing determinant. Then E becomes $C^{1+\alpha}$ for $|z-z_0|\neq\delta$ and satisfies

$$\mathsf{E}_{\bar z}-\mathsf{Q}\mathsf{E}_z=\begin{cases}-\mathsf{CE} & \text{for } |z-z_0|<\delta\\ 0 & \text{for } |z-z_0|>\delta\end{cases} \tag{5.3.22}$$

according to Theorem 5.2.2. Defining

$$\boldsymbol{\phi}\equiv\mathsf{E}^{-1}\mathbf{w}$$

and inserting $\mathbf{w}=\mathsf{E}\boldsymbol{\phi}$ into

$$\mathbf{w}_{\bar z}-\mathsf{Q}\mathbf{w}_z=-\mathsf{C}\mathbf{w}$$

one finds

$$\mathsf{E}\boldsymbol{\phi}_{\bar z}-\mathsf{QE}\boldsymbol{\phi}_z=0 \quad \text{for } |z-z_0|<\delta$$

implying (5.3.19).

In the case of lower triangular matrices a, b,

$$a_{kl}=b_{kl}=0 \quad \text{for } l>k, \tag{5.3.23}$$

C (5.3.20) is also lower triangular and the integral equation (5.3.21) can be solved in $\bar{\mathsf{G}}$ uniquely; i.e. replace $|z-z_0|\leq\delta$ by $\bar{\mathsf{G}}$ and use Theorem 5.2.6 and the triangular shape of C. This is the way Kühn proves in [44] the following.

Theorem 5.3.4 *For* a, b *satisfying* (5.3.23) *the solutions of* (5.3.17) *admit the representation* (5.3.18) *throughout* $\bar{\mathsf{G}}$ *and* (5.3.19) *holds also throughout* G.

Let us omit the detailed proof here.

In the case $\mathsf{Q}\equiv 0$ Theorem 5.3.3 coincides with Pascali's result in [54] which is also a *local* similarity principle due to the work of Habetha [34, Section 4] who gave an example of a Pascali system for which the representation (5.3.18) cannot be true throughout $\bar{\mathsf{G}}$. Since for $\mathsf{Q}\equiv 0$ all the components of $\boldsymbol{\phi}$ are holomorphic, $\mathbf{w}$ *has only isolated zeros of finite order.* This is:

Theorem 5.3.5 (*Carleman's Theorem*)† *Every solution* $\mathbf{w}\not\equiv 0$ *of Pascali's system* (5.3.17) *with* $\mathsf{Q}\equiv 0$ *has only isolated zeros of finite order.*

† Habetha claims in [34, Theorem 3] that Carleman's Theorem also holds for systems with Q purely diagonal. But the constant c_1 on p. 51 in his proof is of the form $c_1=c'\rho^m$ with $\rho=\max_{z\in K}|\zeta_k(z)|/|\zeta_j(z)|>1$ so that his proof is not valid.

Remark Carleman's Theorem for special systems with $\mathsf{Q} \equiv 0$ can also be found in the work of Heinz and Hildebrandt [36].

With a careful investigation of the Hilbert transform (2.4.7) operating on z^m and $\bar{z}^m$, Kühn proves for the system (5.3.1) with $C = 0$ the following.

Theorem 5.3.6 [44, Section 2.1] *Every solution of*

$$\mathsf{D}W + \sum_{j=0}^{n-1} \mathrm{e}^j \sum_{k=0}^{j} (a_{jk} w_k + b_{jk} \bar{w}_k) = 0 \tag{5.3.24}$$

with $w_0 \not\equiv 0$ *admits a local representation*

$$W(z) = \sum_{j=0}^{n-1} \mathrm{e}^j \left\{ \sum_{k=0}^{j-1} \left(\gamma_{jk} + \sum_{l=0}^{k} \alpha_{kjl} \left(\frac{\bar{z} - \bar{z}_0}{z - z_0} \right)^l \right) + \gamma_{jj} \right\} \cdot (z - z_0)^{mj} \exp [\omega_j]$$

$$\text{for} \quad z \in U(z_0) \tag{5.3.25}$$

where the $\gamma_{jk}(z)$ *and* $\omega_j(z) \in C^\alpha$ *and the* α_{kjl} *are constants and where* $\prod_{j=0}^{n-1} \gamma_{jj} \neq 0$.

This theorem also contains Carleman's Theorem thus showing that for $W \not\equiv 0$ the zeros are isolated and of finite order, which can be found in [24].

As a consequence of Carleman's Theorem we find that a solution $\mathbf{w}$ of the corresponding homogeneous system which vanishes on an arc $\gamma \subset \mathsf{G}$ has to vanish identically throughout G. But this property, namely uniqueness of the Cauchy problem has also been shown for another class of elliptic systems by Hile and Protter [37]. As Habetha points out in [34] the uniqueness of the Cauchy problem is not valid for all elliptic systems since the famous counterexample by Pliš corresponds to a first-order elliptic system with $n = 4$. Hence it seems that the only known cases of elliptic systems with *uniqueness of the Cauchy problem* are those listed in the following.

Theorem 5.3.7 *Let* $\mathbf{w} \in C^1(\mathsf{G})$, *respectively* $\mathbf{u} \in C^1(\mathsf{G})$ *be a solution of one of the following homogeneous systems or inequalities:*

(i) *the homogeneous Pascali system*

$$\mathbf{w}_{\bar{z}} + \mathbf{a}\mathbf{w} + \mathbf{b}\bar{\mathbf{w}} = 0 \quad \textit{with } \mathbf{a}, \mathbf{b} \in L^p, p > 2 \tag{5.3.26}$$

(ii) *the generalized hyperanalytic system with triangular* a, b

$$\mathsf{D}\left(\sum_{j=0}^{n-1} \mathrm{e}^j w_j \right) + \sum_{j=0}^{n-1} \mathrm{e}^j \sum_{k=0}^{j} (a_{jk} w_k + b_{jk} \bar{w}_k) = 0 \quad \textit{with } \mathsf{a}, \mathsf{b} \in C^\alpha(\mathsf{G}) \tag{5.3.27}$$

(iii) *the pointwise estimate,* (*or equation* (4.1.1) *with* $\mathbf{f}=0$),

$$|\mathbf{u}_x+\mathsf{B}\mathbf{u}_y|\leq K|\mathbf{u}| \quad \text{in } \mathsf{G} \tag{5.3.28}$$

where $\mathsf{B} \in C^1$ *has no real eigenvalues and is normal,* i.e.

$$\mathsf{BB}^*=\mathsf{B}^*\mathsf{B} \dagger$$

(iv) *the pointwise estimate,* (*or equation* (5.3.17),

$$|\mathbf{w}_{\bar{z}}-\mathsf{Q}\mathbf{w}_z|\leq K|\mathbf{w}| \quad \text{in } \mathsf{G} \quad \text{with } \mathsf{Q} \in C^1 \text{ and} \tag{5.3.29}$$

$$(\mathsf{I}-\mathsf{Q})^{-1}(\mathsf{I}+\mathsf{Q})(\mathsf{I}+\mathsf{Q}^*)(\mathsf{I}-\mathsf{Q}^*)^{-1} = (\mathsf{I}-\mathsf{Q}^*)^{-1}(\mathsf{I}+\mathsf{Q}^*)(\mathsf{I}+\mathsf{Q})(\mathsf{I}-\mathsf{Q})^{-1}. \tag{5.3.30}$$

Then we have:
If $\mathbf{w}|_\gamma=0$ *respectively* $\mathbf{u}|_\gamma=0$ *on a smooth arc* $\gamma\subset\mathsf{G}$ *then* $\mathbf{w}$ *respectively* $\mathbf{u}$ *vanishes identically throughout* G.

It is easily shown that the uniqueness of the Cauchy problem is equivalent to the unique continuation property:

Corollary 5.3.8 *The unique continuation property: If* $\mathbf{w}$ *respectively* $\mathbf{u} \in C^1(\mathsf{G})$ *satisfies one of the conditions* (i)–(iv) *in Theorem* 5.3.7 *and vanishes in some open disk in* G *it vanishes identically throughout* G.

Remark 5.3.9 The case (iv) corresponds to (iii) rewriting (5.3.29) into the form (5.3.28). Note that (5.3.30) is satisfied if Q is purely diagonal. This case corresponds to systems in the normal form (3.1.30) with $\alpha_{jk}=\beta_{jk}\equiv 0$ and is the same system as in [34, Theorem 3]. Hile and Protter prove Theorem 5.3.7 in the case (iii) under the weaker conditions $\mathbf{u} \in C^0(\bar{\mathsf{G}})$, $\mathbf{u}_x, \mathbf{u}_y \in C^0(\mathsf{G})$ and uniformly bounded and $\gamma\subset\bar{\mathsf{G}}$. Other results on the uniqueness of the Cauchy problem were given by Lopatinskiĭ [47], Muramutu [51], Hayashida [35] and Hörmander [38, VIII].

For the purely generalized hyperanalytic hypercomplex case

$$DW+AW+B\overline{W}=0$$

with hypercomplex functions A, $B \in L^p$, $p>2$, Goldschmidt showed in [28, Hilfssatz 5.5] the unique continuation property.

As we have already seen in (5.2.48) and (5.2.86) or (5.2.87), solutions of homogeneous elliptic systems can be generated by Q-holomorphic or hyperanalytic functions. Such systems of solutions usually are given in a bigger domain than G and the question arises whether such systems of solutions can approximate *every* solution in G which itself might not be extendable to a smooth solution in the bigger domain.

† B^* denotes the matrix adjoint to B

Definition *Solutions of an homogeneous equation*

$$\mathbf{w}_{\bar{z}} - \mathsf{Q}\mathbf{w}_z + \mathsf{a}\mathbf{w} + \mathsf{b}\bar{\mathbf{w}} = 0 \tag{5.3.31}$$

are said to have the Runge approximation property if, whenever $\mathsf{G} \subset \mathsf{G}_0$, *any solution in* G *can be approximated uniformly on compact subsets of* G *by solutions of* (5.3.31) *in* G_0; i.e. *for* $\mathbf{w}$ *solving* (5.3.31) *in* G, *any compact* $\mathsf{K} \subset \mathsf{G}$ *and any* $\varepsilon > 0$ *there exists a solution* $\mathbf{w}_\varepsilon$ *of* (5.3.31) *in* G_0 *with*

$$\|\mathbf{w} - \mathbf{w}_\varepsilon\|_{C^0(\mathsf{K})} < \varepsilon.$$

It is shown in [11, Section 5] how the uniqueness of the Cauchy problem is connected with Runge's property for second-order elliptic equations. It seems that this approach can be carried over to the first-order elliptic systems and that the systems for which the Theorem 5.3.7 is true also have the Runge property. But this is yet to be done. For generalized analytic functions, $n = 1$, Bers [6] showed Runge's property. These results have been extended to systems with $n = 1$ and coefficients in $L^{p,2}(\mathbb{C})$ by Gusman [31] and by Gusman and Abdulajeva to such systems on Riemann manifolds [32]. For only L^p_{loc} coefficients Goldschmidt [29] extended Gusman's result. All these results (except those on Riemann surfaces) are contained in the following Theorem by Goldschmidt [28, Satz 5.5]:

Theorem 5.3.10 *The purely hypercomplex generalized hyperanalytic system*

$$\mathsf{D}W + AW + B\bar{W} = 0$$

with coefficients $A, B \in L^p_{\mathrm{loc}}$, $p > 2$ *has the Runge approximation property.*

5.4 Remarks on the Analytic Theory

In case of real analytic coefficients B, C and real analytic $\mathbf{f}$ the solution $\mathbf{u}$ of (4.1.1) is also real analytic according to John [41], Garabedian [19] and hence possesses many properties which are not valid in the nonanalytic case. Hence, the analytic theory of elliptic systems has been developed extensively, see for instance the books by Garabedian [20], Gilbert [21] and Colton [11], to mention only a few. Therefore, it cannot be the aim of this chapter to give any complete survey of the properties of analytic systems (4.1.1). But we shall take up some special questions which already arose in the preceding chapters and shall review a few more problems. We shall begin with Carleman's theorem following Habetha's presentation [34]. We then extend Vekua's generating integral operator to generalized hyperanalytic functions. For the case $n=1$ Vekua [65], by treating the analytically extended elliptic system in $\mathbb{C}^2$ like an hyperbolic system, constructed an integral operator mapping complex analytic functions onto generalized analytic functions. We shall show how this method can be extended to the generalized hyperanalytic systems with n complex unknowns following the work by Gilbert and Wendland [25]. This approach gives the natural generalization of Vekua's method and Pascali's generalized analytic vectors [54] leading to the representation formula which maps holomorphic vectors onto the solutions. As a corollary we find again Carleman's theorem. Following Colton [11, Section 11] we use this generalization of Vekua's representation formula for presenting Lewy's reflection principle for the corresponding systems generalizing results by Chung Ling Yu [72]. The main part of the proof of this new result is due to H. Löffler.

Let us begin with Carleman's theorem following Habetha [34]. He points out by the use of an argument of Bers [5], that the solution of

$$\mathbf{u}_x + \mathsf{B}\mathbf{u}_y + \mathsf{C}\mathbf{u} = 0 \tag{5.4.1}$$

with a zero $z_0=0$ *of finite* order behaves near $z_0=0$ like

$$\mathbf{u}(z) = \mathbf{P}_N(z, \bar{z}) + \boldsymbol{\chi}(z, \bar{z}) \quad \text{with } |\boldsymbol{\chi}| \leq K|z|^{N+\varepsilon} \qquad \varepsilon > 0$$

where $\mathbf{P}_N$ is a homogeneous polynomial of order N and is a solution of the 'osculating equations'

$$\mathbf{P}_{Nx} + \mathsf{B}(x_0, y_0)\mathbf{P}_{Ny} = 0$$

with *constant coefficients.* The osculating equations possess a normal form (3.1.31) in the whole plane and the transformed $\mathbf{S}$ of $\mathbf{P}_N$ satisfies (3.1.31),

$$S_{1\bar{\zeta}_1} = 0, \qquad S_{2\bar{\zeta}_1} + \varepsilon_1 S_{1y} = 0, \ldots,$$

where the ε_k are constants. Hence the zero of the first nonidentically vanishing component is *isolated* and we have as an immediate consequence Carleman's theorem.

Theorem 5.4.1 *The zeros of any solution of an elliptic system* (5.4.1) *with real analytic coefficients* B *and* C *are isolated and of finite order.*

Of course, this provides uniqueness of the Cauchy problem as well as the unique continuation property.

Now let us consider Vekua's generating operator for a system

$$\mathsf{D}W + \sum_{j=0}^{n-1} \sum_{k=0}^{n-1} e^j (a_{jk} w_k + b_{jk} \bar{w}_k) = f \tag{5.4.2}$$

assuming that *all coefficients* in $\alpha(x, y)$ in (5.2.66), $a_{jk}(x, y)$, $b_{jk}(x, y)$, and

$f(x, y)$ *are real analytic* in G. As a preliminary let us consider some properties of analytic functions of several variables.

Let $F(z_1, \ldots, z_n)$ be analytic in $G \subset \mathbb{C}^n$. We denote by

$$G^* \equiv \{(\zeta_1, \ldots, \zeta_n) \mid (\bar{\zeta}_1, \ldots, \bar{\zeta}_n) \in G\}$$

the *reflected domain* and associate with F the function

$$F^*(\zeta_1, \ldots, \zeta_n) := \overline{F(\bar{\zeta}_1, \ldots, \bar{\zeta}_n)}, \qquad (\zeta_1, \ldots, \zeta_n) \in G^*$$

defined by the complex conjugation which is analytic in G^* and is called the *conjugate function* to F. Now let $\phi(x_1, y_1, \ldots, x_n, y_n)$ be a given analytic function of the real variables $x_1, \ldots, y_n$. If we introduce

$$z_k = x_k + \mathrm{i} y_k, \ \zeta_k = x_k - \mathrm{i} y_k,$$

then ϕ becomes an analytic function of the new arguments z_k, ζ_k in some domain $D \subset \mathbb{C}^{2n}$. This function shall also be denoted by

$$\phi(z_1, \zeta_1, \ldots, z_n, \zeta_n).$$

If we continue $\overline{\phi(x_1, \ldots, x_n, y_n)}$ into the domain of complex arguments, we obtain ϕ^*.

Now let F be given as an integral of the form

$$F(z_1, \ldots, z_n, t_1, \ldots, t_n) = \int_{t_1}^{z_1} \cdots \int_{t_n}^{z_n} \phi(z_1, \ldots, z_n, \xi_1, \ldots, \xi_n) \, \mathrm{d}\xi_n \ldots \mathrm{d}\xi_1.$$

Then F^* is given by

$$F^*(\zeta_1, \ldots, \zeta_n, \tau_1, \ldots, \tau_n) = \int_{\tau_1}^{\zeta_1} \cdots \int_{\tau_n}^{\zeta_n} \phi^*(\zeta_1, \ldots, \zeta_n, \eta_1, \ldots, \eta_n) \mathrm{d}\eta_n \ldots \mathrm{d}\eta_1.$$

Since all functions in (5.4.2) are real analytic, they can all be extended to analytic functions of the variables z, ζ. Let us suppose that all given functions in (5.4.2) admit an analytic extension to all of the bicylindrical domain $G \times G^* \subset \mathbb{C}^2$. Then G is called a *fundamental domain* for (5.4.2). There (5.4.2) takes the form

$$\left(\frac{\partial}{\partial \zeta} + \mathrm{e}\alpha \frac{\partial}{\partial z}\right) W + \sum_{j=0}^{n-1} \sum_{k=0}^{n-1} \mathrm{e}^j (a_{jk} w_k + b_{jk} w_k^*) = f. \tag{5.4.3}$$

In order to further simplify the equations (5.4.3) we introduce the Douglis generating variable Ξ by (5.2.63) and introduce the differentiation

$$\frac{\partial}{\partial \bar{t}} = \frac{\Xi_z}{\Xi_{\bar{z}} \Xi_z - \Xi_{\bar{z}} \Xi_z} \left(\frac{\partial}{\partial \bar{z}} + \mathrm{e}\alpha \frac{\partial}{\partial z}\right) \tag{5.4.4}$$

which is well-defined because the Jacobian appearing in (5.4.4) is of the form $1+e\{\cdots\}$; hence its reciprocal must exist. Now (5.4.3) may be rewritten in the form

$$\frac{\partial W}{\partial \bar{t}}+\sum_{j=0}^{n-1}\sum_{k=0}^{n-1} e^j(A_{jk}w_k+B_{jk}w_k^*)=F, \tag{5.4.5}$$

with

$$A_{jk}=\frac{a_{jk}\Xi_z}{\Xi_{\bar z}\Xi_z-\Xi_{\bar z}\Xi_z}, \qquad B_{jk}=\frac{b_{jk}\Xi_z}{\Xi_{\bar z}\Xi_z-\Xi_{\bar z}\Xi_z}, \qquad F=\frac{f\Xi_z}{\Xi_{\bar z}\Xi_z-\Xi_{\bar z}\Xi_z}.$$

For simplicity, we consider the special case where *all these coefficients are entire,* holomorphic functions of x, y and, hence, entire functions of $z=x+iy$, $\zeta=x-iy$. Hence the generating variable Ξ which must be entire, can easily be constructed as shown below.

With (5.2.63) the equations (5.2.9) for the generating variable

$$\Xi(z,\zeta)=z+\sum_{k=1}^{r-1} t_k(z,\zeta)\,e^k=z+T(z,\zeta) \tag{5.4.6}$$

are equivalent to the system

$$\frac{\partial \Xi}{\partial \zeta}+e\alpha\frac{\partial \Xi}{\partial z}=0 \text{ or } \frac{\partial t_k}{\partial \zeta}=-\left(\gamma\frac{\partial t_{k-1}}{\partial \zeta}+\beta\frac{\partial t_{k-1}}{\partial z}\right), \quad k=1,\ldots,n-1;\ t_0=z$$

where $\alpha=\beta(1+e\gamma)^{-1}$.

Therefore, we may define the t_k successively by

$$t_1(z,\zeta)=-\int_0^{\zeta}\beta(z,\xi_1)\,d\xi_1,$$

$$t_k(z,\zeta)=(-1)^k\int_0^{\zeta}\left(\beta(z,\xi_1)\frac{\partial}{\partial z}+\gamma(z,\xi_1)\frac{\partial}{\partial \xi_1}\right)\int_0^{\xi_1} \tag{5.4.7}$$

$$\cdots\left(\beta(z,\xi_{k-1})\frac{\partial}{\partial z}+\gamma(z,\xi_{k-1})\frac{\partial}{\partial \xi_{k-1}}\right)\int_0^{\xi_{k-1}}\beta(z,\xi_k)\,d\xi_k\,d\xi_{k-1}\cdots d\xi_1.$$

For convenience, let us further denote the generating variable by

$$t\equiv\Xi. \tag{5.4.8}$$

By construction t becomes holomorphic wherever $\alpha(z,\zeta)$ is, and hence is entire in $\mathbb{C}^2$. Besides t, we introduce the conjugate function

$$\tau\equiv\zeta+T^*(\zeta,z).$$

The extension of the differential equations, (5.4.3) can now be written as

$$\frac{\partial W}{\partial \tau}+\sum_{j=0}^{n-1}\sum_{k=0}^{n-1} e^j(A_{jk}w_k+B_{jk}w_k^*)=F. \tag{5.4.9}$$

This equation becomes similar to the equation investigated by Vekua [65, p. 66] when we interpret τ as an independent variable. To clarify this idea we consider the transformation

$$\begin{aligned} t &= z + T(z, \zeta), \\ \tau &= \zeta + T^*(\zeta, z). \end{aligned} \tag{5.4.10}$$

which defines a one-to-one mapping from $\mathbb{C}^2$ onto a complex two-dimensional submanifold $A \subset \mathbb{C}^{2n}$. We remark that when we replace z and ζ by hypercomplex variables then this defines a mapping from $\mathbb{C}^{2n}$ into $\mathbb{C}^{2n}$. This may be seen by using hypercomplex z and ζ in the power series of T or T^* and majorising, Douglis [15 Lemma 4.2]. In order to interpret the functions in (5.4.9) as functions of t, τ we need the inverse transformation

$$z = z(\tau, t), \zeta = \zeta(\tau, t) \tag{5.4.11}$$

defined in some $\mathbb{C}^{2n}$ neighbourhood of A. We denote by F, F^* the functions

$$F(t, z, \zeta) \equiv t - z - T(z, \zeta), F^*(\tau, \zeta, z) \equiv \tau - \zeta - T^*(\zeta, z) \tag{5.4.12}$$

and observe that $\partial F/\partial z \neq 0$. Cauchy's formula, for holomorphic functions with values in a Banach space [14, IX 10], may be used here. To this end we supply the space of continuous and bounded hypercomplex valued functions of the form

$$W(\eta) = \sum_{k=0}^{n-1} e^k w_k(\eta), \; w_k(\eta) \in \mathbb{C}, \; \eta \in G,$$

with the sup-norm

$$\|W\|_0 = \sum_{k=0}^{n-1} \sup_{\eta \in G} |w_k(\eta)| \tag{5.4.13}$$

according to (5.2.59).

Then there is a local representation $\hat{z} = \hat{z}(t, \zeta)$ of the function $\hat{z}$ implicitly defined by $F = 0$, namely

$$z = \hat{z}(t, \zeta) = \frac{1}{2\pi i} \int_{|z| = \varepsilon_1 > 0} z \frac{F_z}{F} \, dz. \tag{5.4.14}$$

Using $\hat{z}$ in the second equation of (5.4.12),

$$\hat{F}(\tau, \zeta, t) \equiv \tau - \zeta - T^*(\zeta, \hat{z}(t, \zeta)); \tag{5.4.15}$$

in the same manner one obtains

$$\zeta = \zeta(\tau, t) = \frac{1}{2\pi i} \int_{|\zeta| = \varepsilon_2 > 0} \zeta \frac{\hat{F}_\zeta}{\hat{F}} \, d\zeta, \tag{5.4.16}$$

(see [21, p. 18]) and $z = z(\tau, t) \equiv \hat{z}(t, \zeta(\tau, t))$. This transformation permits a reduction of (5.4.9) to

$$\frac{\partial W}{\partial \tau} + \sum_{j=0}^{n-1} \sum_{k=0}^{n-1} e^j (a_{jk}(t, \tau) w_k + b_{jk}(t, \tau) w_k^*(\tau, t) = f(t, \tau), \tag{5.4.17}$$

where $(t, \tau) \in G$ and G is a suitable domain in $\mathbb{C}^{2n} \times \mathbb{C}^{2n}$ containing A. If $(t, \tau) \in A$ then z, ζ are complex valued and consequently, a_{jk}, b_{jk} and w_k are also complex valued. We remark, however, that for general $(t, \tau) \in G$ all these functions become *hypercomplex valued.*

Example: In the special case

$$\alpha \equiv 1, \quad n = 2, \quad \text{i.e., for } e^2 = 0, \tag{5.4.18}$$

the generating solution t is given by

$$t = z + e\zeta, \tau = \zeta + ez, \tag{5.4.19}$$

Therefore, the inverse transformation (5.4.11) is

$$z = t - e\tau, \zeta = \tau - et, \tag{5.4.20}$$

which is defined everywhere in $\mathbb{C}^2 \times \mathbb{C}^2$. If $A(z, \zeta)$ is given as an entire function in $\mathbb{C}^2$, then the continuation

$$a(t, \tau) = A(t - e\tau, \tau - et)$$

is clearly defined for all $(t, \tau) \in \mathbb{C}^2 \times \mathbb{C}^2$. The manifold A is defined by $t - e\tau \in \mathbb{C}$ and $\tau - et \in \mathbb{C}$.

Now let us consider first the purely hypercomplex systems which can be written as a *single hypercomplex equation,*

$$\frac{\partial W}{\partial \tau} + A(t, \tau) W(t, \tau) + B(t, \tau) W^*(\tau, t) = F(t, \tau) \tag{5.4.21}$$

since in this case Vekua's approach can be simply carried over.

Because z and ζ are thought of as independent variables, t and τ are also independent. Consequently, $A(t, \tau)$ and $B(t, \tau)$ are hyperanalytic in both the t and τ variables. If $W(z, \zeta)$ is a holomorphic solution of the original system, *then the extension satisfies equation* (5.4.21) *and its conjugate equation*

$$\frac{\partial W^*}{\partial t} + A^*(\tau, t) W^*(\tau, t) + B^*(\tau, t) W(t, \tau) = F^*(\tau, t). \tag{5.4.22}$$

Formal integration of (5.4.21) with respect to τ and of (5.4.22) with respect to t leads to a system of Volterra integral equations for W and W^*. Therefore, we are interested in making use of Vekua's approach in [65] to simplify (5.4.21), see also Gilbert [21, Section 3.4 ff.]. For this purpose we introduce as a generalization of the exponential function a particular solution of the differential equation

$$\frac{\partial E}{\partial \tau} = -AE \tag{5.4.23}$$

having nonvanishing complex part. E can be found by Picard's successive approximation method, i.e.

$$E^{(m+1)}(t, \tau) \equiv 1 - \int_{\tau_0}^{\tau} A(t, \sigma) E^{(m)}(t, \sigma)\, d\sigma,\ m = 0, 1, 2, \ldots \text{ with}$$
$$E^{(0)} \equiv 1. \tag{5.4.24}$$

For convenience, we choose τ_0 in A, e.g. $\zeta_0 + T^*(\zeta_0, z_0)$, where $z_0 = z_0 + iy_0$, $\zeta_0 = x_0 - iy_0$ with $x_0, y_0 \in R$. If $(t, \tau) \in A$ also and, moreover, x and y are real, then the integral can be expressed as a simple line integral as shown in Douglis [15], i.e.

$$dt = \frac{t_x}{1 - e\alpha}\{dz - e\alpha\, d\bar{z}\}; \tag{5.4.25}$$

for the hypercomplex differential $d\tau$ we have the corresponding complex conjugate expression. From Douglis [15, 276] we know that when $\mathsf{D}W = 0$, (for x, y real), then the line integral

$$\int_{z_0}^{z} W(z)\, dt(z)$$

is path independent, and indeed defines a *unique, single-valued hyperanalytic function* in simply connected regions. It is a simple computation using the chain rule to see that this integral is also hyperanalytic for x and y hypercomplex valued in simply connected regions. A similar remark holds for integration with respect to the hypercomplex $d\tau(\zeta)$ integration as in (5.4.25).

Let us return to (5.4.24). By induction it follows from the hyperanalyticity of A and $E^{(0)}$ in both variables that every $E^{(m)}$ is hyperanalytic. Therefore, the path of integration in (5.4.24) can be perturbed without changing the value of the integral. If τ belongs to a suitable neighbourhood of τ_0, we can choose straight lines between τ_0 and τ in $\mathbb{C}^n$. From this we have the local estimates

$$|E^{(k)} - E^{(k-1)}| \leq \|A^k\|_0 \frac{|\tau - \tau_0|^k}{k!} \tag{5.4.26}$$

when τ belongs to a sufficiently small closed ball around τ_0. ($|\cdot|$ is defined as in (5.2.59)). We may conclude that $E(t, \tau)$ is locally majorised by the series for $\exp\{\|A\|_0 |\tau - \tau_0|\}$. Hence, E is defined in that ball, and the *limit function becomes hyperanalytic.* This is easy to show using the following result which follows directly from the work of Douglis [15].

If $\{\phi\}$ is a family of hyperanalytic functions in D, which is locally uniformly bounded in D, then this family is normal in D.

By using the iterative procedure (5.4.24) in a compact covering of A where $z(t, \tau)$, $\zeta(\tau, t)$ are defined, we obtain similar estimates to (5.4.26). Hence, E is clearly well-defined in a $\mathbb{C}^{2n}$-neighbourhood of A.

In order to show that E has an inverse everywhere on A we notice that the first component of (5.4.23) is

$$E_{0\zeta} = -\hat{A}_0 E_0(z, \zeta), \tag{5.4.27}$$

an equation containing only the complex part E_0 of E. Consequently, upon integration we have

$$E_0(z, \zeta) = \exp\left(-\int_{\hat{\zeta}(\tau_0, z)}^{\zeta} A_0(z, \xi)\, d\xi\right) \neq 0, \tag{5.4.28}$$

where $\zeta(\tau_0, z)$ is an element on the manifold defined by $\hat{F}^*(\tau_0, \zeta(\tau_0, z), z) = 0$. Since the complex part of $E(z, \zeta)$ does not vanish, the existence of E^{-1} is assured.

By the substitution

$$W = EV \tag{5.4.29}$$

we are able to reduce (5.4.21) to

$$\frac{\partial V}{\partial \tau} = \tilde{B} V^* + \tilde{F}, \tag{5.4.30}$$

where

$$\tilde{B} = -E^{-1} B E^*, \tilde{F} = E^{-1} F. \tag{5.4.31}$$

Hence, $V(t, \tau)$ satisfies the Volterra integral equation

$$V(t, \tau) = \int_{\tau_0}^{\tau} \tilde{B}(t, \sigma) V^*(\sigma, t)\, d\sigma + \Phi(t) + \int_{\tau_0}^{\tau} \tilde{F}(t, \sigma)\, d\sigma, \tag{5.4.32}$$

where $\Phi(t)$ *is a hyperanalytic function.* Corresponding to (5.4.22) we have, in the same manner, the adjoint equation,

$$V^*(\tau, t) = \int_{t_0}^{t} \tilde{B}^*(\tau, s) V(s, \tau)\, ds + \int_{t_0}^{t} \tilde{F}^*(\tau, s)\, ds + \Phi^*(\tau); \tag{5.4.33}$$

here $\Phi^*(\tau)$ is a holomorphic function of τ. Again, we choose τ_0 and t_0 in A. As in (5.4.24), the integrals become hyperanalytic in either variable τ and t for x and y hypercomplex valued.

The special form of the equations (5.4.32), (5.4.33) allows us to combine them into the *single equation*

$$V(t,\tau)=\int_{\tau_0}^{\tau}\tilde{B}(t,\sigma)\int_{t_0}^{t}\tilde{B}^*(\sigma,s)V(s,\sigma)\,ds\,d\sigma+\Phi(t,\tau), \tag{5.4.34}$$

where

$$\Phi(t,\tau)\equiv\Phi(t)+\int_{\tau_0}^{\tau}\tilde{B}(t,\sigma)\Phi^*(\sigma)\,d\sigma+\int_{\tau_0}^{\tau}F(t,\sigma)\,d\sigma \\ +\int_{\tau_0}^{\tau}\int_{t_0}^{t}\tilde{B}(t,\sigma)\tilde{F}^*(\sigma,s)\,ds\,d\sigma. \tag{5.4.35}$$

Now (5.4.34) can be solved using the Picard iteration procedure

$$V_n(t,\tau)\equiv\int_{\tau_0}^{\tau}\int_{t_0}^{t}\tilde{B}(t,\sigma)\tilde{B}^*(\sigma,s)V_{n-1}(s,\sigma)\,ds\,d\sigma+\Phi(t,\tau),\ V_0\equiv\Phi, \tag{5.4.36}$$

as can be seen from the local estimate

$$|V_n-V_{n-1}|\leqslant\|\tilde{B}\|_0^k\|\tilde{B}^*\|_0^k\frac{|t-t_0|^k}{k!}\frac{|\tau-\tau_0|^k}{k!}\|\Phi\|_0, \tag{5.4.37}$$

which holds in a neighbourhood of t_0, τ_0. The family $\{V_n\}$ is clearly a normal family of hyperanalytic functions; consequently the limit function $V(t,\tau)$ exists and is hyperanalytic too. It has a representation in terms of the resolvent kernel Γ, belonging to (5.4.34), namely

$$V(t,\tau)=\Phi(t,\tau)+\int_{t_0}^{t}\int_{\tau_0}^{\tau}\Gamma(t,\tau,s,\sigma)\Phi(s,\sigma)\,d\sigma\,ds. \tag{5.4.38}$$

The resolvent satisfies the hypercomplex equation

$$\Gamma(t,\tau,s,\sigma)=\tilde{B}(t,\sigma)\tilde{B}^*(\sigma,s) \\ +\int_{\sigma}^{\tau}\int_{s}^{t}\tilde{B}(\xi,\sigma)\tilde{B}^*(\sigma,s)\Gamma(t,\tau,\xi,\eta)\,d\xi\,d\eta, \tag{5.4.39}$$

which again can be solved by the Picard iteration procedure. Following Vekua's investigation of the complex case [65, Chap. I, Section 15] we introduce the so-called kernels of 1st and 2nd kind

$$\Gamma_1(t,\tau,s,\sigma)\equiv\int_{\sigma}^{\tau}\Gamma(t,\tau,s,\eta)\,d\eta, \tag{5.4.40}$$

$$\Gamma_2(t,\tau,s,\sigma)\equiv\tilde{B}(t,\sigma)+\int_{\sigma}^{\tau}\int_{s}^{t}\Gamma(t,\tau,\xi,\eta)\tilde{B}(\xi,\sigma)\,d\xi\,d\eta. \tag{5.4.41}$$

Our hypercomplex algebra allows us to establish the following representation theorem using arguments similar to those of Vekua for the complex case.

Theorem 5.4.2 *Every solution of* (5.4.30) *can be represented by a hypercomplex function* $V(t, \tau)$, *where*

$$\begin{aligned} V(t,\tau) = \phi(t) &+ \int_{t_0}^{t} \Gamma_1(t, \tau, s, \tau_0)\phi(s)\, ds \\ &+ \int_{\tau_0}^{\tau} \Gamma_2(t, \tau, t_0, \eta)\phi^*(\eta)\, d\eta + \int_{\tau_0}^{\tau} \tilde{F}(t, \eta)\, d\eta \\ &+ \int_{\tau_0}^{\tau}\int_{t_0}^{t} \Gamma_1(t, \tau, s, \sigma)\tilde{F}(s, \sigma)\, ds\, d\sigma \\ &+ \int_{\tau_0}^{\tau}\int_{t_0}^{t} \Gamma_2(t, \tau, s, \sigma)\tilde{F}^*(\sigma, s)\, ds\, d\sigma \end{aligned} \tag{5.4.42}$$

and where $\phi(t)$ *is the hyperanalytic Goursat data given by*

$$\phi(t) = V(t, \tau_0). \tag{5.4.43}$$

On the other hand, every hyperanalytic function $\phi(t)$ *generates a solution* $V(t, \tau)$ *by* (5.4.42).

As a Corollary we find

Corollary 5.4.3 *Any* W *satisfying the homogeneous hyperanalytic system*

$$\frac{\partial W}{\partial \tau} + AW + B\bar{W} = 0 \tag{5.4.44}$$

has only isolated zeros of finite order.

The proof is similar to Vekua's proof for the complex case [65]. By using the exponential function E, the hypercomplex function W may be represented as

$$W = EV,$$

from which it is clear that W and V have exactly the same zeros. V can be represented by (5.4.42), where we choose $t_0 = \tau_0 = z_0 = 0$, namely

$$\begin{aligned} V(t, \bar{t}) = \phi(t) &+ \int_0^t \Gamma_1(t, \bar{t}, s, 0)\phi(s)\, ds \\ &+ \int_0^t \Gamma_2(t, \bar{t}, 0, \sigma)\phi^*(\sigma)\, d\sigma, \end{aligned} \tag{5.4.45}$$

$\phi(t)$ is hyperanalytic and has, therefore, the form

$$\phi(t)=t^k\phi_0(t) \quad \text{with } \phi_0(0)\neq 0. \tag{5.4.46}$$

Furthermore, for the hypercomplex generating solution it is clear that

$$t=r\theta(t,\bar{t}), \qquad \theta\neq 0, \qquad r=|z|. \tag{5.4.47}$$

We designate the ray from 0 to t by

$$s=pr\theta, \qquad 0\leq p\leq 1, \quad \text{where } \theta \text{ is fixed.} \tag{5.4.48}$$

Using (5.4.46–48) in (5.4.45) one obtains the representation

$$V=r^k\theta^k\phi_0(t)+\int_0^1 \Gamma_1(t,\bar{t},pr\theta,0)(pr\theta)^k\phi_0 r\theta\,\mathrm{d}p$$
$$+\int_0^1 \Gamma_2(t,\bar{t},0,pr\theta^*)(pr\theta^*)^k\phi_0^* r\theta^*\,\mathrm{d}p=r^k\theta^k\phi_0(t)+0(r^{k+1}),$$

from which the desired result follows.

The Bergman–Vekua operator (5.4.22) provides also a proof of the so called *reflection principle* on the lines of Lewy's classical proof [45]. For this purpose let us consider the solution W in the simply connected domain G of the (x, y)-plane whose boundary contains a segment σ of the x-axis with the origin as the interior point and such that G contains the portion $y<0$ of a neighbourhood of each point of σ (Fig. 4).

Then the solution $W(x+\mathrm{i}y, x-\mathrm{i}y)$ of (5.4.44) in G can be 'reflected' at σ; more precisely, W can be continued analytically to $G\cup\sigma\cup G^*$ satisfying everywhere (5.4.44). To this end let us assume that $G\cup\sigma\cup G^*$ is a fundamental domain for (5.4.44). Then $W(x+\mathrm{i}y, x-\mathrm{i}y)$ first can be extended analytically to $W(z,\zeta)$ for $z\in G$ and $\zeta\in G^*$. Next, this extension can be considered to be a function $W(t,\tau)$ of the generating variables if we insert the inverse transformation (5.4.11) for $t\in t(G\times G^*)$ and $\tau\in \tau(G^*\times G)$ where for any domain D we define

$$\begin{aligned} t(D\times D^*)&\equiv\{t=t(z,\zeta) \quad \text{with} \quad z\in D \quad \text{and} \quad \zeta\in D^*\},\\ \tau(D^*\times D)&\equiv\{\tau=\tau(\zeta,z) \quad \text{with} \quad z\in D \quad \text{and} \quad \zeta\in D^*\}. \end{aligned} \tag{5.4.49}$$

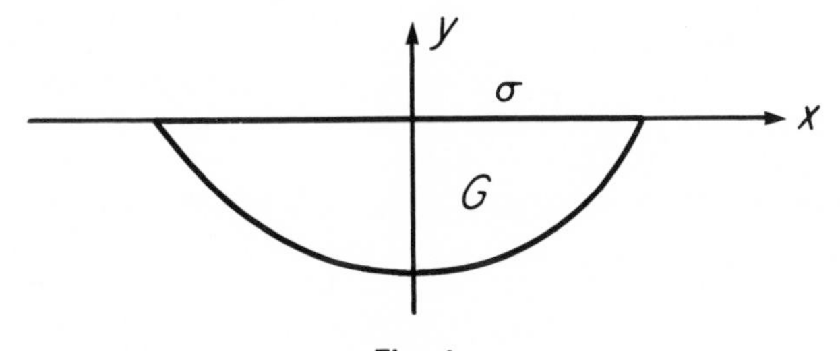

Fig. 4

Now let us consider the first boundary condition, $\rho(x) = \text{Re } W$, on σ and its analytic continuation

$$2\rho(x) = W(x, x) + W^*(x, x) \tag{5.4.50}$$

Let us assume that $\rho(z)$ *is analytic* for all $z \in \mathsf{G} \cup \sigma \cup \mathsf{G}^*$. Now we are in the position to formulate the proposed reflection theorem which is essentially due to H. Löffler.

Theorem 5.4.4 *Let* $W(x+\mathrm{i}y, x-\mathrm{i}y)$ *be a solution of* (5.4.4) *in* G *with real analytic coefficients in* $\mathsf{G} \cup \sigma \cup \mathsf{G}^*$, $W \in C^1(\mathsf{G}) \cap C^0(\mathsf{G} \cup \sigma)$, *let* $\mathsf{G} \cup \sigma \cup \mathsf{G}^*$ *be a fundamental domain for* (5.4.44) *such that integral representation* (5.4.42) *holds for* $t \in t(\mathsf{G} \cup \sigma \cup \mathsf{G}^* \times \mathsf{G} \cup \sigma \cup \mathsf{G}^*)$, $\tau \in \tau(\mathsf{G} \cup \sigma \cup \mathsf{G}^* \times \mathsf{G} \cup \sigma \cup \mathsf{G}^*)$, *and suppose that on* σ *we have* (5.4.50) *where* $\rho(x)$ *can be extended to a regular holomorphic function* $p(z)$ *for* $z \in \mathsf{G} \cup \sigma \mathsf{G}^*$. *Then* W *can be continued analytically into all of* $\mathsf{G} \cup \sigma \cup \mathsf{G}^*$.

Proof Here we follow essentially the presentation in [11, Section 11] but for the more complicated equations (5.4.44). Let us consider $V(t, \tau)$, $t \in t(\mathsf{G} \times \mathsf{G}^*)$, $\tau \in \tau(\mathsf{G}^* \times \mathsf{G})$, obtained from $W(t, \tau)$ by (5.4.29). In a first step let us show that V can be extended continuously for $t \in t(\mathsf{G} \cup \sigma \times \mathsf{G}^* \cup \sigma)$, $\tau \in \tau(\mathsf{G} \cup \sigma \times \mathsf{G}^* \cup \sigma)$. Analogously to the argumentation that led to the Volterra integral equation for solutions of (5.4.30) we integrate (5.4.44) from $\bar{t} \in \tau(\mathsf{G}^* \times \mathsf{G})$ (instead of τ_0) to τ,

$$V(t, \tau) = V(t, \bar{t}) + \int_{\bar{t}}^{\tau} \tilde{B}(t, \sigma) V^*(\sigma, t)\, \mathrm{d}\sigma.$$

Similarly we find the adjoint equation

$$V^*(\tau, t) = V^*(\tau, \bar{\tau}) + \int_{\bar{\tau}}^{t} \tilde{B}^*(\tau, s) V(s, \tau)\, \mathrm{d}s.$$

The two integral equations can be combined to a new Volterra integral equation for $V(t, \tau)$

$$V(t, \tau) = \phi(t, \tau) + \int_{\bar{t}}^{\tau} \int_{\bar{\sigma}}^{t} \tilde{B}(t, \sigma) \tilde{B}^*(\sigma, s) V(s, \sigma)\, \mathrm{d}s\, \mathrm{d}\sigma \tag{5.4.51}$$

with

$$\phi(t, \tau) = V(t, \bar{t}) + \int_{\bar{t}}^{\tau} \tilde{B}(t, \sigma) V^*(\sigma, \bar{\sigma})\, \mathrm{d}\sigma,$$

For $x, y \in \mathsf{G} \cup \sigma$ we already know that

$$V(t, \bar{t}) = V(t(x+\mathrm{i}y, x-\mathrm{i}y), \quad \overline{t(x+\mathrm{i}y, x-\mathrm{i}y)}) = V(x, y) \in C^0(\mathsf{G} \cup \sigma).$$

Hence, $\phi(t, \tau)$ can be continuously extended for $t \in t(\mathbf{G} \cup \sigma \times \mathbf{G}^* \cup \sigma)$, $\tau \in (\mathbf{G} \cup \sigma \times \mathbf{G}^* \cup \sigma)$. Integral equation (5.4.51) can be solved by the successive approximation,

$$V_0(t, \tau) \equiv \Phi(t, \tau)$$

$$V_n(t, \tau) \equiv \Phi(t, \tau) + \int_{\bar{\tau}}^{\tau} \int_{\bar{\sigma}}^{t} \tilde{B}(t, \sigma) \tilde{B}^*(\sigma, s) V_{n-1}(s, \sigma)\, ds\, d\sigma$$

From the local estimates

$$|V_n(t, \tau) - V_{n-1}(t, \tau)| \leq \|\Phi\|_0 \|\tilde{B}\|_0^k \|\tilde{B}^*\|_0^k \frac{|t-\tau|^{2k}}{(2k)!}$$

we find that the limit function $V(t, \tau)$ is hyperanalytic wherever $\Phi(t, \tau)$ is, i.e. for $t \in t(\mathbf{G} \times \mathbf{G}^*)$, $\tau \in \tau(\mathbf{G} \times \mathbf{G}^*)$ and continuous wherever $\Phi(t, \tau)$ is, i.e., for $t \in t(\mathbf{G} \cup \sigma \times \mathbf{G}^* \cup \sigma)$, $\tau \in \tau(\mathbf{G} \cup \sigma \times \mathbf{G}^* \cup \sigma)$.
Thus the integral representation (5.4.42) can be extended and is still valid for $t_0 = \tau_0 = 0$.

In a next step we shall extend the function $V(t, 0)$ analytically into $t \in t(\mathbf{G} \cup \sigma \cup \mathbf{G}^* \times \mathbf{G} \cup \sigma \cup \mathbf{G}^*)$ and also $V(\tau, 0)$ into $\tau \in \tau(\mathbf{G} \cup \sigma \cup \mathbf{G}^* \times \mathbf{G} \cup \sigma \cup \mathbf{G}^*)$. Once these extensions are known then the formula (5.4.42) provides an analytic extension of V to $V(t, \tau)$ for $t \in t(\mathbf{G} \cup \sigma \cup \mathbf{G}^* \times \mathbf{G} \cup \sigma \cup \mathbf{G}^*)$ and $\tau \in \tau(\mathbf{G} \cup \sigma \cup \mathbf{G}^* \times \mathbf{G} \cup \sigma \cup \mathbf{G}^*)$.

Up to now, $V(t, 0) = \phi(t)$ is known for $t \in t(\mathbf{G} \cup \sigma \times \mathbf{G}^* \cup \sigma)$. Since $\phi(t) = \sum_{l=0}^{n-1} e^l \phi_l(t(z, \zeta))$ is a hyperanalytic function, it satisfies the equations

$$\frac{\partial \phi}{\partial \tau} = \frac{\partial \phi}{\partial \zeta} + e\alpha \frac{\partial \phi}{\partial z} = 0, \tag{5.4.52}$$

i.e.

$$\phi_{0\zeta} = 0, \qquad \phi_{k\zeta} = -\gamma \phi_{k-1\zeta} - \beta \phi_{k-1z}, \qquad k = 1, \ldots, n-1.$$

By integration as in (5.4.7), these equations yield

$$\phi_0(z, \zeta) = \phi_0(z),$$

$$\phi_0(z, \zeta) = \phi_k(z, 0) - \int_0^{\zeta} \left\{ \gamma(z, \xi) \frac{\partial}{\partial \xi} + \beta(z, \xi) \frac{\partial}{\partial z} \right\} \phi_{k-1}(z, \xi)\, d\xi,$$
$$k = 1, \ldots, n-1.$$

Thus $\phi(t) = V(t, 0)$ can be further extended to $t \in t(\mathbf{G} \cup \sigma \times \mathbf{G}^* \cup \sigma \cup \mathbf{G})$ by the above integration. In particular the values $V(t(z, z), 0)$ are known for $z \in \mathbf{G} \cup \sigma$ and, correspondingly, $V^*(\tau(z, z), 0)$ for $z \in \mathbf{G}^* \cup \sigma$.

Now let us consider the boundary condition on σ in connection with the representation (5.4.42). The latter holds also on σ due to continuity. This provides the equation

$$\rho(x)=W(t(x,x),\tau(x,x))+W^*(\tau(x,x),t(x,x)),$$

$$\begin{aligned}\rho(x)=E^{-1}(t(x,x),\tau(x,x))\Big[&V(t(x,x),0)\\&+\int_0^{t(x,x)}\Gamma_1(t(x,x),\tau(x,x),s,0)V(s,0)\,ds\\&+\int_0^{\tau(x,x)}\Gamma_2(t(x,x),\tau(x,x),0,\eta)V^*(\eta,0)\,d\eta\Big]\\+E^{-1*}(\tau(x,x),t(x,x))\Big[&V^*(\tau(x,x),0)\\&+\int_0^{\tau(x,x)}\Gamma_1^*(\tau(x,x),t(x,x),\eta,0)V^*(\eta,0)\,d\eta\\&+\int_0^{t(x,x)}\Gamma_2^*(\tau(x,x),t(x,x),0,s)V(s,0)\,ds\Big].\end{aligned}\tag{5.4.53}$$

Let us consider first $x\in\mathbf{G}^*\cup\sigma$. Here we know the values of $V^*(\tau(x,x),0)$ and (5.4.53) can serve as a Volterra integral equation for the yet unknown function $V(t(x,x),0)$ for these x. Now $V(t(x,x),0)$ is known in $x\in\mathbf{G}\cup\sigma$ and in $x\in\mathbf{G}^*\cup\sigma$ and V is analytic in $\mathbf{G}$ and in $\mathbf{G}^*$, respectively. Since the difference of the boundary values on σ satisfies the homogeneous Volterra integral equation corresponding to (5.4.53), this difference vanishes identically. Hence the above defined $V(t(x,x),0)$ is continuous across σ and, hence, even analytic in $\mathbf{G}\cup\sigma\cup\mathbf{G}^*$ due to analyticity in $\mathbf{G}$ and in $\mathbf{G}^*$. In the same manner we define the analytic continuation of $V^*(\tau(x,x),0)$ from $\mathbf{G}^*$ into $\mathbf{G}\cup\sigma\cup\mathbf{G}^*$.

It remains to obtain the hyperanalytic function $V(t,0)=\phi(t)$ for arbitrary $t\in t(\mathbf{G}\cup\sigma\cup\mathbf{G}^*\times\mathbf{G}\cup\sigma\cup\mathbf{G}^*)$. To this end we integrate (5.4.52) once more, now from $z\in\mathbf{G}\cup\sigma\cup\mathbf{G}^*$ to $\zeta\in\mathbf{G}\cup\sigma\cup\mathbf{G}^*$:

$$\phi_0(z,\zeta)=\phi_0(z,z)$$

$$\phi_k(z,\zeta)=\phi_k(z,z)-\int_z^\zeta\left[\gamma(z,\xi)\frac{\partial}{\partial\xi}+\beta(z,\xi)\frac{\partial}{\partial z}\right]\phi_{k-1}(z,\xi)\,d\xi$$

$$k=1,\ldots,n-1$$

where $\phi(z,z)=V(t(z,z),0)$ is already known.

As the result we find $\phi(z,\zeta)=V(t(z,\zeta),0)$ and correspondingly $V^*(\tau(\zeta,z),0)$ for $z,\zeta\in\mathbf{G}\cup\sigma\cup\mathbf{G}^*$. Now (5.4.42) provides the desired extensions of V, V^* and via (5.4.29) those of W, W^* proving Theorem 5.4.4.

In the following let us extend the representation Theorem 5.4.2 to the more general system (5.4.9)

$$\frac{\partial W}{\partial \tau}+\sum_{j=0}^{n-1}\sum_{k=0}^{n-1} \mathrm{e}^j(A_{jk}w_k+B_{jk}\bar{w}_k)=F$$

which in general *cannot* be written solely in terms of the hypercomplex unknown W. Nevertheless (5.4.9) together with the conjugate equation leads to the system of integral equations:

$$W(t,\tau)=-\int_{\tau_0}^{\tau}\sum_{j,k=0}^{n-1} \mathrm{e}^j(A_{jk}w_k(t,\sigma)+B_{jk}w_k^*(\sigma,t))\,\mathrm{d}\sigma + \int_{\tau_0}^{\tau} F(t,\sigma)\,\mathrm{d}\sigma+\phi(t), \tag{5.4.54}$$

$$W^*(\tau,t)=-\int_{t_0}^{t}\sum_{j,k=0}^{n-1} \mathrm{e}^j(A_{jk}(\tau,s)w_k^*(\tau,s)+B_{jk}^*w_k(s,\tau))\,\mathrm{d}s + \int_{t_0}^{t} F^*(\tau,s)\,\mathrm{d}s+\phi^*(\tau), \tag{5.4.55}$$

where as before $W(t,\tau_0)=\phi(t)$, $W^*(\tau,t_0)=\phi^*(\tau)$ are the hyperanalytic Goursat data. But these integral equation systems have to be treated in a manner differently than before. To this end let us denote by $\hat{P}_k$ the projection of the hypercomplex number α,

$$\alpha=\sum_{j=0}^{n-1}\alpha_j\mathrm{e}^j,$$

on its kth component:

$$\alpha_k=\hat{P}_k\alpha.$$

Because e is a nilpotent, this projection from hypercomplex numbers into purely complex numbers *cannot* be written as an algebraic operation in the field of hypercomplex numbers.

If W is a solution of the system (5.4.5) then for x and y real the functions w_k are purely complex valued. But in the *extensions* the $w_k(t,\tau)$ become *hypercomplex valued*. Therefore, in order to write (5.4.54, 55) in a way that permits an iteration procedure, we have to define an operation decomposing W into the proper w_k-components.

Now let $V(t,\tau)$ be given as a hypercomplex valued function. Then in particular V can be decomposed for $(t,\tau)\in A$ and x, y real. If V is hyperanalytic then each component $\hat{P}_kV=V_k$ for x, y real becomes an analytic function of x, y. This function can uniquely be continued as a holomorphic function of z and ζ. Replacing z, ζ by its power series in t, τ (5.4.11), we get a hyperanalytic function $V_k(t,\tau)$. *We designate this operation by*

$$P_k V = V_k(t, \tau).$$

It can be prescribed by the following diagram:

$$V(t, \tau) \to \{V(t(x, y), \tau(x, y)) \quad \text{with} \quad (x, y) \in R^2\} \to \tilde{V}_k(x, y) = \hat{P}_k V \to$$

$$V_k(t, \tau) \equiv \tilde{V}_k\left(\tfrac{1}{2}(z(t, \tau) + \zeta), \frac{1}{2\mathrm{i}}(z - \zeta(t, \tau))\right) \tag{5.4.56}$$

$$V_k(t, \tau) \equiv P_k V(t, \tau).$$

Now (5.4.54, 55) can be written in the form,

$$\begin{aligned} W(t, \tau) &= -\int_{\tau_0}^{\tau} \sum_{j,k=0}^{n-1} \mathrm{e}^j (A_{jk}(t, \sigma) P_k W(t, \sigma) + B_{jk} P_k W^*(\sigma, t))\, \mathrm{d}\sigma \\ &\quad + \int_{\tau_0}^{\tau} F(t, \sigma)\, \mathrm{d}\sigma + \phi(t), \\ W^*(\tau, t) &= -\int_{t_0}^{t} \sum_{j,k=0}^{n-1} \mathrm{e}^j (A^*_{jk}(\tau, s) P_k W^* + B^*_{jk} P_k W(s, \tau))\, \mathrm{d}s \\ &\quad + \int_{t_0}^{t} F^*(\tau, s)\, \mathrm{d}s + \phi^*(\tau), \end{aligned} \tag{5.4.57}$$

where the right-hand sides define a linear operator of Volterra's type. The solution can be found by successive approximation, namely

$$\begin{aligned} W^{(m+1)}(t, \tau) &= -\int_{\tau_0}^{\tau} \sum_{j,k=0}^{n-1} \mathrm{e}^j (A_{jk}(t, \sigma) P_k W^{(m)}(t, \sigma) + B_{jk} P_k W^{*(m)}(\sigma, t))\, \mathrm{d}\sigma \\ &\quad + \int_{\tau_0}^{\tau} F(t, \sigma)\, \mathrm{d}\sigma + \phi(t), \\ W^{*(m+1)}(\tau, t) &= -\int_{t_0}^{t} \sum_{k=0}^{n-1} \mathrm{e}^j (A_{jk}(\tau, s) P_k W^{*(m)}(\tau, s) + B^*_{jk} P_k W^{(m)}(s, \tau))\, \mathrm{d}s \\ &\quad + \int_{t_0}^{t} F^*(\tau, s)\, \mathrm{d}s + \phi^*(\tau), \\ W^{(-1)} &= W^{*(-1)} = 0. \end{aligned} \tag{5.4.58}$$

We remark that the operator P_k does not commute with integration. Hence, we can not expect to obtain a simple expression of the resolvent kernel as before. Moreover, P_k contains the continuation from A to G and it is not obvious that (5.4.58) converges at all. Nevertheless, we shall prove the following.

Theorem 5.4.5 *The approximation converges uniformly in any compact subdomain of G and has an exponential majorant. The recursively defined family* $\{W^{(m)}(t,\tau)\}$ *is normal.*

The operator P_k maps hyperanalytic functions into hyperanalytic functions. Hence the limit function W of the family $W^{(m)}$ is hyperanalytic in each variable and so is the solution of (4.5.9).

Proof The proof is based on a representation of the sequence $W^{(m)}$ other than (5.4.58). To this end let (t,τ) and also the path of integration in (5.4.58) be restricted to A. With (5.4.6), then the formulas (5.4.58) can be written as

$$W^{(m+1)}(t,\tau)=\sum_{l=0}^{n-1} U_l^{(m+1)}\,\mathrm{e}^l$$
$$=\int_{\sigma=\tau_0}^{\tau}\left\{\sum_{j,k=0}^{n-1}\mathrm{e}^j(A_{jk}W_k^{(m)}(t,\sigma)+B_{jk}W_k^{*(m)}(\sigma,t),\right.$$
$$\left.+\sum_{j=0}^{n-1}\mathrm{e}^jF_j(t,\sigma)\right\}$$
$$\times\left\{\sum_{l=0}^{n-1}\mathrm{e}^l\left(\frac{\partial t_l}{\partial\zeta}\,\mathrm{d}\zeta(\sigma,t)+\frac{\partial t_l}{\partial z}\,\mathrm{d}z(t,\sigma)\right)\right\}+\sum_{l=0}^{n-1}\phi_l\mathrm{e}^l$$

and the conjugate equation for $W^{*(m+1)}$. Hence on A the following equations hold:

$$W_l^{(m+1)}(t,\tau)=-\int_{\sigma=\tau_0}^{\tau}\sum_{\substack{j+p=l\\0\leqslant j,\,p\leqslant l}}\left\{\sum_{k=0}^{n-1}\{A_{jk}W_k^{*(m)}(t,\sigma)\right.$$
$$\left.+B_{jk}W_k^{*(m)}(\sigma,t)\}+F_j(t,\sigma)\right\}$$
$$\times\left\{\frac{\partial t_p^*}{\partial\zeta}(\zeta(\sigma,t),z(\sigma,t))\,\mathrm{d}\zeta(\sigma,t)\right.$$
$$\left.+\frac{\partial t_p^*}{\partial z}(\zeta(\sigma,t),z(\sigma,t))\,\mathrm{d}z(\sigma,t)\right\}+\phi_l(t)$$
$$W_l^{*(m+1)}(\tau,t)=-\int_{s=t_0}^{t}\sum_{\substack{j+p=l\\0\leqslant j,\,p\leqslant l}}\left\{\sum_{k=0}^{n-1}\{A_{jk}^*W_k^{*(m)}(\tau,s)\right.$$
$$\left.+B_{jk}W_k^{(m)}(s,\tau)\}+F_j^*(\tau,s)\right\}$$
$$\times\left\{\frac{\partial t_p}{\partial\zeta}(z(s,\tau),\zeta(\tau,s))\,\mathrm{d}\zeta(\tau,s)\right.$$
$$\left.+\frac{\partial t_p}{\partial z}(z(s,\tau),\zeta(s,\tau))\,\mathrm{d}z(s,\tau)\right\}+\phi_l^*(\tau).\qquad(5.4.59)$$

Using (5.4.14, 16) in the right-hand sides, these can be extended to a hyperanalytic function in G by replacing σ, t, τ, s by *hypercomplex values*. The system (5.4.59) is a system of Volterra integral equations. As in the iteration (5.4.36), the approximation (5.4.59) provides an exponential function as a majorant which shows the uniform convergence in compact subsets of G. Consequently, the hypercomplex valued functions

$$\tilde{W}^{(m)} \equiv \sum_{l=0}^{n-1} W_l^{(m)} \mathrm{e}^l \tag{5.4.60}$$

converge there also uniformly. The construction implies $W^{(m)} = \tilde{W}^{(m)}$ on A. *But the hyperanalytic continuation is unique* according to Bochner and Martin [7, Section 36, Theorem 7]. Hence, $\tilde{W}^{(m)}$ coincides with $W^{(m)}$ all over G. Consequently, with the $W^{(m)}$ of (5.4.59), the equations (5.4.58) can be written as

$$W^{(m+1)}(t, \tau) = -\int_{\tau_0}^{\tau} \sum_{j,k=0}^{n-1} \mathrm{e}^j (A_{jk} W_k^{(m)}(t, \sigma) + B_{jk} W_k^{*(m)}(\sigma, t)) \, \mathrm{d}\sigma$$

$$+ \int_{\tau_0}^{\tau} F(t, \sigma) \, \mathrm{d}\sigma + \phi(t)$$

and the corresponding equation for $W^{*(m+1)}$.

Hence, the uniform convergence of the $W_l^{(m)}$ implies the uniform convergence of $W^{(m)}$ and the exponential majorants for $W_l^{(m)}$ define by (5.4.60) an exponential majorant for $W^{(m)} = \tilde{W}^{(m)}$.

The representation (5.4.57) and the method of convergent successive approximations provides again a *proof of Carlemen's Theorem* for solutions W of the homogeneous system (5.4.9) (with $F = 0$).

Let us choose $z_0 = 0$ and $\tau_0 = t_0 = 0$. From (5.4.57) it follows that $\phi(0) = 0$, and

$$\phi(t) = t^k \phi_0(t) \quad \text{with} \quad \phi_0(0) \neq 0, t = r\theta.$$

Using this property of ϕ in (5.4.58) yields

$$W^{(2)} = -\int_0^t \sum_{j,k=0}^{n-1} \mathrm{e}^j (A_{jk}(t, \sigma) \phi_{0k}(t) t^k + B_{jk} \phi_{0k}(\sigma) \sigma^k) \, \mathrm{d}\sigma + t^k \phi_0(t)$$

$$= t^k \phi_0(t) + 0(r^{k+1})$$

and consequently, we have by induction,

$$W^{(m)} = t^k \phi_0(t) + 0(r^{k+1}). \tag{5.4.61}$$

Let us recall that the $W^{(m)}$ form a normal family in G which contains 0. Consequently, $W^{(m)}$ converge uniformly and all derivatives as well in any compact subset of G and the limit function W also has the property

(5.4.61), namely

$$W = t^k \phi_0(t) + 0(r^{k+1}),$$

which is the desired result.

Proposition 5.4.6 *The mappings defined by* (5.4.42) *in Theorem* 5.4.2 *or by the solution of* (5.4.59) *corresponding to Theorem* 5.4.5 *can be used for generating families of solutions to* (5.4.21) *or* (5.4.9) *by choosing holomorphic vectors* $\mathbf{\Phi}(\xi)$ *and inserting* $t(z)$ *for* ξ. *In case of equation* (5.4.21) *one finds complete families by using complete holomorphic families,* e.g. *polynomials* $\{t(z)^k\}_{k=0}^{\infty}$. *The corresponding W can be found by the successive approximations* (5.4.59).

The completeness of the above families is an immediate consequence of the Runge approximation property for (5.4.21), Theorem 5.3.10. Note that (5.4.59) gives a constructive method if one selects from the solution family sequences approximating boundary values and side conditions.

Remark 5.4.7 The proof of the reflection principle Theorem 5.4.4 can be carried over to systems (5.4.9) by using sequences defined by (5.4.59) instead of (5.4.50–53).

5.5 Remarks on the Additional References

In [73] general linear elliptic first-order systems are treated by the use of singular integral equations in plane and in higher dimensional domains. In [84] one finds the theory of singular integral equations and their connections with boundary value problems.

In [76], [103], [109] and [112] generalized Cauchy formulas are obtained and these can be used to derive singular integral equations for solving boundary value problems.

Hyperanalytic functions and hypercomplex algebras are used in [81], [90] and [107], and are applied to problems in mechanics in [94]. The similarity principle and maximum principle have been investigated in [92], [93], [105] and [111]. The Runge approximation property can also be found in [115].

Function theoretic methods are developed in [80], [82], [85], [87], [88], [106] and [119]; especially for the equations with singular coefficients, which can be obtained from rotational symmetric problems, in [95], [98], [99] and [104]. The use of such methods to study transonic flow problems can be found in [75]. Bergman and Vekua operators and the generation of complete systems of solutions can be found in [74], [78], [79], [83], [91],

[96], [97], [100], [101] and [111–114]; their use for numerical methods can be found in [86]. The analyticity of fundamental solution has been investigated in [89]; the reflection principle and analytic continuation in [117–119]. Singularities of solutions have been characterized especially with respect to their location in [77], [102] and [110]. Discontinuities of the solution are investigated in [108] and [116].

References

1 Agmon, S., Multiple layer potentials and the Dirichlet problem for higher order elliptic equations in the plane. *Comm. Pure Appl. Math.* **10,** 179–239, 1975.

2 Atkinson, K., *A survey of numerical methods for the solution of Fredholm integral equations of the second kind.* SIAM, Philadelphia, 1976.

3 Begehr, H. and Gilbert, R. P., Randwertaufgaben ganzzahliger Charakteristik für verallgemeinerte hyperanalytische Funktionen. *Appl. Analysis,* **6,** 189–206, 1977.

4 Bergman, S., *Integral operators in the theory of linear partial differential equations.* Springer, Berlin, Heidelberg, New York, 1960.

5 Bers, L., Local behaviour of solutions of general linear elliptic equations. *Comm. Pure Appl. Math.* **8,** 473–496, 1955.

6 Bers, L., An outline of the theory of pseudo-analytic functions. *Bull. Am. Math. Soc.* **62,** 291–331, 1956.

7 Bochner, S. and Martin, W. T., *Functions of several complex variables.* Princeton University Press, 1948.

8 Bojarski, B., Einige Randwertaufgaben für Systeme von 2m Gleichungen des elliptischen Typs in der Ebene. *Dokl. Akad. Nauk SSSR.* **124,** 15–18, 1958 (Russian).

9 Bojarski, B., Theory of generalized analytic vectors. *Ann. Polon. Math.* **17,** 281–320, 1966 (Russian).

10 Carleman, T., Sur un problème d'unicité pour les systèmes d'équations aux dérivées partielles à deux variables indépendentes. *Ark. Mat. Astr. Fys.* **26,** B No 17, 1–9, 1939.

11 Colton, D., *Partial differential equations in the complex domain.* Pitman, London, San Franciso, Melbourne, 1976.

12 Colton, D., *Solution of boundary value problems by the method of integral operators.* Pitman, London, San Francisco, Melbourne, 1976.

13 Daniljuk, I. I., *Nonregular boundary value problems in the plane.* Izdat. Nauka, Moscow, 1975.

14 Dieudonné, G., *Foundations of modern analysis.* Academic Press, New York, London, 1960.

15 Douglis, A., A function theoretic approach to elliptic systems of equations in two variables. *Comm. Pure Appl. Math.* **6,** 259–289, 1953.

16 Fichera, G., Linear elliptic equations of higher order in two independent variables and singular integral equations. *Proc. Conf. Part. Diff. Equ. and Conf. Mech.*, University Wisconsin Press, 1961.

17 Fichera, G. and Ricci, P., The single layer potential approach in the theory of boundary value problems for elliptic equations. *In Lecture Notes* No. 561, pp. 39–50, Springer, Berlin.
18 Gaier, D., Integralgleichungen erster Art und konforme Abbildung. *Math. Zeitschr.* **147,** 113–129, 1976.
19 Garabedian, P. R., Analyticity and reflection principle for plane elliptic systems. *Comm. Pure Appl. Math.* **14,** 315–322, 1961.
20 Garabedian, P. R., *Partial differential equations.* John Wiley, New York, London, Sydney, 1964.
21 Gilbert, R. P., *Function theoretic methods in partial differential equations.* Academic Press, New York, 1969.
22 Gilbert, R. P., Constructive methods for elliptic equations. *Lecture Notes* No. 365, Berlin, Springer, 1974.
23 Gilbert, R. P., Nonlinear boundary value problems for elliptic systems in the plane. *Proc. Internat. Conf. Nonlinear Systems Appl.*, Techn. Report G3, University of Delaware, 1976.
24 Gilbert, R. P. and Hile, G., Generalized hyperanalytic function theory. *Trans. Am. Math. Soc.*, **195,** 1–29, 1974.
25 Gilbert, R. P. and Wendland, W., Analytic, generalized, hyperanalytic function theory and an application to elasticity. *Proc. Roy. Soc. Edinburgh* **73** A, 317–331, 1975.
26 Gohberg, I. C. and Krupnik, N., *Einführung in die Theorie der eindimensionalen singulären Integraloperatoren.* Birkhäuser, Basel, Stuttgart, to appear 1977/78.
27 Gohberg, I. C. and Krupnik, N. J., Singular integral operators with piecewise continuous coefficients and their symbols. *Math. USSR Izv.* **5,** 955–980, 1971; Orig. : *Izv. Akad. Nauk SSSR* **35,** 1971.
28 Goldschmidt, B., Cousin-Probleme bei partiellen Differentialgleichungen. *Dissertation.* Martin-Luther-Universität Halle-Wittenberg, 1975.
29 Goldschmidt, B., Funktionentheoretische Eigenschaften der Lösungen der Vekuaschen Differentialgleichung mit Koeffizienten in $L_{p,\mathrm{loc}}(G)$, 1977, to appear.
30 Grisvard, P., Behavior of the solution of an elliptic boundary value problem in a polygonal or polyhedral domain. *In* B. Hubbard. (Ed.), *Numerical solution of partial differential equations* III, pp. 207–274, Synspade 1975, Academic Press, New York, 1976.
31 Gusman, S. J., On the uniform approximation of generalized analytic functions. *Dokl. Akad. Nauk SSSR* **144,** 706–708, 1962 (Russian).
32 Gusman, S. J. and Abdulajeva, I. I., On the uniform approximation of generalized analytic functions on Riemann manifolds. *Bull. Acad. Sci. Georgian SSR* **78,** 21–24, 1975.
33 Haack, W. and Wendland, W., *Lectures on partial and pfaffian differential equations.* Pergamon Press, Oxford, 1971.
34 Habetha, K., On zeros of elliptic systems of first order in the plane. *In* R. P. Gilbert and R. J. Weinacht (Eds.), *Research Notes in Mathematics*, 8, pp. 45–62, Pitman, London, San Francisco, Melbourne, 1976.
35 Hayashida, K., Unique continuation theorem of elliptic systems of partial differential equations. *Proc. Japan. Acad.* **38,** 630–635, 1962.

36 Heinz, E. and Hildebrandt, St., Some remarks on minimal surfaces in Riemannian manifolds. *Comm. Pure Appl. Math.* **23,** 371–377, 1970.
37 Hile, G. H. and Protter, M. H., Unique continuation and the Cauchy problem for first order systems of partial differential equations. *Comm. Partial Diff Eqns* **1,** 437–465, 1976.
38 Hörmander, L., *Linear partial differential operators.* Springer, Berlin, Heidelberg, New York, 1969.
39 Hsiao, G. and MacCamy, R., Solution of boundary value problems by integral equations of the first kind. *SIAM Rev.* **15,** 687–705, 1973.
40 Hsiao, G. C., and Wendland, W. L., A finite element method for some integral equations of the first kind. *J. Math. Analysis Appl.* **58**, 449–481, 1977.
41 John, F., Plane waves and spherical means applied to partial differential equations. *Tracts in Pure and Applied Mathematics* 2, Interscience, London, New York, 1955.
42 Kondratiev, V. A., Boundary value problems for elliptic equations in domains with conical or angular points. *Trudy Mosk. Mat. Obshtshestva* **16,** 209–292, 1967; *Trans. Moscow Math. Soc.* 227–313, 1967.
43 Kopp, P., Über eine Klasse von Randwertproblemen mit verletzter Lopatinski-Bedingung für elliptische Systeme erster Ordnung in der Ebene. *Dissertation.* T. H., Darmstadt, D17, 1977.
44 Kühn, E., Über die Funktionentheorie und das Ähnlichkeitsprinzip einer Klasse elliptischer Differentialgleichungen in der Ebene. *Dissertation.* Dortmund, 1974.
45 Lewy, H., On the reflection laws of second order differential equations in two independent variables. *Bull. Am. Math. Soc.* **65,** 37–58, 1959.
46 Lopatinskiĭ, J. B., On a method of reduction of boundary value problems for systems of elliptic differential equations to regular integral equations. *Ukrain. Matem. J.* **5,** 123–151, 1953; *Am. Math. Soc. Transl.* II **89,** 149–183, 1970.
47 Lopatinskiĭ, V. B., Uniqueness of the solution of the Cauchy problem for a class of elliptic equations. *Dopovidi Akad. Nauk Ukrain. SSR*, 689–693, 1958 (Ukrainian).
48 Meister, E., *Singular integral equations.* Pitman Press, London, San Francisco, Melbourne, to appear.
49 Mikhlin, S. G., *Advanced course of mathematical physics.* North-Holland, Amsterdam, London, 1970.
50 Miranda, C., *Partial differential equations of elliptic type.* Springer, Berlin, Heidelberg, New York, 1970.
51 Muramutu, T., On the uniqueness of the Cauchy problem for elliptic systems. *Sci. Papers College Gen. Ed. Univ. Tokyo* **11,** 13–23, 1961.
52 Muschelischwili, N., *Singuläre Integralgleichungen.* Akademie Verlag, Berlin, 1965.
53 Noble, Ben., Error analysis of colocation methods for solving Fredholm integral equations. *In* J. H. Miller (Ed.), *Tropics in numerical analysis,* Academic Press, London, 1972.
54 Pascali, D., Vecteurs analytiques généralisés. *Rev. Roum. Math. pura appl.* **10,** 779–808, 1965.

55 Pettineo, B., Nuove dimonstrazione dei teoremi di esistenza per i problemi al contorno regolari relativi alle equazioni lineari a derivate parziali di tipo ellittico. *Rend. Acc. Lincei* **23,** 32–38, 1957.
56 Pliš, A., A smooth linear elliptic differential equation without any solution in a sphere. *Comm. Pure Appl. Math.* **14,** 599–617, 1951.
57 Pröβdorf, S., Über eine Klasse singulärer Integralgleichungen nicht normalen Typs. *Math. Ann.* **183,** 130–150, 1969.
58 Prößdorf, S., *Einige Klassen singulärer Gleichungen.* Birkhäuser, Basel, Stuttgart, 1974.
59 Ricci, P., Sui potenziale di semplice strato per le equazioni ellittiche di ordine superiore in due variabili. *Rend. Mat.* **7,** 1–39, 1974.
60 le Roux, M. N., Method d'elements finis par la resolution numérique de problèmes exterieurs en dimension deux. *Rev. Francaise d'Automatique,* Inf. Rech. Opérationelle, to appear.
61 Schüppel, B., Regularisierung singulärer Integralgleichungen vom nicht normalen Typ mit stückweise stetigen Koeffizienten (Technische Hochschule Darmstadt Fachbereich Mathematik, Preprint **293,** 1976), to appear.
62 Tovmasjan, N. E., On a method for solving boundary value problems for elliptic systems of differential equations of the second order in the plane. *Mat. Sbornik* **89,** 131, 599–615, 1972.
63 Tovmasjan, N. E., An effective method for solving the Dirichlet problem for elliptic systems of differential equations of the second order with constant coefficients in plane bounded ellipses. *Differentialnye Uravnenija* **5,** 60–71, 1969.
64 Vekua, I. N., *Verallgemeinerte Analytische Funktionen.* Akademie Verlag, Berlin, 1963.
65 Vekua, I., *New methods for solving elliptic equations.* North Holland, Amsterdam, 1967.
66 Vekua, N. P., *Systems of singular integral equations.* Noordhoff, Groningen, 1967.
67 Vol'pert, A., On the index and normal solvability of boundary value problems for elliptic systems of differential equations in the plane. *Trudy Mosk. Mat. Obštš.* **10,** 41–87, 1961 (Russian).
68 Wendland, W., Bemerkungen über die Fredholmschen Sätze. *In* B. Brosowski and M. Martensen (Eds.), *Methoden und Verfahren der Mathematischen Physik* **3,** pp. 141–176, BI-Mannheim Hochschulskripten 722/722a*, 1970.
69 Wendland, W., On a class of semilinear boundary value problems for certain elliptic systems in the plane, to appear.
70 Wolska-Bochenek, J., On some generalized nonlinear problem of the Hilbert type. *Zeszyty Nauk. Politechn. Warszawskiej* **183**; *Matematyka* **14,** 15–32, 1968.
71 Wolska-Bochenek, J., A compound non-linear boundary value problem in the theory of pseudo-analytic functions. *Demonstratio Mathematica* **4,** 105–117, 1972.
72 Chung Ling Yu, Reflection principle for systems of first order elliptic

equations with analytic coefficients. *Trans. Am. Math. Soc.* **164,** 489–501, 1972.

Additional References

73 Avantaggiati, A., Nuovi contributi allo studio dei problemi al contorno per i sistemi ellittici del primo ordine. *Ann. Matem.* **69,** 107–170, 1965.

74 Bauer, K. W. and Florian, H., Bergman-Operatoren mit Polynomerzeugenden. *In* R. P. Gilbert and R. J. Weinacht (Eds.), *Function theoretic methods in differential equations.* Res. Notes 8, pp. 85–93, Pitman, London, San Francisco, Melbourne, 1976.

75 Bauer, F., Garabedian, P. and Korn, D., A theory of supercritical wing sections, with computer programs and examples. *Lecture Notes in Economics and Mathematical Systems* No. 66, Springer, Berlin, 1972.

76 Bojarski, B., The general representation of solutions to elliptic systems of 2m equations in the plane. *Dokl. Akad. Nauk* **122,** 543–546, 1958.

77 Colton, D. and Gilbert, R. P., Singularities of solutions to elliptic partial differential equations with analytic coefficients. *Quart. J. Math. Oxford* **19,** 391–396, 1968.

78 Colton, D., Integral operators for elliptic equations in three independent variables, I, II. *Applicable Analysis* **4,** 77–95, 1974 and **4,** 283–295, 1975.

79 Colton, D. and Gilbert, R. P., Integral operators and complete families of solutions for $\Delta_{p+2}^2 u(\underset{\sim}{x}) + A(r^2)\Delta_{p+2} u(\underset{\sim}{x}) + B(r^2)u(\underset{\sim}{x}) = 0$. *Arch. Rat. Mech. Anal.* **43,** 62–78, 1971.

80 Danyljuk, I. I., On the oblique derivative problem for the general quasilinear system of the first order. *Dokl. Akad. Nauk SSSR* **127,** 953–956, 1959.

81 Delanghe, R. and Brackx, F., Hyperkomplex function theory and Hilbert modules with reproducing kernel, to appear.

82 Douglis, A., Function theoretic properties of certain elliptic systems of first order linear equations. *Lectures on functions of a complex variable.* University of Michigan, 335–340, 1955.

83 Farooqui, A. S., A complete set of orthonormal harmonic functions. *SIAM J. Math. Anal.,* **4** (2), 309–313, 1973.

84 Fichera, G., Una introduzione alla teoria delle equazioni integrali singolari. Ediz. Cremonese, Roma, 1958.

85 Garabedian, P. R., Lectures on function theory and partial differential equations. *Rice University Studies* **49,** No. 4, 1963.

86 Gilbert, R. P., Integral operator methods for approximating solutions of Dirichlet problems. *Iterationsverfahren, Numerische Mathematik, Approximationstheorie,* Birkhäuser Basel, ISNM, **15,** 1970.

87 Gilbert, R. P. and Hile, G. N., Hilbert function modules with reproducing kernels, to appear.

88 Gilbert, R. P. and Hile, G. N., Hypercomplex function theory in the sense of L. Bers. *Math. Nachr.* **72,** 187–200, 1976.

89 Habetha, K., Über lineare elliptische Differentialgleichungen mit analytischen Koeffizienten. *Ber. Ges. Math. Datenverarb. Bonn* **77,** 65–89, 1973.

90 Habetha, K., Eine Bemerkung zur Funktionentheorie in Algebren. *In Function theoretic methods for partial differential equations,* Springer Lecture Notes 561, 502–509, 1976.

91 Hile, G. N., Representations of solutions of a special class of first order systems, *J. Differential Equations* to appear.

92 Hile, G. N. and Protter, M. H., Maximum principles for a class of first order elliptic systems, *J. Differential Equations* **24,** 136–151, 1977.

93 Koohora, A., Similarity principle of the generalized Cauchy–Riemann equations for several complex variables. *J. Math. Soc. Jap.* **23,** 213–249, 1971.

94 Krawietz, A., Ein Beitrag zur Behandlung ebener Elastizitätsprobleme. *Techn. Report* Techn. University, Berlin, 1975.

95 Kühnau, R., Über drehstreckungssymmetrische Potentiale. *Math. Nachr.* **56,** 201–205, 1973.

96 Lanckau, E., Konstruktive Methoden zur Lösung von elliptischen Differentialgleichungen mittels Bergman-Operatoren. *In* Anger (Ed.), *Elliptische Differentialgleichungen* I, pp. 67–76, 1970.

97 Lanckau, E., Zur Lösung gewisser partieller Differentialgleichungen mittels parameterabhängiger Bergman-Operatoren. *Nova Acta Leopoldina* **201,** Bd. 36, 1971.

98 Lanckau, E., Beitrag zur Theorie axialsymmetrischer Potentiale. *Wiss. Zeitschr. Techn. Hochschl. Karl-Marx-Stadt* **15,** 55–64, 1973.

99 Lanckau, E., Über die Differentialgleichungen der Torsion von Rotationskörpern. *Beiträge z. Numerischen Math.* **4** 147–155, 1975.

100 Lanckau, E., Über ein strömungsmechanisches Problem. *Wiss. Zeitschr. Techn. Hochsch. Karl-Marx-Stadt* **17,** 777–785, 1975.

101 Lanckau, E., Zur Integration der Differentialgleichungen ebener kompressibler Unterschallströmungen. *ZAMM* **56,** T 450–T 453, 1976.

102 Millar, R. F., Singularities of solutions to linear, second order analytic elliptic equations in two independent variables I. The completely regular boundary. *Appl. Anal.* **1,** 101–121, 1971.

103 Nicolau, A., Die Zurückführung von Verallgemeinerten Dirichletschen und Neumannschen Problemen für verallgemeinerte analytische Funktionen auf Fredholmsche Integralgleichungen. *St.Čerč. Mat. Bucaresti* **23,** 1413–1457, 1971.

104 Quinn, D. W. and Weinacht, R. J., Boundary value problems in generalized biaxially symmetric potential theory. *J. Differential Equations* **21,** 113–133, 1976.

105 Sarkisjan, S. C., Properties of solutions of Cauchy-Riemann systems with non-linear right-hand sides. *Akad. Nauk Armjansk. SSR Dokl.* **36,** 141–146, 1963.

106 Schubert, H., Über das Randwertproblem von Poincaré in der Ebene. *Wiss. Zeitschr. Univ. Rostock, math.-naturw.* **19,** 397–404, 1970.

107 Snyder, H. H., *A hyperkomplex function theory associated with Laplace's equation.* Verl. Dt. Wiss., Berlin, 1968.

108 Sovin, J. A., Elliptic boundary value problems for plane domains with angles and discontinuities reaching the boundary. *Soviet Math. Dokl.* **10,** 985–988, 1969.

109 Tutschke, W., Die Cauchysche Integralformel für morphe Funktionen mehrerer komplexer Variabler. *Math. Nachr.* **54,** 385–391, 1972.

110 Tutschke, W., Konstruktion von globalen Lösungen mit vorgeschriebenen Singularitäten bei partiellen komplexen Differentialgleichungssystemen. *Sitzber. d. sächsische Akad. Wiss. Leipzig, Math.-nath. Kl.* **109,** 1–24, 1972.

111 Tuschke, W., A new application of I. N. Vekua's proof to Carleman's theorem. *Dokl. Akad. Nauk, Soviet Math. Dokl.* **15,** 374–378, 1974.

112 Vološina, M. S., The generalized Cauchy formula for a first-order elliptic system. *Soviet Math. Dokl.* **4,** 1784–1787. 1963.

113 Watzlawek, W., Über Zusammenhänge zwischen Fundamental systemen, Riemann-Funktion und Bergmann-Operatoren. *J. reine angewandte Math.* **251,** 200–211, 1972.

114 Watzlawek, W., Zur Lösungsdarstellung bei gewissen linearen partiellen Differentialgleichungen zweiter Ordnung. *Monatsh. Math.* **73,** 461–472, 1969.

115 Weck, N., Über die Lösung von Außenraumaufgaben durch Approximation der Randdaten, to appear.

116 Wigley, N. M., Mixed boundary value problems in plane domains with corners. *Math. Z.* **115,** 33–52, 1970.

117 Chung Ling Yu, Reflection principle for solutions of higher order elliptic equations with analytic coefficients. *SIAM J. Appl. Math.* **20,** 358–363, 1971.

118 Chung Ling Yu, Cauchy problem for systems of first order analytic equations in the plane. *SIAM J. Math. Anal.* To appear.

119 Chung Ling Yu, Cauchy problem and analytic continuation for systems of first order elliptic equations with analytic coefficients. *Trans. Am. Math. Soc.* **185,** 429–443, 1973.

B

Computational Methods

6

Numerical Methods Using Integral Equations

In Sections 1.1–1.3 we saw that the general boundary value problems for $n=1$ could be solved by using some solutions of the *first* boundary value problem. With the Green and the Neumann functions G^I, G^{II}, which are explicitly known from conformal mapping, the first boundary value problem is equivalent to the system of integral equations,

$$u(\zeta)=\iint_{\mathsf{G}}\{(Au+Bv)G^I_x+(\tilde{A}u+\tilde{B}v)G^I_y\}[dx, dy] + \iint_{\mathsf{G}}(CG^I_x+\tilde{C}G^I_y)[dx, dy]-\oint_{\dot{\mathsf{G}}}\psi\, d_nG^I$$

$$v(\zeta)=-\iint_{\mathsf{G}}\{(Au+Bv)G^{II}_y-(\tilde{A}u+\tilde{B}v)G^{II}_x\}[dx, dy] - \iint_{\mathsf{G}}(CG^{II}_y-\tilde{C}G^{II}_x)[dx, dy]+\oint_{\dot{\mathsf{G}}}\psi\, dG^{II}+\kappa. \tag{6.0.1}$$

In order to find approximate solutions to (6.0.1), we approximate G^I, G^{II}, G^I_x, G^I_y, G^{II}_x, G^{II}_y. For this it is convenient to approximate the conformal mapping ϕ taking G onto the unit disk, as well as ϕ'.

Hence, Section 6.1 begins with constructive methods for conformal mapping by the use of polynomials and the Ritz method following Gaier [7]. In the integral equations appear boundary integrals corresponding to harmonic functions with given Dirichlet or Neumann data. In Section 6.2 it is shown how the boundary integrals can be evaluated. For these two sections I have to thank Dr. W. F. Moss who revised the first version and who made many valuable remarks. In particular, the error analysis for the approximating conformal mappings in connection with (6.1.15) is due to him [15].

In Section 6.3 we consider the discretization of (6.0.1) and the convergence properties depending on the decreasing mesh width. Here we use quadrature formulas following [9, Section 14] and the concept of Anselone [1]. Since the discretized integral equations lead to linear equations with completely filled matrices one needs iteration procedures for their solution. One such procedure is presented in Section 6.4. This method hinges on the fact that the real spectrum of a Fredholm equation of the second kind coincides with the real eigenvalues λ of

$$w_{\bar{z}} = \lambda a w + \lambda b \bar{w}, \qquad \text{Re}\, w_{|\dot{G}} = 0, \qquad \oint_{\dot{G}} \text{Im}\, w\sigma \, ds = 0.$$

The latter do not exist and, hence, the integral equations do not have real eigenvalues at all. This allows the use of an iteration procedure based on a mapping of the spectrum and the integral operator [4]. In Section 6.5 we investigate an integral equation method using single layer potentials. This method does not need the Green and Neumann functions but only the well known fundamental solution $\log|z-\zeta|$ of the Laplacian. The integral equations of the first kind on the boundary curve will have logarithmic principal part. These integral equations include the equations of Fichera and Ricci (5.2.110). Following [10] we shall review their treatment with Galerkin–Ritz methods and present the corresponding asymptotic error analysis. By a slight extension, the method can also be used for solving the first boundary value problem [24]. Also in this case, the Galerkin–Ritz method can be applied and we obtain convergence and asymptotic estimates.

6.1 Conformal Mapping

We shall use an approximation of ϕ by suitable polynomials which originated with Carleman and Szegö. In this method we can use integrals over G or over $\dot{G}$. But because the one-dimensional integrals over $\dot{G}$ are numerically easier to handle, we will use here only boundary integrals. The following considerations are from the books of Golusin [8, Chapters IX and X], and Gaier [7, pp. 126 ff]. The approximation of the exterior conformal mappings is due to Moss [15] who presents also numerical results. As before, let G be a bounded domain whose boundary $\dot{G}$ is a closed, rectifiable, Jordan curve, and let $0 \in G$. Let $w = f(z)$ be the conformal mapping of G onto the disk Δ_R of radius R, centred at the origin. Let f have the normalization $f(0) = 0$, $f'(0) = 1$. R is called the *conformal radius* of G. It is well-known that $f(z)$ is the unique solution to the problem of minimizing

$$\sup_{z \in G} |F(z)|$$

over the class of functions $F(z)$, holomorphic in G with $F(0)=0$ and $F'(0)=1$, and that

$$R=\sup_{z\in G}|f(z)|.$$

The function ϕ which we wish to approximate satisfies

$$\phi(z)=\frac{f(z)}{R}, \qquad R=\frac{1}{\phi'(0)}.$$

There are several known extremal characterizations of $f(z)$. We now introduce another of these which leads easily to approximations of $f(z)$ and hence to approximations of $\phi(z)$. For this we introduce the class $L_2(\dot{G})$.

Let $z=g(w)$ denote the *inverse* of $f(z)$. Because the boundary of G is a closed, rectifiable, Jordan curve it follows that $g(w)$ is *continuous* in the closed disk $\bar{\Delta}_R$ and is *absolutely continuous* on the circle $|w|=R$. Furthermore,

$$\sup_{0\leq r<R}\int_0^{2\pi}|g'(re^{i\theta})|\,d\theta<\infty. \tag{6.1.1}$$

Now (6.1.1) implies that there exists $h\in L_1([0,2\pi))$ such that

$$\lim_{w\to Re^{i\theta}} g'(w)=h(\theta)$$

for almost all θ, where w approaches $Re^{i\theta}$ along any nontangential path. We will denote $h(\theta)$ by $g'(Re^{i\theta})$, i.e. the nontangential boundary values of $g'(w)$. It also follows that $dg(Re^{i\theta})/d\theta=Rie^{i\theta}g'(Re^{i\theta})$ for almost all θ, and that the length $S(\theta', \theta'')$ of the arc, $z=g(Re^{i\theta})$, $\theta'\leq\theta\leq\theta''$, is given by

$$S(\theta',\theta'')=\int_{\theta'}^{\theta''}|g'(Re^{i\theta})|\,d\theta. \tag{6.1.2}$$

Now let

$$S(\theta)=\int_0^{\theta}|g'(Re^{i\theta})|\,d\theta,$$

$$l=\int_0^{2\pi}|g'(Re^{i\theta})|\,d\theta,$$

and let s denote Lebesgue measure on $[0, l)$. If $F(z)$ is holomorphic in G, then the integral

$$\int_{C_r}|F(z)|^2\,ds=r\int_0^{2\pi}|F(g(re^{i\theta}))\sqrt{g'(re^{i\theta})}|^2\,d\theta,$$

where $C_r=g(|w|=r)$, is a monotonic increasing function of r. We say that

$F(z)\in L_2(\dot{G})$ if this integral is bounded for $0\leqslant r<R$. Now $\dot{G}$ has a tangent for almost all $0\leqslant s<l$ and if $F(z)\in L_2(\dot{G})$ there exists $\tilde{h}\in L_1([0,l))$ such that

$$\lim_{z\to z(s)\in\dot{G}} F(z)=\tilde{h}(s)$$

for almost all s, where z approaches $z(s)\in\dot{G}$ along any nontangential path. We will denote $F(z(s))$ by $\tilde{h}(s)$, i.e. the nontangential boundary values of $F(z)$. We can introduce an inner product on $L_2(\dot{G})$ by

$$(F,H)\equiv\lim_{r\to R}\int_{C_r} F(z)\overline{H(z)}\,ds=\int_0^l F(z(s))\overline{H(z(s))}\,ds$$

and denote by $\|\cdot\|$ the corresponding norm. With this inner product it can be shown that $L_2(\dot{G})$ is a complex Hilbert space.

Now let $L\equiv\{F\mid F(z)$ holomorphic in G with $F_{|\dot{G}}\in L_2(\dot{G})$ and $F(0)=1\}$. The next theorem relates $f(z)$ to the unique solution to a certain minimum problem over L.

Theorem 6.1.1 (Julia) *The function $F_0(z)=\sqrt{f'(z)}$ and only this function minimizes the integral*

$$\int_0^l |F(z(s))|^2\,ds \tag{6.1.3}$$

over the class L. This minimum is equal to $2\pi R$.

Proof For $F\in L$

$$\begin{aligned}\int_0^l |F(z(s))|^2\,ds&=\lim_{r\to R}\int_{C_r}|F(z)|^2\,ds\\&=\lim_{r\to R} r\int_0^{2\pi}|F(g(re^{i\theta}))\sqrt{g'(re^{i\theta})}|^2\,d\theta\\&=\lim_{r\to R} r\int_0^{2\pi}|H(re^{i\theta})|^2\,d\theta,\end{aligned}$$

when $H(w)=F(g(w))\sqrt{g'(w)}$ and $H(0)=1$. Now $H(w)=\sum_{n=0}^{\infty}a_nw^n$ with $a_0=1$ so that

$$\int_0^l |F(z(s))|^2\,ds=2\pi\sum_{n=0}^{\infty}|a_n|^2R^{2n+1}\geqslant 2\pi R,$$

where equality holds precisely when $a_n=0$, $n>0$, in which case

$$H(w)\equiv 1,\quad F(z)=\sqrt{f'(z)},\quad\text{and}\quad f(z)=\int_0^z [F_0(\xi)]^2\,d\xi.$$

This completes the proof.

The following lemma will be useful in what follows.

Lemma 6.1.2 *The function $F_0(z)$ is orthogonal to each function $\mathsf{P} \in L_2(\dot{\mathsf{G}})$ with $\mathsf{P}(0)=0$, i.e.*

$$(F_0, \mathsf{P}) = \int_0^l F_0(z(s))\overline{\mathsf{P}(z(s))}\,\mathrm{d}s = 0. \tag{6.1.4}$$

Proof For each $\varepsilon > 0$ and $0 \leqslant s < l$, $F_0(z) + \varepsilon e^{2\pi i s/l}\mathsf{P}(z) \in \mathsf{L}$ and

$$\begin{aligned}\int_0^l |F_0(z(s))|^2\,\mathrm{d}s &\leqslant \int_0^l |F_0(z(s)) + \varepsilon e^{2\pi i s/l}\mathsf{P}(z(s))|^2\,\mathrm{d}s \\ &= \int_0^l |F_0(z(s))|^2\,\mathrm{d}s \\ &\quad + 2\varepsilon \operatorname{Re}\left\{e^{-2\pi i s/l}\int_0^l F_0(z(s))\overline{\mathsf{P}(z(s))}\,\mathrm{d}s\right\} \\ &\quad + \varepsilon^2 \int_0^l |\mathsf{P}(z(s))|^2\,\mathrm{d}s.\end{aligned}$$

Division by ε yields

$$2\operatorname{Re}\left\{e^{-2\pi i s/l}\int_0^l F_0(z(s))\overline{\mathsf{P}(z(s))}\,\mathrm{d}s\right\} + \varepsilon \int_0^l |\mathsf{P}(z(s))|^2\,\mathrm{d}s \geqslant 0. \tag{6.1.5}$$

If (6.1.4) were false, by suitable choice of s and $\varepsilon > 0$, the left hand side of (6.1.5) could be made negative, a contradiction. The proof is complete.

We now introduce the *Ritz method* for the approximation of $F_0(z)$. Because of (6.1.1) all functions holomorphic and bounded in G are in class $L_2(\dot{\mathsf{G}})$. *Let L_n denote the set of all polynomials $P(z)$ of degree less than or equal to n with $P(0) = 1$.*

Theorem 6.1.3 *For each $n \geqslant 1$, there exists a polynomial $P_n(z)$ in class L_n which uniquely minimizes*

$$\int_0^l |P(z(s))|^2\,\mathrm{d}s$$

over L_n.

Proof For a fixed n let

$$\inf_{P(z) \in \mathsf{L}_n} \int_0^l |P(z(s))|^2\,\mathrm{d}s = \mu.$$

Because the polynomial $P(z)\equiv 1$ is in class L_n, it is sufficient to consider only polynomials $P(z)=1+\alpha_1 z+\cdots+\alpha_n z^n$ for which

$$\int_0^l |P(z(s))|^2\,\mathrm{ds}\leq l. \tag{6.1.6}$$

Such polynomials are locally uniformly bounded in every disk $\bar{\Delta}_{r'}\subseteq \mathsf{G}$ due to the following. If $\{|z-z_0|\leq r'\}\subseteq \mathsf{G}$ and $G(z)\in L_2(\dot{\mathsf{G}})$, then applying Cauchy's formula to $F^2(z)$ it follows that

$$|F(z_0)|^2\leq\frac{1}{2\pi r'}\int_{C_r}|F(z)|^2\,ds\leq\frac{1}{2\pi r'}\int_0^l|F(z(s))|^2\,\mathrm{ds} \tag{6.1.7}$$

where r is sufficiently close to R so that $\{|z-z_0|\leq r'\}$ is contained in the interior of C_r. Thus, there exists a constant $M_k>0$ such that the kth coefficient α_k of any polynomial $P(z)$ in L_n, for which (6.1.6) holds, satisfies

$$|\alpha_k|\leq M_k.$$

We can, therefore, find a sequence of polynomials $P^{(m)}(z)=1+\alpha_1^{(m)}z+\cdots+\alpha_n^{(m)}z^n$ so that

$$\int_0^l |P^{(m)}(z(s))|^2\,\mathrm{ds}\rightarrow\mu$$

and $\alpha_k^{(m)}\rightarrow a_k$ as $m\rightarrow\infty$, $1\leq k\leq n$. But then

$$\lim_{m\rightarrow\infty}\int_0^l |P^{(m)}(z(s))|^2\,\mathrm{ds}=\int_0^l |P_n(z(s))|^2\,\mathrm{ds}=\mu$$

where $P_n(z)=1+a_1z+\ldots+a_nz^n$.

To show that $P_n(z)$ is unique, we first remark that $P_n(z)$ is orthogonal to each polynomial $Q(z)$ of degree less than or equal to n with $Q(0)=0$, i.e.

$$(P_n, Q)=\int_0^l P_n(z(s))\overline{Q(z(s))}\,\mathrm{ds}=0. \tag{6.1.8}$$

The proof of this is analogous to the proof of Lemma 6.1.2. If $R(z)\in\mathsf{L}_n$ and $R(z)\neq P_n(z)$, then since $(P_n,\ R-P_n)=0$, we have

$$\begin{aligned}(R, R)&=(P_n+[R-P_n], P_n+[R-P_n])\\&=(P_n, P_n)+(R-P_n, R-P_n)>(P_n, P_n),\end{aligned}$$

or

$$\int_0^l |R(z(s))|^2\,\mathrm{ds}>\int_0^l |P_n(z(s))|^2\,\mathrm{ds}.$$

Thus, we have shown that the minimum polynomial $P_n(z)$ is *uniquely determined.* Furthermore, we see that $P_n(z)$ is completely characterized by property (6.1.8).

The coefficients of $P_n(z)=1+a_1z+\cdots+a_nz^n$ are characterized by the fact that

$$\int_0^l \left(\sum_{k=0}^n a_k z^k(s)\right)\overline{(z^i(s))}\,\mathrm{ds}=0, \quad a_0=1, \quad i=1,\ldots,n.$$

Let

$$\beta_{ik}=\int_0^l z^k(s)\overline{z^i(s)}\,\mathrm{ds}, \quad i,k=0,1,\ldots,n.$$

The coefficients of $P_n(z)$ must satisfy the linear system

$$\sum_{k=0}^n \beta_{ik}a_k=0, \quad a_0=1, \quad i=1,2,\ldots,n. \tag{6.1.9}$$

Hence (6.1.8) must have a unique solution.

In Berlin at the Hahn–Meitner Institute the best results for computing the β_{ik} were obtained using Gaussian numerical integration, and the computation time was short. For piecewise analytic boundaries, the convergence is very rapid. Because the matrix $((\beta_{ik}))$ has very bad conditioning for large n, it is advised to use double length of the machine numbers for the determination of $a_1,\ldots,a_n$.

Because of Lemma 6.1.2, for an arbitrary polynomial $P(z)$ in class L_n, we have $(F_0,P)=(F_0,P-F_0)+(F_0,F_0)=(F_0,F_0)$, and hence $\|F_0-P\|^2=\|P\|^2-\|F_0\|^2$, so that $P_n(z)$ *has the additional minimum property that*

$$\int_0^l |F_0(z(s))-P(z(s))|^2\,\mathrm{ds}$$

is minimal in L_n for $P(z)=P_n(z)$.

Now $\mathsf{L}_{n+1}\supset\mathsf{L}_n$ and thus $\|F_0-P_n\|$ is monotone decreasing. The question remains under what hypotheses on $\dot{\mathsf{G}}$ is it true that $\|F_0-P_n\|\to 0$ as $n\to\infty$? If the set of all polynomials is dense in $L_2(\dot{\mathsf{G}})$ then $\|F_0-P_n\|\to 0$ as $n\to\infty$. We now state a necessary and sufficient condition for the set of all polynomials to be dense in $L_2(\dot{\mathsf{G}})$.

Theorem 6.1.4 *The set of polynomials is dense in $L_2(\dot{\mathsf{G}})$ if and only if the function* $\log|g'(w)|$ *which is harmonic in $|w|<R$ can be represented by its Poisson integral, i.e.*

$$\log|g'(re^{i\theta})|=\frac{1}{2\pi}\int_0^{2\pi}\log|g'(Re^{i\phi})|\frac{R^2-r^2}{R^2+r^2-2Rr\cos(\theta-\phi)}\,d\phi. \tag{6.1.10}$$

If (6.1.10) *holds,* $\dot{\mathsf{G}}$ *is said to satisfy condition* (S), *after Smirnov.*

Because $\dot{G}$ is a closed, rectifiable, Jordan curve $\log|g'(Re^{i\theta})| \in L^1([0, 2\pi))$; however, in general, the right hand side of (6.1.10) is a *subharmonic* function. When the right hand side of (6.1.10) is harmonic then (6.1.10) holds. There exist domains G with $\dot{G}$ a closed, rectifiable, Jordan curve such that $\dot{G}$ does *not* satisfy condition (S). If G is a domain with $\dot{G}$ a closed, rectifiable Jordan curve, then any one of the following conditions is sufficient for $\dot{G}$ to satisfy condition (S):

(a) G *is convex, or starlike with respect to a point of* G,

(b) $\dot{G}$ *is piecewise smooth, and the smooth arcs are joined with interior angles* $\neq 0$ (*i.e. no cusps are allowed*),

(c) *there exists* $M > 0$ *such that for all* $0 \leq s_1 < s_2 < l$

$$(s_2 - s_1)/|z(s_2) - z(s_1)| < M.$$

Even if $\dot{G}$ is a closed, rectifiable, Jordan curve which does not satisfy condition (S), it follows that

$$\|P_n\|^2 \to 2\pi R \cdot D,$$

where

$$D = \exp\left\{\frac{1}{2\pi}\int_0^{2\pi} \log|g'(Re^{i\theta})|\,d\theta\right\}.$$

Under assumptions (a), (b), or (c) one has, therefore, $\|F_0 - P_n\| \to 0$ and because of (6.1.7), we also have that $P_n(z)$ converges uniformly to $F_0(z)$ on compact subsets of G. Hence the sequence of polynomials

$$\pi_{2n+1}(z) = \int_0^z [P_n(v)]^2\,dv$$

converges uniformly to $f(z)$ *on compact subsets of* G.

Actually, we can show more. In fact, we can show that $\pi_{2n+1}(z)$ converges *uniformly on* $\bar{G}$ to $f(z)$ and that

$$\max_{\bar{G}} |f(z) - \pi_{2n+1}(z)|$$

can be estimated by $\|F_0 - P_n\|$. For this we need

Lemma 6.1.5 *Let* $F \in L_2(\dot{G})$. *Then for each* $z \in G$,

$$\left|\int_0^z F(\xi)\,d\xi\right| \leq \frac{1}{2}\int_0^l |F(z(s))|\,ds. \tag{6.1.11}$$

Proof The path of integration for $\int_0^z$ can be chosen so that its image in the w-plane is the line segment joining 0 and $w=f(z)$. Then

$$\begin{aligned}\left|\int_0^z F(\xi)\,\mathrm{d}\xi\right| &= \left|\int_0^w F(g(r))g'(r)\,\mathrm{d}r\right| \\ &\leqslant \int_0^w |F(g(r))g'(r)|\,|\mathrm{d}r| \\ &\leqslant \frac{1}{2}\int_{|r|=|w|} |F(g(r))g'(r)|\,|\mathrm{d}r| = \frac{1}{2}\int_{C_{|w|}} |F(z(s))|\,\mathrm{d}s \\ &\leqslant \frac{1}{2}\int_0^l |F(z(s))|\,\mathrm{d}s.\end{aligned}$$

Here we have used the inequality of Fejér-Riesz and also the fact that

$$\int_{C_r} |F(z)|\,\mathrm{d}s$$

is monotone increasing.

Lemma 6.1.5 leads to

Theorem 6.1.6 *Suppose $P\in L_n$, and let*

$$p(z)=\int_0^z P^2(\xi)\,\mathrm{d}\xi.$$

Then

$$\max_{\bar{G}} |f(z)-p(z)| \leqslant \sqrt{2\pi R}\,\|F_0-P\| + \tfrac{1}{2}\|F_0-P\|^2. \tag{6.1.12}$$

Proof For $z\in G$, we have

$$\begin{aligned}|f(z)-p(z)| &= \left|\int_0^z [f'(\xi)-P^2(\xi)]\,\mathrm{d}\xi\right| \\ &\leqslant \frac{1}{2}\int_0^l |f'(z(s))-P^2(z(s))|\,\mathrm{d}s.\end{aligned}$$

This last integral is

$$\begin{aligned}&\int_0^l |\sqrt{f'(z(s))}-P(z(s))|\,|\sqrt{f'(z(s))}+P(z(s))|\,\mathrm{d}s \\ &\qquad\leqslant \left[\int_0^l |\sqrt{f'(z(s))}-P(z(s))|^2\,\mathrm{d}s\right]^{1/2} \\ &\qquad\quad\times\left[\int_0^l |\sqrt{f'(z(s))}+P|^2\,\mathrm{d}s\right]^{1/2} \\ &\qquad= \|F_0-P\|\,\|F_0+P\|.\end{aligned}$$

Using Lemma 6.1.2 it follows that

$$\begin{aligned}\|F_0+P\|^2 &= 4\,\|F_0\|^2+\|F_0-P\|^2\\ &\leq 4\cdot 2\pi R+\|F_0-P\|^2.\end{aligned}$$

Hence $\|F_0+P\|\leq 2\sqrt{2\pi R}+\|F_0-P\|$, and the proof is complete.

In particular Theorem 6.1.6 implies that

$$\max_{\bar{G}}|f(z)-\pi_{2n+1}(z)|\leq\sqrt{2\pi R}\,\|F_0-P_n\|+\tfrac{1}{2}\,\|F_0-P_n\|^2. \tag{6.1.13}$$

For brevity of notation, as in (1.1.32), let

$$\|\cdot\|_0\equiv\sup_{\bar{G}}|\cdot|$$

and let $R_n=\|P_n\|^2/2\pi$.

Let $\delta>0$ be given and suppose $\tilde{\beta}_{ik}$ is a numerical approximation for

$$\beta_{ik}=\int_0^l z^k(s)\overline{z^i(s)}\,\mathrm{d}s,\quad 1\leq i,\,k\leq n,$$

such that $|\beta_{ik}-\tilde{\beta}_{ik}|\leq\delta$. Suppose $\tilde{\alpha}_i$, $1\leq i\leq n$, is an approximate solution to the linear system

$$\sum_{k=0}^n \beta_{ik}\alpha_k=0,\quad \alpha_0=1,\quad i=1,\ldots,n. \tag{6.1.14}$$

Then $\tilde{P}_n(z)=1+\tilde{\alpha}_1 z+\cdots+\tilde{\alpha}_n z^n$ is an approximation to $P_n(z)$ and

$$2\pi\hat{R}_n=\sum_{i,j=0}^n \tilde{\alpha}_i\bar{\tilde{\alpha}}_j\tilde{\beta}_{ji}$$

is an approximation to $\|\tilde{P}_n\|^2$. Let $\tilde{R}_n=\|\tilde{P}_n\|^2/2\pi$. Then

$$\begin{aligned}|\tilde{R}_n-\hat{R}_n| &= \frac{1}{2\pi}\sum_{i,j=0}^n \tilde{\alpha}_i\bar{\tilde{\alpha}}_j(\beta_{ji}-\tilde{\beta}_{ji})\\ &\leq\frac{\delta}{2\pi}\sum_{i,j=0}^n|\tilde{\alpha}_i\tilde{\alpha}_j|\equiv\varepsilon.\end{aligned}$$

Furthermore, let

$$\tilde{p}_{2n+1}(z)=\int_0^z \tilde{P}_n^2(\xi)\,\mathrm{d}\xi.$$

Then

$$\left|\phi(z)-\frac{\tilde{p}_{2n+1}(z)}{\hat{R}_n}\right|=\left|\frac{f(z)}{R}-\frac{\tilde{p}_{2n+1}(z)}{\hat{R}_n}\right|$$

$$\leq \|f\|_0\left|\frac{1}{R}-\frac{1}{\hat{R}_n}\right|+\frac{1}{\hat{R}_n}\|f-\tilde{p}_{2n+1}\|_0$$

$$=\frac{(\hat{R}_n-R)}{\hat{R}_n}+\frac{1}{\hat{R}_n}\|f-\tilde{p}_{2n+1}\|_0.$$

Now if $P \in \mathsf{L}$, $\|F_0-P\|^2=\|P\|^2-\|F_0\|^2$. Using this fact, together with the previous theorem, we have that

$$\|\phi-\tilde{p}_{2n+1}/\hat{R}_n\|_0 \leq \frac{(\hat{R}_n-R)}{\hat{R}_n}+\frac{1}{\hat{R}_n}[\sqrt{2\pi R}\sqrt{2\pi(\tilde{R}_n-R)}+\pi(\tilde{R}_n-R)].$$

Since $\tilde{R}_n \leq \hat{R}_n+\varepsilon$ we have

$$\|\phi-\tilde{p}_{2n+1}/\hat{R}_n\|_0 \leq \frac{(\hat{R}_n-R)}{\hat{R}_n}+\frac{\pi}{\hat{R}_n}[2\sqrt{R(\hat{R}_n+\varepsilon-R)}+(\hat{R}_n+\varepsilon-R)]. \tag{6.1.15}$$

If it is possible to calculate a lower bound R_l for R, then (6.1.15), with R replaced by R_l, *yields an a posteriori error bound.* In any case, we take $\tilde{p}_{2n+1}(z)/\hat{R}_n$ *to be our approximation to* $\phi(z)$.

Lower bounds for R can be obtained by the following analysis due to W. F. Moss [15]. Let $I(z)=1/z$, and let G^e denote the exterior of G. Furthermore, let $\mathsf{G}^*=I(\mathsf{G}^e)$. If ρ is the conformal radius of G^*, then there exists a function $w=F(z)$ which maps G^e conformally onto $|w|>1/\rho$ and satisfies

$$F(\infty)=\infty, \qquad \lim_{z\to\infty}\frac{F(z)}{z}=1. \tag{6.1.16}$$

We call $1/\rho$ *the exterior mapping radius* of G^e. It follows from $\mathsf{G}=(\mathsf{G}^*)^*$ that $1/R$ is the exterior mapping radius of G^{*e}. Thus, upper bounds for $1/R$ yield lower bounds for R.

Now let $w=F(z)$ be the conformal mapping of G^{*e} onto $\{|w|>\nu\}$ satisfying (6.1.16) and let $z=H(w)$ be the inverse of $F(z)$. For $r>\nu$ let $C_r=H(|w|=r)$. Let M denote the class of functions, holomorphic in G^{*e} and continuous on $\overline{\mathsf{G}^{*e}}$ having an expansion

$$1+\frac{a_{-1}}{z}+\frac{a_{-2}}{z^2}+\cdots.$$

Clearly, M contains all polynomials in $1/z$ with the constant term 1. Then we have the following version of Julia's theorem:

Theorem 6.1.7 *The function* $\mathsf{F}_0(z)=\sqrt{F'(z)}$ *and only this function minimizes the integral*

$$\oint_{\dot{\mathsf{G}}} |F(z)|^2 \, \mathrm{d}s' \tag{6.1.17}$$

over the class M. *This minimum is equal to* $2\pi\nu$.

Since the proof is very similar to that of Theorem 6.1.1 let us omit the details here. Similarly to Lemma 6.1.2 we also have

Lemma 6.1.8 *The function* $\mathsf{F}_0(z)$ *is orthogonal to every function* $H(z)$ *which is holomorphic in* G^{*e}, *continuous in* $\overline{\mathsf{G}^{*e}}$ *and has an expansion at* ∞,

$$H(z)=\frac{a_{-1}}{z}+\frac{a_{-2}}{z^2}+\ldots,$$

i.e.

$$(\mathsf{F}_0, H)=\oint_{\dot{\mathsf{G}}^*} \mathsf{F}_0(z)\overline{H(z)} \, \mathrm{d}s'=0. \tag{6.1.18}$$

Again, the proof is quite the same as for Lemma 6.1.2; we leave it to the reader.

We now formulate the *Ritz method* for the approximation of $\mathsf{F}_0(z)$. Let M_n denote the class of elements of M of the form

$$1+\sum_{\nu=1}^{n} a_{-\nu}z^{-\nu}.$$

Then we have

Theorem 6.1.9 *For each* $n \geqslant 1$, *there exists a function* $\mathsf{P}_n(z)$ *in class* M_n *which uniquely minimizes*

$$\int_{\dot{\mathsf{G}}^*} |P(z)|^2 \, \mathrm{d}s'$$

over M_n.

Replacing G by G^{*e} and using Laurent's theorem instead of Cauchy's theorem, the proof can be obtained in the same manner as that for Theorem 6.1.3 so that we omit the details here.

The coefficients of $\mathsf{P}_n(z)=1+a_{-1}z^{-1}+\cdots+a_{-n}z^{-n}$ are characterized by the fact that

$$\int_{\dot{\mathsf{G}}^*} \sum_{k=0}^{n} a_{-k}z^{-k}\overline{z^{-i}} \, \mathrm{d}s'=0, \quad a_0=1, \quad i=1,\ldots,n. \tag{6.1.19}$$

Let

$$\gamma_{ik}=\int_{\dot{\mathsf{G}}^*} z^{-k}\overline{z^{-i}}\,\mathrm{d}s', \quad i,\ k=0,1,\ldots,n.$$

Then the coefficients of $\mathsf{P}_n(z)$ must satisfy the linear system

$$\sum_{k=0}^{n} \gamma_{ik}a_{-k}=0, \quad a_0=1, \quad i=1,2,\ldots,n. \tag{6.1.20}$$

Hence, (6.1.20) must have a unique solution.

Because of Lemma 6.1.8, for an arbitrary function $\mathsf{P}(z)$ in class M_n, we have $(\mathsf{F}_0,\mathsf{P})=(\mathsf{F}_0,\mathsf{P}-\mathsf{F}_0)+(\mathsf{F}_0,\mathsf{F}_0)=(\mathsf{F}_0,\mathsf{F}_0)$, and hence $\|\mathsf{F}_0-\mathsf{P}\|^2=\|\mathsf{P}\|^2-\|\mathsf{F}_0\|^2$, so that $\mathsf{P}_n(z)$ has the additional minimum property that

$$\int_{\dot{\mathsf{G}}^*} |\mathsf{F}_0(z)-\mathsf{P}(z)|^2\,\mathrm{d}s'$$

is minimal in M_n for $\mathsf{P}(z)=\mathsf{P}_n(z)$.

Now $\mathsf{M}_{n+1}\supset\mathsf{M}_n$ and thus $\|\mathsf{F}_0-\mathsf{P}_n\|$ is monotone decreasing. Let $\zeta=1/z$ and let d denote the diameter of G^*. Furthermore, suppose that $\dot{\mathsf{G}}$ as well as $\dot{\mathsf{G}}^*$ satisfy condition (S). Let $\tilde{\mathsf{F}}_0(\zeta)=\mathsf{F}_0(1/z)$. Then $\tilde{\mathsf{F}}_0(\zeta)$ is holomorphic in G, continuous on $\bar{\mathsf{G}}$, and has an expansion at the origin of the form

$$1+a_{-1}\zeta+a_{-2}\zeta^2+\ldots.$$

Hence $\tilde{\mathsf{F}}_0(\zeta)\in L_2(\dot{\mathsf{G}})$. Thus, given $\varepsilon>0$, there exists a polynomial $P(\zeta)$ such that

$$\int_{\dot{\mathsf{G}}} |\tilde{\mathsf{F}}_0(\zeta)-P(\zeta)|^2\,|\mathrm{d}\zeta|<\varepsilon/d^2.$$

But then

$$\frac{1}{d^2}\int_{\dot{\mathsf{G}}^*}\left|\mathsf{F}_0(z)-P\left(\frac{1}{z}\right)\right|^2\mathrm{d}s'\leqslant\int_{\dot{\mathsf{G}}^*}\left|\mathsf{F}_0(z)-P\left(\frac{1}{z}\right)\right|^2\frac{\mathrm{d}s'}{|z|^2}$$
$$=\int_{\dot{\mathsf{G}}}|\tilde{\mathsf{F}}_0(\zeta)-P(\zeta)|^2\,|\mathrm{d}\zeta|<\varepsilon/d^2.$$

Let $\mathsf{P}(z)=P(1/z)$. Then we have that $\|\mathsf{F}_0-\mathsf{P}\|^2<\varepsilon$ and $\mathsf{P}(z)$ is in class M_n for some n depending on ε. It follows that $\|\mathsf{F}_0-\mathsf{P}_n\|\to 0$ as $n\to\infty$.

Let $\delta>0$ be given and suppose $\tilde{\gamma}_{ik}$ is a numerical approximation for

$$\gamma_{ik}=\int_{\dot{\mathsf{G}}^*} z^{-k}\overline{z^{-i}}\,\mathrm{d}s', \quad i,\ k=-,1,\ldots,n, \tag{6.1.21}$$

such that $|\gamma_{ik}-\tilde{\beta}_{ik}|\leqslant\delta$. Furthermore, suppose $\tilde{\alpha}_{-i}$, $1\leqslant i\leqslant n$, is an approximate solution to the linear system

$$\sum_{k=0}^{n} \gamma_{ik}\alpha_{-k}=0, \quad \alpha_0=1, \quad i=1,\ldots,n. \tag{6.1.22}$$

Then $\tilde{\mathsf{P}}_n(z)=1+\tilde{\alpha}_{-1}z^{-1}+\cdots+\tilde{\alpha}_{-n}z^{-n}$ is an approximation to $\mathsf{P}_n(z)$ and

$$2\pi\hat{\nu}_n = \sum_{i,j=0}^{n} \tilde{\alpha}_{-1}\bar{\tilde{\alpha}}_{-j}\tilde{\gamma}_{ji}$$

is an approximation to $\|\tilde{\mathsf{P}}_n\|^2$. Let $\tilde{\nu}_n=\|\tilde{\mathsf{P}}_n\|^2/2\pi$. Then

$$\begin{aligned}|\tilde{\nu}_n-\hat{\nu}_n| &= \frac{1}{2\pi}\left|\sum_{i,j=0}^{n} \tilde{\alpha}_{-i}\bar{\tilde{\alpha}}_{-j}(\gamma_{ji}-\tilde{\gamma}_{ji})\right| \\ &\leqslant \frac{\delta}{2\pi}\sum_{i,j=0}^{n} |\tilde{\alpha}_{-i}\tilde{\alpha}_{-j}| \equiv \varepsilon.\end{aligned}$$

Hence $\nu<\tilde{\nu}_n<\hat{\nu}_n+\varepsilon$, and we have the desired upper bound for ν.

If we apply the above analysis, then we obtain *an upper bound for* $1/R$, *yielding a lower bound for* R. Putting this lower bound into (6.1.15), we obtain the desired *a posteriori error estimate.*

Suppose $\dot{\mathsf{G}}$ and $\dot{\mathsf{G}}^*$ satisfy condition (S) and $\dot{\mathsf{G}}=\{z=\lambda(t):0\leqslant t\leqslant 1\}$. Then we must numerically approximate

$$\beta_{ik}=\int_0^1 \lambda^k(t)\overline{\lambda^i(t)}\,|\lambda'(t)|\,\mathrm{d}t, \quad i,\ k=0,\ 1,\ldots,n,$$

to obtain an upper bound for R, and we must numerically approximate

$$\begin{aligned}\gamma_{ik} &= \int_0^1 \lambda^k(t)\overline{\lambda^i(t)}\frac{|\lambda'(t)|}{|\lambda^2(t)|}\,\mathrm{d}t \\ &= \int_0^1 \lambda^{k-1}(t)\overline{\lambda^{i-1}(t)}\,|\lambda'(t)|\,\mathrm{d}t, \quad i,\ k=0,\ 1,\ \ldots,n,\end{aligned} \tag{6.1.23}$$

to obtain a lower bound for R. *However,* $\beta_{i-1,k-1}=\gamma_{ik}$, i, $k=1,\ 2,\ldots,n$, so that *only* γ_{ok} *and* γ_{ko}, $k=0,\ 1,\ldots,n$, *need be approximated in addition to the* β_{ik}, i, $k=0,1,\ldots,n$.

Again, suppose that $\dot{\mathsf{G}}$ and $\dot{\mathsf{G}}^*$ satisfy condition (S). Then we may use the above procedure to obtain upper bounds for the exterior mapping radius ν of G^{*e} and for the conformal radius $1/\nu$ of G. Hence, we can obtain improvable upper and lower bounds for ν. Since ν is also equal to the capacity of $\bar{\mathsf{G}}^*$ and the transfinite diameter of $\bar{\mathsf{G}}^*$ or $\dot{\mathsf{G}}^*$, we have a procedure for estimating these quantities.

Conjecture $\dot{\mathsf{G}}$ *satisfies condition* (S) *if and only if* $\dot{\mathsf{G}}^*$ *satisfies condition* (S).

The following theorem leads to estimates for

$$\max_{z\in G}|\sqrt{f'(z)}-P_n(z)|.$$

Theorem 6.1.10 (Suetin [19, (8)]) *Let $\dot{G}\in C^2$ and let the function $\psi(z)$ be holomorphic in* G and $\psi\in C^{k+\beta}(\bar{G})$. *Then the Fourier expansion of ψ with respect to z^j and the L_2 scalar product $(.,.)$ on $\dot{G}$ converges as follows:*

$$\left|\psi(z)-\sum_{j=0}^{n} a_j z^j\right|\leq cn^{-k-\beta}\log n \quad \text{for} \quad z\in\bar{G}. \tag{6.1.24}$$

Here c depends only on G but not on ψ nor n.

Applying this theorem to the function $\sqrt{f'(z)}$ we find the following

Assertion 6.1.11 *Let the boundary $\dot{G}$ be $C^{k+\beta}$ with $k\geq 2$ and, hence, $f\in C^{k+\beta}(\bar{G})$. Then there exists a constant $M>0$ such that*

$$\max_{\bar{G}}|\sqrt{f'(z)}-P_n(z)|\leq Mn^{-k-\beta+1}\log n, \quad n\geq 2. \tag{6.1.25}$$

This estimate implies that there exists a constant $c'>0$ such that the estimates

$$\max_{\bar{G}}|f'(z)-p_n(z)|\leq c'n^{-k-\beta+1}\log n$$

and

$$\max_{\bar{G}}|f(z)-\pi_{2n+1}(z)|\leq c'n^{-k-\beta+1}\log n$$

hold. This means for the approximation $\tilde{\phi}$ of ϕ, defined by

$$\tilde{\phi}(z)\equiv\tilde{R}^{-1}\pi_{2n+1}(z)=2\pi\left|\sum_{k=0}^{n}\beta_{0k}a_k\right|^{-1}\pi_{2n+1}(z), \tag{6.1.26}$$

the corresponding estimates,

$$\max_{\bar{G}}\left|\phi'(z)-2\pi\left|\sum_{k=0}^{n}\beta_{0k}a_k\right|^{-1}p_n(z)\right|=\max_{\bar{G}}|\phi'(z)-\tilde{\phi}'(z)|$$
$$\leq c''n^{-k-\beta+1}\log n \tag{6.1.27}$$

and

$$\max_{\bar{G}}|\phi(z)-\tilde{\phi}(z)|\leq c''n^{-k-\beta+1}\log n \tag{6.1.28}$$

are valid.

Remark Let $\{q_j(z)\}_{j=0}^{\infty}$ denote the orthonormal sequence of polynomials obtained by applying the Schmidt orthogonalization procedure to $\{z^j\}_{j=0}^{\infty}$ with respect to the above inner product of $L_2(\dot{\mathsf{G}})$. Then it can be shown that

$$P_n(z) = \frac{\sum_{j=0}^{n} \overline{q_j(0)} q_j(z)}{\sum_{j=0}^{n} |q_j(0)|^2} \tag{6.1.29}$$

and

$$\|P_n\|^2 = \sum_{j=0}^{n} |q_j(0)|^2 \tag{6.1.30}$$

The $q_j(z)$ can be determined strictly by integrations and no linear system needs to be solved. But the integrals β_{jk} have to be evaluated also in this case and the Schmidt orthogonalization needs as much numerical effort as the solution of the linear equations. For further discussion see [7].

The whole Ritz method for the conformal mapping is also closely related to expansions of the Bergman kernel function for G. This together with explicit computations can be found in [5].

6.2 Boundary Integrals

Besides the surface integrals two boundary integrals,

$$h(\zeta) = -\oint_{\dot{\mathsf{G}}} \psi \, \mathrm{d}_n G^I \tag{6.2.1}$$

and

$$\tilde{h}(\zeta) = \oint_{\dot{\mathsf{G}}} \chi \, \mathrm{d} G^{II}, \tag{6.2.2}$$

appear in the integral equations (6.0.1). Both functions are harmonic on $\mathbb{R}^2 \backslash \dot{\mathsf{G}}$, the first has Dirichlet data ψ and the second Neuman data χ on $\dot{\mathsf{G}}$. Although the Green functions depend on z and ζ we shall evaluate these integrals here *explicitly* since the conformal mapping ϕ can be considered to be known and can be computed by an approximate polynomial. This has been done numerically in [14] providing surprisingly good numerical results. To this end both integrals (6.2.1), (6.2.2) have been integrated by parts. Then in (6.2.1) the harmonic measure appears.

If $a(z)$ is any holomorphic function then it follows from (1.1.8) that

$$\mathrm{d}_n a(z) = -\mathrm{i} \, \mathrm{d} \, a(z), \qquad \mathrm{d}_n \operatorname{Re} a(z) = \mathrm{d} \operatorname{Im} a(z).$$

Since

$$G^I(z, \zeta) = -\frac{1}{2\pi} \log \left| \frac{\phi(z) - \phi(\zeta)}{1 - \phi(z)\overline{\phi(\zeta)}} \right| = -\frac{1}{2\pi} \operatorname{Re} \log \frac{\phi(z) - \phi(\zeta)}{1 - \phi(z)\overline{\phi(\zeta)}} \tag{6.2.3}$$

we conclude that

$$d_n G^I(z,\zeta) = -\frac{1}{2\pi}\, d\,\mathrm{Im}\log\frac{\phi(z)-\phi(\zeta)}{1-\phi(z)\overline{\phi(\zeta)}} = -\frac{1}{2\pi}\, d\arg\frac{\phi(z)-\phi(\zeta)}{1-\phi(z)\overline{\phi(\zeta)}}. \tag{6.2.4}$$

Note that the total differential appearing in (6.2.4) is *periodic*, while the argument function is *not*. From (6.2.4) we have

$$h(\zeta) = \frac{1}{2\pi}\oint_{\dot{G}} \psi\, d\arg\frac{\phi(z)-\phi(\zeta)}{1-\phi(z)\overline{\phi(\zeta)}}. \tag{6.2.5}$$

This last integral is a principal value Cauchy integral. But if $\psi \in C^1(\dot{G})$, this formula has the advantage that it can be integrated by parts. Noting that for $z \in \dot{G}$, $1/\phi(z) = \overline{\phi(z)}$, we have

$$d\left(\arg\frac{\phi(z)-\phi(\zeta)}{1-\phi(z)\overline{\phi(\zeta)}}\right)\bigg|_{z\in\dot{G}} = d\left(\arg\frac{\phi(z)-\phi(\zeta)}{\overline{\phi(z)}-\overline{\phi(\zeta)}}\cdot\overline{\phi(z)}\right)\bigg|_{z\in\dot{G}}$$

$$= d\left(\arg\frac{\phi(z)-\phi(\zeta)}{\overline{\phi(z)}-\overline{\phi(\zeta)}}\cdot\frac{\overline{\phi(z)}}{\overline{\phi(z_0)}}\cdot\frac{\overline{\phi(z_0)-\phi(\zeta)}}{\phi(z_0)-\phi(\zeta)}\right)\bigg|_{z\in\dot{G}}$$

$$= d\left(\arg\frac{\phi(z)-\phi(\zeta)}{\phi(z_0)-\phi(\zeta)}\cdot\frac{\overline{\phi(z_0)-\phi(\zeta)}}{\overline{\phi(z)-\phi(\zeta)}}\cdot\frac{\overline{\phi(z)}}{\overline{\phi(z_0)}}\right)\bigg|_{z\in\dot{G}}$$

$$= d\left(2\arg\frac{\phi(z)-\phi(\zeta)}{\phi(z_0)-\phi(\zeta)} - \arg\frac{\phi(z)}{\phi(z_0)}\right)\bigg|_{z\in\dot{G}},$$

where $z_0 = z(0)$. Setting

$$M(\zeta,s) = \frac{1}{2\pi}\left\{2\arg\frac{\phi(z(s))-\phi(\zeta)}{\phi(z(0))-\phi(\zeta)} - \arg\frac{\phi(z(s))}{\phi(z(0))}\right\}, \tag{6.2.6}$$

we have

$$h(\zeta) = \frac{1}{2\pi}\int_0^l \psi(s)\, dM(\zeta,s). \tag{6.2.7}$$

For fixed s, $M(\xi+i\eta, s)$ is harmonic for $(\xi,\eta)\in G$. Its two terms

$$\alpha = \arg\frac{\phi(z(s))-\phi(\zeta)}{\phi(z(0))-\phi(\zeta)}, \qquad \beta = \arg\frac{\phi(z(s))}{\phi(z(0))}$$

define two angles in the unit circle as indicated in Fig. 5. For fixed s and variable ζ, M has the boundary values

$$1 \text{ on } \Gamma_1 \equiv \{\zeta(\sigma) \mid 0<\sigma<s\} \quad \text{and} \quad 0 \text{ on } \Gamma_2 \equiv \{\zeta(\tau) \mid s<\tau<l\}.$$

M is the so called *harmonic measure of* Γ_1 *in* G.

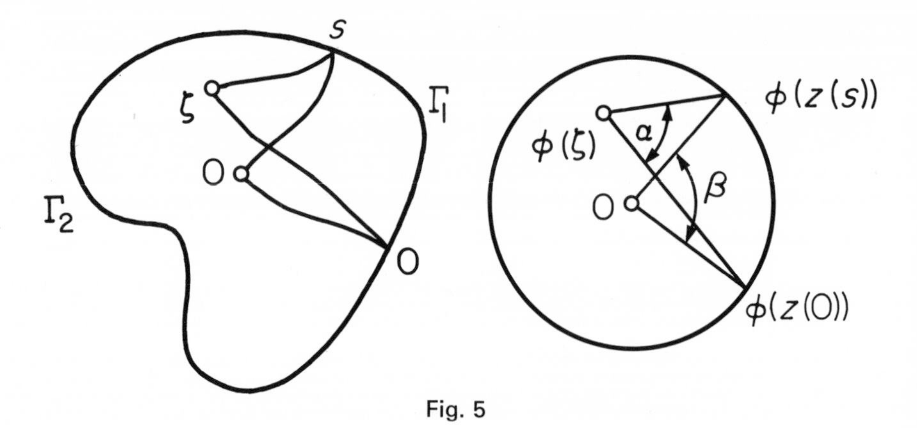

Fig. 5

For fixed $\zeta \in \mathrm{G}$ and variable s, M is not periodic in s since

$$\lim_{\varepsilon \to 0} M(\zeta, l-\varepsilon) - M(\zeta, 0) = 1.$$

Setting $M(\zeta, l) = 1$, we have $M \in C^1([0, l])$. Integrating (6.2.7) by parts, we obtain for $\psi \in C^1(\dot{\mathrm{G}})$

$$h(\zeta) = \psi(0) - \int_0^l \psi'(s) M(\zeta, s)\, \mathrm{d}s. \tag{6.2.8}$$

This integral has a *bounded integrand* for $\zeta \in \bar{\mathrm{G}}$ and hence it is better adapted for numerical integration than is the integral in (6.2.5).

The second harmonic integral in (6.2.2) can also be integrated by parts for $\chi \in C^1([0, l])$ and we have

$$\tilde{h}(\zeta) = -\int_0^l \chi'(s) G^{II}(\zeta, z(s))\, \mathrm{d}s. \tag{6.2.9}$$

Now, we can use the polynomial approximation for ϕ and Gaussian numerical integration to obtain approximations to $h(\zeta)$ and $\tilde{h}(\zeta)$. In case ψ is not given explicitly or $\psi \notin C^1([0, l])$, we first approximate ψ by smooth functions.

Some numerical computations for $\dot{\mathrm{G}}$ an ellipse and a triangle can be found in [14].

6.3 The Approximation and Discretization of Fredholm Integral Equations of the Second Kind

Using (6.0.1) in order to compute the solution w of the first BVP (1.1.1) we are forced to two different approximations; the approximation of the conformal mapping ϕ by $\tilde{\phi}$ and the discretization of the continuous integral equation by some kind of quadrature procedures. For simplicity, in this section we shall restrict our presentation to approximations with respect to uniform convergence and maximum norm.

Let us begin with the influence of $\tilde{\phi}$. Using $\tilde{\phi}$ instead of ϕ in (1.1.10) and (1.1.14) the integral operator K, (1.1.23), in (6.0.1) becomes a perturbed new operator $\tilde{\mathsf{K}}$. Its difference from K can be estimated by the pointwise inequality,

$$
\begin{aligned}
|\mathsf{K}w-\tilde{\mathsf{K}}w|^2 \leqslant & \left\{\iint_{\mathsf{G}} (|A|\,|u|+|B|\,|v|)\,|G_x^I-\tilde{G}_x^I|\,[\mathrm{d}x,\mathrm{d}y]\right.\\
&\left.+\iint_{\mathsf{G}} (|\tilde{A}|\,|u|+|\tilde{B}|\,|v|)\,|G_y^I-\tilde{G}_y^I|\,[\mathrm{d}x,\mathrm{d}y]\right\}^2\\
&+\left\{\iint_{\mathsf{G}} (|A|\,|u|+|B|\,|v|)\,|G_y^{II}-\tilde{G}_y^{II}|\,[\mathrm{d}x,\mathrm{d}y]\right.\\
&\left.+\iint_{\mathsf{G}} (|\tilde{A}|\,|u|+|\tilde{B}|\,|v|)\,|G_x^{II}-\tilde{G}_x^{II}|\,[\mathrm{d}x,\mathrm{d}y]\right\}^2.
\end{aligned}
\tag{6.3.1}
$$

Using the norm $\|w\|_0 \equiv \sup_{\bar{\mathsf{G}}}\{|u|^2+|v|^2\}^{1/2}$ we find with

$$
M \equiv \max\{\|A+\mathrm{i}B\|_0, \|\tilde{A}+\mathrm{i}\tilde{B}\|_0\}
\tag{6.3.2}
$$

for the operator norm induced by $\|\cdot\|_0$, the inequality

$$
\begin{aligned}
\|\mathsf{K}-\tilde{\mathsf{K}}\|_0^2 \leqslant M^2\cdot\max\Bigg(&\left\{\iint_{\mathsf{G}} (|G_x^I-\tilde{G}_x^I|+|G_y^I-\tilde{G}_y^I|)[\mathrm{d}x,\mathrm{d}y]\right\}^2\\
&+\left\{\iint_{\mathsf{G}} (|G_x^{II}-\tilde{G}_x^{II}|+|G_y^{II}-\tilde{G}_y^{II}|)[\mathrm{d}x,\mathrm{d}y]\right\}^2\Bigg).
\end{aligned}
\tag{6.3.3}
$$

Using (1.1.10) we find for the Green function and its approximation

$$
\begin{aligned}
|G_x^I-\tilde{G}_x^I| \leqslant \frac{1}{2\pi}\Bigg\{&\left|\frac{\mathrm{d}\phi}{\mathrm{d}z}\right|\left|\frac{1}{\phi(z)-\phi(\zeta)}-\frac{1}{\tilde{\phi}(z)-\tilde{\phi}(\zeta)}\right|+\left|\frac{\phi'(z)-\tilde{\phi}'(z)}{\tilde{\phi}(z)-\tilde{\phi}(\zeta)}\right|\\
&+|\phi'|\left|\frac{\overline{\phi(\zeta)}}{1+\phi(z)\overline{\phi(\zeta)}}-\frac{\overline{\tilde{\phi}(\zeta)}}{1-\tilde{\phi}(z)\overline{\tilde{\phi}(\zeta)}}\right|\\
&+\left|\frac{\overline{\tilde{\phi}(\zeta)}\,|\phi'(z)-\tilde{\phi}'(z)|}{1-\tilde{\phi}(z)\overline{\tilde{\phi}(\zeta)}}\right|\Bigg\}.
\end{aligned}
\tag{6.3.4}
$$

Hence the estimates for $\mathsf{K}-\tilde{\mathsf{K}}$ hinge on estimates for $\phi-\tilde{\phi}$, (6.1.27) and (6.1.28). Since the first derivatives $\tilde{\phi}'$ converge uniformly and $\phi'\neq 0$ in $\bar{\mathsf{G}}$ we can assume without any loss of generality that

$$\left|\frac{\tilde{\phi}(z)-\tilde{\phi}(\zeta)}{z-\zeta}\right|\geqslant\kappa>0 \tag{6.3.5}$$

holds for all $n\geqslant n_0$. Furthermore, since

$$|R-\tilde{R}|\leqslant c'''n^{-k-\beta+1}\log n \tag{6.3.6}$$

holds according to (6.1.25) and Theorem 6.1.1, we may replace $\tilde{R}$ in (6.1.26) by a bigger constant such that the estimates (6.1.27), (6.1.28) and (6.3.5) remain valid and, in addition, the estimate

$$|\tilde{\phi}(z)|\leqslant 1 \quad \text{for all} \quad z\in\bar{\mathsf{G}} \tag{6.3.7}$$

holds true. This implies the inequality

$$\left|\frac{\tilde{\phi}(z)-\tilde{\phi}(\zeta)}{1-\tilde{\phi}(z)\overline{\tilde{\phi}(\zeta)}}\right|\leqslant 1 \quad \text{for all} \quad z\in\bar{\mathsf{G}} \tag{6.3.8}$$

according to the properties of the Möbius transform.

Now we are in the position to further estimate (6.3.4). Let us begin with the first term on the right hand side of (6.3.4) finding

$$\begin{aligned}\left|\frac{1}{\phi(z)-\phi(\zeta)}-\frac{1}{\tilde{\phi}(z)-\tilde{\phi}(\zeta)}\right| &\leqslant \frac{|\phi(z)-\tilde{\phi}(z)-\phi(\zeta)+\tilde{\phi}(\zeta)|^{\gamma}}{|\phi(z)-\phi(\zeta)|^{\gamma}\,|\tilde{\phi}(z)-\tilde{\phi}(\zeta)|^{\gamma}}\\ &\quad\times\left\{\frac{1}{|\phi(z)-\phi(\zeta)|^{1-\gamma}}+\frac{1}{|\tilde{\phi}(z)-\tilde{\phi}(\zeta)|^{1-\gamma}}\right\}\\ &\leqslant 2^{\gamma}\|\phi-\tilde{\phi}\|_{\gamma}\cdot 2\frac{1}{\kappa^{1+\gamma}}\frac{1}{|z-\zeta|^{1+\gamma}}\end{aligned} \tag{6.3.9}$$

where γ can be any constant with $0<\gamma<1$. Using (6.3.5) in the next term of (6.3.4) and (6.3.8) in the further terms we find after similar calculations an estimate

$$\begin{aligned}|G_x^I-\tilde{G}_x^I| &\leqslant \|\phi-\tilde{\phi}\|_{\gamma}\cdot c(\mathsf{G},\kappa,\gamma,\|\phi'\|_0)\,|z-\zeta|^{-1-\gamma}\\ &\quad+\frac{1}{\pi\kappa}\|\phi'-\tilde{\phi}'\|_0|z-\zeta|^{-1}\end{aligned} \tag{6.3.10}$$

where the constant c depends on the indicated quantities but is independent of n. In the same way we find estimates for $G_y^I-\tilde{G}_y^I$ and $G_x^{II}-\tilde{G}_x^{II}$, $G_y^{II}-\tilde{G}_y^{II}$ of the same kind as (6.3.10). Substituting them into (6.3.3) we arrive at the following theorem.

Theorem 6.3.1 *If the approximating conformal mappings $\tilde{\phi}(z)$ satisfy the assumptions* (6.3.5) *and* (6.3.7) *then the corresponding integral operators in*

(6.0.1). K *and* $\tilde{\mathsf{K}}$, *satisfy an inequality*

$$\|\mathsf{K}-\tilde{\mathsf{K}}\|_0 \leq c(\mathsf{G}, \gamma, \kappa, M)\{\|\phi-\tilde{\phi}\|_\gamma + \|\phi'-\tilde{\phi}'\|_0\} \tag{6.3.11}$$

with respect to the operator norm associated to $C^0(\bar{\mathsf{G}})$ *with any* γ, $0<\gamma<1$ *where* c *depends on the indicated quantities.*

As an immediate consequence we have from (6.1.27) and (6.1.28) the following:

Corollary 6.3.2 *If* $\tilde{\mathsf{K}}$ *is constructed with* $\tilde{\phi}$ *from* (6.1.26) *as indicated above then we have an estimate*

$$\|\mathsf{K}-\tilde{\mathsf{K}}\|_0 \leq c'(\mathsf{G}, \gamma, \kappa, M)\{n^{-k-\beta+1}\log n\}^\gamma \tag{6.3.12}$$

for every γ, $0<\gamma<1$ *or, as a simple conclusion, to every* $\varepsilon>0$ *there exists a constant* c'' *independent of* n *such that*

$$\|\mathsf{K}-\tilde{\mathsf{K}}\|_0 \leq c'' n^{1+\varepsilon-k-\beta}. \tag{6.3.13}$$

As we have already seen in Section 1.1 and, especially in the proof of Theorem 1.1.2, the Fredholm integral equation (6.0.1) of the second kind is uniquely solvable e.g. in the space of continuous functions. Hence, $(\mathsf{I}-\mathsf{K})^{-1}$ exists as a bounded linear operator in $C^0(\bar{\mathsf{G}})$. Since (6.3.13) implies

$$\lim_{n\to\infty} \|\mathsf{K}-\tilde{\mathsf{K}}\|_0 = 0, \tag{6.3.14}$$

every operator approximating $\mathsf{I}-\mathsf{K}$ also has an inverse for $n \geq n_0$ with n_0 big enough, and

$$\|(\mathsf{I}-\mathsf{K})^{-1}-(\mathsf{I}-\tilde{\mathsf{K}})^{-1}\|_0 \leq c''' n^{1+\varepsilon-k-\beta} \tag{6.3.15}$$

due to a well known argument (see e.g. [1, p. 3ff.]). Consequently, the solutions of the approximate integral equations converge uniformly to the desired w as n tends to infinity.

Replacing ϕ by $\tilde{\phi}$ in (6.0.1) the kernels are explicitly known but now we still have to replace (6.0.1) by discrete equations. Here this will be done in two different ways. First we shall replace the unknown w by piecewise constant functions. Although this is a very crude approximation, the corresponding discrete equations have the advantage yielding a sequence of operators that converge to K (or $\tilde{\mathsf{K}}$) with respect to the operator norm. Once the approximating step functions are found, they might even be smoothed by a further application of K (see e.g. [4]). The other method is the well known quadrature formula method [1, Chapter II] where we just cancel the diagonal terms which involve the singularities in (6.0.1).

For the discretizations we follow essentially the presentations in [9, Section 14.3, 14.4], [23] and [12]. Let us cover $\bar{G}$ by n squares defined by a net of meshwidth $h>0$ and let us denote the midpoints of the squares by $\zeta_j = \xi_j + i\eta_j$. We further consider only the squares q_j for which

$$g_j \equiv \iint_{q_j \cap \bar{G}} [dx, dy] > 0 \quad \text{for} \quad j = 1, \ldots, n.$$

The collocation of the discretized equations will be based on points z_j, taking z_j to be the centre of the square q_j if q_j lies entirely within $\bar{G}$. If, however, q_j contains a part of the boundary $\dot{G}$ of positive length $s_j^+ - s_j^-$ we choose z_j to be different from ζ_j defining

$$z_j \equiv \begin{cases} \zeta_j & \text{for} \quad g_j = h^2 \\ x((s_j^+ + s_j^-)/2) + iy((s_j^+ + s_j^-)/2) & \text{for} \quad g_j < h^2. \end{cases} \tag{6.3.16}$$

To make this definition reasonable we assume that h is chosen so small that $q_j \cap \dot{G}$ is a connected curve. Now let us consider (6.0.1) at z_j. Then we use the rectangular rule for every q_r with $r \neq j$ and some suitable other approximation near the singularity at q_j. Writing $H(j) = H(x_j, y_j)$ the discretization is defined by

$$\begin{aligned} \tilde{u}(j) = {\sum_{\substack{r=1\\ r\neq j}}^{n}}{}' \, g_r[(A\tilde{u} + B\tilde{v})(r) G_x^I(r, j) + (\tilde{A}\tilde{u} + \tilde{B}\tilde{v})(r) G_y^I(r, j)] \\ + (A\tilde{u} + B\tilde{v})(j) p_1^I(j) + (\tilde{A}\tilde{u} + \tilde{B}\tilde{v})(j) p_2^I(j) + F(j), \\ \tilde{v}(j) = {\sum_{\substack{r=1\\ r\neq j}}^{n}}{}' \, g_r[-(A\tilde{u} + B\tilde{v})(r) G_y^{II}(r, j) + (\tilde{A}\tilde{u} + \tilde{B}\tilde{v})(r) G_x^{II}(r, j)] \\ -(A\tilde{u} + B\tilde{v})(j) p_2^{II}(j) + (\tilde{A}\tilde{u} + \tilde{B}\tilde{v})(j) p_1^{II}(j) + H(j), \quad j = 1, \ldots, n. \end{aligned} \tag{6.3.17}$$

Here F and H are defined by the *given* parts in (6.0.1). From (1.1.10) and (1.1.14) one easily finds the expressions

$$\begin{aligned} G_x^I - iG_y^I &= -\frac{\phi'(z)}{2\pi} \left\{ \frac{\overline{\phi(\zeta)}}{1 - \phi(z)\overline{\phi(\zeta)}} + \frac{1}{\phi(z) - \phi(\zeta)} \right\}, \\ G_x^{II} - iG_y^{II} &= \frac{\phi'(z)}{2\pi} \left\{ \frac{\overline{\phi(\zeta)}}{1 - \phi(z)\overline{\phi(\zeta)}} - \frac{1}{\phi(z) - \phi(\zeta)} \right\}. \end{aligned} \tag{6.3.18}$$

These are used in (6.3.17) for $r \neq j$. Hence, (6.3.17) corresponds to a certain quadrature formula evaluated at the grid points [1, Section 2.3,

2.4]. For $r=j$ we approximate the weight

$$\iint_{q_j}(G_x^I-\mathrm{i}G_y^I)[\mathrm{d}x,\mathrm{d}y]=-\frac{1}{2\pi}\iint_{q_j}\phi'(z)\left\{\frac{1}{\phi(z)-\phi(\zeta)}+\frac{\overline{\phi(z)}}{1-\phi(z)\overline{\phi(\zeta)}}\right\}\times[\mathrm{d}x,\mathrm{d}y]$$

by using the Taylor expansion for the integrand and we choose

$$p_1^I(j)-\mathrm{i}p_2^I(j)\equiv-\frac{h^2}{2\pi}\phi'(j)\frac{\overline{\phi(j)}}{1-\phi(j)\overline{\phi(j)}},$$
$$p_1^{II}(j)-\mathrm{i}p_2^{II}(j)\equiv\frac{h^2}{2\pi}\phi'(j)\frac{\overline{\phi(j)}}{1-\phi(j)\overline{\phi(j)}}, \tag{6.3.19}$$

for the gridpoints in G, where $g_j=h^2$. For gridpoints $\zeta_j\in\dot{\mathrm{G}}$, on the boundary, we replace $\mathrm{G}\cap q_j$ by a polygonial $\hat{q}_j$ where $\dot{\mathrm{G}}\cap q_j$ is replaced by the tangent line as indicated in Fig. 6. Then the integral

$$\iint_{\hat{q}_j}\frac{[\mathrm{d}x,\mathrm{d}y]}{z-\zeta_j}=\int_{\alpha=\vartheta}^{\vartheta+\pi}\int_0^{r(\alpha)}\frac{r\,\mathrm{d}r\,\mathrm{d}\alpha}{r(\cos\alpha+\mathrm{i}\sin\alpha)} \tag{6.3.20}$$

can be evaluated explicitly [9, pp. 338–339] where ϑ denotes the angle as on Fig. 6. Since $G^I{}_{|\dot{\mathrm{G}}}=0$ and $\phi(\zeta)\overline{\phi(\zeta)}=1$ on $\dot{\mathrm{G}}$, we choose the remaining weights as

$$p_1^I(j)-\mathrm{i}p_2^I(j)=0,$$
$$p_1^{II}(j)-\mathrm{i}p_2^{II}(j)=-\frac{1}{\pi}\iint_{\hat{q}_j}\frac{[\mathrm{d}x,\mathrm{d}y]}{z-\zeta_j}\quad\text{for}\quad\zeta_j\in\dot{\mathrm{G}}. \tag{6.3.21}$$

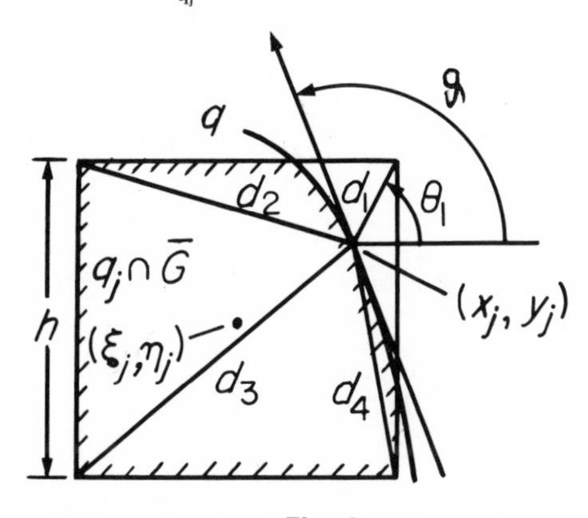

Fig. 6

With (6.3.18), (6.3.19) and (6.3.21) introduced into (6.3.17), the latter form a system of $2n$ linear equations for the $2n$ unknowns $\tilde{u}(j)$ and $\tilde{v}(j)$. For every small enough $h>0$ these equations will be uniquely solvable and provide approximating solutions which converge to the desired w as will be indicated below. In order to formulate this convergence we need to consider function spaces containing continuous functions as well as piecewise constant functions which are still equipped with the maximum norm. Thus, let B denote the linear space of all continuous functions and all functions of the first Baire's class over $\bar{\mathsf{G}}$ considered as a linear space over the field of reals. Then B supplied with the norm $\|w\|_0 = \sup_{z\in\bar{\mathsf{G}}}|w(z)|$ is a Banach space. Of course, K is by (1.1.23) also well defined in B and K maps the unit ball of B into a class of equi-Hölder-continuous functions. Hence the extended K is also compact in B. The extension of the discrete equations (6.3.17) to an operator acting in B is also quite natural. Let $k_1(z, \zeta)$, $k_2(z, \zeta)$ denote the kernels of (6.0.1). Then we define K_h, the extension, by the step function

$$
\begin{aligned}
\mathsf{K}_h w(\zeta) \equiv {\sum_{\substack{r=1\\ r\neq j}}^{n}}' \, & k_1(r, j) \iint_{q_j\cap\bar{\mathsf{G}}} u(z)[\mathrm{d}x, \mathrm{d}y] + k_2(r, j) \iint_{q_j\cap\bar{\mathsf{G}}} v(z)[\mathrm{d}x, \mathrm{d}y] \\
&+ \iint_{q_j\cap\bar{\mathsf{G}}} \Big\{(A(j)u(z)+B(j)v(z))\Big[G_x^I(z, j) \\
&+ \frac{1}{g_j}\Big(p_1^I(j) - \iint_{q_j\cap\bar{\mathsf{G}}} G_x^I(\hat{z}, j)[\mathrm{d}\hat{x}, \mathrm{d}\hat{y}]\Big)\Big] \\
&+ (\tilde{A}(j)u(z)+\tilde{B}(j)v(z))\Big[G_y^I(z, j) \\
&+ \frac{1}{g_j}\Big(p_2^I(j) - \iint_{q_j\cap\bar{\mathsf{G}}} G_y^I(\hat{z}, j)[\mathrm{d}\hat{x}, \mathrm{d}\hat{y}]\Big)\Big]\Big\}[\mathrm{d}x, \mathrm{d}y] \\
&+ \mathrm{i} \iint_{q_j\cap\bar{\mathsf{G}}} \Big\{-(A(j)u(z)+B(j)v(z))\Big[G_y^{II}(z, j) \\
&+ \frac{1}{g_j}\Big(p_2^{II}(j) - \iint_{q_j\cap\bar{\mathsf{G}}} G_y^{II}(\hat{z}, j)[\mathrm{d}\hat{x}, \mathrm{d}\hat{y}]\Big)\Big] \\
&+ (\tilde{A}(j)u(z)+\tilde{B}(j)v(z))\Big[G_x^{II}(z, j) \\
&+ \frac{1}{g_j}\Big(p_1^{II}(j) - \iint_{q_j\cap\bar{\mathsf{G}}} G_x^{II}(\hat{z}, j)[\mathrm{d}\hat{x}, \mathrm{d}\hat{y}]\Big)\Big]\Big\}[\mathrm{d}x, \mathrm{d}y]
\end{aligned}
\qquad (6.3.22)
$$

for all $\zeta \in q_j \cap \bar{G}$ with $j = 1, \ldots, n$. Hence choosing for the right hand sides step functions as

$$(F^h + iH^h)(\zeta) \equiv F(j) + iH(j) \quad \text{for all} \quad \zeta \in q_j \cap \bar{G}$$

the system of linear equations (6.3.17) coincides with the approximate operator equation in B,

$$w^h = K_h w^h + F^h + iH^h. \tag{6.3.23}$$

Let $\|K_h\|_0$ denote the operator norm associated with the sup-norm in B. Then we have:

Theorem 6.3.3 *To* $\bar{G}$ *and the coefficients of the system* (1.1.1) *there exists a constant* $\hat{c}$ *independent of the solution and of* h *such that*

$$\|K - K_h\|_0 \leq \hat{c} h^\alpha \tag{6.3.24}$$

holds. Here $\alpha > 0$ *is the Hölder index of the coefficients.*

The proof is based on the weak singularity of the kernel in (6.0.1) and requires several elementary but lengthy estimates. Since it can be found in [9, p. 341 ff.] let us omit this here.

As a consequence, from the existence of $(I - K)^{-1}$ we have immediately the following *stability property* [1, p. 3]: *There exist constants* C *and* $h_0 > 0$ *such that* $(I - K_h)^{-1}$ *exists and*

$$\|(I - K_h)^{-1}\|_0 \leq C \quad \text{for all} \quad 0 < h \leq h_0. \tag{6.3.25}$$

Hence, for these h the linear equations (6.3.17) are uniquely solvable. As is well known, the stability property together with (6.3.24) implies from

$$w - w^h = (I - K_h)^{-1}\{(K - K_h)(I - K)^{-1}(F + iH) + (F - F^h) + i(H - H^h)\} \tag{6.3.26}$$

the error estimate [4, Section 4]

$$\|w - w^h\|_0 \leq C\{c' \cdot h^\alpha \, \|F + iH\|_0 + \|F - F^h + i(H - H^h)\|_0\} \tag{6.3.27}$$

yielding the uniform convergence of the step functions w^h to w for decreasing meshwidth $h \to 0$.

For the quadrature formula method we choose a partition of $\bar{G} = \bigcup_{j=1}^n F_j$ where $F_j \cap F_l = \emptyset$ for $j \neq l$ and where the gridpoints $p_j \in F_j$ are chosen rather arbitrarily. Now let h be some measure of the fineness of the partition,

$$h \equiv \max_{j=1,\ldots,n} \operatorname{diam} F_j \quad \text{and} \quad |F_j| \equiv \iint_{F_j} [dx, dy]. \tag{6.3.28}$$

To the above choice of the partition let us assign a quadrature formula having weights $\gamma_j > 0$ such that if $f \in C^0(\bar{\mathsf{G}})$,

$$\iint_{\mathsf{G}} f(x, y)[dx, dy] = \lim_{h \to 0} \sum_{j=0}^{n} f(p_j)\gamma_j. \tag{6.3.29}$$

Whereas for continuous kernels no more assumptions are needed in order to prove convergence of the discrete solutions [1, Chap. 2], the weak singularity of our kernels requires either a special quadrature formula or further restrictions on the partitions in connection with the weights. Therefore, we further assume that there exists a constant $M' \geq 1$ such that

$$0 < \gamma_j \leq M' |F_j| \quad \text{and} \quad h \leq M' |z - p_j| \quad \text{for} \quad z \notin F_j \quad \text{and for all } j \text{ and all } h > 0. \tag{6.3.30}$$

Similarly to (6.3.17), we define approximate equations by

$$\begin{aligned} \tilde{u}(\zeta) = \sum_{r=1}^{n} \gamma_r [&(A\tilde{u} + B\tilde{v})(r) G_x^I(r, \zeta) \\ &+ (\tilde{A}\tilde{u} + \tilde{B}\tilde{v})(r) G_y^I(r, \zeta)] \chi_h(|\zeta - p_r|) + F(\zeta), \\ \tilde{v}(\zeta) = \sum_{r=1}^{n} \gamma_r [&-(A\tilde{u} + B\tilde{v})(r) G_y^{II}(r, \zeta) \\ &+ (\tilde{A}\tilde{u} + \tilde{B}\tilde{v})(r) G_x^{II}(r, \zeta)] \chi_h(|\zeta - p_r|) + H(\zeta) \end{aligned} \tag{6.3.31}$$

where χ_h is a cut off function,

$$\chi_h(\rho) \equiv \begin{cases} M'^2 h^{-2} \rho^2 & \text{for} \quad M'\rho \leq h, \\ 1 & \text{for} \quad M'\rho > h. \end{cases}$$

Clearly, (6.3.31) is completely solved if the corresponding system of linear equations obtained with $\zeta = p_j$ is solved.

The right hand side of (6.3.31) defines a continuous function of ζ since all terms are continuous and, hence, it defines a continuous operator K_h from $C^0(\bar{\mathsf{G}})$ into itself. Clearly, (6.3.31) may also be written as an equation in C^0:

$$\tilde{w}(\zeta) = \mathsf{K}_h \tilde{w} + (F(\zeta) + iH(\zeta)). \tag{6.3.32}$$

Although now the approximation takes place within the continuous functions the norm convergence of $\|\mathsf{K} - \mathsf{K}_h\|_0$ to zero is *not* true anymore. Nevertheless a stability property (6.3.25) can be proved by using Anselone's concept of *collectively compact operators*. The family of operators

$\{\mathsf{K}_h\}$ is called *collectively compact* provided the set of functions

$$C \equiv \{v = \mathsf{K}_h w \text{ with } \|w\|_0 \leq 1\} \tag{6.3.33}$$

is relatively compact in $C^0(\bar{\mathsf{G}})$ [1, p. 3].

Lemma 6.3.4 *For* $\dot{\mathsf{G}} \in C^{1+\alpha}$ *and* $A, \ldots, \tilde{B} \in C^\alpha(\bar{\mathsf{G}})$ *the above family* K_h *is collectively compact.*

Proof The proof is very similar to that of Theorem 6.3.3 given in [9, p. 342 ff.]. We shall use here some of the inequalities from [9]. Parts of the proof can also be found in [12].

Let ζ, ζ' be any given points. Then $\zeta \in F_j$ and $\zeta' \in F_l$ with suitable j, l and for any $w \in C^0(\bar{\mathsf{G}})$ with $\|w\|_0 \leq 1$ we find from (6.3.31) with (6.3.2) and (6.3.18) the inequality

$$\begin{aligned}
|(\mathsf{K}_h w)(\zeta) - (\mathsf{K}_h w)(\zeta')| \leq M \sum_{r=1}^{n} \gamma_r \{ & |G_x^I(r,\zeta)\chi_h(|p_r - \zeta|) \\
& - G_x^I(r,\zeta')\chi_h(|p_r - \zeta'|)| + \cdots + |G_y^{II}(r,\zeta) \\
& \times \chi_h(|p_r - \zeta|) - G_y^{II}(r,\zeta')\chi_h(|p_r - \zeta'|)\} \\
\leq M'' \sum_{\substack{r \neq j \\ r \neq l}} \gamma_r \Bigg\{ & \left| \frac{1}{|\phi(p_r) - \phi(\zeta)|} - \frac{1}{|\phi(p_r) - \phi(\zeta')|} \right| \\
& + \left| \frac{1}{|1 - \phi(p_r)\overline{\phi(\zeta)}|} - \frac{1}{|1 - \phi(p_r)\overline{\phi(\zeta')}|} \right| \Bigg\} \\
+ M'' \Bigg\{ \gamma_j & \left| \frac{\chi_h(|p_j - \zeta|)}{|\phi(p_j) - \phi(\zeta)|} - \frac{\chi_h(|p_j - \zeta'|)}{|\phi(p_j) - \phi(\zeta')|} \right| + \cdots \\
+ \gamma_l & \left| \frac{\chi_h(|p_l - \zeta|)}{|1 - \phi(p_l)\overline{\phi(\zeta)}|} - \frac{\chi_h(|p_l - \zeta'|)}{|1 - \phi(p_l)\overline{\phi(\zeta')}|} \right| \Bigg\}.
\end{aligned} \tag{6.3.34}$$

In the first sum the $\chi_h(|p_j - \zeta|)$ are identically equal to 1 thanks to the assumption (6.3.30).

In the first term of (6.3.34) let us use the inequalities [9, (14.4.9)],

$$\begin{aligned}
& \big| |\phi(p_r) - \phi(\zeta)|^{-1} - |\phi(p_r) - \phi(\zeta')|^{-1} \big| \\
& \quad + \big| |1 - \phi(p_r)\overline{\phi(\zeta)}|^{-1} - |1 - \phi(p_r)\overline{\phi(\zeta')}|^{-1} \big| \\
& \leq |\phi(\zeta) - \phi(\zeta')|^\alpha \{ |\phi(p_r) - \phi(\zeta)|^{-\alpha} \cdot |\phi(p_r) - \phi(\zeta')|^{-1} \\
& \quad + |\phi(p_r) - \phi(\zeta')|^{-\alpha} \, |\phi(p_r) - \phi(\zeta)|^{-1} \} \\
& \leq \text{const} \cdot |\zeta - \zeta'|^\alpha \{ |p_r - \zeta|^{-1-\alpha} + |p_r - \zeta'|^{-1-\alpha} \}
\end{aligned} \tag{6.3.35}$$

where the constant depends only on the conformal mapping ϕ. In the last terms of (6.3.34) the differences belong to piecewise differentiable functions where the gradients are bounded independent of j and ζ.

$$\left|\nabla_\zeta \frac{\chi_h(|p_j-\zeta|)}{|\phi(p_j)-\phi(\zeta)|}\right|, \qquad \left|\nabla_\zeta \frac{\chi_h(|p_j-\zeta|)}{|1-\phi(p_j)\overline{\phi(\zeta)}|}\right| \leqslant \frac{\text{const.}}{h^2}. \tag{6.3.36}$$

Thus, (6.3.34) yields with the above estimates

$$\begin{aligned} |(\mathsf{K}_h w)(\zeta)-(\mathsf{K}_h w)(\zeta')| &\leqslant |\zeta-\zeta'|\, M''' \frac{1}{h^2}(\gamma_j+\gamma_l) \\ &\quad + |\zeta-\zeta'|^\alpha M''' \sum_{\substack{r\neq j\\ r\neq l}} \gamma_r\{|p_r-\zeta|^{-1-\alpha}+|p_r-\zeta'|^{-1-\alpha}\}. \end{aligned} \tag{6.3.37}$$

In (6.3.37) we use (6.3.28) and (6.3.30), yielding $\gamma_j+\gamma_l \leqslant 4\pi h^2$ and

$$\begin{aligned} \gamma_r\,|p_r-\zeta|^{-1-\alpha} &\leqslant m' \iint_{F_r} \frac{[\mathrm{d}x,\mathrm{d}y]}{|p_r-\zeta|^{1+\alpha}} \\ &\leqslant M'(1+M')^{1+\alpha} \iint_{F_r} \frac{[\mathrm{d}x,\mathrm{d}y]}{|z-\zeta|^{1+\alpha}} \end{aligned}$$

since $\zeta \notin F_r$,

$$\begin{aligned} |(\mathsf{K}_h w)(\zeta)-(\mathsf{K}_h w)(\zeta')| &\leqslant |\zeta-\zeta'|^\alpha M^+ \iint_{\mathsf{G}\setminus(F_j\cup F_l)} \left\{\frac{1}{|z-\zeta|^{1+\alpha}}+\frac{1}{|z-\zeta'|^{1+\alpha}}\right\}[\mathrm{d}x,\mathrm{d}y] \\ &\quad + 4\pi M'''|\zeta-\zeta'| \leqslant \tilde{M}|\zeta-\zeta'|^\alpha \end{aligned}$$

where the constant $\tilde{M}$ is independent of w and ζ, ζ'. Hence (6.3.33) is a family of uniformly bounded equi-Hölder-continuous functions which is relatively compact in $C^0(\bar{\mathsf{G}})$ due to the well known Arzela–Ascoli Theorem.

Since K is weakly singular, from (6.3.29) with (6.3.30) it can easily be shown that for every continuous function $v \in C^0(\bar{\mathsf{G}})$ we have

$$\lim_{h\to 0} \|\mathsf{K}_h v - \mathsf{K} v\|_0 = 0, \tag{6.3.38}$$

i.e. the family of operators K_h converges 'elementwise' to K. Since $(\mathsf{I}-\mathsf{K})^{-1}$ exists, the collective compactness of $\{\mathsf{K}_h\}$ and the convergence (6.3.38) imply by [1, Theorem 1.11] or [4, Sections 3 and 4] the following:

Proposition 6.3.5 *For* $\dot{\mathsf{G}} \in C^{1+\alpha}$ *and* $A, \ldots, \tilde{B} \in C^\alpha(\bar{\mathsf{G}})$, *to the approximate equations* (6.3.31) *there exist constants* C *and* $h_0>0$ *such that*

(6.3.31) *is uniquely solvable for every* $0<h\leq h_0$, *and the inverses are stable,*

$$\|(\mathsf{I}-\mathsf{K}_h)^{-1}\|_0\leq C \quad \textit{for all} \quad 0<h\leq h_0. \tag{6.3.39}$$

The solutions of (6.3.31) *converge uniformly to* w, i.e.

$$\|w_h-w\|_0\leq C\|(\mathsf{K}-\mathsf{K}_h)w\|_0\to 0 \quad \textit{for} \quad h\to 0. \tag{6.3.40}$$

6.4 A Successive Approximation Method

As we have seen in Section 6.3 the discretized equations (6.3.17) or (6.3.31) define suitable approximations of the integral equations (6.0.1) which are equivalent to finite systems of linear equations for the values $\tilde{u}(j)$, $\tilde{v}(j)$ at the grid points. But for an increasing number of grid points a new difficulty arises from the numerical solution of these discrete equations since, in general, the corresponding matrices are completely filled up. Hence iteration procedures are extremely welcome for solving the discrete equations. In this context successive approximation is the most convenient procedure yielding the Neumann series for the solution. It surely converges for small coefficients $A, \ldots, \tilde{B}$. But if they are big the convergence will no longer hold. Nevertheless, by using a suitable rearrangement of the terms in the Neumann series, we shall find a procedure which converges also for big coefficients due to an estimate for the resolvent set of K. The idea for this method can be found in [11, II Section 2] and the following presentation is based on [4]. The procedure will be applicable to both the original and the discretized equations.

Let us denote by $\mathsf{A}(\lambda)$ and $\mathsf{A}_h(\lambda)$ the parameter dependent operators,

$$\mathsf{A}(\lambda)\equiv\mathsf{I}-\lambda\mathsf{K}, \qquad \mathsf{A}_h(\lambda)\equiv\mathsf{I}-\lambda\mathsf{K}_h. \tag{6.4.1}$$

Since K and K_h are compact the spectra of A and A_h consist of isolated eigenvalues only [20]. An estimate for the resolvent set of A, ρ_{A}, can be found with the *a priori* estimate (1.2.55); for convenience let us write

$$\gamma_1\equiv(\hat{\gamma}+\gamma(\phi_0)\|\phi_0\|_0), \qquad \gamma_2\equiv 5\gamma'. \tag{6.4.2}$$

Note that γ_1, γ_2 depend only on the domain G and the weight function σ. The latter coincides with $|\phi'|_{|\dot{\mathsf{G}}}$ in this section.

Lemma 6.4.1 *The resolvent set* ρ_{A} *includes the set of* $\lambda\in\mathbb{C}$ *satisfying*

$$|\mathrm{Im}\,\lambda|<\frac{1}{2\gamma_1 M}\exp\left(-\gamma_2 M\,|\mathrm{Re}\,\lambda|\right) \tag{6.4.3}$$

where M *is given by* (6.3.2).

Proof Let $\lambda=\lambda_1+i\lambda_2$ be an eigenvalue of (6.0.1). Then to λ we have *complex valued* eigenfunctions u and v satisfying

$$u=\lambda\iint_G\{(Au+Bv)G_x^I+(\tilde{A}u+\tilde{B}v)G_y^I\}[dx, dy],$$
$$v=\lambda\iint_G\{-(Au+Bv)G_y^{II}+(\tilde{A}u+\tilde{B}v)G_x^{II}\}[dx, dy]. \tag{6.4.4}$$

Hence, the real and imaginary parts of $u=u_1+iu_2$ and $v=v_1+iv_2$ satisfy the following homogeneous boundary value problem for a Pascali system,

$$\begin{aligned}
u_{1x}-v_{1y} &= \lambda_1(Au_1+Bv_1)-\lambda_2(Au_2+Bv_2)\\
u_{1y}+v_{1x} &= \lambda_1(\tilde{A}u_1+\tilde{B}v_1)-\lambda_2(\tilde{A}u_2+\tilde{B}v_2)\\
u_{2x}-v_{2y} &= \lambda_2(Au_1+Bv_1)+\lambda_1(Au_2+Bv_2)\\
u_{2y}+v_{2x} &= \lambda_2(\tilde{A}u_1+\tilde{B}v_1)+\lambda_1(\tilde{A}u_2+\tilde{B}v_2)
\end{aligned} \tag{6.4.5}$$

$$u_1\,|_{\dot{G}}=0,\quad u_2\,|_{\dot{G}}=0,\qquad \oint_{\dot{G}} v_1\,|\phi'|\,ds=\oint_{\dot{G}} v_2\,|\phi'|\,ds=0.$$

With $h_1=u_1+iv_1$ and $h_2=u_2+iv_2$ and the complex coefficients (1.1.6) it can also be written as

$$h_{1\bar{z}}=\lambda_1(\bar{a}h_1+b\bar{h}_1)+c_1,\qquad \operatorname{Re} h_{1|\dot{G}}=0,\qquad \oint_{\dot{G}} \operatorname{Im} h_1\,|\phi'|\,ds=0,$$
$$h_{2\bar{z}}=\lambda_1(\bar{a}h_2+b\bar{h}_2)+c_2,\qquad \operatorname{Re} h_{2|\dot{G}}=0,\qquad \oint_{\dot{G}} \operatorname{Im} h_2\,|\phi'|\,ds=0 \tag{6.4.6}$$

where

$$c_1=-\lambda_2(\bar{a}h_2+b\bar{h}_2)\quad\text{and}\quad c_2=\lambda_2(\bar{a}h_1+b\bar{h}_1). \tag{6.4.7}$$

Hence, for both h_1, h_2, Theorem 1.2.5 is valid and (1.2.55) with (6.4.2) yields the estimates

$$\begin{aligned}
\|h_1\|_0 &\leq \gamma_1 M\,|\lambda_2|\,\|h_2\|_0 \exp(M\gamma_2\,|\lambda_1|),\\
\|h_2\|_0 &\leq \gamma_1 M\,|\lambda_2|\,\|h_1\|_0 \exp(M\gamma_2\,|\lambda_1|).
\end{aligned} \tag{6.4.8}$$

Since $\|u\|_0+\|v\|_0\neq 0$, at least one of the numbers $\|h_1\|_0$ or $\|h_2\|_0$ is positive. Hence, inserting one of the inequalities of (6.4.8) into the remaining leads to the estimate

$$1\leq \gamma_1^2M^2\,|\lambda_2|^2\,(\exp(\gamma_2 M\,|\lambda_1|))^2 \tag{6.4.9}$$

for any eigenvalue of A. Thus, the complementary complex λ, satisfying (6.4.3), belong to ρ_{A}.

Lemma 6.4.1 provides a strip-like region including the real λ-axis belonging to ρ_{A}. The following theorem on the approximation of integral equations shows that, in some sense, this domain also belongs to ρ_{A_h}.

Theorem 6.4.2 ([1, *Theorem* 4.7], [4, *Satz* 5]) *Let Λ be a compact subset of ρ_{A}, $\Lambda \subset \rho_{\mathsf{A}}$. Then there exist $C>0$ and $h_0>0$ such that*

$$\Lambda \subset \rho_{\mathsf{A}_h} \quad \textit{and} \quad \|\mathsf{A}_h(\lambda)^{-1}\|_0 \leqslant C \quad \textit{for all } \lambda \in \Lambda \textit{ and } h \leqslant h_0. \tag{6.4.10}$$

This is exactly Anselone's Theorem 4.7 [1, p. 64] if we use Theorem 6.3.3 and Lemma 6.3.4, respectively, which shows the convergence $(\mathsf{K}-\mathsf{K}_h)\, w$ for every w and the collective compactness of $\{\mathsf{K}-\mathsf{K}_h\}$ since K is compact in both cases. For the proof we refer to [1], [4].

Now let us choose some simply connected compact set Λ in the domain (6.4.3) containing the real segment $0 \leqslant \lambda_1 \leqslant 1$, $\lambda_2 = 0$ in its open interior. Let $\eta(\lambda)$ be the conformal mapping from Λ onto the unit disk having the properties

$$\eta(0)=0, \quad \eta_0 = \eta(1) > 0 \; (\eta_0 \in \mathbb{R}), \tag{6.4.11}$$

and let $\lambda(\eta)$ be the inverse mapping (Fig. 7). Since we shall only need $0 \in \mathring{\Lambda}$ and $1 \in \mathring{\Lambda}$, the compact Λ can be chosen in such a way that $\lambda(\eta)$ is a *polynomial,*

$$\lambda(\eta) = \sum_{\nu=1}^{M} c_\nu \eta^\nu. \tag{6.4.12}$$

Moreover, since Λ can be chosen as to be symmetric to the real axis, the coefficients c_ν in (6.4.12) can be chosen as to be reals,

$$c_\nu \in \mathbb{R}. \tag{6.4.13}$$

Since $\lambda(\eta)$ is holomorphic in the unit disk up to $|\eta|=1$ so are $\mathsf{A}(\lambda(\eta))$ and $\mathsf{A}(\lambda(\eta))^{-1}$ holomorphic operator valued functions [6]. Both possess Taylor expansions at $\eta=0$ converging for $|\eta| \leqslant 1$,

$$\mathsf{A}(\lambda(\eta)) = \mathsf{I} - \sum_{\nu=1}^{M} c_\nu \eta^\nu \mathsf{K}, \tag{6.4.14}$$

$$\mathsf{A}(\lambda(\eta))^{-1} = \mathsf{I} + \sum_{j=1}^{\infty} \eta^j \mathsf{T}_j. \tag{6.4.15}$$

Then the identity

$$\mathsf{I} = \left(\mathsf{I} - \sum_{\nu=1}^{M} c_\nu \eta^\nu \mathsf{K}\right)\left(\mathsf{I} + \sum_{j=1}^{\infty} \eta^j \mathsf{T}_j\right) \tag{6.4.16}$$

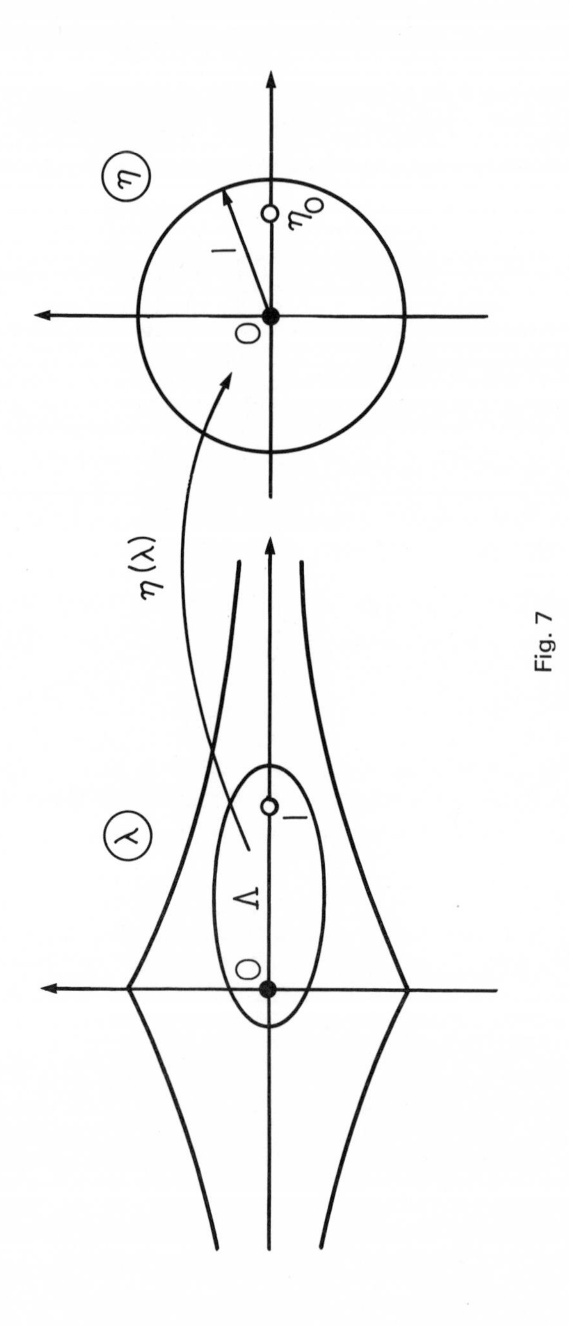

Fig. 7

for all $|\eta| \leq 1$ implies the recursion formula

$$\mathsf{T}_\sigma = \mathsf{K} \sum_{\nu=1}^{\sigma} c_\nu T_{\sigma-\nu}, \qquad \mathsf{T}_0 = \mathsf{I}, \qquad c_{M+1} = c_{M+2} = \cdots = 0, \sigma = 1, 2, \ldots. \tag{6.4.17}$$

Using (6.4.15) at η_0 for solving (6.0.1), the solution w can be obtained by the iteration

$$g_0 \equiv f \equiv (F + \mathrm{i}H), \qquad g_j \equiv \mathsf{T}_j f = \mathsf{K} \sum_{\nu=1}^{\min\{j,M\}} c_\nu g_{j-\nu}, \; j = 1, 2, \ldots \tag{6.4.18}$$

or with the approximate terms

$$w^{(0)} \equiv f, \qquad w^{(m)} \equiv \sum_{j=0}^{m} \eta_0^j g_j, \tag{6.4.19}$$

using (6.4.18) for $m < M$ and for $m \geq M$ using

$$\begin{aligned}\sum_{j=0}^{m} \eta_0^j g_j &= w^{(0)} + \mathsf{K} \sum_{\nu=1}^{m} \sum_{\nu=1}^{M} \eta_0^\nu \eta_0^{j-\nu} c_\nu g_{j-\nu} \\ &= w^{(0)} + \mathsf{K} \sum_{\nu=1}^{M} \eta_0^\nu c_\nu \sum_{s=0}^{m-\nu} \eta_0^s g_s,\end{aligned}$$

we find for all $m \geq 1$

$$w^{(m)} \equiv f + \mathsf{K} \sum_{\nu=1}^{\mathrm{Min}\{m,M\}} c_\nu \eta_0^\nu w^{(m-\nu)}. \tag{6.4.20}$$

Using the above Λ of Theorem 6.4.2 we have for the discrete equations and $h \leq h_0$ the uniform boundedness of $\mathsf{A}_h(\lambda(\eta))^{-1}$ in the unit disk which allows us to use the same iterations for the discrete equations,

$$\begin{aligned}&\tilde{w}^{(0)} \equiv F_h + \mathrm{i}H_h, \\ &\tilde{w}^{(m)} \equiv \tilde{w}^{(0)} + \mathsf{K}_h \sum_{\nu=1}^{\min\{m,M\}} c_\nu \eta_0^\nu \tilde{w}^{(m-\nu)}.\end{aligned} \tag{6.4.21}$$

The convergence of (6.4.21) is equivalent to the convergence of the operator series

$$\mathsf{A}_h(\lambda(\eta_0))^{-1} = \sum_{j=0}^{\infty} \eta_0^j \mathsf{T}_{hj}. \tag{6.4.22}$$

The convergence of the latter follows from the Cauchy formula [6],

$$\mathsf{T}_{hj} = \frac{1}{2\pi \mathrm{i}} \oint_{|\eta|=1} \mathsf{A}_h(\lambda(\eta))^{-1} \eta^{-j-1} \, \mathrm{d}\eta, \tag{6.4.23}$$

providing with Theorem 6.4.2 the *uniform estimate*

$$\|T_{hj}\|_0 \leq C \quad \text{for all } h \leq h_0 \text{ and all } j = 0, 1, \ldots. \tag{6.4.24}$$

The immediate consequence of (6.4.24) is then the:

Proposition 6.4.3 *The iteration* (6.4.21) *converges for* $h \leq h_0$ *and we have an error estimate*

$$\|\tilde{w}^{(m)} - \tilde{w}\|_0 \leq C(1-\eta_0)^{-1}\eta_0^{m+1}\|\tilde{w}^{(0)}\|_0 \tag{6.4.25}$$

where C *is independent of* h, m *and* $\tilde{w}^{(0)}$.

6.5 Layer Methods

The previous sections were devoted to the computational solution of the integral equations (6.0.1) hinging on G^I and G^{II} and, hence, on the conformal mapping ϕ. In this section we shall consider the first kind integral equations (5.2.110) which are defined *without the knowledge of* ϕ. Moreover, these equations can also be used for elliptic systems with a more general principal part than (1.1.1) and with $n \geq 1$. The presentation follows essentially [10] where special higher order equations were treated with Galerkin's method using finite element functions as trial functions. Since the principal part of (5.2.110) is the integral operator of the first kind,

$$Vg(\xi) \equiv -\oint_{\dot{G}} \log|z(s)-z(\xi)|\, g(s)\, ds = f(\xi), \quad z(\xi) \in \dot{G}, \tag{6.5.1}$$

with an inverse being a Fourier integral operator like a differentiation, its discretization leads to discrete equations with condition tending to ∞ with an increasing number of grid points. Although in practice for (6.5.1) collocation methods are extensively used, for those in connection with (6.5.1) no error analysis is available yet. But for Galerkin's method an asymptotic error analysis can be made and it leads with the finite element functions to an optimal rate of convergence.†

If the differential equations include linear linking terms without derivatives then one can append surface potentials to the integral representation without changing the fundamental solution. For the simplest case of the equations (1.1.1) this yields a system of Fredholm integral equations with some operators of the first and some of the second kind. The presentation follows [24]. Then using again Galerkin's method we shall again find an optimal rate of convergence.

Let us begin with the Galerkin method for (6.5.1). Let r be real and let

† In [61] this method is implemented numerically providing high accuracy and extremely short computational time.

us denote by $H^r(\dot{\mathsf{G}})$ the Sobolev spaces as well as their interpolation spaces on $\dot{\mathsf{G}}$ for noninteger r with the understanding that if $r<0$, $H^r(\dot{\mathsf{G}})$ is the dual of $H^{-r}(\dot{\mathsf{G}})$ (see e.g. [13], p. 34 ff.]). The norms are denoted by $\|\cdot\|_r$ as long as misunderstandings are not possible.

Since V in (6.5.1) may have eigensolutions we suppose in the following that the domain G has

$$\text{diameter }(\mathsf{G})<1. \tag{6.5.2}$$

Then it is easy to verify that V has a *positive kernel*. Furthermore, if we define the scalar product $\langle , \rangle_0$ by

$$\langle f, g\rangle_0 = \mathrm{Re} \oint_{\dot{\mathsf{G}}} f(s)\bar{g}(s)\, \mathrm{d}s,$$

considering $L^2(\dot{\mathsf{G}})$ now as a *Hilbert space over the field* $\mathbb{R}$. We see that V is self-adjoint with respect to $\langle , \rangle_0$. The fundamental properties of V can be summarized in the following theorems:

Theorem 6.5.1 *There exists a constant* $\gamma>0$ *depending only on* $\dot{\mathsf{G}}$ *such that*

$$\langle Vg, g\rangle_0 \geq \gamma \|Vg\|^2_{1/2} \tag{6.5.3}$$

holds for all functions $g \in L^2(\dot{\mathsf{G}})$.

Proof The operator V has a weakly singular kernel and hence it is a Hilbert–Schmidt operator mapping $L^2(\dot{\mathsf{G}})$ compactly into $L^2(\dot{\mathsf{G}})$. It suffices to prove (6.5.3) for smooth g only. Now let e be the natural layer of $\dot{\mathsf{G}}$ with $\oint_{\mathsf{G}} e(s)\, \mathrm{d}s = 1$ [9, (4.12.6)]. Then if $\dot{\mathsf{G}} \in C^{\rho+\alpha}$, $0<\alpha<1$ and $\rho>0$, an integer, we have the properties:

$$e \in C^{\rho+\alpha-1}(\dot{\mathsf{G}}), \quad e>0 \quad \text{and} \quad Ve=E=\text{const}>0 \text{ on } \dot{\mathsf{G}}.$$

In terms of e, one may express g in the form:

$$g = g_0 + e\omega, \tag{6.5.4}$$

where $\oint_{\dot{\mathsf{G}}} g_0\, \mathrm{d}s = 0$ and $\omega \equiv \oint_{\dot{\mathsf{G}}} g\, \mathrm{d}s$. With this representation, the left hand side of (6.5.3) reads

$$\langle g, Vg\rangle_0 = \langle g_0, Vg_0\rangle_0 + E\omega^2. \tag{6.5.5}$$

The simple layer potential

$$Vg_0(\zeta) \equiv -\oint_{\dot{\mathsf{G}}} \log|z(s)-\zeta|\, g_0(s)\, \mathrm{d}s, \qquad \zeta \in \mathbb{C}, \tag{6.5.6}$$

vanishes at infinity; hence one can apply Green's formula to Vg_0 in G as well as in $\mathbb{C}\backslash\mathsf{G}$. This leads to

$$\langle Vg_0, g_0\rangle_0 = \frac{2}{\pi} \iint_{\mathbb{C}\backslash\dot{\mathsf{G}}} |(Vg_0)_{\bar{z}}|^2[\mathrm{d}x, \mathrm{d}y] \tag{6.5.7}$$

provided one makes use of the jump relation

$$g_0 = \frac{1}{2\pi}\left(\frac{\partial Vg_0}{\partial n_i} - \frac{\partial Vg_0}{\partial n_a}\right) \quad \text{on} \quad \dot{\mathsf{G}}.$$

Here $\partial/\partial n_i$ and $\partial/\partial n_a$ denote, respectively, the limits of the outer normal derivatives from the inside and outside. Hence, from (6.5.7), (6.5.5) can be estimated as

$$\langle g, Vg\rangle_0 \geqslant \frac{2}{\pi} \iint_{\mathsf{G}} |(Vg_0)_{\bar{z}}|^2[\mathrm{d}x, \mathrm{d}y] + E\omega^2. \tag{6.5.8}$$

Now from the representation of g in (6.5.4) and the fact that $Ve = \text{const}$ in G, it follows that

$$(Vg)_{\bar{z}} = (Vg_0)_{\bar{z}} + \omega(Ve)_{\bar{z}} = (Vg_0)_{\bar{z}}.$$

Also from the self-adjointness of V, we see that

$$\frac{1}{E}\langle e, Vg\rangle_0 = \frac{1}{E}\langle Ve, g\rangle_0 = \langle 1, g\rangle_0 = \omega.$$

Consequently, (6.5.8) reads

$$\langle g, Vg\rangle_0 \geqslant \frac{2}{\pi} \iint_{\mathsf{G}} |(Vg)_{\bar{z}}|^2[\mathrm{d}x, \mathrm{d}y] + \frac{1}{E}\left(\int_{\dot{\mathsf{G}}} eVg\, \mathrm{d}s\right)^2. \tag{6.5.9}$$

The right hand side is equivalent to $\|Vg\|^2_{H^1(\mathsf{G})}$ ([21], p. 382, Theorem 28.2), and hence there exists a constant $\tilde{\gamma} > 0$, depending only on $\dot{\mathsf{G}}$ and e such that

$$\langle g, Vg\rangle_0 \geqslant \tilde{\gamma}\|Vg\|^2_{H^1(\mathsf{G})}. \tag{6.5.10}$$

The Trace Theorem ([13], p. 41 with $\mu = 0$, $s = 1$) implies the desired estimate

$$\langle g, Vg\rangle_0 \geqslant \tilde{\gamma} \inf_{F|_{\dot{\mathsf{G}}} = Vg} \|F\|^2_{H^1(\mathsf{G})} \geqslant \gamma\|Vg\|^2_{1/2} \tag{6.5.11}$$

Theorem 6.5.2 *There exists a constant $\kappa > 0$ depending only on $\dot{\mathsf{G}}$ such that*

$$\kappa^{-1}\|g\|_{-1/2} \leqslant \|Vg\|_{1/2} \leqslant \kappa\|g\|_{-1/2} \tag{6.5.12}$$

holds for all distributions $g \in H^{-1/2}(\dot{\mathsf{G}})$.

Here we shall not present the complete proof but only the ideas. The complete proof can be found in [10]. First we notice that the relation

$$\langle Vg, \phi\rangle_0 = \langle g, V\phi\rangle_0 \tag{6.5.13}$$

for any test functions ϕ leads to a natural extension of V operating on distributions g.

The right inequality in (6.5.12) follows from Theorem 6.5.1:

$$\|Vg\|_{1/2}^2 \leqslant \gamma^{-1}\langle g, Vg\rangle_0 \leqslant \gamma^{-1}\|g\|_{-1/2}\|Vg\|_{1/2}. \tag{6.5.14}$$

For proving the left inequality in (6.5.12), one can use interior and exterior harmonic extensions of functions on $\dot{\mathsf{G}}$ to show that $V: H^{-1/2}(\dot{\mathsf{G}}) \to H^{1/2}(\dot{\mathsf{G}})$ is a *bijective* mapping. It is continuous from (6.5.14). Then Banach's Theorem [20] implies the continuity of the inverse $V^{-1}: H^{1/2}(\dot{\mathsf{G}}) \to H^{-1/2}(\dot{\mathsf{G}})$ which is equivalent to the left inequality.

As a consequence of Theorems 6.5.1 and 6.5.2, we have the following corollary.

Corollary 6.5.3 *Let* $\{\,,\}$ *be the inner product defined by*

$$\{g, h\} := \langle Vg, h\rangle_0 \quad \text{for} \quad g, h \in H^{-1/2}(\dot{\mathsf{G}}). \tag{6.5.15}$$

Then, $\{g, g\}$ *is equivalent to* $\|g\|_{-1/2}^2$:

$$\nu^{-1}\|g\|_{-1/2}^2 \leqslant \{g, g\} \leqslant \nu\|g\|_{-1/2}^2 \tag{6.5.16}$$

for some constant $\nu > 0$.

Proof Indeed, (6.5.12) and (6.5.3) lead to

$$\gamma\kappa^{-2}\|g\|_{-1/2}^2 \leqslant \|Vg\|_{1/2}^2 \leqslant \langle g, Vg\rangle_0 = \{g, g\} \leqslant \|g\|_{-1/2}\|Vg\|_{1/2} \leqslant \kappa\|g\|_{-1/2}^2$$

As a generalization of Theorem 6.5.2 we also have:

Theorem 6.5.4 *Let* $\dot{\mathsf{G}} \in C^{\{|r|+1\}}$ *for any real* r, *where* $\{r\} \equiv -[-r]$ *and* $[r]$ *stands for the greatest integer* $\leqslant r$. *Then there exists a positive constant* $\beta > 0$, *depending only on* r *and* $\dot{\mathsf{G}}$, *such that*

$$\beta^{-1}\|g\|_{r-1} \leqslant \|Vg\|_r \leqslant \beta\|g\|_{r-1} \tag{6.5.17}$$

for all $g \in H^{r-1}(\dot{\mathsf{G}})$.

For $r = -\frac{1}{2}$, (6.5.17) coincides with (6.5.12). For $r > 1$ the inequality can be proved by solving the Dirichlet problem in G with boundary values Vg and a corresponding Dirichlet problem in the exterior. Then regularity properties in connection with the jump condition of the normal derivatives of (6.5.6) yield (6.5.17). For the details see [10].

Now let us formulate Galerkin's method for solving equation (6.5.1). We adopt here the finite elements for approximating periodic functions and functions on the closed curve $\dot{\text{G}}$. To this end let h, $0<h\leq 1$ be a parameter of meshwidth and let $\tilde{H}(h)$ be a sequence of finite dimensional subspaces of H^m with m a nonnegative integer such that $\cup_{h>0}\tilde{H}(h)$ is dense in H^m and $\tilde{H}(h)\subset\tilde{H}(h')$ for $h'<h$.

Furthermore, we assume $\tilde{H}$ to be a so called *regular finite element space* satisfying the following conditions:

Convergence property

(C) *For* $k\leq r$ *with* $-m-1\leq k\leq m$, $-m\leq r\leq m+1$ *and for any* $u\in H^r(\dot{\text{G}})$ *there exist a constant* c_{rk} *independent of* h *and* u, *and an* $\bar{u}\in\tilde{H}$ *such that*

$$\|u-\bar{u}\|_k\leq c_{rk}h^{r-k}\|u\|_r$$

for all k *with the above restrictions.*
(see Bramble and Schatz [3], p. 659).

Stability property

(S) *For* $k\leq r$ *with* $|k|,|r|\leq m$ *there exists a constant* M_{rk} *independent of* h *and of* $\chi\in\tilde{H}$ *such that*

$$\|\chi\|_r\leq M_{rk}h^{k-r}\|\chi\|_k$$

for all $\chi\in\tilde{H}$ (see Nitsche [17]).

As a preparation for considering the equations (5.2.110) let us consider first Galerkin's method for solving (6.5.1) which can be formulated as the problem of finding a finite element function $\tilde{g}\in\tilde{H}$ satisfying the *Galerkin equations*

$$\langle V\tilde{g},\tilde{\phi}\rangle_0=\langle f,\tilde{\phi}\rangle_0 \quad \textit{for all } \tilde{\phi}\in\tilde{H}. \tag{6.5.18}$$

Alternatively, let us introduce the orthogonal projection P_h of $L^2=H^0$ onto $\tilde{H}$, that is,

$$P_h: H^0\to\tilde{H}\subset H^0. \tag{6.5.19}$$

From (C) and (S) it follows easily that

$$\|u-P_hu\|_k\leq c'_{rk}h^{r-k}\|u\|_r \tag{6.5.20}$$

holds for all u and k as in (C). Then, using P_h, the Galerkin method (6.5.18) is identical with the problem of finding the approximate solution $\tilde{g}\in\tilde{H}$ which satisfies

$$P_hVP_h\tilde{g}=P_hf\equiv\tilde{f}=P_hVg. \tag{6.5.21}$$

Note that (6.5.21) is just the Ritz equation for the operator $V^{1/2}$, the square root of V, which is well defined in H^0, since V is a self-adjoint, semidefinite compact operator on H^0 according to Theorem 6.5.4 and the compact imbedding $H^1 \to H^0$ (see e.g. [20, Section 6.5]).

For illustration, in the following let us consider a special treatment of (6.5.18). Our approach is a variation of technique due to Aubin ([2, Chap. 4]) for treating functions of one variable on the real axis. It should be emphasized that the results obtained below are true in general; they do not depend on the particular choice of special finite elements as long as properties (C) and (S) are fulfilled.

To begin, let us identify functions on $\dot{G}$ with L-periodic functions on the real axis:

$$x(s+L)=x(s), \qquad f(s+L)=f(s) \quad \text{etc. for all } s\in\mathbb{R}. \tag{6.5.22}$$

Let N and m be integers with $N\geq m\geq 0$. We consider the grid points x_j on $\dot{G}$ defined by

$$x_j\equiv x(jh), \qquad j=0,\pm 1,\ldots, \quad \text{where} \quad h\equiv\frac{L}{N+1}, \quad L\equiv\oint_{\dot{G}} ds,$$

and periodic grid functions, u_h, with values $u_h^j=u_h(x_j)$ such that

$$u_h^{j+(N+1)}=u_h^j \quad \text{for all} \quad j\in\mathbb{Z}, \text{ the set of all integers.} \tag{6.5.23}$$

We assume that $\mu\in H^m(\mathbb{R})$ is a given (non-periodic) function on the real axis with

$$\operatorname{supp}(\mu)=[a,b]\subset\mathbb{R} \quad \text{and} \quad \int_{-\infty}^{\infty}\mu(s)\,ds=1; \tag{6.5.24}$$

moreover, μ is assumed to satisfy the criterion of the *m-convergence* [2, p. 131] and the stability property [2, p. 135]. The criterion of m-convergence states that the Fourier transform $\hat{\mu}$ of μ and its derivatives $\hat{\mu}^{(j)}$ $(0\leq j\leq m$ with $\hat{\mu}^{(0)}=\hat{\mu})$ satisfy the relation:

$$\hat{\mu}^{(j)}(2k\pi L)=0 \quad \text{for all} \quad k=\pm 1,\pm 2,\ldots; \tag{6.5.25}$$

while by the stability property, we mean that for every $y\in(-\pi L,\pi L)$ there exists an integer $j\in\mathbb{Z}$ such that

$$\hat{\mu}(y+2\pi Lj)\neq 0. \tag{6.5.26}$$

We also assume that $\lambda\in L^\infty(\mathbb{R})$ is a given (non-periodic) function with

$$\operatorname{supp}(\lambda)\subset[0,L] \quad \text{and} \quad \int_{-\infty}^{+\infty}\lambda(s)\,ds=1; \tag{6.5.27}$$

in addition, λ is related to μ by the double integral

$$\int_{-\infty}^{+\infty}\int_{-\infty}^{+\infty}\mu(s)\lambda(\sigma)(s-\sigma)^k\,ds\,d\sigma=\begin{cases}1 & \text{if } k=0\\ 0 & \text{if } 1\leq k\leq m\end{cases} \tag{6.5.28}$$

(see [2, p. 128]). Now with the given μ and λ, we define periodic prolongations by

$$p_h u_h(s) \equiv \sum_{-\infty}^{+\infty} u_h^j \mu\left(\frac{s}{h} - j\right) \tag{6.5.29}$$

and periodic restrictions by

$$(r_h u)^j \equiv h^{-1} \int_{-\infty}^{+\infty} \lambda\left(\frac{s}{h} - j\right) u(s)\,\mathrm{d}s. \tag{6.5.30}$$

Here p_h and r_h are referred to as the prolongation and restriction operators; u is L-periodic as in (6.5.22) and $u_h = \{u_h^j\}$ stands for a periodic grid function with values u_h^j at x_j given by (6.5.23).

Remark The spline approximations are special finite element approximations of this kind, In this case μ is given by $(m+1)$-fold convolution of the characteristic function $\chi_{[0,L]}$ on $\mathbb{R}$, namely,

$$\mu \equiv \chi_{[0,L]} * \chi_{[0,L]} * \cdots * \chi_{[0,L]} \tag{6.5.31}$$

and λ is a special polynomial on $[0, L]$, defined by

$$\lambda(x) \equiv \chi_{[0,L]}(x) \sum_{j=0}^{m} \frac{b_j}{j!} x^j \tag{6.5.32}$$

where the b_j's are determined by a suitable system of linear equations [2, p. 124]. All approximate functions $p_h u_h$ are piecewise polynomials of order m in class $C^{m-1}(\dot{\mathrm{G}})$.

We notice that for any fixed N and for any periodic grid function u_h, the approximate functions $p_h u_h$ in (6.5.29) form the $(N+1)$-dimensional subspace $\tilde{H}$. Hence a basis $\{\tilde{\phi}_k\}_{k=0}^N$ of $\tilde{H}$ can be written explicitly in the form:

$$\tilde{\phi}_k(s) = \sum_{l=-\infty}^{+\infty} \mu\left(\frac{s}{h} - k - l(N+1)\right), \quad k = 0, \ldots, N. \tag{6.5.33}$$

The Galerkin method (6.5.18) can then be formulated as to find a finite element function $\tilde{g}(s) = \sum_{j \in Z} \gamma_h^j \mu((s/h) - j)$ with periodic γ_h^j satisfying (6.5.18). By a summation with modulo $N+1$ it follows easily that (6.5.18) is equivalent to the $N+1$ linear algebraic equations

$$-\sum_{j=0}^{N} \gamma_h^j \int_{\mathbb{R}} \int_{\mathbb{R}} \log|x(s) - x(\sigma)| \mu\left(\frac{s}{h} - j\right) \mu\left(\frac{\sigma}{h} - k\right) \mathrm{d}s\,\mathrm{d}\sigma$$

$$= \int_R f(s) \mu\left(\frac{s}{h} - k\right) \mathrm{d}s, \quad k = 0, \ldots, N, \tag{6.5.34}$$

for the unknowns γ_h^j. The method in [61] is based on these equations

It is clear from Corollary 6.5.3 that equations (6.5.18) are uniquely solvable for any N. The coefficient matrix is positive definite and symmetric. Supplying all spaces with the norm $\{,\}^{1/2}$ defined in (6.5.15), we see that the method converges with respect to this norm, and hence, by Corollary 6.5.3, converges in $H^{-1/2}(\dot{G})$ as $N \to \infty$. For the numerical treatment, however, convergence is more involved and further information concerning the coefficient matrix is needed.

In what follows, we consider the sequence $h = L/n$ for $n = 1, 2, \ldots$ and denote by $\tilde{H}_n$ the corresponding finite element spaces $\tilde{H}(h)$. Without loss of generality we assume that $\dim \tilde{H}_n = n$. Now, if we supply the chain of finite dimensional spaces $\tilde{H}_{N+1} \subset \tilde{H}_{2N+2} \subset \ldots \tilde{H}_{lN+l} \subset \tilde{H}^{-1/2}$ with bases $\tilde{\phi}_1, \ldots, \tilde{\phi}_{N+1}; \tilde{\phi}_{N+2}, \ldots, \tilde{\phi}_{2N+2}, \ldots,$ so that the sequence $\tilde{\phi}_1, \ldots, \tilde{\phi}_{lN+l}$ forms an orthonormal (with respect to $\{,\}$) basis for $\tilde{H}_{lN+l}$, then it is clear by Corollary 6.5.3 that $\tilde{g}$ is just the partial sum of the Fourier expansion of g in $H^{-1/2}$ with respect to $\{\phi_j\}_{j=1}^{lN+l}$ under the inner product $\{,\}$. It is easy to show that the $\tilde{\phi}_j$'s are complete in $H^{-1/2}$. Hence, we can conclude the following lemma.

Lemma 6.5.5 *If $f \in H^{1/2}$ then*

$$\|\tilde{g} - g\|_{-1/2} \to 0 \quad \text{as} \quad N \to \infty, \quad \text{and} \quad \|\tilde{g}\|_{-1/2} \leq \nu' \|g\|_{-1/2} \tag{6.5.35}$$

with some constant $\nu' = \nu'(\dot{G}) > 0$.

It is also clear that for $f \notin H^{1/2}$ (e.g. $f \in H^0 \backslash H^{1/2}$) the method will no longer converge to the solution g because then $g \notin H^{-1/2}$ whereas the Galerkin method just coincides with the Fourier expansion in $H^{-1/2}$. *For convergence we have to demand at least $f \in H^{1/2}(\dot{G})$.* For smoother f we will see that the Galerkin method converges with *optimal order*, i.e. $\tilde{g} \to g$ with the same order as $\bar{g} \to g$ (cf. the property (C)).

To this end let us introduce the *Galerkin projection* Γ_h defined by

$$\Gamma_h g \equiv \tilde{g} = (P_h V P_h)^{-1} (P_h V) g. \tag{6.5.36}$$

Observe that for χ in $\tilde{H}$ we have

$$\Gamma_h \chi = \chi. \tag{6.5.37}$$

From Corollary 6.5.3, Theorem 6.5.4, (6.5.20) and (6.5.21) we find the simple inequality

$$\begin{aligned} \|\tilde{g}\|_{-1/2}^2 \leq \nu \langle V\tilde{g}, \tilde{g} \rangle_0 = \nu \langle P_h V\tilde{g}, \tilde{g} \rangle_0 = \nu \langle P_h V g, \tilde{g} \rangle_0 &\leq c'\nu \|Vg\|_{1/2} \|\tilde{g}\|_{-1/2} \\ &\leq c'\nu\beta \|g\|_{-1/2} \|\tilde{g}\|_{-1/2} \end{aligned}$$

yielding the following stability property.

Proposition 6.5.6 *The Galerkin projections Γ_h from $H^{-1/2}$ into $H^{-1/2}$ are uniformly bounded, that is*

$$\|\Gamma_h\|_{-1/2,-1/2} \equiv \sup_{\|g\|_{-1/2} \leq 1} \|\Gamma_h g\|_{-1/2} \leq \nu'. \tag{6.5.38}$$

Of course, (6.5.38) corresponds to (6.5.35). As a consequence we have

Theorem 6.5.7 *Let* $f \in H^r(\dot{\mathrm{G}})$; $1/2 \leqslant s+1 \leqslant r \leqslant m+2$, $s \leqslant m$. *Then the solution* $\tilde{g}$ *of the Galerkin equation* (6.5.18) *converges to* g *in* $H^s(\dot{\mathrm{G}})$ *with optimal rate of convergence*:

$$\|\tilde{g}-g\|_s \leqslant c_1(r, s, \dot{\mathrm{G}}) h^{r-s-1} \|f\|_r. \tag{6.5.39}$$

For the inverses of the Galerkin equations (6.5.21) *there holds the following estimate for their condition numbers as* $h \to 0$:

$$\|(P_h V P_h)^{-1} P_h f\|_0 \leqslant c \cdot h^{-1} \|f\|_0 \quad \textit{for} \quad f \in H^0 \tag{6.5.40}$$

with c *independent of* f *and* h.

Proof The estimate (6.5.39) follows from (6.5.36), (6.5.37), (C), (S), (6.5.20) and (6.5.17):

$$\begin{aligned}\|\tilde{g}-g\|_s &= \|\Gamma_h g - \Gamma_h P_h g + P_h g - g\|_s \leqslant \|\Gamma_h(g-P_h g)\|_s + \|P_h g - g\|_s \\ &\leqslant h^{-1/2-s} M \|\Gamma_h(g-P_h g)\|_{-1/2} + c h^{r-s-1} \|g\|_{r-1} \\ &\leqslant h^{-1/2-s} M \nu' \|g-P_h g\|_{-1/2} + c h^{r-s-1} \|g\|_{r-1} \\ &\leqslant c' h^{r-s-1} \|Vf\|_{r-1} \leqslant c_1 h^{r-s-1} \|f\|_r.\end{aligned}$$

The estimate (6.5.40) follows from (S), (6.5.38) and (6.5.12),

$$\begin{aligned}\|(P_h V P_h)^{-1} P_h f\|_0 &= \|\Gamma_h V^{-1} P_h f\|_0 \\ &\leqslant M h^{-1/2} \|\Gamma_h V^{-1} P_h f\|_{-1/2} \leqslant \nu' M h^{-1/2} \|V^{-1} P_h f\|_{-1/2} \\ &\leqslant \kappa \nu' M h^{-1/2} \|P_h f\|_{1/2} \leqslant c_1 h^{-1} \|P_h f\|_0 \leqslant c_1 h^{-1} \|f\|_0.\end{aligned}$$

Now let us consider the more general equations (5.2.110) including side conditions for $\oint_{\dot{\mathrm{G}}} \boldsymbol{\sigma}\, ds$:

$$V\boldsymbol{\sigma} + \pi \mathbf{k} + \mathrm{K}\boldsymbol{\sigma} = \mathbf{f} \quad \text{on } \dot{\mathrm{G}} \quad \text{and} \quad \oint_{\dot{\mathrm{G}}} \boldsymbol{\sigma}\, ds = \mathbf{A}. \tag{6.5.41}$$

Here $\mathbf{f}(s)$ and the constant $\mathbf{A}$ are given $\{\mathbf{A}=0$ in (5.2.110)$\}$ and the density vector $\boldsymbol{\sigma}(s)$ together with $\mathbf{k}$ are the unknowns. Although (6.5.41) has principal part V and, hence, defines a Fredholm mapping of index zero, in general it might still admit eigensolutions and additional solvability conditions for $\mathbf{f}$ and $\mathbf{A}$. For excluding this merely formal difficulty let us suppose the following additional assumption:

(U) *Any solution of* (6.5.41) *is unique.*

In the following each component σ_j of $\boldsymbol{\sigma}$ will belong to $H^r(\dot{\mathrm{G}})$; therefore we write simply $H^r(\dot{\mathrm{G}})$ instead of $(H^r(\dot{\mathrm{G}}))^n$ and $\tilde{H}$ instead of $\tilde{H}^n$. The corresponding inner products are defined in the natural manner,

$$\langle \mathbf{f}, \mathbf{g}\rangle_0 \equiv \sum_{j=1}^{n} \langle f_j, g_j\rangle_0, \tag{6.5.42}$$

and similarly for the norms and dualities. In addition we denote by $|\mathbf{k}|$ the Euclidian norm of $\mathbf{k}$.

Based on the continuity properties of K for (6.5.41) the same approximation properties hold as for (6.5.1).

Lemma 6.5.8 *Let* $\dot{\mathsf{G}} \in C^{[|r|+t+1]}$, $1 \leq t \leq 3$. *Then* K *maps* $H^r \to H^{r+t}$ *continuously.*

This proposition follows straightforwardly from the continuity properties of H in (5.2.103) and of $\partial/\partial n_t \log|t-z|\,|_{\dot{\mathsf{G}}}$ which can be found in [18] and [10, Appendix].

Theorem 6.5.9 *Suppose* (U) *and let* $\dot{\mathsf{G}} \in C^{[|r-1|+3]} \cap C^{\{|r|+1\}}$. *Then there exists a constant* $\beta > 0$ *independent of* $\boldsymbol{\sigma}$ *and* $\mathbf{k}$ *such that*

$$\beta^{-1}\{\|\boldsymbol{\sigma}\|_{r-1} + |\mathbf{k}|\} \leq \{\|V\boldsymbol{\sigma} + \mathsf{K}\boldsymbol{\sigma} + \pi\mathbf{k}\|_r + |\oint_{\dot{\mathsf{G}}} \boldsymbol{\sigma}\, \mathrm{d}s|\} \leq \beta\{\|\boldsymbol{\sigma}\|_{r-1} + |\mathbf{k}|\}. \tag{6.5.43}$$

Proof Indeed, let us write (6.5.41) after multiplication by V^{-1} in the form

$$\mathsf{T}(\boldsymbol{\sigma}, \mathbf{k}) \equiv \begin{cases} (\mathsf{I} + V^{-1}\mathsf{K})\boldsymbol{\sigma} + \pi(V^{-1}1)\mathbf{k} = V^{-1}\mathbf{f}, \\ \oint \boldsymbol{\sigma}\, \mathrm{d}s \qquad\qquad\qquad\qquad = \mathbf{A}. \end{cases} \tag{6.5.44}$$

Lemma 6.5.8 and Theorem 6.5.4 yield that the mapping $V^{-1}\mathsf{K}: H^{r-1} \to H^r$ is continuous and by the compactness of the embedding $H^r \to H^{r-1}$ is completely continuous [2, p. 203]. Hence, (6.5.44) forms a system of second kind Fredholm equations for which the classical Fredholm alternative holds [20]. Since we assume uniqueness (U), (6.5.44) defines a bijective linear mapping of $H^{r-1} \times \mathbb{R}^n$ into itself. Hence by Banach's theorem there exists $\gamma > 0$ such that

$$\gamma^{-1}\{\|\boldsymbol{\sigma}\|_{r-1} + |\mathbf{k}|\} \leq \{\|V^{-1}\mathbf{f}\|_{r-1} + |\mathbf{A}|\} \leq \gamma\{\|\boldsymbol{\sigma}\|_{r-1} + |\mathbf{k}|\}. \tag{6.5.45}$$

Using (6.5.17) on both sides and replacing $\mathbf{f}$ and $\mathbf{A}$ by (6.5.41) we find (6.5.43).

The Galerkin equations for (6.5.41) now read

$$\langle \boldsymbol{\chi}, V\tilde{\mathbf{g}} + \mathsf{K}\tilde{\mathbf{g}}\rangle_0 + \pi\langle \boldsymbol{\chi}, 1\rangle_0 \tilde{\mathbf{k}} = \langle \boldsymbol{\chi}, \mathbf{f}\rangle_0 \quad \textit{for all } \boldsymbol{\chi} \in \tilde{H}, \quad \langle \tilde{\mathbf{g}}, 1\rangle_0 = \mathbf{A} \tag{6.5.46}$$

or

$$P_h V P_h \tilde{\mathbf{g}} + P_h \mathsf{K} P_h \tilde{\mathbf{g}} + \pi(P_h 1)\tilde{\mathbf{k}} = P_h \mathbf{f} = P_h V\{\boldsymbol{\sigma} + V^{-1}\mathsf{K}\boldsymbol{\sigma} + \pi(V^{-1}1)\mathbf{k}\},$$

$$\oint_{\dot{\mathsf{G}}} \tilde{\mathbf{g}}\, ds = \mathbf{A} = \oint_{\dot{\mathsf{G}}} \boldsymbol{\sigma}\, ds. \quad (6.5.47)$$

Again the *Galerkin projection* can be introduced for (6.5.47), mapping the solution $(\boldsymbol{\sigma}, \mathbf{k})$ onto the solution of the Galerkin equations:

$$\underset{\sim}{\Gamma}_h : (\boldsymbol{\sigma}, \mathbf{k}) \mapsto (\tilde{\mathbf{g}}, \tilde{\mathbf{k}}). \quad (6.5.48)$$

Lemma 6.5.10 The Galerkin projections $\underset{\sim}{\Gamma}_h : H^{-1/2} \to H^{-1/2}$ are uniformly bounded:

$$\|\underset{\sim}{\Gamma}_h(\boldsymbol{\sigma}, \mathbf{k})\|_{-1/2} = \|\tilde{\mathbf{g}}\|_{-1/2} + |\tilde{\mathbf{k}}| \leq \nu'\{\|\boldsymbol{\sigma}\|_{-1/2} + |\mathbf{k}|\}. \quad (6.5.49)$$

Proof Multiplying (6.5.47) by $(P_h V P_h)^{-1}$ we find with (6.5.36)

$$\mathsf{T}_h(\tilde{\mathbf{g}}, \tilde{\mathbf{k}}) \equiv \begin{cases} \{\mathsf{I} + \Gamma_h V^{-1} P_h \mathsf{K} P_h\}\tilde{\mathbf{g}} + \pi(\Gamma_h V^{-1}1)\tilde{\mathbf{k}} \\ \qquad = \Gamma_h[\{\mathsf{I} + V^{-1}\mathsf{K}\}\boldsymbol{\sigma} + \pi(V^{-1}1)\mathbf{k}], \\ \oint_{\dot{\mathsf{G}}} \tilde{\mathbf{g}}\, ds = \oint_{\dot{\mathsf{G}}} \boldsymbol{\sigma}\, ds. \end{cases} \quad (6.5.50)$$

Since $\mathsf{K} : H^{-1/2} \to H^{1/2}$ is a *compact* mapping, $P_h \to \mathsf{I}$ in $H^{-1/2}$ and $H^{1/2}$ due to (6.5.20) and $\Gamma_h \to \mathsf{I}$ due to (6.5.38), elementwise, we find

$$\|\mathsf{T}_h - \mathsf{T}\|_{-1/2} \to 0 \quad \text{as} \quad h \to 0 \quad (6.5.51)$$

in the corresponding operator norm [4, Hilfssatz 30)], [1, p. 8]. Since T^{-1} exists, (6.5.51) yields also the existence of T_h^{-1} and the uniform boundedness

$$\|\mathsf{T}_h^{-1}\|_{-1/2,-1/2} \leq C \quad \text{for all } h \leq h_0 \quad (6.5.52)$$

with $h_0 > 0$ and C independent of h [1, Proposition 1.4 p. 3]. Hence, multiplying (6.5.50) by T_h^{-1} we find

$$(\tilde{\mathbf{g}}, \tilde{\mathbf{k}}) = \mathsf{T}_h^{-1}(\Gamma_h\{\{\mathsf{I} + V^{-1}\mathsf{K}\}\boldsymbol{\sigma} + \pi(V^{-1}1)\mathbf{k}\}, \oint_{\dot{\mathsf{G}}} \boldsymbol{\sigma}\, ds) = \underset{\sim}{\Gamma}_h(\boldsymbol{\sigma}, \mathbf{k})$$

which leads with (6.5.52) and (6.5.38) to the desired estimate (6.5.49).

From Lemma 6.5.10 one obtains the following theorem in exactly the same way as in the proof of Theorem 6.5.7.

Theorem 6.5.11 Let $\mathbf{f} \in H^r(\dot{\mathsf{G}})$, $\dot{\mathsf{G}} \in C^{[|r-1|+3]} \cap C^{\{|r|+1\}}$, $1/2 \leq s+1 \leq r \leq m+2$. Then the solutions $\tilde{\mathbf{g}}, \tilde{\mathbf{k}}$ of the Galerkin equations (6.5.46) *converge* *to* $\boldsymbol{\sigma}, \mathbf{k}$ *with optimal rate of convergence:*

$$\begin{aligned} \|\tilde{\mathbf{g}}-\boldsymbol{\sigma}\|_s &\leq c_1' h^{r-s-1}\{\|\mathbf{f}\|_r+|\mathbf{A}|\}, \\ |\tilde{\mathbf{k}}-\mathbf{k}| &\leq c'' h^r\{\|\mathbf{f}\|_r+|\mathbf{A}|\}. \end{aligned} \tag{6.5.53}$$

We omit the proof.

Corollary 6.5.12 *For* $2\leq r\leq m+1$ *and the same assumptions as in Theorem* 6.5.11, $\tilde{\mathbf{g}}$ *converge uniformly:*

$$\max_{\dot{G}} |\tilde{\mathbf{g}}(s)-\boldsymbol{\sigma}(s)|\leq c_2' h^{r-(3/2)}(\|\mathbf{f}\|_r+|\mathbf{A}|). \tag{6.5.54}$$

Proof This follows from the Schwarz inequality,

$$\begin{aligned} |\mathbf{g}(s)|^2 &\leq 2\left|\int_{s_0}^{s} |\mathbf{g}'\mathbf{g}(\sigma)|\,\mathrm{d}\sigma\right|+|\mathbf{g}(s_0)|^2 \\ &\leq h^{-1}\|\mathbf{g}\|_0^2+h\|\mathbf{g}\|_1^2+L^{-1}\oint|\mathbf{g}(s)|^2\,\mathrm{d}s \\ &\leq (1+L^{-1})\{h^{-\frac{1}{2}}\|\mathbf{g}\|_0+h^{\frac{1}{2}}\|\mathbf{g}\|_1\}^2 \end{aligned} \tag{6.5.55}$$

where s_0 denotes the minimal point of the continuous function $|\mathbf{g}(s)|^2$. Inserting $\mathbf{g}=\tilde{\mathbf{g}}-\boldsymbol{\sigma}$ and (6.5.53) (with $s=0$ and with $s=1$) we find (6.5.54).

Now let us consider an integral equation method for the standard problem

$$\begin{aligned} &w_{\bar{z}}=\bar{a}w+b\bar{w}+c \quad \text{in } \mathrm{G}, \\ &\operatorname{Re} w=\psi \text{ on } \dot{\mathrm{G}} \quad \text{and} \quad \oint_{\dot{\mathrm{G}}} \operatorname{Im} w\sigma\,\mathrm{d}s=\kappa. \end{aligned} \tag{6.5.56}$$

It is based on:

Lemma 6.5.13 *Every solution* $w\in C^{1+\alpha}(\bar{\mathrm{G}})$ *of* (6.5.56) *admits a representation*

$$w(\zeta)=\iint_{\mathrm{G}} p(z)\log|z-\zeta|[\mathrm{d}x,\mathrm{d}y]+\oint_{\dot{\mathrm{G}}} q(s)\log|z(s)-\zeta|\,\mathrm{d}s, \tag{6.5.57}$$

where the layers $p=p_1+\mathrm{i}p_2\in C^\alpha(\bar{\mathrm{G}})$ *and* $q=q_1+\mathrm{i}q_2\in C^\alpha(\dot{\mathrm{G}})$ *are uniquely determined.*

Proof Differentiation of (6.5.56) with $w\in C^{1+\alpha}$ implies $w_{\bar{z}z}\in C^\alpha(\bar{\mathrm{G}})$

and, hence, the Green theorem provides the representation

$$w(\zeta)=\frac{2}{\pi}\iint\limits_{G} w_{\bar{z}z}\log|\zeta-z|[dx, dy]-\frac{1}{2\pi}\oint\limits_{\dot{G}}\log|\zeta-z|\frac{\partial w}{\partial n_z}ds_z$$

$$+\frac{1}{2\pi}\oint_{\dot{G}} w\frac{\partial}{\partial n_z}\log|\zeta-z|\,ds_z. \qquad (6.5.58)$$

Comparing (6.5.58) and (6.5.57) we find $p=(2/\pi)w_{\bar{z}z}$ and

$$\phi_j(\xi)=-\oint_{\dot{G}} q_j\log|z(s)-\xi|\,ds=Vq_j \quad \text{for } \xi\in\dot{G} \qquad (6.5.59)$$

where ϕ_j denote the boundary values of the two harmonic boundary layer integrals in (6.5.58). As we have already seen in this section, (6.5.59) can be solved uniquely. Since $w\in C^{1+\alpha}(\dot{G})$, one finds $\phi\in C^{1+\alpha}(\dot{G})$ and hence $q\in C^{\alpha}(\dot{G})$ following the arguments of Muskelishvili [16, p. 180 ff.].

For convenience, let us use the following abbreviations for the potentials:

$$Pp\equiv\iint\limits_{G} p(\tilde{z})\log|z-\tilde{z}|[d\tilde{x}, d\tilde{y}],$$

$$Qq\equiv\oint_{\dot{G}} q(\tilde{z})\log|z-\tilde{z}|\,ds_{\tilde{z}},$$

$$Kp\equiv(\partial/\partial z)Pp=(1/2)\iint\limits_{G} p(\tilde{z})(z-\tilde{z})^{-1}[d\tilde{x}, d\tilde{y}], \qquad (6.5.60)$$

$$Sq\equiv(\partial/\partial z)Qq=(1/2)\oint_{\dot{G}} q(\tilde{z})(z-\tilde{z})^{-1}\,ds_{\tilde{z}}.$$

If the potential (6.5.57) solves the boundary value problem (6.5.56), then it will satisfy the following system of integral equations:

$$\begin{aligned} &Kp+Sq-\bar{a}(Pp+Qp)-b(P\bar{p}+Q\bar{q})=c \quad \text{in } G,\\ &Pp_1+Qq_1=\psi \quad \text{on } \dot{G},\\ &-\oint\limits_{\dot{G}} p_2\tilde{\sigma}\,ds-\iint\limits_{G} q_2\tilde{\sigma}[dx, dy]=\kappa \end{aligned} \qquad (6.5.61)$$

where $\tilde{\sigma}$ is defined by the potential

$$\tilde{\sigma}(z)=-P\sigma. \qquad (6.5.62)$$

$\tilde{\sigma}$ is positive in $\bar{G}$ since $\log|z-\tilde{z}|<0$ for all $(z,\tilde{z})\in\bar{G}\times\bar{G}$. If we differentiate the first equation in (6.5.61) with respect to z then it becomes

$$(\pi/2)p=\bar{a}_z(Pp+Qq)+b_z(P\bar{p}+Q\bar{q})$$
$$+\bar{a}(\overline{K\bar{p}}+\overline{S\bar{q}})+b(\overline{Kp}+\overline{Sq})+c_z. \tag{6.5.63}$$

This is an equation of the second kind and in the special case $\bar{a}=b=0$, it determines p uniquely. Then q_1 is determined by the second equation of (6.5.61) but q_2 is not. Therefore, we need a new equation for q_2. It can be found from (6.5.63), if z approaches boundary points. Using (6.5.61) on the boundary and the special combination

$$w_z\dot{z}+w_{\bar{z}}\dot{\bar{z}}=2\,\mathrm{Re}\,w_{\bar{z}}\dot{\bar{z}}$$

where $z\to\dot{G}$ and $\dot{z}$ approaches the tangent at the boundary point, we obtain with the jump conditions for S, the equation:

$$\pi q_2-\oint_{\dot{G}}\{q_2(\mathrm{d}/\mathrm{d}n_z)\log|z-\tilde{z}|\}\,\mathrm{d}s_{\tilde{z}}$$
$$=\iint_G p_2\{(\mathrm{d}/\mathrm{d}n_z)\log|z-\tilde{z}|\}[\mathrm{d}\tilde{x},\mathrm{d}\tilde{y}]+\dot{\psi}$$
$$-(\bar{a}\dot{\bar{z}}+\bar{b}\dot{z})(Pp+Qq)-(a\dot{z}+b\dot{\bar{z}})(P\bar{p}+Q\bar{q})-(c\dot{\bar{z}}+\bar{c}\dot{z}). \tag{6.5.64}$$

The left hand side defines on the boundary the 'Neumann integral operator', which has one eigensolution and is therefore not invertible. This can be overcome by the elimination of this eigenvalue through the use of a degenerate kernel. We use the last equation of (6.5.61) and replace the equation (6.5.64) by

$$\pi Nq_2\equiv\pi\Big(q_2(z)-\frac{1}{\pi}\oint_{\dot{G}}q_2(\tilde{z})\{((\mathrm{d}/\mathrm{d}n_z)\log|z-\tilde{z}|)-\tilde{\sigma}(\tilde{z})\}\,\mathrm{d}s_{\tilde{z}}\Big)$$
$$=-\kappa-\iint_G p_2\tilde{\sigma}[\mathrm{d}\tilde{x},\mathrm{d}\tilde{y}]+\iint_G p_2(\mathrm{d}/\mathrm{d}n_z)\log|z-\tilde{z}|\,[\mathrm{d}\tilde{x},\mathrm{d}\tilde{y}]+\dot{\psi}$$
$$-(\bar{a}\dot{\bar{z}}+\bar{b}\dot{z})(Pp+Qq)-(a\dot{z}+b\dot{\bar{z}})(P\bar{p}+Q\bar{q})-(c\dot{\bar{z}}+\bar{c}\dot{z}). \tag{6.5.65}$$

In [24, Lemma 4] it is shown that N possesses a continuous inverse which in fact can be found by the Neumann series.

Collecting the above computations we find the following system of integral equations for the layers p, q:

$$p(z)-\frac{1}{\pi}(\bar{a}(z)+b(z))\oint_{\dot{G}}\frac{q_1(\tilde{z})\,ds_{\tilde{z}}}{(z-\tilde{z})}$$

$$=\frac{2}{\pi}\Bigg\{\bar{a}_z\Bigg(\iint_G p\log|z-\tilde{z}|\,[d\tilde{x}, d\tilde{y}]+\oint_{\dot{G}} q\log|z-\tilde{z}|\,ds_{\tilde{z}}\Bigg)$$

$$+b_z\Bigg(\iint_G \bar{p}\log|z-\tilde{z}|\,[d\tilde{x}, d\tilde{y}]+\oint_{\dot{G}} \bar{q}\log|z-\tilde{z}|\,ds_{\tilde{z}}\Bigg)$$

$$+\frac{1}{2}\bar{a}\Bigg(\iint_G \frac{p}{(z-\tilde{z})}[d\tilde{x}, d\tilde{y}]+i\oint_{\dot{G}}\frac{q_2}{(z-\tilde{z})}ds_{\tilde{z}}\Bigg)$$

$$+\frac{1}{2}b\Bigg(\iint_G \frac{\bar{p}}{(z-\tilde{z})}[d\tilde{x}, d\tilde{y}]-i\oint_{\dot{G}}\frac{q_2}{(z-\tilde{z})}ds_{\tilde{z}}\Bigg)\Bigg\}+\frac{2}{\pi}c_z$$

$$\equiv T_1(p, q)+d \quad \text{in } G, \tag{6.5.66}$$

$$-\oint_{\dot{G}} q_1(\tilde{z})\log|z-\tilde{z}|\,ds_{\tilde{z}} = Vq_1 = \iint_G p_1\log|z-\tilde{z}|\,[d\tilde{x}, d\tilde{y}]-\psi \text{ on } \dot{G}$$

$$\equiv VT_2(p_1)-\psi, \tag{6.5.67}$$

$$q_2=\frac{1}{\pi}\oint_{\dot{G}} q_2(\tilde{z})\Big\{\frac{\partial}{\partial n_z}\log|z-\tilde{z}|-\tilde{\sigma}(\tilde{z})\Big\}ds_{\tilde{z}}$$

$$+\frac{1}{\pi}\iint_G p_2\Big\{\frac{\partial}{\partial n_z}\log|z-\tilde{z}|-\tilde{\sigma}(\tilde{z})\Big\}[d\tilde{x}, d\tilde{y}]$$

$$-2\mathrm{Re}\,\Big\{\frac{1}{\pi}(\bar{a}\dot{\bar{z}}+\bar{b}\dot{z})\iint_G p(\tilde{z})\log|z-\tilde{z}|\,[d\tilde{x}, d\tilde{y}]+\oint_{\dot{G}} q\log|z-\tilde{z}|\,ds_{\tilde{z}})\Big\}$$

$$+\frac{1}{\pi}(-\kappa-(c\dot{\bar{z}}+\bar{c}\dot{z})+\dot{\psi})$$

$$\equiv T_3(p, q)+\phi. \tag{6.5.68}$$

From [24] we have the following two theorems which we shall quote here without proofs:

Theorem 6.5.14 *Every solution (p, q) of the above system of integral* *equations* (6.5.66–68) *generates by* (6.5.57) *the solution of the boundary value problem* (6.5.56).

Theorem 6.5.15 *The mapping* $(p, q) \mapsto (d, \psi, \phi)$ *defined by* (6.5.66–68) *maps* $C^{\alpha}(\bar{G}) \times C^{\alpha}(\dot{G})$ *onto* $C^{\alpha}(\bar{G}) \times C^{1+\alpha}(\dot{G}) \times C^{\alpha}(\dot{G})$ *bijectively and continuously.*

In [24] one can also find a successive approximation for (6.5.66–68) which converges for small a and b and which is also related to Section 6.4. Moreover, there it is pointed out how to use this method for solving the semilinear problems (1.4.1) with $\tau \equiv 0$. Some numerical computations with collocation methods are also presented.

In the following let us solve (6.5.66–68) by using the Galerkin method again. To this end we need to approximate the boundary layers q and also the surface layer p. Hence, for the latter we need *two-dimensional* finite element functions $\chi \in \tilde{H}(\bar{G})$. Here $\tilde{H}(\bar{G})$ denotes a regular finite element space having also the

convergence property:

(C) *For* $k \leq r$ *with* $-m-1 \leq k \leq m$, $-m \leq r \leq m+1$ *and for any* $v \in H^r(G)$ *there exists an* $\tilde{v} \in \tilde{H}(\bar{G})$ *such that*

$$\|v - \tilde{v}\|_{H^k(G)} \leq c_{rk} h^{r-k} \|v\|_{H^r(G)}$$

holds where $\tilde{v}$ *is independent of* k *and* c_{rk} *is independent of* v, $\tilde{v}$ *and* h [3, p. 659];

and the stability property:

(S) *For* $k \leq r$ *with* $|k|, |r| \leq m$ *there exists* $M_{rk} > 0$ *independent of* h *and of* $\chi \in \tilde{H}(\bar{G})$ *such that*

$$\|\chi\|_{H^r(G)} \leq M_{rk} h^{k-r} \|\chi\|_{H^k(G)} \quad \textit{for all } \chi \in \tilde{H}(\bar{G}).$$

Here $H^r(G)$ denote the Sobolev spaces and corresponding interpolation spaces on G [13, p. 34 ff.]. The construction of such finite element spaces is by no means trivial; we shall come back to this point in Chapter 8. For the Galerkin equations let us denote the scalar product $(\,,\,)_0$ by

$$(v, w)_0 = \operatorname{Re} \iint_G v\bar{w}[dx, dy], \tag{6.5.69}$$

considering $L^2(G)$ now as a *Hilbert space over the field* $\mathbb{R}$.

Similar to the treatment of (6.5.41) we investigate first the principal part of (6.5.66–68) for which we show:

Theorem 6.5.16 *There exist two constants* $\gamma_1, \gamma_2 > 0$ *independent of* p

and q such that

$$(p,p)_0-\frac{1}{\pi}\left((\bar a+b)\oint_{\dot{G}}\frac{q_1(\tilde z)}{(\cdot-\tilde z)}\,ds_{\tilde z},p\right)_0+\gamma_1\langle Vq_1,q_1\rangle_0+\langle q_2,q_2\rangle_0$$

$$\geqslant \gamma_2\{\|p\|^2_{L^2(G)}+\|q_1\|^2_{H^{-1/2}(\dot G)}+\|q_2\|_{L^2(\dot G)}\} \quad (6.5.70)$$

holds for all $p\in L^2(G)$, $q_1\in H^{-1/2}(\dot G)$, $q_2\in L^2(\dot G)$.

Proof Since the Theorems 6.5.1 and 6.5.2 are available it remains only to show that

$$\left|\frac{1}{\pi}\left((\bar a+b)\oint_{\dot G}\frac{q_1(\tilde z)\,ds_{\tilde z}}{(\cdot-\tilde z)},p\right)_0\right|\leqslant \text{const}\,\|q_1\|_{H^{-1/2}(\dot G)}\|p\|_{L^2(G)} \quad (6.5.71)$$

holds where p and q_1 can even be chosen to be C^∞ functions. Therefore the order of integration can be changed and we find that

$$p\mapsto\frac{1}{\pi}\iint_G\frac{(\bar a(z)+b(z))\bar p(z)}{(z-\tilde z)}[dx,dy]$$

defines a continuous mapping $L^2(G)\to H^1(G)$ [22, I Section 9.2] and the trace Theorem can be applied [13, Theorem 9.4] yielding a continuous mapping $L^2(G)\to H^{1/2}(\dot G)$. Thus, the bilinear form

$$\text{Re}\,\frac{1}{\pi}\iint_G(\bar a+b)\oint_{\dot G}\frac{q_1(\tilde z)\,ds_{\tilde z}}{(z-\tilde z)}\bar p(z)[dx,dy]$$

$$=\left\langle q_1,\frac{1}{\pi}\iint_G\frac{(\bar a(z)+b(z))\bar p(z)[dx,dy]}{(z-\cdot)}\right\rangle_0$$

is continuous for $q_1\in H^{-1/2}(\dot G)$ and $p\in L^2(G)$ and (6.5.71) holds. Choosing $\gamma_1>0$ with $\gamma_1\gamma\kappa^{-2}\geqslant$const., the desired inequality (6.5.70) is now a simple consequence of the triangle inequality.

Using the properties of the two-dimensional logarithmic and Newtonian potentials obtained by Vekua [22, I Section 9.2], the trace Theorem [13, Theorem 9.4] and compact embeddings of the Sobolev spaces, for the remaining operators in (6.5.66–68) the following properties can easily be shown:

Proposition 6.5.17 *The following mappings T_j defined in* (6.5.66–68) *are compact from* $p=p_1+ip_2\in L^2(G)$, $q=q_1+iq_2$ with $q_1\in H^{-1/2}(\dot G)$, $q_2\in L^2(\dot G)$ *into the indicated spaces*

$$T_1(p,q)\in L^2(G),\qquad T_2(p)\in H^{-1/2}(\dot G),\qquad T_3(p,q)\in L^2(\dot G).$$

be holomorphic in G and $\psi\in C^{k+\beta}(\bar G)$ then the Fourier expansion of ψ with respect to z^j and the L_2 scalar product $(.,.)_0$ on $\dot G$ converges as follows:

Now we are in the position to formulate the Galerkin equations and to prove the uniform boundedness of the corresponding Galerkin projections for (6.5.66–68). The Galerkin equations for $\tilde{p} \in \tilde{H}(\bar{\mathsf{G}})$ and $\tilde{q} \in \tilde{H}(\dot{\mathsf{G}})$ read

$$\left(\tilde{p} - \frac{1}{\pi}(\bar{a}+b)\oint_{\dot{\mathsf{G}}} \frac{\tilde{q}_1(\tilde{z})\,\mathrm{d}s_{\tilde{z}}}{(\cdot - \tilde{z})}, \tilde{\phi}\right)_0 = (T_1(\tilde{p}, \tilde{q}), \tilde{\phi})_0 + \frac{2}{\pi}(c_z, \tilde{\phi})_0 \qquad \text{for all } \tilde{\phi} \in \tilde{H}(\bar{\mathsf{G}}),$$

$$\langle V\tilde{q}_1, \tilde{\chi}\rangle_0 = \langle VT_2(\tilde{p}_1), \tilde{\chi}\rangle_0 - \langle \psi, \tilde{\chi}\rangle_0 \tag{6.5.72}$$

and

$$\langle \tilde{q}_2, \tilde{\chi}\rangle_0 = \langle T_3(\tilde{p}, \tilde{q}), \tilde{\chi}\rangle_0 + \langle -\kappa - (c\dot{\bar{z}} + \bar{c}\dot{z}) + \psi, \chi\rangle_0 \qquad \text{for all } \tilde{\chi} \in \tilde{H}(\dot{\mathsf{G}}).$$

Since the original equations (6.5.66–68) are uniquely solvable so are the Galerkin equations for $h \leq h_0$ where $h_0 > 0$ is a suitable constant. Introducing the orthogonal projection $\tilde{P}_h$ of $L^2(G)$ onto $\tilde{H}(\bar{\mathsf{G}})$, that is

$$\tilde{P}_h : L^2(\mathsf{G}) \to \tilde{H}(\bar{\mathsf{G}}) \subset L^2(\mathsf{G}) \tag{6.5.73}$$

the equations (6.5.72) are equivalent to

$$\begin{aligned} &\tilde{p} - \tilde{P}_h\left\{\frac{1}{\pi}(\bar{a}+b)\oint_{\dot{\mathsf{G}}} \frac{P_h\tilde{q}_1\,\mathrm{d}s_{\tilde{z}}}{(\cdot - \tilde{z}(s))}\right\} = \tilde{P}_h T_1(\tilde{P}_h\tilde{p}, P_h\tilde{q}) + \frac{2}{\pi}\tilde{P}_h c_z \\ &\tilde{q}_1 = \Gamma_h T_2(\tilde{P}_h\tilde{p}_1) - \Gamma_h V^{-1}\psi, \\ &\tilde{q}_2 = P_h T_3(\tilde{P}_h\tilde{p}, P_h\tilde{q}) + P_h\{-\kappa - (c\dot{\bar{z}} + \bar{c}\dot{z}) + \dot{\psi}\} \end{aligned} \tag{6.5.74}$$

in $L^2(\mathsf{G}) \times H^{-1/2}(\dot{\mathsf{G}}) \times L^2(\dot{\mathsf{G}})$ where the right-hand side operators converge with respect to the corresponding operator norms,

$$\begin{aligned} &\|\tilde{P}_h T_1(\tilde{P}_h\,\cdot, P_h\cdot) - T_1(\cdot,\cdot)\| \to 0, \\ &\|\Gamma_h T_2(\tilde{P}_h\cdot) - T_2(\cdot)\| \to 0, \\ &\|P_h T_3(\tilde{P}_h\cdot, P_h\cdot) - T_3(\cdot,\cdot)\| \to 0 \quad \text{as} \quad h \to 0 \end{aligned}$$

since all the limits T_j are known as to be compact. The left-hand sides in (6.5.74) are essentially the identities except the first term, and are definite having uniformly bounded inverses due to (6.5.70). Hence, using the same arguments as in the proof of Lemma 6.5.10 we find for the Galerkin projection to (6.5.74), mapping the unique solution (p, q_1, q_2) of (6.5.66–68) onto $(\tilde{p}, \tilde{q}_1, \tilde{q}_2)$, that of (6.5.74),

$$\mathbf{G}(p, q_1, q_2) \equiv (\tilde{p}, \tilde{q}_1, \tilde{q}_2), \tag{6.5.75}$$

the following lemma.

Lemma 6.5.18 *The Galerkin projections*

$$\mathbf{G} : L^2(\mathsf{G}) \times H^{-1/2}(\dot{\mathsf{G}}) \times L^2(\dot{\mathsf{G}}) \to L^2(\mathsf{G}) \times H^{-1/2}(\dot{\mathsf{G}}) \times L^2(\dot{\mathsf{G}})$$

With this inner product it can be shown that $L_2(\dot{\mathsf{G}})$ is a Hilbert space.

Now let $\mathsf{L} \equiv \{F = F(z)$ holomorphic in G with $F_{\dot{\mathsf{G}}} \in L_2(\dot{\mathsf{G}})$ and $F(0) = 1\}$.

are uniformly bounded, i.e.

$$\|\tilde{p}\|_{L^2(\mathrm{G})}+\|\tilde{g}_1\|_{H^{-1/2}(\dot{\mathrm{G}})}+\|\tilde{q}_2\|_{L^2(\dot{\mathrm{G}})} \leq \nu''\{\|p\|_{L^2(\mathrm{G})}+\|q_1\|_{H^{-1/2}(\dot{\mathrm{G}})}+\|q_2\|_{L^2(\dot{\mathrm{G}})}\}. \quad (6.5.76)$$

Since the proof is very similar to that of Lemma 6.5.10 we leave it to the reader.

Using $\tilde{p}$ and $\tilde{q}$ in (6.5.57) we find the desired approximation $\tilde{w}$ for the solution of the original boundary value problem. As a consequence of Lemma 6.5.18, we have then the following theorem.

Theorem 6.5.19 *Let the solution w be in $H^{r+2}(\mathrm{G})$,*

$$\dot{\mathrm{G}}\in C^{[|r+1/2|+3]}\cap C^{\{|r+3/2|+1\}}, -1\leq s\leq r\leq m+1/2.$$

Then the Galerkin approximations $\tilde{w}$ converge to w with optimal rate of convergence:

$$\|w-\tilde{w}\|_{H^{s+2}(\mathrm{G})}\leq ch^{r-s}\|w\|_{H^{r+2}(\mathrm{G})}. \quad (6.5.77)$$

Proof Since the proof is very similar to that of Theorem 6.5.7 we shall indicate here only the additional ideas. Let (p, q_1, q_2) be the layers representing the given solution w by (6.5.57). Then $p\in H^r(\mathrm{G})$, $q_1, q_2\in H^{r+1/2}(\dot{\mathrm{G}})$ according to the trace Theorem and the properties of the logarithmic potentials. Now let us decompose w into

$$w=w_1+w_2+w_3 \quad (6.5.78)$$

with

$$w_1\equiv\iint_{\mathrm{G}} p\log|z-\zeta|\,[\mathrm{d}x, \mathrm{d}y],$$

$$w_2\equiv\oint_{\dot{\mathrm{G}}} q_1\log|z-\zeta|\,\mathrm{d}s_z,$$

$$w_3\equiv\oint_{\dot{\mathrm{G}}} q_2\log|z-\zeta|\,\mathrm{d}s_z.$$

Since $\mathbf{G}$ is a linear operator we can introduce approximations corresponding to the above decomposition,

$$\tilde{w}_1=\iint_{\mathrm{G}} \tilde{p}\log|z-\zeta|\,[\mathrm{d}x, \mathrm{d}y] \quad \text{with} \quad \tilde{p}=\mathbf{G}(p, 0, 0)$$

$$\tilde{w}_2=\oint_{\dot{\mathrm{G}}} \tilde{q}_1\log|z-\zeta|\,\mathrm{d}s_z \quad \text{with} \quad \tilde{q}_1=\mathbf{G}(0, q_1, 0),$$

$$\tilde{w}_3=\oint_{\dot{\mathrm{G}}} \tilde{q}_2\log|z-\zeta|\,\mathrm{d}s_z \quad \text{with} \quad \tilde{q}_2=\mathbf{G}(0, 0, q_2),$$

finding

$$\tilde{w} = \tilde{w}_1 + \tilde{w}_2 + \tilde{w}_3.$$

Estimating each term separately we find

$$\begin{aligned}\|w_1 - \tilde{w}_1\|_{H^{s+2}(G)} &\leq \|p - \mathbf{G}(p, 0, 0)\|_{H^s(G)} \\ &\leq \|p - \tilde{P}_h p\|_{H^s(G)} + Mh^{-s}\|\mathbf{G}(\tilde{P}_h p - p, 0, 0)\|_{L^2(G)} \\ &\leq ch^{r-s}\|p\|_{H^r(G)} + Mh^{-s}\|\tilde{P}_h p - p\|_{L^2(G)} \\ &\leq c'h^{r-s}\|p\|_{H^r(G)} \leq c''h^{r-s}\|w_1\|_{H^{r+2}(G)}\end{aligned}$$

for the first term and corresponding inequalities for the remaining two terms. The triangle inequality yields (6.5.77).

6.6 Remarks on the Additional References

General questions of the approximate solution of integral equations of the second kind are investigated and collected in [25–27]. The convergence of the approximation of the conformal mapping can be found in [52]. The integral equations of the second kind in connection with acoustics are treated in [45], with conformal mapping in [60], and for electromagnetic theory and two-dimensional fluid flow problems, in [51]. The approximation of corresponding eigenvalue problems can be found in [41]. General methods for the treatment of Fredholm integral equations of the first kind are investigated in [42, 46] and [53]. The equation of the first kind on $\dot{G}$ with logarithmic kernel was investigated in [29–31], [36], [37], [39], [57] and [58]. In [33] and [59] one finds this equation in connection with conformal mapping. Applications of equations of the first kind to problems in mechanics are given in [28], [34], [38], [40]. In [28] they are used for solving a mixed boundary value problem. In [34] one also finds many generalizations to other Fredholm equations of the first kind and applications. In [50] equations of the first kind on an interval are studied. The solvability theory in Sobolev spaces and finite element methods are presented in [47–49]. The corresponding integral equations in $\mathbb{R}^3$ are investigated in [32], [54–56] where also applications and computations can be found. For Cauchy Riemann equations the integral equation methods can often be replaced by methods which provide the computation of the harmonic conjugate for which one finds suitable approximations in [43, 44].

References

1 Anselone, P. M., *Collectively compact operator approximation theory.* Prentice-Hall Int. Inc., London, 1971.

2 Aubin, J. P., *Approximation of elliptic boundary-value problems.* Wiley-Interscience, New York, 1972.

3 Bramble, J. and Schatz, A., Rayleigh–Ritz–Galerkin methods for Dirichlet's problem using subspaces without boundary conditions. *Comm. Pure Appl. Math.* **23**, 653–675, 1970.

4 Bruhn, G. and Wendland, W., Über die näherungsweise Lösung von linearen Funktionalgleichungen. *ISNM* **7,** 136–164, 1967, Birkhäuser, Basel, Stuttgart.

5 Burbea, J., A procedure for conformal maps of simply connected domains by using the Bergman function. *Math. Computation* **24,** 821–829, 1970.

6 Dieudonné, G., *Foundations of modern analysis.* Academic Press, New York, 1960.

7 Gaier, D., *Konstruktive Methoden der konformen Abbildung.* Springer, Berlin, 1964.

8 Golusin, G., *Geometrische Funktionentheorie.* Deutscher Verlag d. Wiss., Berlin, 1957.

9 Haack, W. and Wendland, W., *Lectures on pfaffian and partial differential equations.* Pergamon Press, Oxford, New York, 1972.

10 Hsiao, G. C. and Wendland, W., A finite element method for some integral equations of the first kind. *J. Math. Anal. Appl.* **58,** 449–481, 1977.

11 Kantorovich, L. V. and Krylov, V. I., *Approximate methods of higher analysis.* Interscience, New York, 1964.

12 Kleinman, R. E. and Wendland, W. L., On Neumann's method for the exterior Neumann problem for the Helmholtz equation. *J. Math. Anal. Appl.* **57,** 170–202, 1977.

13 Lions, J. L. and Magenes, E., *Non-homogeneous boundary value problems and applications,* Vol. I, Springer, Berlin, 1972.

14 Löhr, K. P. and Wendland, W., Die numerische Lösung der ersten und zweiten Randwertaufgabe unter Verwendung elementarer Greenscher Funktionen. *Elektronische Datenverarbeitung* **4,** 166–172, 1967.

15 Moss, W. F., Approximation of exterior conformal mappings, *Annales Polonici Mathematici* **35,** 1978.

16 Muskhelishvili, N. L., *Singular integral equations.* Noordhoff, Groningen, The Netherlands, 1953.

17 Nitsche, J., Zur Konvergenz von Näherungsverfahren bezüglich verschiedener Normen. *Num. Math.* **15,** 224–228, 1970.

18 Ricci, P. E., Sui potenziali di semplice strato per le equazioni ellittiche di ordine superiore in due variabili. *Rendiconti di Mat.* **7,** Ser. VI, 1–39, 1974.

19 Suetin, P., On the degree of approximation of analytic functions by partial sums of series of orthogonal polynomials. *Izv. Akad. Nauk Armjan. SSR* **15,** 81–86, 1962.

20 Taylor, A. E., *Functional analysis* (2nd Edn), John Wiley, New York, 1958 (1967).

21 Triebel, H., *Höhere Analysis.* Deutscher Verlag d. Wiss., Berlin, 1972.

22 Vekua, I. N., *Generalized analytic functions.* Pergamon Press, Oxford, New York, 1962.

23 Wendland, W., Zur numerischen Behandlung der Randwertaufgaben für

elliptische Systeme. *ISNM* **9,** 187–207, 1968, Birkhäuser Basel · Stuttgart.
24 Wendland, W., An integral equation method for generalized analytic functions. *In* D. Colton and R. P. Gilbert. (Eds.), *Proc. Conf. Constr. Comp. Methods Differential and Integral Equations Lecture Notes* No. 430, 414–452, 1974, Berlin, Springer.

Additional references

25 Atkinson, K. A., *A survey of numerical methods for the solution of Fredholm integral equations of the second kind.* SIAM, Philadelphia, 1976.
26 Ben Noble, Error analysis of collocation methods for solving Fredholm integral equations. *In* John J. H. Miller (Ed.), Topics in numerical analysis. *Proc. Royal Irish Academy Conference Num. Analysis.* Academic Press, London, 1972.
27 Ben Noble, *A bibliography on 'Methods for solving integral equations'*. Math. Research Center, University of Wisconsin, Madison, Wisconsin, MCR Reports 1176 and 1177, 1971.
28 Bischoff, H., Die Berechnung von Potentialfeldern mit der Randintegralmethode dargestellt am Beispiel der ebenen stationären Grundwasserbewegungen. *Dissertation.* T.H., Darmstadt, 1977 (D17).
29 Christiansen, S., Integral equations without a unique solution can be made useful for solving some plane harmonic problems. *J. Inst. Math. Appl.* **16,** 143–159, 1975.
30 Christiansen, S. and Rasmussen, H., Numerical solution for two-dimensional annular electrochemical machining problems. *Report* No. **53,** Technical University of Denmark, 1973.
31 Cyrlin, L. E., On a method of solving integral equations of the first kind in potential theory problems. *Z. Vyčisl. Mat. i Mat. Fiz.* (1967) (Russian) translated in *USSR Comput. Mat. Phys.* **9,** 324–328, 1969.
32 Djaoua, M., *Methode d'éléments finis pour la resolution d'un problème exterieur dans R^3.* Centre de Mathematiques Appliquees, Ecole Polytechnique, Palaiseau, France, Rapport Int. No. 3, 1975.
33 Gaier, D., Integralgleichungen erster Art und konforme Abbildungen. *Math. Zeitschr.* **147,** 113–129, 1976.
34 Giroire, J., *Formulation variationnelle par equations intégrales de problèmes aux limites exterieurs.* Centre de Mathematiques Appliquees, Ecole Polytechnique, Palaiseau, France, Rapport Int. No. 6, 1976.
35 Howland, J. L., The numerical solution of an induced potential problem. *J. Math. Anal. Appl.* **8,** 245–257, 1964.
36 Howland, J. L., Symmetrizing kernels and the integral equations of first kind of classical potential theory. *Proc. Am. Math. Soc.* **19,** 1–7, 1968.
37 Howland, J. L. and Vaillancourt, R., Series expansions of induced potentials. *J. Math. Anal. Appl.* **16,** 385–395, 1966.
38 Hsiao, G. C. and MacCamy, R. C., Solution of boundary value problems by integral equations of the first kind. *SIAM Review* **15,** 687–705, 1973.

39 Jaswon, M. A., Integral equation methods in potential theory I. *Proc. Roy. Soc. London*, Series A, **275,** 23–32, 1963.

40 Jaswon, M. A., Maiti, M. and Symm, G. T., Numerical biharmonic analysis and some applications. *Int. J. Solids Structures* **3,** 309–322, 1967.

41 Jeggle, H. and Wendland, W. L., On the discrete approximation of eigenvalue problems with holomorphic parameter dependence. *Proc. Roy. Soc. Edinburgh*, **78A,** 1–29, 1977.

42 Kammerer, W. J. and Nashed, M. Z., Iterative methods for best approximate solutions of linear integral equations of the first and second kinds. *J. Math. Anal. Appl.* **40,** 547–574, 1972.

43 Knauff W. und Kreß, R., Optimale Approximation linearer Funktionale auf periodischen Funktionen. *Num. Math.* **22,** 187–205, 1974.

44 Kreß, R., Über die numerische Berechnung konjugierter Funktionen. *Computing* **10,** 177–187, 1972.

45 Kussmaul, R. and Werner, P., Fehlerabschätzungen für ein numerisches Verfahren zur Auflösung linearer Integralgleichungen mit schwach singulären Kernen. *Computing* **3,** 22–46, 1968.

46 Landweber, L., An iteration formula for Fredholm integral equations of the first kind. *Am. J. Math.* **LXXIII,** 615–624, 1951.

47 Le Roux, M. N., Résolution numérique du problème du potential dans le plan par une méthode variationelle d'éléments finis. *Thèse,* L'université de Rennes 1974, Serie A, No. d'ordre 347, No. de série 38.

48 Le Roux, M. N., Equations intégrales pur le problème du potential électrique dans le plan. *C. R. Acad. Sci. Paris,* **278,** A541, 1974.

49 Le Roux, M. N., Mèthod d'éléments finis pur la resolution numérique de problèmes exterieurs en dimension deux. *Revue Franc. Automatique Inf. Rech. Opérationnelle,* to appear.

50 MacCamy, R. C., On singular integral equations with logarithmic or Cauchy kernels. *J. Math. Mech.* **7,** 355–376, 1958.

51 Martensen, E., *Potentialtheorie.* Teubner, Stuttgart, 1968.

52 Mergelian, S., On best approximation in a complex domain. *Acta Sci. Math. Szeged* **12,** A 198–212, 1950.

53 Nashed, M. Z., Approximate regularized solutions to improperly posed linear integral and operator equations. *In Proc. Conference on Constructive and Computational Methods for Differential and Integral Equations,* Lecture Notes in Math. No. 430, 289–332, Springer, Berlin 1974.

54 Nedelec, J. C., Curved finite element methods for the solution of singular integral equations on surfaces in R^3. *Comp. Meth. Appl. Mech. Engin.* **8,** 61–80, 1976.

55 Nedelec, J. C., *Computation of eddy currents on a surface in R^3 by finite element methods.* Centre de Mathematiques, Ecole Polytechnique, Palaiseau, France. Rapport Int. No. 10, 1976.

56 Nedelec, J. C. and Planchard, J., Une méthode variationnelle d'éléments finis pour la résolution numerique d'un problème exterieurs dans R^3. *Revue Franc. Automatique, Inf. Rech. Opérationnelle* **R3,** 105–129, 1973.

57 Siddalingaiah, H., A constructive solution of the induced potential problem. *Research Report* 73–108, University of Pittsburg, 1973.

58 Symm, G. T., Integral equation methods in potential theory II. *Proc. Roy. Soc. London*, Ser. A **275,** 33–46, 1963.
59 Symm, G. T., An integral equation method in conformal mapping. *Num. Math.* **9,** 250–258, 1966.
60 Warschawski, S. E., On the solution of the Lichtenstein-Gershgorin integral equation in conformal mapping: I theory. *Nat. Bureau of Standards, Applied Math. Series* **42,** 7–29, 1955.
61 Kopp, P. and Wendland, W., Numerische Behandlung von Integralgleichungen 1. Art bei zähen Strömungen und in der ebenen Elastizitätstheorie, to appear (Fachbereich Math., T. H., Darmstadt, Preprint 411, 1978).

7

Finite Difference Methods

Since the formulation and numerical solution of the finite difference equations replacing the original boundary value problem is much simpler than the solution of integral equations, finite difference methods are very popular for numerical computations. But in contradiction to their simple formulation, the corresponding error analysis is by no means trivial. For second order Dirichlet problems with one unknown function the maximum principle for both the discrete and nondiscrete equations has been the main tool which has been connected with *a priori* estimates by Stummel [27] and Hainer [18] yielding asymptotic convergence for rather general Dirichlet problems. But for other boundary conditions such as the Neumann condition and also for systems of equations the asymptotic error analysis has been established only recently. Here it will be obtained by estimating the differences of corresponding Green and Neumann functions and their discrete counterparts. For difference equations corresponding to variational problems Scott [25], Frehse [15] and Natterer [24] developed such estimates for the Dirichlet problem, and Scott [26] and Höhn [19] for Neumann problems with homogeneous boundary data. But here we shall need inhomogeneous boundary data and we shall define the discrete boundary conditions explicitly and not by the natural conditions for a variational problem. Since the above estimates are available only for strongly elliptic second-order operators with diagonally dominant principal part as the Laplacian and for Dirichlet or Neumann boundary conditions, we shall restrict here the difference approximation to the first boundary value problem (1.1.1) for systems in the Hilbert normal form with point condition,

$$\begin{aligned} &u_x - v_y = Au + Bv + C, \\ &u_y + v_x = \tilde{A}u + \tilde{B}v + \tilde{C} \quad \text{in } \mathsf{G} \quad \text{and} \\ &u = \psi \text{ on } \dot{\mathsf{G}} \quad \text{and } v(p_0) = v_0, \quad p_0 \in \mathsf{G}. \end{aligned} \tag{7.0.1}$$

Of course, this is only a very special case of first-order problems. But the general boundary value problems for generalized analytic functions have been solved in Chapter 1 by solving standard problems of the form (7.0.1) consecutively.

The generalization of our approach to corresponding problems for Pascali systems (3.2.1) can be done without difficulties, whereas more general boundary conditions raise new problems.

Differentiating the first equation in (7.0.1) with respect to x, the second with respect to y and adding, we find an equation with Δu, and similarly by differentiating the second with respect to x and the first with respect to y and subtracting we find another equation for Δv. Hence, we have the second order boundary value problem

$$\begin{aligned}
&\Delta u = (Au + Bv + C)_x + (\tilde{A}u + \tilde{B}v + \tilde{C})_y,\\
&\Delta v = -(Au + Bv + C)_y + (\tilde{A}u + \tilde{B}v + \tilde{C})_x \quad \text{in } \mathsf{G},\\
&u = \psi \quad \text{and}\\
&\mathrm{d}_n v = -\mathrm{d}\psi + (Au + Bv + C)\,\mathrm{d}x + (\tilde{A}u + \tilde{B}v + \tilde{C})\,\mathrm{d}y \text{ on } \dot{\mathsf{G}},\\
&v(p_0) = v_0.
\end{aligned} \tag{7.0.2}$$

For these equations (and some semilinear systems) we shall formulate the difference equations. In order to find error estimates we consider first the case $A = B = \tilde{A} = \tilde{B} \equiv 0$.

Here we present well known difference equations for the Dirichlet and Neumann problems, their corresponding Green and Neumann functions, and show the order of convergence $0(h^2)$ and $0((h \log h)^2)$, respectively, following the lines of Bramble, Hubbard, Thomée and Ciarlet [2–9]. These results together with the behaviour of the original Green and Neumann functions and the fundamental solutions of the discrete and nondiscrete Laplacian allow estimates for the differences between the discrete and nondiscrete Green and Neumann functions on the grids. These estimates are rather useful. Writing the difference equations formally with discrete Green and Neumann functions, they become discretized versions of Fredholm equations of the second kind (6.0.1). The estimates yield convergence of the discretized operators. This and collective compactness implies the solvability of the discrete equations and the convergence of their solutions. This is our main result. If A, B, $\tilde{A}$, $\tilde{B}$ vanish at $\dot{\mathsf{G}}$ then we obtain convergence of order $h^2 |\log h|$ which is optimal due to Höhn's results [19]. But for more general coefficients $A, \ldots, \tilde{B}$ we can only show convergence of order $h^{1/2-\varepsilon}$ (with arbitrary small $\varepsilon > 0$).

In Section 7.1 we follow Bramble and Hubbard [4] introducing matrices of positive type which provide a discrete maximum principle.

In Section 7.2 the discrete Laplacian is introduced in the grid domain and near the boundary according to [14].

In Section 7.3 we use the classical ideas by Courant, Friedrichs and Lewy [11] as in [5] and [9] introducing the discrete Green functions. Referring to Bramble, Hubbard and Thomée [7], the discrete analogue to the Green representation formula is established. Further, the basic estimates for the Green functions and the $0(h^2)$ convergence is shown following Ciarlet [9]. Referring to Bramble and Thomée [8] we present pointwise estimates for the Green function.

Section 7.4 is devoted to the pointwise and L^1 estimates between the discrete and nondiscrete Green functions. They are based on the discrete fundamental solution introduced by McCrea and Whipple [23] and the explicit representation of the Green function in terms of conformal mappings and fundamental solutions.

In Section 7.5 we introduce the discrete Neumann function appearing in the second equation of (7.0.2). According to Bramble and Hubbard [5, 6], the discrete normal derivative is introduced in such a way that the finite difference equations belonging to the discrete Neumann problem admit a matrix of positive type providing the maximum principle. For the discrete Neumann function the basic estimates are shown yielding the well known convergence of order $h^2 |\log h|$ for the discrete solutions of the Neumann problem for the Laplacian.

In Section 7.6 we obtain pointwise and L^1 estimates between the discrete and nondiscrete Neumann functions using the explicit form of the special G^{II} in (1.1.14). Here we need some slight extensions of the asymptotic estimates by McCrea and Whipple [23].

In Section 7.7 we formulate the difference equations of the Neumann problem with boundary condition

$$\oint_{\dot{G}} v\sigma \, ds = \kappa = \sum_{p_j \in \partial G_h} \gamma_j v(p_j)\sigma(p_j) + \text{order}\,(h^2)$$

instead of the point condition and introduce the corresponding discrete Neumann function.

In Section 7.8 the main results are presented. (7.0.2) is replaced by suitable difference equations. Rewriting them by means of the discrete Green and Neumann functions they become equivalent to discrete integral equations for piecewise constant functions which we consider as equations in B, the space of bounded Baire functions of first category equipped with the supnorm. The L^1 estimates for the Green and Neumann functions imply that the discrete equations form an 'approximation regular' family in the framework of Stummel's discrete convergence theory [28, 29] or, by using our embedding in B, they define a collectively compact perturbation of the corresponding integral equations of the original problem in the sense of Anselone's theory [1]. These properties yield inverse stability for the system of difference equations if the meshwidth h is small enough. Finally the order of consistency in the difference equations implies convergence of the difference solutions. For our system, which is in general *not* the Euler equation to a definite variational problem, this seems to be a new result.

We conclude this chapter with some remarks on the treatment of the difference equations and on semilinear problems.

For the first sections of this chapter I have to thank Mr. G. Gray who revised the first version. For the whole chapter I have to thank Dr. K. Merten who has given me most of the valuable references.

7.1 Matrices of Positive Type

The following concept allows the use of discrete analogues to the maximum principle for elliptic difference equations which will be needed for the Dirichlet problem as well as for the Neumann problem. It can be found in Varga's book [32]; here we follow the presentation by Bramble and Hubbard [4] which is general enough for our purposes.

Definition An $N \times N$ matrix $\mathsf{B} = ((b_{ij}))$ is said to be of *positive type* if the following conditions are satisfied

$$b_{ij} \leq 0 \qquad \text{for all } i \neq j, \tag{7.1.1}$$

$$\sum_{k=1}^{N} b_{ik} \geq 0 \quad \text{for all } i, \tag{7.1.2}$$

$$\sum_{k=1}^{N} b_{ik} > 0 \quad \text{for all } i \in J(\mathsf{B}) \subset \{1, 2, \ldots, N\} \tag{7.1.3}$$

with $J \neq \varnothing$ and,

If $i \notin J$, then there exists a finite sequence of nonzero elements $\{b_{ik_1}, b_{k_1k_2}, \ldots, b_{k_rj}\}$ where $j \in J(\mathsf{B})$. Such a sequence is called a *connection* from i to $J(\mathsf{B})$. (7.1.4)

For such matrices the following lemma provides a monotonicity property.

Lemma 7.1.1 *If* B *is a matrix of positive type, then* B *is nonsingular and its inverse has nonnegative entries.*

Proposition 7.1.2 *The inequalities*

$$\sum_{j=1}^{N} b_{ij} x_j \geq 0 \quad \text{for} \quad i = 1, \ldots, N \tag{7.1.5}$$

imply

$$x_j \geqslant 0 \quad \text{for} \quad j=1, \ldots, N, \tag{7.1.6}$$

if (and only if) $\mathbf{B}$ *is of positive type.*

Proof Since we shall later need only Lemma 7.1.1 let us restrict the proof to the necessity of (7.1.5), (7.1.6). For the remaining part see [4, 32]. If for any i, $b_{ii}=0$, then for all j not equal to this i, using (7.1.1) and (7.1.2), we have $b_{ij}=0$. Thus $i \notin J$. This, however, violates the connection property. *Thus,* $b_{ii}>0$ *for every* i. We now prove that $\mathbf{B}$ is of monotone type. Let us assume that x_k are given satisfying (7.1.5). Using $b_{ii}>0$ and (7.1.1) for $j \neq i$, we obtain

$$x_i \geqslant \sum_{j \neq i} \left|\frac{b_{ij}}{b_{ii}}\right| x_j \tag{7.1.7}$$

with

$$-\sum_{k \neq i} \frac{b_{ik}}{b_{ii}} \begin{cases} <1 & \text{for} \quad i \in J \\ =1 & \text{for} \quad i \notin J. \end{cases} \tag{7.1.8}$$

from (7.1.2), (7.1.3).

Now let us suppose that one of the x_i was negative, i.e. that

$$\bar{x}=x_i \equiv \min_{j=1, \ldots, N} x_j<0. \tag{7.1.9}$$

In this case for $i \in J(\mathbf{B})$ we would find

$$\bar{x}<\sum_{k \neq i}\left|\frac{b_{ik}}{b_{ii}}\right| \bar{x} \leqslant \sum_{k \neq i}\left|\frac{b_{ik}}{b_{ii}}\right| x_k \leqslant x_i=\bar{x}, \tag{7.1.10}$$

a contradiction. On the other hand, if $i \notin J(\mathbf{B})$ there exists, by the connection property (7.1.4), an index k_1 with $b_{ik_1} \neq 0$. Here, for $\bar{x}<x_{k_1}$,

$$\bar{x}=\sum_{k \neq i}\left|\frac{b_{ik}}{b_{ii}}\right| \bar{x}<\sum_{i \neq k_1}\left|\frac{b_{ik}}{b_{ii}}\right| x_k \leqslant x_i=\bar{x} \tag{7.1.11}$$

would hold, a contradiction. Thus $\bar{x}=x_{k_1}$. Next we replace i by k_1 and repeat the proof. Then we replace k_1 by k_2. After a finite number of steps we find a $j \in J$ with $\bar{x}=x_j$. But with $j \in J(\mathbf{B})$ the inequality (7.1.10) holds, i.e. we have arrived at a contradiction.

Therefore, there *cannot* be a negative minimum of the x_i-s, or $x_i \geqslant 0$ *for every* i. Hence (7.1.6) holds true. Moreover,

$$\mathbf{By}=\mathbf{0} \quad \text{implies} \quad y_j \geqslant 0 \quad \text{and} \quad -y_j \geqslant 0 \quad \text{for} \quad j=1, \ldots, N.$$

Thus $\mathbf{y}=0$ and so $\mathbf{B}$ is non-singular. This and Proposition 7.1.2 implies Lemma 7.1.1.

7.2 The Approximation Scheme for the Laplacian

Following Bramble, Hubbard and Ciarlet [2, 5, 9] let us introduce a square mesh on G having mesh width $h>0$ and let us use the following notation.

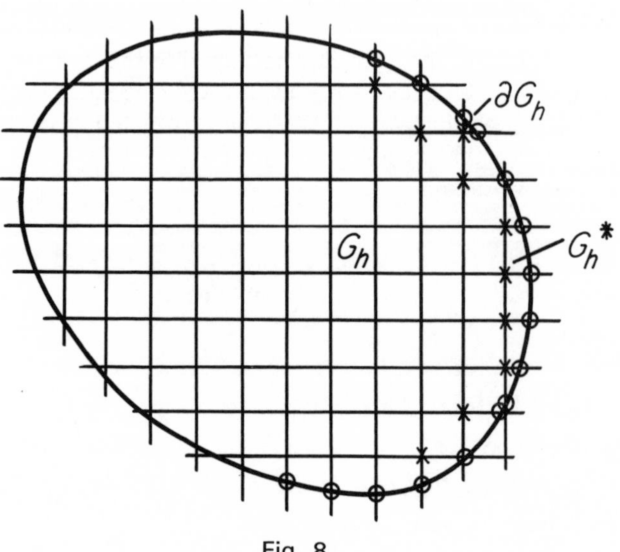

Fig. 8

All intersections of the meshlines with $\dot{G}$ define the point set ∂G_h. The interior netpoints forming G_h are those mesh points in G whose four nearest neighbours also belong to the domain G. The remaining mesh points in G will be denoted by G_h^*.

If F is any given mesh function then we define for all $z \in G_h$ the discrete Laplacian by

$$\Delta_h F \equiv h^{-2}\{F(x+h, y)+F(x-h, y)+F(x, y+h)+F(x, y-h) \\ -4F(x, y)\} \tag{7.2.1}$$

For a smooth function, $F \in C^4(G)$, this is an approximation of order h^2 in interior points. The number of these points is of order h^{-2} whereas the number of points belonging to G_h^* near the boundary is only of order h^{-1}. This is the reason why in G_h^* only an approximation of order h can be chosen. An $0(h)$ approximation is given by the following construction introduced by Shortley and Weller (see [14], p. 201):

$$\Delta_h F := \frac{2}{h_1(h_1+h_3)} F(p_1) + \frac{2}{h_2(h_2+h_4)} F(p_2) + \frac{2}{h_3(h_3+h_1)} F(p_3) + \frac{2}{h_4(h_4+h_2)} F(p_4) - \left\{\frac{2}{h_1 h_3} + \frac{2}{h_2 h_4}\right\} F(p)$$

for $p \in \mathsf{G}_h^*$. (7.2.2)

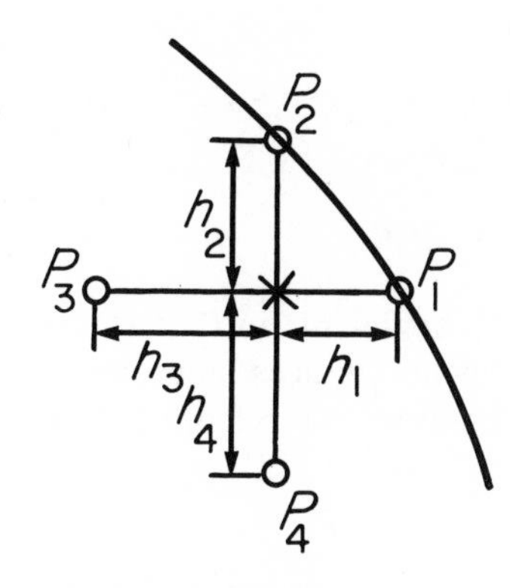

Fig. 9

7.3 The Discrete Green Function

In a similar manner as the Green function G^I provides the solution of the Dirichlet problem to the Poisson equation by an explicit integral formula, Courant, Friedrichs and Lewy [11] introduced in their famous paper a discrete Green function which provides the solution of the corresponding difference equations by an explicit summation formula. Here we follow the presentations of Bramble and Hubbard [2, 3, 4] and Ciarlet [9] defining $G_h(p, q)$, the discrete Green function depending on the pair of grid points p, q by the solution of the following systems of linear equations:
For $q \in \mathsf{G}_h \cup \mathsf{G}_h^*$ solve

$$\begin{aligned} -\Delta_{h,p} G_h(p, q) &= h^{-2}\delta(p, q) \quad \text{for all } p \in \mathsf{G}_h \cup \mathsf{G}_h^*, \\ G_h(p, q) &= 0 \qquad\qquad\quad \text{for } p \in \partial G_h \end{aligned} \qquad (7.3.1)$$

and for $q \in \partial \mathsf{G}_h$ solve

$$\begin{aligned} -\Delta_{h,p} G_h(p, q) &= 0 \qquad \text{for all } p \in \mathsf{G}_h \cup \mathsf{G}_h^*, \\ G_h(p, q) &= h^{-1}\delta(p, q) \quad \text{for all } p \in \partial \mathsf{G}_h \end{aligned} \tag{7.3.2}$$

where the index at $\Delta_{h,p}$ denotes that the discrete Laplacian (7.2.1) operates with respect to p and $\delta(p, q)$ denotes the Kronecker symbol,

$$\delta(p, q) = \begin{cases} 1 & \text{for} \quad p = q \quad \text{and} \\ 0 & \text{for} \quad p \neq q. \end{cases}$$

The following lemma guarantees the existence of G_h (see e.g. Collatz [10], p. 46).

Lemma 7.3.1 *For $h > 0$, if $\mathsf{G}_h \cup \mathsf{G}_h^* \neq \varnothing$, the discrete Green function G_h exists and is uniquely determined by* (7.3.1), (7.3.2). *Moreover,*

$$0 \leq G_h(p, q) \quad \textit{for all} \quad p, q. \tag{7.3.3}$$

Proof The matrices of coefficients in (7.3.1) and (7.3.2) are denoted by $((b_{pt}))$ where we omit further indices since the grid points can be numbered for $h > 0$ fixed. Then we have from (7.2.1) and (7.2.2)

$$b_{pt} \leq 0 \quad \text{for all} \quad t \neq p, \tag{7.3.4}$$

$$\begin{cases} \sum_t b_{pt} = 0 & \text{for interior grid points } p \in \mathsf{G}_h \text{ and} \\ \sum_t b_{pt} \geq 0 & \text{for grid points } p \in \mathsf{G}_h^*, \end{cases} \tag{7.3.5}$$

and

$$\sum_t b_{pt} = \sum_t \delta(p, t) = b_{pp} = 1 \quad \text{for all boundary points } p \in \partial \mathsf{G}_h. \tag{7.3.6}$$

Hence $\partial \mathsf{G}_h \subset J(\mathsf{B}) \neq \varnothing$. Moreover, every grid point $p \in \mathsf{G}_h \cup \mathsf{G}_h^*$ can be connected with a boundary point $r \in \partial \mathsf{G}_h$ in the sense of (7.1.4). To this end, e.g. consider the sequence of grid points on the horizontal mesh line with increasing x-coordinates including the first boundary point r which surely exists.

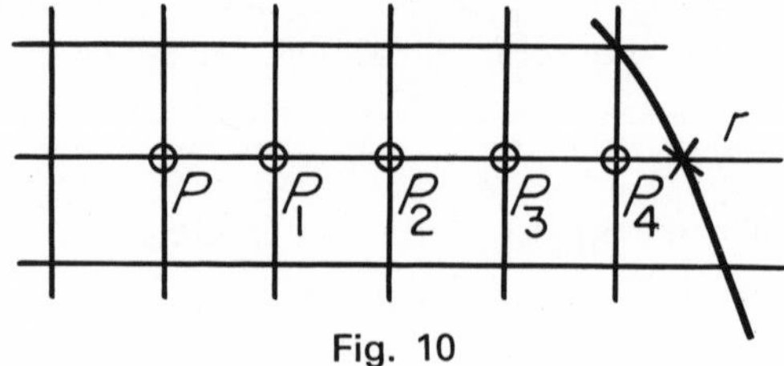

Fig. 10

Then from (7.2.1) and (7.2.2) we have

$$b_{pp_1} = b_{p_1p_2} = \cdots = b_{p_3p_4} = -\frac{1}{h^2} \neq 0 \quad \text{and} \quad b_{p_4r} = -\frac{2}{h_1(h_1+h_3)} \neq 0$$

Consequently, the matrix B corresponding to (7.3.1), (7.3.2) *is of positive type*. Hence Lemma 7.1.1 yields the unique solvability of (7.3.1) and (7.3.2). Since the right-hand sides are always ≥ 0, the Proposition 7.1.2 yields (7.3.3).

The next lemma provides the announced representation formula for mesh functions corresponding to Green's integral formula ([17, (4.3.2)]).

Lemma 7.3.2 [7] *For every given mesh function $F(p)$ the following representation formula holds,*

$$F(p) = -h^2 \sum_{q \in \mathsf{G}_h{}^* \cup \mathsf{G}_h} G_h(p,q)\Delta_h F(q) + h \sum_{q \in \partial\mathsf{G}_h} G_h(p,q)F(q). \tag{7.3.7}$$

Proof For given $F(p)$, let us denote by $R(p)$ the right-hand side in (7.3.7). Then the equations

$$\begin{aligned}\Delta_{h,p}(F-R)(p) &= \Delta_h F(p) + h^2 \sum_{q \in \mathsf{G}_h \cup \mathsf{G}_h{}^*} \Delta_{h,p} G_h(p,q)\Delta_h F(q) \\ &\quad - h \sum_{q \in \partial\mathsf{G}_h} \Delta_{h,p} G_h(p,q)F(q) \\ &= \Delta_h F(p) - \Delta_h F(p) = 0 \quad \text{for all } p \in \mathsf{G}_h \cup \mathsf{G}_h^*\end{aligned}$$

and

$$\begin{aligned}(F-R)(p) &= F(p) + h^2 \sum_{q \in \mathsf{G}_h \cup \mathsf{G}_h{}^*} G_h(p,q)\Delta_h F(q) - h \sum_{q \in \partial\mathsf{G}_h} G_h(p,q)F(q) \\ &= F(p) - F(p) = 0 \quad \text{for all } p \in \partial\mathsf{G}_h\end{aligned}$$

follow from (7.3.1) and (7.3.2). But this homogeneous system of equations is uniquely solvable since it has a coefficient matrix of positive type as we have seen in the proof of Lemma 7.3.1. Hence, $F-R$ vanishes identically, i.e. (7.3.7).

Next we prove several properties of G_h by using the representation formula (7.3.7).

Every $r \in \mathsf{G}_h^*$ has at least one—and for h small enough—at most two neighbours on $\partial\mathsf{G}_h$ belonging to the difference star of $\Delta_h F(r)$ in (7.2.2). Let h_r denote their smallest distance to r; further let us denote

$$\bar{\mathsf{G}}_h \equiv \mathsf{G}_h \cup \mathsf{G}_h^* \cup \partial\mathsf{G}_h.$$

Lemma 7.3.3 [3] *G_h has the following properties:*

$$h \sum_{q \in \partial \mathsf{G}_h} G_h(p, q) = 1 \quad \text{for all } p \in \bar{\mathsf{G}}_h. \tag{7.3.8}$$

$$h^2 \sum_{q \in \mathsf{G}_h \cup \mathsf{G}_h^*} G_h(p, q) \leqslant C \quad \text{for all } p \text{ and uniformly in } h, \tag{7.3.9}$$

$$\sum_{r \in \mathsf{G}_h^*} G_h(p, r) \frac{h}{h_r} \leqslant \sum_{r \in \mathsf{G}_h^*} G_h(p, r) \leqslant 1 \quad \text{for all } p \in \bar{\mathsf{G}}_h \tag{7.3.10}$$

Proof The equation (7.3.8) follows from (7.3.7) by the use of $F \equiv 1$. To prove (7.3.9) we use the special function

$$F(x, y) \equiv -e^{\alpha x} \tag{7.3.11}$$

where $\alpha > 0$ is chosen big enough in order to assure

$$\Delta F = -\alpha^2 e^{\alpha x} \leqslant -2 \quad \text{in } \bar{\mathsf{G}}. \tag{7.3.12}$$

Since $F \in C^\infty$ we have

$$|\Delta F - \Delta_h F| = 0(h^2)$$

uniformly for all grid points in $\bar{\mathsf{G}}$ [14] and, hence

$$\Delta_h F \leqslant -1 \quad \text{uniformly for all } h \leqslant h_0 \tag{7.3.13}$$

where h_0 is a suitable constant depending only on $\bar{\mathsf{G}}$. Now (7.3.7) for F with $G_h \geqslant 0$ and (7.3.13) yields

$$F(p) \geqslant h^2 \sum_{q \in \mathsf{G}_h \cup \mathsf{G}_h^*} G_h(p, q) + h \sum_{q \in \partial \mathsf{G}_h} G_h(p, q) F(q). \tag{7.3.14}$$

Since F is bounded in $\bar{\mathsf{G}}$ we find from (7.3.14)

$$h^2 \sum_{q \in \mathsf{G}_h \cup \mathsf{G}_h^*} G_h(p, q) \leqslant \max_{\bar{\mathsf{G}}} |F| \left\{ 1 + h \sum_{q \in \partial \mathsf{G}_h} G_h(p, q) \right\}. \tag{7.3.15}$$

For the last expression we can use (7.3.8), which has already been proved, and then (7.3.15) yields (7.3.9).

To prove (7.3.10) we define a special mesh function

$$F(p) \equiv \begin{cases} 1 & \text{for } p \in \partial \mathsf{G}_h \quad \text{and} \\ 0 & \text{for } p \in \mathsf{G}_h \cup \mathsf{G}_h^*. \end{cases} \tag{7.3.16}$$

$F(p)$ can be represented by (7.3.7),

$$\left.\begin{matrix} 0 \\ 1 \end{matrix}\right\} = -F(p) + h \sum_{r \in \partial \mathsf{G}_h} G_h(p, r) = h^2 \sum_{r \in \mathsf{G}_h^*} G_h(p, r) \Delta_{hr} F(r) \tag{7.3.17}$$

where (7.3.8) has been used. The definition for Δ_{hr}, (7.2.2) implies for F

(7.3.16) the estimate

$$\frac{1}{hh_r} \leq \frac{2}{h_r(h+h_r)} \leq \Delta_{hr}F(r) \quad \text{for } r \in \mathsf{G}_h^*. \tag{7.3.18}$$

Hence, (7.3.18) in (7.3.17) yields

$$\sum_{r \in \mathsf{G}_h^*} G_h(p, r) \frac{h^2}{hh_r} \leq \sum_{r \in \mathsf{G}_h^*} G_h(p, r)\, \Delta_{hr}F(r)h^2 \leq 1,$$

the desired inequality (7.3.10).

The summation properties in Lemma 7.3.3 can already be used for proving error estimates for the Dirichlet problem to the Poisson equation.

Theorem 7.3.4 *Let U be the solution of the discrete equations*

$$\begin{aligned} &\Delta_h U = F_h \quad \text{in } \mathsf{G}_h \cup \mathsf{G}_h^* \quad \textit{and} \\ &U|_{\partial \mathsf{G}_h} = \psi_h. \end{aligned} \tag{7.3.19}$$

Let u, the solution of the original problem satisfy the estimates

$$\begin{cases} |\Delta_h u - F_h| \leq c_u h^2 & \text{in } \mathsf{G}_h, \\ |\Delta_h u - F_h| \leq c_u & \text{in } \mathsf{G}_h^*, \\ \quad |u - \psi_h| \leq c_u h^2 & \text{on } \partial \mathsf{G}_h. \end{cases} \tag{7.3.20}$$

Then

$$|u - U| \leq c_u(C+2)h^2 \quad \textit{in } \bar{\mathsf{G}}_h \tag{7.3.21}$$

where C is the constant from (7.3.9).

Proof The representation formula (7.3.7) for $u-U$ with (7.3.19) reads as

$$\begin{aligned}(u-U)(p) = h^2 &\sum_{q \in \mathsf{G}_h} G_h(p, q)(F_h(q) - \Delta_h u(q)) \\ &+ h^2 \sum_{q \in \mathsf{G}_h^*} G_h(p, q)(F_h(q) - \Delta_h u(q)) \\ &+ h \sum_{q \in \partial \mathsf{G}_h} G_h(p, q)(u(q) - \psi_h(q)).\end{aligned}$$

Using the estimates (7.3.20) together with (7.3.3) and (7.3.8–10) the desired estimate (7.3.21) follows immediately.

As a simple consequence we have:

Proposition 7.3.5 *Let u be the solution of the Dirichlet problem*

$$\Delta u = F, \qquad u|_{\dot{\mathsf{G}}} = \psi \tag{7.3.22}$$

with $u \in C^4(\bar{G})$. *Then the estimates*

$$|\Delta_h u - \Delta u| \leq \frac{h^2}{6} \|u\|_{C^4(\bar{G})} \quad \text{for } p \in G_h \tag{7.3.23}$$

and

$$|\Delta_h u - \Delta u| \leq \tfrac{2}{3} h \|u\|_{C^3(\bar{G})} \quad \text{for } p \in G_h^* \tag{7.3.24}$$

hold [14, p. 201]. *Choosing* $\psi_h(q) = \psi(q)$ *and* $F_h(p) = F(p)$ *we find the error estimate*

$$|u - U| \leq h^2 \left\{ \frac{C}{6} \|u\|_{C^4(\bar{G})} + \tfrac{2}{3} \|u\|_{C^3(\bar{G})} \right\}. \tag{7.3.25}$$

For more general equations and also for the Neumann problem we need more estimates for the Green function. Pointwise estimates of this type (for a much bigger class of equations) have been proved by Bramble and Thomée [8] and McAllister and Sabotka [22]. Here we follow [8].

Theorem 7.3.6 [8] *The discrete Green function satisfies*

$$0 \leq G_h(p, q) \leq C \log \frac{C}{|p-q| + h} \quad \text{for all } p, q \in \bar{G}_h \tag{7.3.26}$$

where C is a suitable constant independent of p, q and h.

Proof Let us consider the function

$$W_q(p) \equiv \log \left\{ \frac{C}{(|p-q|^2 + \mu^2 h^2)^{1/2} + \mu h} \right\} \tag{7.3.27}$$

where the positive constants C and μ will be chosen later on. Using the abbreviations

$$\rho_0 \equiv (|p-q|^2 + \mu^2 h^2)^{1/2}, \ \rho \equiv \rho_0 + \mu h \tag{7.3.28}$$

we find with

$$\rho_x = \rho_{0x} = (x - \xi)\rho_0^{-1}, \ p = x + iy, \ q = \xi + i\eta,$$

for the derivatives of W the relations

$$\begin{aligned} W_{qx} &= -(x-\xi)\rho^{-1}\rho_0^{-1}, \\ W_{qxx} &= -\rho^{-1}\rho_0^{-1} + (x-\xi)^2\{\rho^{-2}\rho_0^{-2} + \rho^{-1}\rho_0^{-3}\}, \\ W_{qxxx} &= -[(x-\xi)(\rho^{-2}\rho_0^{-2} + \rho^{-1}\rho_0^{-3}) \\ &\quad + (x-\xi)^3\{2\rho^{-3}\rho_0^{-3} + 3\rho^{-2}\rho_0^{-4} + 3\rho\rho_0^{-5}\}] \end{aligned} \tag{7.3.29}$$

and similar expressions for the other derivatives. Thus, for the third derivatives we have estimates

$$|D^3 W_q| \leq c_1 \rho_0^{-3} \leq c_2 \rho^{-2} \rho_0^{-1} \tag{7.3.30}$$

with constants c_1, c_2 independent of p, q, C, μ, h since

$$\begin{aligned}4\rho_0^2 &\geq |p-q|^2+2\mu^2h^2+2\rho_0(|p-q|^2+\mu^2h^2)^{1/2}\\ &\geq |p-q|^2+2\mu^2h^2+2\mu h\rho_0=\rho^2\end{aligned} \tag{7.3.31}$$

holds. For the Laplacian we find

$$\begin{aligned}-\Delta_p W_q(p) &= \{\rho^2\rho_0^2-\rho\rho_0^3+\mu^2h^2(\rho\rho_0+\rho^2)\}\rho^{-3}\rho_0^{-3} \geq (\rho-\rho_0)\rho^{-2}\rho_0^{-1}\\ &= \mu h\rho^{-2}\rho_0^{-1}.\end{aligned} \tag{7.3.32}$$

Hence, for the discrete Laplacian, we find from (7.3.30) with

$$|\Delta W_q-\Delta_h W_q| \leq h\tfrac{2}{3}c_2\rho^{-2}\rho_0^{-1} \tag{7.3.33}$$

the inequality

$$-\Delta_h W_q \geq -\Delta W_q-\tfrac{2}{3}hc_2\rho^{-2}\rho_0^{-1} \geq \mu h\rho^{-2}\rho_0^{-1}\tfrac{1}{2} \tag{7.3.34}$$

if we choose μ as

$$\mu \equiv \tfrac{4}{3}c_2. \tag{7.3.35}$$

With this μ we can choose C in such a way that

$$W_q(p) \geq 0 \quad \text{for all } p, q \text{ in } \bar{\mathrm{G}} \text{ and all } h \leq 1. \tag{7.3.36}$$

With $W_q(p)$ let us define an auxiliary function to G_h by introducing

$$\lambda(q) \equiv -(h^2\,\Delta_h W_q(q))^{-1}, \qquad Z(p, q) \equiv \lambda(q)\,W_q(p), \tag{7.3.37}$$

Then

$$0<\lambda(q) \leq \left.\frac{2\rho^2\rho_0}{\mu h^3}\right|_{p=q} = 8\mu^2. \tag{7.3.38}$$

Now from (7.3.34), (7.3.36) and (7.3.38) it follows easily that the inequalities

$$\left.\begin{aligned}&-\Delta_{hp}(Z(p, q)-G_h(p, q)) \geq 0 \quad \text{for } p \in \mathrm{G}_h \cup \mathrm{G}_h^* \quad \text{and}\\ &Z-G_h \geq 0 \qquad\qquad\qquad\qquad\quad\ \text{for } p \in \partial\mathrm{G}_h\end{aligned}\right. \tag{7.3.39}$$

hold. Since the linear difference equations in (7.3.39) are of positive type, Lemma 7.1.1 implies

$$0 \leq Z-G_h$$

or

$$\begin{aligned}G_h \leq \lambda(q)\,W_q(p) &\leq 8\mu^2 \log\left\{\frac{C}{(|p-q|^2+\mu^2h^2)^{1/2}+\mu h}\right\}\\ &\leq 8\mu^2 \log\left\{\frac{C}{|p-q|+\mu h}\right\},\end{aligned}$$

the desired estimate.

Since $G_h(p,q)$ vanishes for $p\in\partial\mathrm{G}_h$ identically we have besides (7.3.10) the estimate:

Lemma 7.3.7 *For* $p\in\bar{\mathrm{G}}_h$

$$\sum_{r\in\mathrm{G}_h\cup\mathrm{G}_h^*\wedge|r-\dot{\mathrm{G}}|\le 4h} G_h(p,r)\le C \tag{7.3.40}$$

with C independent of p and h.

Proof The proof is very similar to the proof of (7.3.10). Choosing

$$F(p)=\begin{cases}1 & \text{for } p\in\mathrm{G}_h^*\cup\partial\mathrm{G}_h\cup\mathrm{G}_h^1\\ 0 & \text{for } p\in\mathrm{G}_h\backslash\mathrm{G}_h^1\end{cases} \tag{7.3.41}$$

where G_h^1 denotes all neighbours of G_h^* in G_h, one finds with the representation formula

$$\sum_{r\in\mathrm{G}_h{}^1} G_h(p,r)\le 1 \quad \text{for all grid points } p\in\bar{\mathrm{G}}_h. \tag{7.3.42}$$

If G_h^2 denotes all neighbours of G_h^1 in $\mathrm{G}_h\backslash\mathrm{G}_h^1$ then choose in the next step

$$F(p)=\begin{cases}1 & \text{for } p\in\mathrm{G}_h^*\cup\partial\mathrm{G}_h\cup\mathrm{G}_h^1\cup\mathrm{G}_h^2\\ 0 & \text{for } p\in\mathrm{G}_h\backslash(\mathrm{G}_h^1\cup\mathrm{G}_h^2)\end{cases} \tag{7.3.43}$$

and find

$$\sum_{r\in\mathrm{G}_h{}^2} G_h(p,r)\le 1. \tag{7.3.44}$$

Proceeding in this manner one finds after a finite number of steps the desired estimate (7.3.40).

7.4 The Discrete and Nondiscrete Green Functions

In order to extend the application of difference equations to more general equations than the Laplacian we need estimates for the difference between the Green functions on the grid. Since G^I is explicitly given by (1.1.10), let us first introduce the discrete counterpart to the fundamental solution following McCrea and Whipple [23, p. 288]:

$$\Gamma_h(p,q)\equiv\frac{1}{2\pi}\int_0^\pi\frac{1-\cos\left(\frac{\lambda}{h}(y-\eta)\right)\exp\left(\frac{-\mu}{h}|x-\xi|\right)}{\sinh\mu}\,d\lambda$$
$$+\frac{1}{2\pi}\log h-\frac{1}{4\pi}(\log 8+2\gamma),\ p=x+iy,\ q=\xi+i\eta \tag{7.4.1}$$

where $\mu = \mu(\lambda)$ is defined by

$$\cosh \mu(\lambda) = 2 - \cos \lambda \quad \text{with} \quad \mu(\lambda) \cdot \lambda^{-1} \to 1 \text{ for } \lambda \to 0$$

and where γ is Euler's constant. For Γ_h we have:

Lemma 7.4.1 *$\Gamma_h(p, q)$ satisfies*

$$\Delta_{hp} \Gamma_h(p, q) = h^{-2} \delta(p, q) \tag{7.4.2}$$

with Δ_h from (7.1.1) for all corresponding regular grid points, i.e. *in $G_h \cup G_h^*$. Moreover, there exists a constant K such that*

$$\left| \Gamma_h(p, q) - \frac{1}{2\pi} \log |p - q| \right| \leq K \frac{h^2}{|p - q|^2} \tag{7.4.3}$$

holds for all the regular gridpoints p, q and all $h > 0$.

The proof can be found in [23]; it is based on the asymptotic series expansion of (7.4.1); we shall omit it here.

We shall need Γ_h not only on the regular grid points $p, q \in G_h \cup G_h^*$ but also on the boundary grid points ∂G_h. To this end we extend Γ_h to ∂G_h such that (7.4.3) remains valid. Let us distinguish two cases for any chosen boundary point $\tilde{p} \in \partial G_h$ according to its definition by a net line $x = \text{const.}$ or $y = \text{const.}$ and let us consider only the latter case. In the other case Γ_h can be extended correspondingly.

Lemma 7.4.2 *Let $\tilde{p} \in \partial G_h$ have the coordinates $\tilde{p} = x_1 + h_1 + iy_1$ with $0 < h_1 \leq h$ and $p_1 = x_1 + iy_1 \in G_h^*$, p_1 being a regular grid point. Then Γ_h defined by the linear combination*

$$\Gamma_h(\tilde{p}, q) \equiv \frac{h - h_1}{h} \Gamma_h(p_1, q) + \frac{h_1}{h} \Gamma_h((x_1 + h + iy_1), q) \tag{7.4.4}$$

satisfies the inequality

$$\left| \Gamma_h(\tilde{p}, q) - \frac{1}{2\pi} \log |\tilde{p} - q| \right| \leq K' \frac{h^2}{|\tilde{p} - q|^2}, \quad \tilde{p} \in \partial G_h \tag{7.4.5}$$

for all $q \in G_h \cup G_h^*$ with $|q - \tilde{p}| \geq 4h$ where K' is independent of h, $\tilde{p}$, q.

Proof For $|q-\tilde{p}|\geq 4h$ we have $|q-p_1|\geq 3h$, $|q-p_2|\geq 3h$ where $p_2=x_1+h+iy_1$. Using the mean value theorem we find from (7.4.4) and (7.4.3)

$$\begin{aligned}\left|\Gamma_h(\tilde{p},q)+\frac{1}{2\pi}\log|\tilde{p}-q|\right| &\leq \frac{h-h_1}{h}K\frac{h^2}{|p_1-q|^2}+\frac{h_1}{h}K\frac{h^2}{|p_2-q|^2}\\ &\quad+\left|\frac{h-h_1}{h}\frac{1}{4\pi}\log\frac{|\tilde{p}-q|^2}{|p_1-q|^2}\right|+\left|\frac{h_1}{h}\frac{1}{4\pi}\log\frac{|\tilde{p}-q|^2}{|p_2-q|^2}\right|\\ &\leq \tfrac{9}{4}K\frac{h^2}{|\tilde{p}-q|^2}+\frac{1}{2\pi h}\left|(h-h_1)\left(-h_1\frac{(x_1+h_1-\xi)}{|\tilde{p}-q|^2}\right)\right.\\ &\quad\left.+h_1(h-h_1)\frac{(x_1+h_1-\xi)}{|\tilde{p}-q|^2}\right|\\ &\quad+C\frac{h^2}{|\tilde{p}-q|^2}\end{aligned}$$

where C does not depend on $\tilde{p}, q$ and h. This is the desired estimate (7.4.5).

Since the Green function for the continuum (1.1.10) for the special domain of the unit disk is completely determined by p, q and their images under the reflection at the unit circle, we shall define a generalization of the reflection procedure to more general domains. This idea is very much related to the expansion about the contiguous circle at $q\in\dot{\mathsf{G}}$ which was used by Scott [26, Appendix].

Lemma 7.4.3 *Let $\phi\in C^{k+\alpha}(\bar{\mathsf{G}})$ be the conformal mapping of G onto the unit disk, $k\geq 1$. Then there exists an extension Φ of ϕ across $\dot{\mathsf{G}}$ with the following properties:*

$$\begin{aligned}&\Phi^{-1}(w) \text{ is bijective for } |w|\leq R_1>1 \quad \text{and}\\ &\Phi^{-1}(w)=\phi^{-1}(w) \quad \text{for} \quad |w|\leq 1.\end{aligned} \tag{7.4.6}$$

For all $q\in\bar{\mathsf{G}}$ with $R_1^{-1}\leq|\phi(q)|\leq 1$, the generalized reflected point

$$q'(q)\equiv\Phi^{-1}\left(\frac{1}{\overline{\phi(q)}}\right) \tag{7.4.7}$$

is well defined satisfying

$$|p-q|\leq 2\,|p-q'(q)| \quad \text{for all } p\in\bar{\mathsf{G}}. \tag{7.4.8}$$

Proof For ϕ given in $\bar{\mathsf{G}}$, the Tietze–Urysohn Theorem [13] provides an extension to a $C^{k+\alpha}$ function Φ on a bigger domain. Of course this extension will not be holomorphic outside of $\bar{\mathsf{G}}$ anymore since there $\phi_{\bar{z}}$

might not vanish, in general. Since the Jacobian is continuous and given by

$$J=|\phi_z|^2-|\phi_{\bar z}|^2 \text{ with } J\neq 0 \text{ in } \bar{\mathsf{G}} \tag{7.4.9}$$

and $\phi=\Phi$ in $\bar{\mathsf{G}}$ bijectively, there exists a radius $R_0>1$ such that Φ maps a closed region (containing $\bar{\mathsf{G}}$ in its interior) bijectively onto the disk $|w|\leq R_0$.

Since (7.4.8) trivially holds for all points with $|p-q|\leq 2\,|p-q'|$ let us consider for a moment the points $p,\ q\in\bar{\mathsf{G}}$ satisfying

$$|p-q|>2\,|p-q'|=2\left|\Phi^{-1}(\phi(p))-\Phi^{-1}\left(\frac{1}{\overline{\phi(p)}}\right)\right| \tag{7.4.10}$$

where $0<1/R_0\leq 1/R_1\leq|\phi(q)|\leq 1$. The latter expression can be estimated by

$$|p-q'|\geq 2\gamma\,|\phi(p)-\overline{(\phi(q))^{-1}}| \tag{7.4.11}$$

with a constant $\gamma>0$ since Φ is bijective and at least $C^{1+\alpha}$. This yields

$$\begin{aligned}|p-q'|&\geq 2\gamma|\phi(q)|^{-1}|\phi(p)\overline{\phi(q)}-1| \quad \text{and with Möbius transform}\\ &\geq 2\gamma|\phi(p)-\phi(q)| \text{ since } \quad |\phi(p)|,|\phi(q)|\leq 1,\\ |p-q'|&\geq 2\gamma^2|p-q|.\end{aligned} \tag{7.4.12}$$

Thus,

$$F(R_1)\equiv \max_{q,p\in\bar{\mathsf{G}}\wedge 1/R_1\leq|\phi(q)|\leq 1}\{|p-q|\,|p-q'(q)|^{-1}\} \tag{7.4.13}$$

is bounded and continuous with

$$F(1)=1 \quad \text{since} \quad q'(q)=q \quad \text{for} \quad q\in\dot{\mathsf{G}}. \tag{7.4.14}$$

Therefore there exists R_1 with $R_0\geq R_1>1$ such that

$$F(R_1)\leq 2 \tag{7.4.15}$$

implying the proposition of Lemma 7.4.3.

Before we prove our proposed estimate for the Green functions let us formulate the following Lemma.

Lemma 7.4.4 *For q with $4h\leq\min_{r'\in\dot{\mathsf{G}}}|r'-q|$ and $R_1^{-1}\leq|\phi(q)|<1$ there hold the estimates*

$$\Delta_{hr}\frac{1}{|r-q'(q)|^2}\geq|r-q'(q)|^{-4} \text{ for all } r\in\mathsf{G}_h, \tag{7.4.16}$$

and

$$\Delta_{hr} |r - q'(q)|^2 \leq 800 \, |r - q'(q)|^{-4} \text{ for all } r \in \mathsf{G}_h \cup \mathsf{G}_h^*. \tag{7.4.17}$$

Proof If $\tilde{r} \in \dot{\mathsf{G}}$ is a point on the straight line between r and $q'(q)$ then the above restrictions imply

$$h \leq \tfrac{1}{4} |\tilde{r} - q| \leq \tfrac{1}{2} |\tilde{r} - q'(q)| \leq \tfrac{1}{2} |r - q'(q)|.$$

For $r \in \mathsf{G}_h$ elementary computations yield

$$\frac{1}{|r-q'|^4} \leq \frac{4|r-q'|^6 - 2\,|r-q'|^2}{|r-q'|^2 \dfrac{17^2}{16^2} |r-q'|^8} \leq \frac{4\,|r-q'|^6 + 8h^6 - 4h^6 - 8h^2\,|r-q'|^4}{|r-q'|^2[|r-q'|^4 + h^4]^2}$$

$$\leq \frac{4\,|r-q'|^6 + 8h^4\,|r-q'|^2 - 4h^6 - 32(x-x')^2(y-y')^2h^2}{|r-q'|^2[|r-q'|^4 + h^4 - 2h^2(x-x')^2][|r-q'|^4 + h^4 - 2h^2(y-y')^2]}$$

$$= \Delta_{hr} \frac{1}{|r-q'|^2}.$$

The second estimate follows from

$$|\Delta_{hr} F(r) - \Delta F(r)| \leq 2h \cdot \max_{|p-r| \leq h} |D^3 F(p)|$$

where D^3 denotes any third order derivative of F and from

$$\Delta_r \, |r - q'|^{-2} = 2|r - q'|^{-4}.$$

Theorem 7.4.5 *Let be $\dot{\mathsf{G}} \in C^{4+\alpha}$, $\alpha > 0$. Then there exist two constants c_1, c_2 independent of p, q and h such that the following estimates are valid:*

$$|G^I(p, q) - G_h(p, q)| \leq \frac{h^2}{|p-q|^2} \quad \textit{for} \quad q \in \mathsf{G}_h \quad \textit{with} \quad |\phi(q)| \leq \frac{1}{R_1} \tag{7.4.18}$$

and

$$|G^I(p, q) - G_h(p, q)| \leq c_1 \frac{h^2}{|p-q|^2}$$
$$+ c_2 h^3 \Big\{ \sum_{r \in \mathsf{G}_h{}^*} G_h(p, r) \frac{1}{h_r} |r - q'|^{-2} + \sum_{r \in \partial \mathsf{G}_h} G_h(p, r) \, |r - q'|^{-2} \Big\}$$
$$\textit{for } q \in \mathsf{G}_h \textit{ with } \frac{1}{R_1} < |\phi(q)| < 1 \quad \textit{and} \quad |q - \dot{\mathsf{G}}| \geq 4h \tag{7.4.19}$$

and for $p \in \bar{\mathsf{G}}_h$ with $|p - q| \geq 8h$.

Corollary 7.4.6 (7.4.19) *yields the weaker estimate*

$$|G^I(p,q)-G_h(p,q)| \leq h^2\left\{\frac{c_1}{|p-q|^2}+\frac{c_2}{\min\limits_{r\in\partial G_h}|r-q|^2}\right\}. \tag{7.4.20}$$

Remarks In the book by Forsythe and Wasow [14, p. 318] one can find a corresponding $0(h)$ estimate. But as we shall see, for our first-order system (7.0.1), or for the corresponding system (7.0.2) we need a higher order estimate. The idea of our proof can be found in Laasonen's work [21], who proved the sharper estimate (7.4.18) with $c_2=0$ for the case of a rectangular domain.

Our estimate (7.4.20) coincides with a result by Natterer [24] which goes back to Scott [25] if one uses special quadrature formulas for Galerkin equations to the Dirichlet problem for the Laplacian with piecewise linear finite element functions as trial functions.

But we shall use (7.4.19) providing estimates up to $\dot{G}$.

Proof of Theorem 7.4.5 The following proof is based on inner estimates for G^I-G_h and on the properties of G^I near the boundary. Using the representation formula (7.3.7) for the function $G_h-G^I-(1/2\pi)\log|p-q|+\Gamma_h$ we find with (7.4.5)

$$\begin{aligned}&\left|G_h(p,q)-G^I(p,q)-\frac{1}{2\pi}\log|p-q|+\Gamma_h(p,q)\right|\\ &\quad\leq h^3\sum_{r\in\partial G_h}G_h(p,r)K'\,|r-q|^{-2}\\ &\qquad+h^2\sum_{r\in G_h{}^*}G_h(p,r)\,|\Delta_{hr}\Gamma_h(r,q)|\\ &\qquad+h^2\sum_{r\in G_h\cup G_h{}^*}G_h(p,r)\left|\Delta_{hr}\left(G^I(r,q)+\frac{1}{2\pi}\log|r-q|\right)\right|.\end{aligned} \tag{7.4.21}$$

Since (7.2.2) applied to any grid function F yields

$$|\Delta_h F(r)| \leq \frac{12}{hh_r}\max|F| \quad \text{for} \quad r\in G_h^*, \tag{7.4.22}$$

the term containing $\Delta_{hr}\Gamma_h$ with $|r-q|\geq 3h$ can be estimated by

$$\begin{aligned}|\Delta_{hr}\,\Gamma_h(r,q)| &\leq \left|\Delta_{hr}\frac{1}{2\pi}\log|r-q|\right|+12K'\,\frac{h}{h_r}|r-q|^{-2}\\ &\leq c\left\{h\,|r-q|^{-3}+\frac{h}{h_r}|r-q|^{-2}\right\}\leq 2c\frac{h}{h_r}|r-q|^{-2} \quad \text{for} \quad r\in G_h^*\end{aligned} \tag{7.4.23}$$

where c is a suitable constant independent of h, h_r, r and q. Thus, (7.4.21) takes the form

$$|G_h(p,q)-G^I(p,q)-\frac{1}{2\pi}\log|p-q|+\Gamma_h(p,q)|$$

$$\leq h^3\sum_{r\in\partial G_h} G_h(p,r)K'\,|r-q|^{-2}+h^3\sum_{r\in G_h^*} G_h(p,r)2c\frac{1}{h_r}|r-q|^{-2}$$

$$+\,h^2\sum_{r\in G_h\cup G_h^*} G_h(p,r)\left|\Delta_{hr}\left(G^I(r,q)+\frac{1}{2\pi}\log|r-q|\right)\right| \quad (7.4.24)$$

In the first case (7.4.18), $|\phi(q)|\leq 1/R_1<1$, we conclude that

$$|r-q|\geq k>0 \quad \text{for} \quad r\in G_h^*\cup\partial G_h \quad (7.4.25)$$

holds where k is independent of h, r, q. For these q, the function $G^I(r,q)+(1/2\pi)\log|r-q|$ is harmonic in $r\in G$ and $C^{4+\alpha}(\bar{G})$. Therefore the discrete Laplacian is of order h^2 and with (7.3.8–10) we find

$$\left|G_h(p,q)-G^I(p,q)-\frac{1}{2\pi}\log|p-q|+\Gamma_h(p,q)\right|\leq ch^2 \quad (7.4.26)$$

which implies with (7.4.3) and (7.4.5) the desired estimate (7.4.18).

In the second case (7.4.19), $1/R_1<|\phi(q)|<1$ we observe that for any boundary point $\tilde{r}\in\dot{G}$ the mapping (7.4.7) yields

$$\tilde{r}'=\tilde{r}. \quad (7.4.27)$$

Therefore the quotient,

$$\frac{|\tilde{r}-q'(q)|}{|r-q|}\leq k', \quad (7.4.28)$$

is uniformly bounded for all $\tilde{r}\in\dot{G}$ and all q in (7.4.19). This yields with $|r-\dot{G}|\leq h_r\leq h$ for $r\in G_h^*$ and $|q-\dot{G}|\geq 4h$ a uniform bound for

$$\frac{|r-q'|}{|r-q|}\leq k'' \quad \text{for all } r\in G_h^*\cup\partial G_h \quad \text{and} \quad (7.4.29)$$

the above q.

The function in the last expression of (7.4.24) can now be written as

$$G^I(r,q)+\frac{1}{2\pi}\log|r-q|=-\frac{1}{2\pi}\log\frac{|\phi(r)-\phi(q)|}{|r-q|}+\frac{1}{2\pi}\log|\phi(q)|$$

$$+\frac{1}{2\pi}\log\frac{|\phi(r)-\Phi(q')|}{|r-q'|}+\frac{1}{2\pi}\log|r-q'|$$

$$=\frac{1}{2\pi}\log|r-q'|+H(r,q) \quad (7.4.30)$$

where $H(r, q)$ together with all derivatives with respect to $r=(x, y)$ up to the fourth order is continuous in $\bar{\mathsf{G}}\times\{q|\,(1/R_1)\leq|\phi(q)|\leq 1\}$. This follows by the use of the Taylor expansion in the logarithmic terms of (7.4.30). Hence, we find

$$\left|\Delta_{hr}\left(G^I(r, q)+\frac{1}{2\pi}\log|r-q|\right)\right|\leq\begin{cases}k''h^2\,|r-q'|^{-4} & \text{for}\quad r\in\mathsf{G}_h,\\ k''h\,|r-q'|^{-3} & \text{for}\quad r\in\mathsf{G}_h^*,\end{cases}\tag{7.4.31}$$

[14, p. 193]. Inserting (7.4.29) and (7.4.31) into (7.4.24) we get the estimate

$$\begin{aligned}&\left|G_h(p, q)-G^I(p, q)-\frac{1}{2\pi}\log|p-q|+\Gamma_h(p, q)\right|\\ &\quad\leq Ch^3\left\{\sum_{r\in\partial\mathsf{G}_h}G_h(p, r)\,|r-q'|^2+\sum_{r\in\mathsf{G}_h{}^*}G_h(p, r)\frac{1}{h_r}|r-q'|^{-2}\right\}\\ &\qquad+kh^4\sum_{r\in\mathsf{G}_h}G_h(p, r)\frac{1}{|r-q'|^4}\end{aligned}\tag{7.4.32}$$

with $h_r\leq|r-q'|$.

For the last sum we use (7.4.16), and then (7.3.7), yielding

$$\begin{aligned}h^4\sum_{r\in\mathsf{G}_h}G_h(p, r)\frac{1}{|r-q'|^4}&\leq h^4\sum_{r\in\mathsf{G}_h}G_h(p, r)\,\Delta_{hr}\frac{1}{|r-q'|^2}\\ &\leq\frac{-h^2}{|p-q'|^2}+\sum_{r\in\partial\mathsf{G}_h}G_h(p, r)\frac{h^3}{|r-q'|^2}-\sum_{r\in\mathsf{G}_h{}^*}G_h(p, r)\,\Delta_{hr}\frac{h^4}{|r-q'|^2}.\end{aligned}$$

In the last sum we use (7.4.22). Then we finally have

$$\begin{aligned}&\left|G_h(p, q)-G^I(p, q)-\frac{1}{2\pi}\log|p-q|+\Gamma_h(p, q)\right|\\ &\quad\leq C'h^3\left\{\sum_{r\in\partial\mathsf{G}_h}G_h(p, r)\,|r-q'|^{-2}+\sum_{r\in\mathsf{G}_h{}^*}G_h(p, r)\frac{1}{h_r}|r-q'|^{-2}\right\}\end{aligned}\tag{7.4.33}$$

with a new constant C'. With (7.4.3), (7.4.5) the inequality (7.4.33) leads to the desired estimate (7.4.19).

The proof of the Corollary 7.4.6 follows in case (i) from (7.4.33) by the use of (7.3.8) and (7.3.10) and in case (ii) as above.

The main estimate for further error analysis is the following L^1-estimate.

Theorem 7.4.7 *Let be* $\dot{\mathsf{G}} \in C^{4+\alpha}$, $\alpha > 0$. *Then there exists a constant C independent of h such that the estimate*

$$\sum_{\substack{q \in \mathsf{G}_h \cup \mathsf{G}_h^* \\ |p-q| \geq h}} h^2 |G^I(p, q) - G_h(p, q)| \leq Ch^2 |\log h|$$

is valid for all grid points $p \in \bar{\mathsf{G}}_h$. (7.4.34)

Proof With Theorem 7.4.5 we find for I, the left hand side of (7.4.34), the following sums,

$$I \leq \sum_{h \leq |p-q| < 8h} h^2(G^I(p, q) + G_h(p, q)) + \sum_{8h \leq |p-q| \wedge q \in \mathsf{G}_h \cup \mathsf{G}_h^*} c_1 h^4 |p-q|^{-2}$$

$$\times c_2 \sum_{\substack{8h \leq |p-q| \\ \wedge \frac{1}{R_1} < |\phi(q)| \wedge |q - \dot{\mathsf{G}}| \geq 4h}} h^5 \Bigg\{ \sum_{r \in \mathsf{G}_h^*} G_h(p, r) \frac{1}{h_r} |r - q'|^{-2}$$

$$+ \sum_{r \in \partial \mathsf{G}_h} G_h(p, r) |r - q'|^{-2} \Bigg\} + h^2 \sum_{8h \leq |p-q| \wedge |q - \dot{\mathsf{G}}| < 4h} |G^I(p, q) - G_h(p, q)|, \tag{7.4.35}$$

where p is fixed and the summation involves q. We shall estimate these four expressions separately.

Since $|p-q| \geq h$ and both Green functions grow as $|\log|p-q||$ for $q \to p$, (1.1.15), (7.3.26), and the fact that the set $h \leq |p-q| < 8h$ has at most 148 points we find

$$\sum_{h \leq |p-q| < 8h} h^2(G^I(p, q) + G_h(p, q)) \leq Ch^2 |\log h|. \tag{7.4.36}$$

In the second sum let us use the monotonicity of $|p-q|^{-2}$ which yields

$$h^2 c_1 \sum_{8h \leq |p-q|} h^2 |p-q|^{-2} \leq c_1 h^2 \iint_{7h \leq |p-z| \cap \bar{\mathsf{G}}} |p-z|^{-2} [\mathrm{d}x, \mathrm{d}y]$$

$$\leq Ch^2 |\log h|. \tag{7.4.37}$$

Similarly, for any $r \in \mathsf{G}_h^* \cup \partial \mathsf{G}_h$ fixed and with the estimate $|q'(q) - \bar{\mathsf{G}}| \geq 2h$ for all q with $|q - \dot{\mathsf{G}}| \geq 4h$ there holds the estimate

$$\sum_{\substack{|q - \dot{\mathsf{G}}| \geq 4h \wedge \\ |\phi(q)| > R_1^{-1}}} h^2 |r - q'(q)|^{-2} \leq C' \iint_{2h \leq |z-r| \leq D} |z - r|^{-2} [\mathrm{d}x, \mathrm{d}y] \leq C'' |\log h| \tag{7.4.38}$$

which is uniformly valid; $D \equiv \max_{|\zeta|=R_1} |\Phi^{-1}(\zeta)| + \text{diam}\,(\mathsf{G})$. Thus, with (7.3.8), (7.3.10) we find

$$c_2 \sum_{\substack{q \in \mathsf{G}_h \\ 8h \leq |p-q| \wedge |q-\mathsf{G}| \geq 4h \wedge |\phi(q)| > R_1^{-1}}} h^2 \Big\{ \sum_{r \in \mathsf{G}_h} G_h(p,r) h_r^{-1} |r-q'|^{-2}$$

$$+ \sum_{r \in \partial \mathsf{G}_h} G_h(p,r) |r-q'|^{-2} \Big\} \leq Ch^2 |\log h|. \quad (7.4.39)$$

For the remaining sum in (7.4.35) we use Lemma 7.3.7 for $G_h(p,q)$ and the estimate

$$|G^I(p,q)| \leq ch\,|p-q|^{-1} \quad \text{for } |q-\dot{\mathsf{G}}| \leq 4h \text{ and } |p-q| \geq 8h \qquad (7.4.40)$$

which follows from (1.1.10) by the use of the mean value theorem. Moreover, for $|p-q|^{-1}$ we can use its monotonicity to show that

$$h^2 \sum_{8h \leq |p-q| \wedge |q-\dot{\mathsf{G}}| < 4h} |G^I(p,q) - G_h(p,q)|$$

$$\leq h^2 \Big\{ \sum_{8h \leq |p-q| \wedge |q-\dot{\mathsf{G}}| < 4h} ch\,|p-q|^{-1} + C \Big\}$$

$$\leq h^2 C + hc' \iint_{|z-\dot{\mathsf{G}}| \leq 4h \wedge |z-p| \geq 7h} |z-p|^{-1} [\mathrm{d}x, \mathrm{d}y]$$

$$\leq h^2 C + c'h \int_{t=0}^{4h} \int_{s=7h}^{L} \frac{1}{s}\, \mathrm{d}s\, \mathrm{d}t \leq C'h^2 |\log h|. \qquad (7.4.41)$$

7.5 An $0(h^2)$ Approximation for the Discrete Neumann Function

For the discrete Neumann problem and the corresponding discrete version of G^{II}, the discrete Neumann function, we shall use the maximum principle of Section 7.1. To this end we need an $0(h^2)$ approximation of the normal derivative which has additional properties yielding matrices of positive type in the discrete equations. Such an approximation was introduced by Bramble and Hubbard [4, 6] as follows. Let us introduce local coordinates with origin at the boundary point $p \in \dot{\mathsf{G}}$ with the tangent vector to $\dot{\mathsf{G}}$ at p and the *inner* normal as the basis. Let K denote the curvature at p and let us denote by $\bar{K}$ its maximum,

$$\bar{K} \equiv \max_{\dot{\mathsf{G}}} |K|. \qquad (7.5.1)$$

Near p we introduce the new coordinates s, t by the implicit equations

$$x(s, t) = x(s) - t\dot{y}(s), \qquad y(s, t) = y(s) + t\dot{x}(s) \tag{7.5.2}$$

where for $t = 0$, s denotes the arc length on $\dot{G}$ with the origin at $p = x(0) + iy(0) = 0$. Hence we have the following relations:

$$\begin{cases} x(0) = 0, & \dot{x}(0) = 1, & \ddot{x}(0) = 0, \\ y(0) = 0, & \dot{y}(0) = 0, & \ddot{y}(0) = (\dot{x}\ddot{y} - \ddot{x}\dot{y})(0) = K \end{cases} \tag{7.5.3}$$

and for every smooth function g,

$$\frac{\partial}{\partial s}(g(x(s, t), y(s, t))) = g_x \cdot (\dot{x} - t\ddot{y}) + g_y \cdot (\dot{y} + t\ddot{x}),$$

$$\frac{\partial}{\partial s}(g(x(s, t), y(s, t)))(p) = g_x(p),$$

$$\frac{\partial^2}{\partial s \partial t}(g(x(s, t), y(s, t))) = -g_x \cdot \ddot{y} + g_y \ddot{x} - g_{xx} \cdot \dot{y}(\dot{x} - t\ddot{y})$$
$$+ g_{xy}(\dot{x}(\dot{x} - t\ddot{y}) - \dot{y}(\dot{y} + t\ddot{x})) + g_{yy} \cdot \dot{x}(\dot{y} + t\ddot{x}), \tag{7.5.4}$$

$$\frac{\partial^2 g}{\partial s \partial t}(p) = g_{xy} - Kg(x(s, t), y(s, t))_s = -\frac{\partial}{\partial s}\left(\frac{\partial g}{\partial n}\bigg|_{\dot{G}}\right). \tag{7.5.5}$$

Inserting (7.5.4) and (7.5.5) into the Taylor expansion of $g \in C^3$ at p, the expansion takes the form

$$g(z) - g(p) = -y\frac{\partial g}{\partial n}(p) + x[1 + yK]\frac{\partial g}{\partial s}(p)$$
$$+ \tfrac{1}{2}(x^2 - y^2)g_{xx}(p) + \frac{y^2}{2}\Delta g(p) - xy\frac{\partial}{\partial s}\left(\frac{\partial g}{\partial n}\right)(p)$$
$$+ 0(|z - p|^3), \qquad z = x + iy. \tag{7.5.6}$$

If we choose three different grid points $\tilde{p}_j = x_j + iy_j = z$ in (7.5.6) and form a linear combination with linear factors α_j then this yields an approximation of the normal derivative $(\partial g/\partial n)(p)$ if we require the linear equations

$$\begin{aligned} \sum_{j=1}^{3} \alpha_j x_j[1 + y_j K(p)] &= 0, \\ \sum_{j=1}^{3} \alpha_j y_j \qquad &= 1, \quad j = 1, 2, 3, \\ \sum_{j=1}^{3} \alpha_j (x_j^2 - y_j^2) \quad &= 0 \end{aligned} \tag{7.5.7}$$

for the points $\tilde{p}_j$ and the weights α_j. Provided that we can ever find such points and weights the desired approximation is then given by

$$\delta_n g(p) \equiv \sum_{j=1}^{3} (g(p) - g(\tilde{p}_j))\alpha_j = \frac{\partial g}{\partial n}(p)$$
$$- \sum_{j=1}^{3} \alpha_j \left\{ \frac{y_j^2}{2} \Delta g(p) - x_j y_j \frac{\mathrm{d}}{\mathrm{d}s} \left(\frac{\partial g}{\partial n} \bigg|_{\dot{G}} \right)(p) \right\} + 0\left(\sum_{j=1}^{3} \alpha_j |\tilde{p}_j - p|^3 \right). \tag{7.5.8}$$

Since the points $\tilde{p}_j$ are to be chosen in a neighbourhood of order h near p, the α_j *should behave exactly like:*

$$c_0 h^{-1} \leqslant |\alpha_j| \leqslant c_1 h^{-1} \quad \text{with} \quad c_0 > 0 \tag{7.5.9}$$

for $j = 1, 2, 3$ and all $0 < h \leqslant h_0$, $h_0 > 0$. We also have in mind that the whole system of difference equations for the Neumann problem defines a *coefficient matrix of positive type* where the boundary equations will be given by (7.5.8). There p traces the boundary points ∂G_h. Hence (7.1.1) and (7.1.2) imply

$$\alpha_j > 0. \tag{7.5.10}$$

Finally, we see that for points in G the coordinates y_j *have to be positive.*

Following Bramble and Hubbard [6] we shall see that suitable points $\tilde{p}_j$ can be found satisfying all the requirements above provided h is small enough. Bramble and Hubbard consider first the case of a *rectilineal* boundary near p which is interpreted as a perturbation of (7.5.7),

$$\begin{aligned} \sum_{j=1}^{3} \bar{\alpha}_j y_j \quad &= 1, \\ \sum_{j=1}^{3} \bar{\alpha}_j x_j \quad &= 0, \\ \sum_{j=1}^{3} \bar{\alpha}_j (x_j^2 - y_j^2) &= 0. \end{aligned} \tag{7.5.11}$$

Here we find the *explicit* solution given by Cramer's rule,

$$\bar{\alpha}_1 = \frac{\overline{D_1}}{\overline{D}} \quad \text{with} \quad \overline{D_1} = x_2(x_3^2 - y_3^2) - x_3(x_2^2 - y_2^2),$$

$$\bar{\alpha}_2 = \frac{\overline{D_2}}{\overline{D}} \quad \text{with} \quad \overline{D_2} = x_3(x_1^2 - y_1^2) - x_1(x_3^2 - y_3^2),$$

$$\bar{\alpha}_3 = \frac{D_3}{\overline{D}} \quad \text{with} \quad \overline{D_3} = x_1(x_2^2 - y_2^2) - x_2(x_1^2 - y_1^2),$$

where

$$\bar{D} = y_1\overline{D_1} + y_2\overline{D_2} + y_3\overline{D_3}.$$

In order to find positive y_j and $\overline{D_j}$ let us choose $h > 0$ small enough satisfying

$$h < \frac{2}{3\sqrt{17}\,\bar{K}}. \tag{7.5.12}$$

Setting

$$\varepsilon \equiv \tfrac{3}{2}h \tag{7.5.13}$$

we define three triangles in G:

$$\begin{aligned} TI&: 2\varepsilon < y_1 + \varepsilon < x_1 < 4\varepsilon, \\ TII&: 2\varepsilon < y_2 + \varepsilon < -x_2 < 4\varepsilon, \\ TIII&: 5\varepsilon + |x_3| < y_3 \leqslant 6\varepsilon. \end{aligned} \tag{7.5.14}$$

With (7.5.14) it is easily shown that each of these triangles contains at least one mesh point $\tilde{p}_j$ and *these points are the ones chosen for the desired difference formula.*

In the following figure the possible choices are indicated by $\otimes$. Moreover, it shows that these points can easily be found by the geometric interpretation corresponding to (7.5.14).

From the construction it is clear that the inequality

$$|\tilde{p}_j - p| = (x_j^2 + y_j^2)^{1/2} < 7\varepsilon = \tfrac{21}{2}h \tag{7.5.15}$$

holds and from (7.5.14) with (7.5.12) we have for the determinants

$$\begin{aligned} \bar{D}_1 &> 2\varepsilon \cdot 25\varepsilon^2 - 6\varepsilon \cdot \varepsilon^2 = 44\varepsilon^3 > 12\varepsilon^3, \\ \bar{D}_2 &> -6\varepsilon^3 + 2\varepsilon \cdot 25\varepsilon^2 = 44\varepsilon^3 > 12\varepsilon^3, \\ \bar{D}_3 &> (y_1+\varepsilon)(y_2 - x_2)(-y_2 - x_2) + (y_2 + \varepsilon)(x_1 - y_1)(x_1 + y_1) \\ &> 6\varepsilon^3 + 6\varepsilon^3 = 12\varepsilon^3. \end{aligned} \tag{7.5.16}$$

In a similar manner one finds inequalities

$$\bar{D}_1 < 159\varepsilon^3, \qquad \bar{D}_2 < 159\varepsilon^3, \qquad \bar{D}_3 < 120\varepsilon^3 \quad \text{and} \quad 148\varepsilon^4 < \bar{D} < 1028\varepsilon^4. \tag{7.5.17}$$

Thus, for a rectilineal part of $\dot{G}$ the above choice yields weights $\bar{\alpha}_j$ satisfying all the requirements (7.5.7), (7.5.9) and (7.5.10). But the *same choice* of points works also for a curved part of $\dot{G}$ if we solve (7.5.7). Let us express the desired α_j by

$$\alpha_j = D_j/D \tag{7.5.18}$$

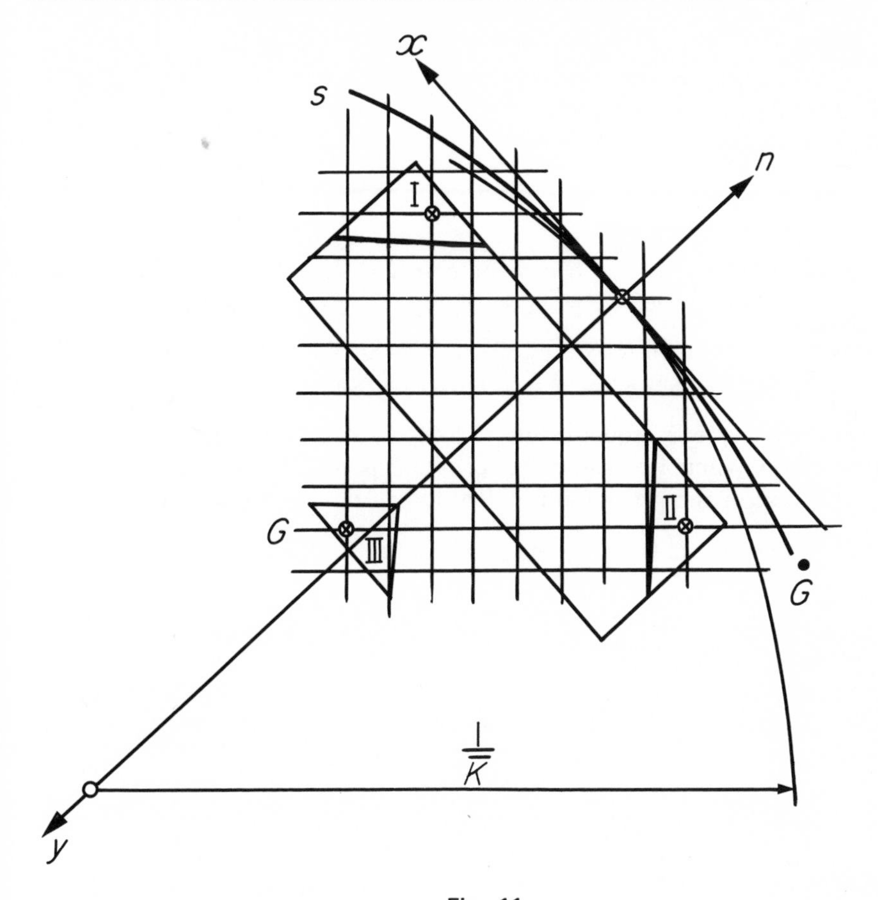

Fig. 11

where the D_j are the cofactors corresponding to (7.5.7). D is the determinant which satisfies

$$D = y_1 D_1 + y_2 D_2 + y_3 D_3. \qquad (7.5.19)$$

For the cofactors one finds the explicit formulas

$$\begin{aligned} D_1 &= x_2[1+y_2K](x_3^2-y_3^2) - x_3[1+y_3K](x_2^2-y_2^2) \\ &= \bar{D}_1 + x_2y_2K(x_3^2-y_3^2) - x_3y_3K(x_2^2-y_2^2), \\ D_2 &= x_3[1+y_3K](x_1^2-y_1^2) - x_1[1+y_1K](x_3^2-y_3^2) \\ &= \bar{D}_2 + x_3y_3K(x_1^2-y_1^2) - x_1y_1K(x_3^2-y_3^2), \\ D_3 &= x_1[1+y_1K](x_2^2-y_2^2) - x_2[1+y_2K](x_1^2-y_1^2) \\ &= \bar{D}_3 + x_1y_1(x_2^2-y_2^2) - x_2y_2K(x_1^2-y_1^2). \end{aligned} \qquad (7.5.20)$$

Using (7.5.14) in (7.5.20), these determinants satisfy the estimates

$$|D_1-\bar{D}_1|\leq 522\bar{K}\varepsilon^4, \qquad |D_2-\bar{D}_2|\leq 522\bar{K}\varepsilon^4, \qquad |D_3-\bar{D}_3|\leq 360\bar{K}\varepsilon^4 \tag{7.5.21}$$

and with (7.5.19)

$$|D-\bar{D}|<5292\bar{K}\varepsilon^5. \tag{7.5.22}$$

Hence with the above choice of points (7.4.14), the linear equations (7.5.7) provide coefficients α_j satisfying

$$\tfrac{2}{3}h\frac{12-348h\bar{K}}{1028+3528h\bar{K}}\leq \alpha_j\leq \tfrac{2}{3}h\frac{12+348h\bar{K}}{1028-3528h\bar{K}}, \tag{7.5.23}$$

which imply our requirements (7.5.9), (7.5.10). If p denotes different boundary points we shall write $\alpha_j(p)$ and $\tilde{p}_j(p)$.

In case of the Neumann problem for v,

$$-\Delta v=F \quad \text{in } \mathsf{G}, \qquad \frac{\partial v}{\partial n}=g \quad \text{on } \dot{\mathsf{G}}$$

$$\text{with} \quad \oint_{\dot{\mathsf{G}}} g\, ds+\iint_{\mathsf{G}} F[dx, dy]=0 \quad \text{and} \quad v(p_0)=v_0, \tag{7.5.24}$$

all the quantities on the right hand side of (7.5.6) are given and, hence, to (7.5.24) there corresponds the system of difference equations

$$\begin{aligned}
&-\Delta_h V=F(p) \quad \text{for} \quad p\in \mathsf{G}_h\cup \mathsf{G}_h^*\backslash\{p_0\},\\
&\delta_n V\equiv \sum_{j=1}^{3} \alpha_j(p)(F(p)-F(\tilde{p}_j(p)))\\
&\quad=g(p)+\sum_{j=1}^{3} \alpha_j\left\{\frac{y_j^2}{2} F(p)+x_jy_j\left(\frac{d}{ds}\, g\right)(p)\right\} \quad \text{for} \quad p\in \partial\mathsf{G}_h,\\
&V(p_0)=v_0.
\end{aligned} \tag{7.5.25}$$

Correspondingly, we define a *discrete Neumann function* N_h by

$$\begin{aligned}
-\Delta_{hp}N_h(p, q)&=h^{-2}\delta(p, q) \quad \text{for all} \quad p\in \mathsf{G}_h\cup \mathsf{G}_h^*-\{p_0\},\\
\delta_{np}N_h(p, q)&=h^{-1}\delta(p, q) \quad \text{for} \quad p\in \partial\mathsf{G}_h \quad \text{and}\\
N_h(p_0, q)&=0 \quad \text{for} \quad q\in \bar{\mathsf{G}}_h
\end{aligned} \tag{7.5.26}$$

following essentially Bramble and Hubbard [4]. For its existence and further properties we shall use that the matrix of coefficients to (7.5.26) is of *positive type*. Let us denote this matrix by B which is given by (7.2.1), (7.2.2) and the $\alpha_j(p)$.

Lemma 7.5.1 *The matrix* B *of* (7.5.25) *is of positive type.*

Proof In case $p=p_0$ we have on the left hand side of (7.5.26),

$$b_{p_0q}=\delta(p_0,q) \quad \text{and, hence,} \quad p_0\in J(\mathbf{B})\neq\emptyset. \tag{7.5.27}$$

In case $p\in \mathrm{G}_h$ we find from (7.2.1)

$$b_{pp}=\frac{4}{h^2} \quad \text{and} \quad b_{pq}=\begin{cases} -h^{-2} & \text{for } p\neq q \text{ and } q \text{ a neighbour,} \\ 0 & \text{for } p\neq q \text{ and } q \text{ not a neighbour,}\end{cases} \tag{7.5.28}$$

satisfying (7.1.1) and (7.1.2) and $p\notin J$.
In case $p\in \mathrm{G}_h^*$, the properties (7.1.1), (7.1.2) and $p\notin J$ follow immediately from (7.2.2).
In case $p\in\partial \mathrm{G}_h$ we have

$$\begin{aligned} &b_{pp}=\sum_{j=1}^{3}\alpha_j>0 \quad \text{and} \\ &b_{pq}=-\alpha_j<0 \quad \text{for} \quad q=\tilde{p}_j(p) \quad \text{and} \\ &b_{pq}=0 \quad \text{for the other points} \quad q\neq p \quad \text{and} \quad q\neq\tilde{p}_j(p). \end{aligned} \tag{7.5.29}$$

Hence (7.1.1), (7.1.2) and $p\notin J$ are obvious. For any $p\neq p_0$ the connection property between p and p_0 is fulfilled since G is simply connected and the two points can be joined by a chain of difference stars with corresponding coefficients (7.2.1) or (7.2.2).

With Lemma 7.1.1 it is clear that the special right hand sides in (7.5.26) provide the following:

Lemma 7.5.2 *The discrete Neumann function N_h is nonnegative,*

$$N_h(p,q)\geq 0 \quad \text{for all} \quad p,q\in \mathrm{G}_h\cup \mathrm{G}_h^*\cup\partial \mathrm{G}_h \tag{7.5.30}$$

and for every mesh function F,

$$F(p)=-h^2\sum_{q\in \mathrm{G}_h\cup \mathrm{G}_h^*-\{p_0\}} N_h(p,q)\,\Delta_h F(q) + h\sum_{q\in\partial \mathrm{G}_h} N_h(p,q)\,\delta_n F(q)+F(p_0). \tag{7.5.31}$$

In order to find error estimates we need estimates for the terms in (7.5.31) involving N_h.

Lemma 7.5.3 [4] *Let* $\dot{\mathsf{G}}$ *be smooth enough,* e.g. $\dot{\mathsf{G}} \in C^{4+\alpha}$. *Then the discrete Neumann function satisfies the following inequalities:*

$$h^2 \sum_{q \in \mathsf{G}_h} N_h(p, q) \leqslant C\,|\log h|, \tag{7.5.32}$$

$$h \sum_{q \in \mathsf{G}_h{}^*} N_h(p, q) \frac{h}{h_q} \leqslant C\,|\log h|, \tag{7.5.33}$$

$$h \sum_{q \in \partial \mathsf{G}_h} N_h(p, q) \leqslant C\,|\log h| \tag{7.5.34}$$

where the constant C *is independent of* h, p *and* q.

Proof For proving the above inequalities we shall use the Green function of *the first kind* and the representation of N_h together with the discrete G_h. From Theorem 7.4.5, (7.4.18), we have

$$|G^I(p, p_0) - G_h(p, p_0)| \leqslant ch^2 \tag{7.5.35}$$

for all $p \in \partial \mathsf{G}_h$ and all the points $p = \tilde{p}_j(p)$ in the discrete Neumann condition provided $h \leqslant h_0$ and $h_0 > 0$ is small enough. As a simple consequence we find for the discrete Neumann data

$$|\delta_{np} G_h(p, p_0) - \delta_{np} G^I(p, p_0)| \leqslant c'h \quad \text{for} \quad p \in \partial \mathsf{G}_h. \tag{7.5.36}$$

Moreover with $G^I(p, p_0)$ being harmonic and smooth near $\dot{\mathsf{G}}$ i.e. $\Delta G^I(p, p_0) = 0$ for $p \neq p_0, p \in \mathsf{G}$ and $G^I(p, p_0) \in C^{4+\alpha}(\bar{\mathsf{G}} - \{p_0\})$ we have from (7.5.8), (7.5.14) and (7.5.23) the estimate

$$\left| \delta_{np} G^I(p, p_0) - \frac{\partial G^I}{\partial n}(p, p_0) \right| \leqslant c''h \quad \text{for} \quad p \in \partial \mathsf{G}_h. \tag{7.5.37}$$

The properties of G^I together with the maximum principle imply that G^I takes its minimum value 0 at every point on the boundary $\dot{\mathsf{G}}$. Then E. Hopf's second lemma [17, p. 45] ensures the inequality

$$\frac{\partial G^I}{\partial n}(p, p_0) \leqslant -\beta < 0 \quad \text{for all} \quad p \in \dot{\mathsf{G}} \tag{7.5.38}$$

with some suitable constant β. Hence there exists an $h_0 > 0$ such that for all $0 < h \leqslant h_0$ (7.5.36–38) imply

$$0 < \delta \leqslant -\delta_{np} G_h(p, p_0) \leqslant \delta^{-1} \quad \text{uniformly in} \quad h \leqslant h_0 \tag{7.5.39}$$

with a suitable constant $\delta > 0$. Using (7.5.31) with $F(p) = G_h(p, p_0)$,

$$G_h(p, p_0) = h \sum_{q \in \partial \mathsf{G}_h} N_h(p, q) \delta_{np} G_h(p, p_0) + G_h(p_0, p_0)$$

we find from (7.5.39) the inequality

$$-\delta h \sum_{q \in \partial G_h} N_h(p, q) \leqslant G_h(p_0, p_0) - G_h(p, p_0) \leqslant \delta^{-1} h \sum_{q \in \partial G_h} N_h(p, q). \tag{7.5.40}$$

Hence, the pointwise estimate (7.3.26) for G_h implies with (7.5.40) one of the desired inequalities, (7.5.34).

For the remaining estimates we use (7.5.34); first with the representation formula (7.5.31) for

$$F(p) \equiv -x^2 - y^2. \tag{7.5.41}$$

Since all third derivatives of F vanish identically we have

$$\Delta_h F = -4 \tag{7.5.42}$$

and moreover

$$|\delta_n F| \leqslant c \tag{7.5.43}$$

independently of h. Hence (7.5.31) implies with (7.5.34)

$$h^2 \sum_{q \in G_h \cup G_h^* - \{p_0\}} N_h(p, q) \leqslant c\, |\log h|. \tag{7.5.44}$$

With the mesh function given by

$$F(p) = \begin{cases} 1 & \text{for} \quad p \in G_h \cup G_h^*, \\ 0 & \text{for} \quad p \in \partial G_h \end{cases} \tag{7.5.45}$$

having the properties

$$-\Delta_h F(p) \begin{cases} \geqslant \dfrac{1}{h h_p} & \text{for} \quad p \in G_h^* \quad \text{and} \\ = 0 & \text{for} \quad p \in \partial G_h, \end{cases} \tag{7.5.46}$$

from (7.2.2) and

$$|\delta_n F| \leqslant c \cdot h, \tag{7.5.47}$$

the representation formula (7.5.31) yields with (7.5.23)

$$F(p) - F(p_0) - h \sum_{q \in \partial G_h} N_h(p, q)\, \delta_n F(q) = -h^2 \sum_{p \in G_h^*} N_h(p, q)\, \Delta_h F(q) \tag{7.5.48}$$

and with (7.5.46), (7.5.47) and (7.5.34) the desired estimate (7.5.33).

The proposed error estimate for the Neumann problem can now be formulated as follows.

Theorem 7.5.4 *Let v be a solution of the Neumann problem* (7.5.24) *and let V be the solution of the discrete equations* (7.5.25) *where*

$$|F+\Delta_h v|\leq c_v h^2 \quad \text{in } \mathsf{G}_h, \qquad |F+\Delta_h v|\leq c_v h \quad \text{in } \mathsf{G}_h^*, \qquad |\delta_n V-\delta_n v|\leq c_v h^2 \quad \text{on } \partial\mathsf{G}_h \quad \text{and} \quad v(p_0)=v_0. \tag{7.5.49}$$

Then

$$|v-V|\leq 3c_v Ch^2\,|\log h| \quad \text{for all} \quad p\in\bar{\mathsf{G}}_h. \tag{7.5.50}$$

Proof The estimate is an immediate consequence of the representation formula (7.5.31) for $v-V$, Lemma 7.5.3 and the estimates (7.5.49).

As a simple consequence we have:

Proposition 7.5.5 *Let v be the solution of the Neumann problem* (7.5.24) *with $v\in C^4(\bar{\mathsf{G}})$. Then the error estimates*

$$|\Delta_h v-\Delta v|\leq\frac{h^2}{6}\|v\|_{C^4(\bar{\mathsf{G}})} \quad \text{for} \quad p\in\mathsf{G}_h \quad \text{and} \tag{7.5.51}$$

$$|\Delta_h v-\Delta v|\leq kh\,\|v\|_{C^3(\bar{\mathsf{G}})} \quad \text{for} \quad p\in\mathsf{G}_h^* \tag{7.5.52}$$

are valid [14]. *Choosing*

$$g_h(p)=g(p)+\sum_{j=1}^{3}\alpha_j(p)\left\{\frac{y_j^2(p)}{2}F(p)+x_j(p)y_j(p)\frac{dg}{ds}(p)\right\} \tag{7.5.53}$$

and $F_h(p)=F(p)$ we find the error estimate for V, the solution of (7.5.25),

$$|v-V|\leq Ch^2\,|\log h|\,\{\tfrac{1}{6}\|v\|_{C^4(\bar{\mathsf{G}})}+(k+k')\,\|v\|_{C^3(\bar{\mathsf{G}})}\} \tag{7.5.54}$$

where k' is given by

$$\left|\frac{\partial v}{\partial n}(p)-g_h(p)\right|\leq k'\|v\|_{C^3(\bar{\mathsf{G}})} \tag{7.5.55}$$

which follows from (7.5.8) *and* (7.5.53).

7.6 The Discrete and the Nondiscrete Neumann Functions

In this section we shall prove an estimate similar to (7.4.18), but now for the Neumann functions. Comparing the representation formula (7.5.31) and the corresponding representation formula with G^{II} (1.1.14) (see [17,

Section 4.8, formula (4.8.3)]) we see that N_h is the discrete counterpart to

$$N(p, q) = G^{II}(q, p) - G^{II}(q, p_0) \tag{7.6.1}$$

where the conformal mapping ϕ satisfies $\phi(p_0)=0$. For the estimates corresponding to (7.4.18) we shall need an estimate similar to (7.4.3) concerning the normal derivatives of Γ_h which can be obtained using McCrea's and Whipple's approach in [23]. Similar methods were used by McAllister and Sabotka [22].

Lemma 7.6.1 *There exists a constant K such that*

$$\left|\delta_{np}\left(\Gamma_h(p, q) - \frac{1}{2\pi}\log|p-q|\right)\right| \leq K\frac{h^2}{|p-q|^3} \tag{7.6.2}$$

holds for all $p \in \partial G_h$ and $q \in G_h \cup G_h^*$ *with* $|q-p| \geq 12h$ *where K is independent of h, p, q.*

Proof First let us consider (7.6.2) for points p belonging to the regular grid. Using (7.4.1) as a function also between the grid points we find for the derivatives,

$$\Gamma_{hy}(p, q) = \frac{1}{2\pi h}\int_0^{\pi} \frac{\lambda \sin\left(\frac{\lambda}{h}(y-\eta)\right)\exp\left(-\frac{\mu(\lambda)}{h}|x-\xi|\right)}{\sinh \mu(\lambda)}\, d\lambda, \tag{7.6.3}$$

$$\Gamma_{hx}(p, q) = \frac{1}{2\pi h}\int_0^{\pi} \frac{\mu(\lambda)}{\sinh \mu(\lambda)} \cos\left(\frac{\lambda}{h}(y-\eta)\right)\exp\left(-\frac{\mu(\lambda)}{h}|x-\xi|\right) d\lambda \tag{7.6.4}$$

if $x-\xi>0$. Since Γ_h is symmetric we may restrict the following estimates to the case

$$r \cdot h = x - \xi \geq y - \eta = s \cdot h; \tag{7.6.5}$$

all other cases follow by symmetry. Following McCrea and Whipple [23] we use an asymptotic expansion together with

$$\begin{aligned} \frac{1}{2\pi}\int_0^{\infty} \sin(s\lambda)\cdot e^{-r\lambda}\, d\lambda &= \frac{1}{2\pi}\frac{s}{s^2+r^2}, \\ \frac{1}{2\pi}\int_0^{\infty} \cos(s\lambda)\cdot e^{-r\lambda}\, d\lambda &= \frac{1}{2\pi}\frac{r}{s^2+r^2} \end{aligned} \tag{7.6.6}$$

and the behaviour of $\mu(\lambda)$, which has an analytic expansion about $\lambda=0$,

$$\mu(\lambda) = \lambda - \frac{1}{12}\lambda^3 + - \ldots. \tag{7.6.7}$$

Using the estimates

$$\left|\int_{r^{-1/3}}^{\pi} \frac{\lambda}{\sinh \mu} \begin{Bmatrix} \sin s\lambda \\ \cos s\lambda \end{Bmatrix} \exp(-r\mu)\, \mathrm{d}\lambda\right| \leq C \exp(-r^{2/3}),$$

$$\left|\int_{r^{-1/3}}^{\infty} \begin{Bmatrix} \sin s\lambda \\ \cos s\lambda \end{Bmatrix} \mathrm{e}^{-r\lambda}\, \mathrm{d}\lambda\right| \leq \frac{1}{r} \exp(-r^{2/3}),$$

$$\left|\int_{0}^{r^{-1/3}} \sin(s\lambda)\left(\mathrm{e}^{-r\lambda} - \mathrm{e}^{-r\mu} \frac{\lambda}{\sinh \mu}\right) \mathrm{d}\lambda\right| \leq C \cdot r^{-3},$$

$$\left|\int_{0}^{r^{-1/3}} \cos(s\lambda)\left(\mathrm{e}^{-r\lambda} - \mathrm{e}^{-r\mu} \frac{\mu}{\sinh \mu}\right) \mathrm{d}\lambda\right| \leq C \cdot r^{-3},$$

elementary computations yield

$$\begin{aligned} \left| h\Gamma_{hy} - \frac{1}{2\pi} \frac{s}{s^2 + r^2} \right| &\leq C \cdot r^{-3} \leq 3C\{r^2 + s^2\}^{-3/2} \\ \left| h\Gamma_{hx} - \frac{1}{2\pi} \frac{r}{s^2 + r^2} \right| &\leq C \cdot r^{-3} \leq 3C\{r^2 + s^2\}^{-3/2} \end{aligned} \tag{7.6.8}$$

where all the constants are independent of r and s.

Inserting (7.6.5) we finally find the desired behaviour for the first derivatives:

$$\left|\Gamma_{hx} - \left(\frac{1}{2\pi} \log|p-q|\right)_x\right| + \left|\Gamma_{hy} - \left(\frac{1}{2\pi} \log|p-q|\right)_y\right| \leq K' \frac{h^2}{|p-q|^3}. \tag{7.6.9}$$

The corresponding estimate on $\partial\mathbf{G}_h$ for Γ_h defined by (7.4.4) follows as in Lemma 7.4.2. Then (7.6.2) is an immediate consequence of (7.6.9) e.g. by the use of

$$|\delta_{np} F| \leq c \cdot \max_{|p'-p| \leq 10h} \{|F_x(p')| + |F_y(p')|\} \tag{7.6.10}$$

which follows from (7.5.14), (7.5.15) and the definition of δ_n.

Lemma 7.6.2 *For the difference between the fundamental solutions we have*

$$\left|\frac{\partial^j}{\partial x^{j-l} \partial y^l} \left(\Gamma_h(p, q) - \frac{1}{2\pi} \log|p-q|\right)\right| \leq K_j \frac{h^2}{|p-q|^{j+2}} \tag{7.6.11}$$

for each $j = 0, 1, \ldots$.

The proof follows by further differentiation of (7.6.3), (7.6.4), (7.6.6) and similar estimates as in the proof of Lemma 7.6.1. We shall omit the details.

In order to prove pointwise estimates for the Neumann function we need a better estimate for $\Delta_{hr}\,\Gamma_h(r, q)$ replacing (7.4.23).

Lemma 7.6.3 *For $|r-q|\geq 12h$ there exists a constant K such that*

$$|\Delta_{hr}\,\Gamma_h|\leq K\frac{h}{|r-q|^3}\quad \textit{holds for all}\quad r, q\in \mathsf{G}_h\cup \mathsf{G}_h^* \textit{ and all}\quad h>0. \tag{7.6.12}$$

Proof The proof follows from (7.6.11) and the approximation of Δ by Δ_h:

$$|\Delta_{hr}\Gamma_h(r, q)|\leq|\Delta_{hr}\Gamma_h-\Delta\Gamma_h|+\left|\Delta\Gamma_h-\Delta\frac{1}{2\pi}\log|r-q|\right|$$

$$\leq h\cdot\max_{|\tilde r-r|\leq h}|D^3\Gamma_h(\tilde r, q)|+2K_2h^2|r-q|^{-4}$$

$$\leq h\cdot\max_{|\tilde r-r|\leq h}\left\{\left|D^3\frac{1}{2\pi}\log|\tilde r-q|\right|+\left|D^3\left(\Gamma_h-\frac{1}{2\pi}\log|\tilde r-q|\right)\right|\right\}$$

$$+2K_2h^2|r-q|^{-4}$$

$$\leq 2h(|r-q|-h)^{-3}+K_2\cdot\tfrac{1}{6}\cdot h|r-q|^{-3}+h^3K_3(|r-q|-h)^{-5}$$
$$\leq Kh\,|r-q|^{-3}.$$

Theorem 7.6.4 *Let $\dot{\mathsf{G}}\in C^{4+\alpha}$, $\alpha>0$. Then there exist two constants c_1', c_2' independent of p, q and h such that the following pointwise estimates hold:*

$$|N(p, q)-N_h(p, q)|\leq c_1'h^2\left\{\frac{1}{|p-q|^2}+\frac{1}{|p_0-q|^2}+|\log h|\right\} \tag{7.6.13}$$

for $q\in\mathsf{G}_h$ with $|\phi(q)|\leq(1/R_1)$ and

$$|N_h(p, q)-N(p, q)|\leq c_1'h^2\left\{\frac{1}{|p-q|^2}+\frac{1}{|p_0-q|^2}\right\}$$

$$+c_2'h^3\left\{\sum_{r\in\partial\mathsf{G}_h}N_h(p, r)|r-q'|^{-3}+\sum_{r\in\mathsf{G}_h{}^*}N_h(p, r)\,|r-q'|^{-3}\right\} \tag{7.6.14}$$

for $q\in\mathsf{G}_h$ with $R_1^{-1}<|\phi(q)|<1$ and $12h\leq|q-\dot{\mathsf{G}}|$ and for $p\in\bar{\mathsf{G}}_h$ with $|p-q|\geq 12h$.

Corollary 7.6.5 (7.6.13) *yields the weaker estimate*

$$|N(p,q)-N_h(p,q)| \leqslant h^2 c_1' \left\{ \frac{1}{|p-q|^2} + \frac{1}{|p_0-q|^2} \right\}$$

$$+ c_2' h^2 |\log h| \frac{1}{\min\limits_{r \in \partial G_h} |r-q|^3}. \qquad (7.6.15)$$

Remark For a rectangular domain, Deeter and Springer have shown an $0(h|\log h|)$ estimate in [12] instead of (7.6.13) if $|p-q| \geqslant \delta > 0$ independently of h. But for the system (7.0.2) we shall need the estimates up to the boundary.

Proof The proof combines inner estimates with the behaviour of $N(p, q)$ near $\dot{G}$ according to (1.1.14). If we use the representation formula (7.5.31) for $F(p) \equiv N_h(p,q) - N(p,q) - (1/2\pi)\log|p-q| + \Gamma_h(p,q) + (1/2\pi)\log|p_0-q| - \Gamma_h(p_0,q)$ then we find with (7.6.2) the estimate

$$\left|N_h(p,q) - N(p,q) - \frac{1}{2\pi}\log|p-q| + \Gamma_h(p,q) + \frac{1}{2\pi}\log|p_0-q| - \Gamma_h(p_0,q)\right|$$

$$\leqslant h^3 \sum_{r \in \partial G_h} N_h(p,q) K |r-q|^{-3}$$

$$+ h^2 \sum_{r \in G_h^*} N_h(p,r) |\Delta_{hr} \Gamma_h(r,q)|$$

$$+ h^2 \sum_{r \in G_h \cup G_h^*} N_h(p,r) \left| \Delta_{hr}\left(N(r,q) + \frac{1}{2\pi}\log|r-q| \right) \right|.$$

Inserting (7.6.12) into the second sum we have with (7.4.3) and (7.4.5)

$$|N_h(p,q) - N(p,q)| \leqslant kh^2 \left\{ \frac{1}{|p-q|^2} + \frac{1}{|p_0-q|^2} \right\}$$

$$+ h^3 \sum_{r \in \partial G_h} N_h(p,r) K |r-q|^{-3} + h^3 \sum_{r \in G_h^*} N_h(p,r) K |r-q|^{-3}$$

$$+ h^2 \sum_{r \in G_h \cup G_h^*} N_h(p,r) \left| \Delta_{hr}\left(N(r,q) + \frac{1}{2\pi}\log|r-q| \right) \right|. \qquad (7.6.16)$$

In the first case (7.6.13), $|\phi(q)| \leqslant (1/R_1) < 1$, we can use (7.4.25), $|r-q| \geqslant k > 0$. Moreover, for these q, the function $N(r,q) + (1/2\pi)\log|r-q|/|p_0-q|$ is harmonic in $r \in G$ and $C^{4+\alpha}(\bar{G})$. Therefore the discrete Laplacian is of order h^2 and with (7.5.32–34) we find (7.6.13), the first of the desired estimates.

In the second case (7.6.14), $R_1^{-1} < |\phi(q)| < 1$, we use (7.4.29) in the terms of (7.6.16) with $r \in \mathsf{G}_h \cup \mathsf{G}_h^*$. In the last expression we use (1.1.14) yielding the representation

$$N(r, q) + \frac{1}{2\pi} \log \frac{|r-q|}{|p_0 - q|} = \frac{1}{2\pi} \log \left\{ \frac{|r-q|\,|\phi(q)|\,|r-q'|\,|\Phi(q')|}{|\phi(r)-\phi(q)|\,|p_0-q|\,|\phi(r)-\Phi(q')|} \right\}$$

$$-\frac{1}{2\pi} \log |r-q'| = -\frac{1}{2\pi} \log |r-q'| + H'(r, q) \tag{7.6.17}$$

where $H'(r, q)$ together with all its derivatives up to the fourth order with respect to $r = (x, y)$ is continuous in $\bar{\mathsf{G}} \times \{q \,|(1/R_1) \le |\phi(q)| \le 1\}$. Hence (7.4.31) holds also for $N + (1/2\pi) \log |r-q|$ implying the estimate

$$|N_h(p, q) - N(p, q)| \le kh^2 \left\{ \frac{1}{|p-q|^2} + \frac{1}{|p_0 - q|^2} \right\}$$

$$+ Ch^3 \left\{ \sum_{r \in \partial \mathsf{G}_h} N_h(p, r)\,|r-q'|^{-3} + \sum_{r \in \mathsf{G}_h{}^*} N_h(p, r)\,|r-q'|^{-3} \right\}$$

$$+ kh^4 \sum_{r \in \mathsf{G}_h - \{p_0\}} N_h(p, r) \frac{1}{|r-q'|^4}. \tag{7.6.18}$$

For the last sum we use (7.4.17), and then (7.5.31),

$$h^4 \sum_{r \in \mathsf{G}_h - \{p_0\}} N_h(p, r) \frac{1}{|r-q'|^4} \le h^2 \{|p-q'|^{-2} + |p_0 - q'|^{-2}\}$$

$$+ h^3 \sum_{r \in \partial \mathsf{G}_h} N_h(p, r) \delta_{nr}(|r-q'|^{-2}) - \sum_{r \in \mathsf{G}_h{}^*} N_h(p, r)\, \Delta_{hr} \frac{h^4}{|r-q'|^2}.$$

Here $\delta_{nr}(|r-q'|^{-2})$ can be replaced by its definition together with (7.4.17). Inserting the resulting estimate into (7.6.18) we find the desired inequality (7.6.14).

The Corollary 7.6.5 follows from (7.6.13) and (7.6.14) with (7.4.8) and (7.5.32), (7.5.33).

Since Theorem 7.6.4 excludes points $|p-q| < 12h$ and $|q - \dot{\mathsf{G}}| < 12h$ we need some estimates for these remaining points. In contrast to Theorem 7.3.6, here we can only show:

Theorem 7.6.6 *The discrete Neumann function satisfies the pointwise inequality*

$$0 \le N_h(p,q) \le C\Big[|\log h| + \sum_{r\in\partial G_h} N_h(p,r)\Big\{\frac{1}{|r-q|} + h^2 \sum_{t\in G_h{}^*} G_h(r,t) h_t^{-1} |t-q'(q)|^{-2} + h^2 \sum_{t\in\partial G_h} G_h(r,t)\,|t-q'(q)|^{-2}\Big\}\Big] \tag{7.6.19}$$

for all $p \in G_h \cup G_h^*$ *and* $q \in G_h$ *with* $|q-\dot{G}| > 12h$.

Proof The representation formula (7.5.31) for the special function $F(p) \equiv G_h(p,q) - G_h(p_0,q)$ (q fixed) yields

$$0 \le N_h(p,q) \le |G_h(p,q) - G_h(p_0,q)| + h \sum_{r\in\partial G_h} N_h(p,r)\,|\delta_{nr} G_h(r,q)|. \tag{7.6.20}$$

For $\delta_{nr} G_h(r,q)$ we use

$$|\delta_{nr} G^I(r,q)| \le C\,|r-q|^{-1} \tag{7.6.21}$$

which follows from $|\delta_{nr} F(r)| \le \max_{|r-p|\le 10h} |D^1 F(p)|$, $|r-q| \ge 12h$ and (1.1.10), together with Theorem 7.4.5,

$$|\delta_{nr} G_h(r,q)| \le C|r-q|^{-1} + c_1 h |r-q|^{-2} + c_2 h^2 \sum_{t\in G_h{}^*} G_h(r,t) h_t^{-1} |t-q'|^{-2} + c_2 h^2 \sum_{t\in\partial G_h} G_h(r,t)\,|t-q'|^{-2}. \tag{7.6.22}$$

In the second term we use $h \le (1/12)\,|r-q|$ and insert the resulting inequality (7.6.22) into (7.6.20). This implies with (7.3.26) the desired inequality (7.6.19).

For the remaining points we have

Lemma 7.6.7 *For all* $p \in G_h \cup G_h^* \cup \partial G_h$

$$h \sum_{q\in G_h\cup G_h{}^* \wedge 0\le|q-\dot{G}|\le 12h} N_h(p,q) \le C\,|\log h| \tag{7.6.23}$$

where C *is independent of* p *and* h.

Since the proof repeats all arguments of the proof to (7.5.33) using the same mesh functions as have been used in the proof of Lemma 7.3.7 we shall omit the details here.

Since all estimates for $N_h(p, q)$ are much weaker than the corresponding estimates for $G_h(p, q)$ we shall also have a weaker main estimate which can only be used if the right hand side of Δv in (7.0.2) vanishes at $\dot{\mathsf{G}}$. For this purpose let us use a function $\Psi \in C^{4+\alpha}(\bar{\mathsf{G}})$ having the properties

$$\Psi(q)=0 \quad \text{on} \quad \dot{\mathsf{G}} \quad \text{and} \quad 0 \leqslant \Psi(q) \leqslant 2\delta^{-1}|q-\dot{\mathsf{G}}| \quad \text{for} \quad q \in \mathsf{G} \quad \text{and}$$
$$\Psi(q)=1 \quad \text{for all} \quad q \in \mathsf{G} \quad \text{with} \quad |q-\dot{\mathsf{G}}| \geqslant \delta. \tag{7.6.24}$$

Here $\delta>0$ is a constant which will be chosen later on. For the weight function Ψ we have:

Lemma 7.6.8 *The following estimates are valid for* Ψ:

$$\Psi(q) \leqslant |r-q| \frac{1}{\delta} \quad \text{for all} \quad r \in \dot{\mathsf{G}} \quad \text{and} \quad q \in \bar{\mathsf{G}}, \tag{7.6.25}$$

$$\Psi(q) \leqslant |r-q| \frac{24}{11\delta} \quad \text{for} \quad r \in \mathsf{G}_h^* \quad \text{and} \quad 12h \leqslant |q-\dot{\mathsf{G}}|, \tag{7.6.26}$$

$$\Psi(q) \leqslant \gamma \cdot \delta^{-1}|t-q'(q)| \quad \text{for all} \quad t \in \bar{\mathsf{G}} \quad \text{and} \quad q \in \bar{\mathsf{G}} \quad \text{with} \quad R_1^{-1} \leqslant |\phi(q)| \leqslant 1, \tag{7.6.27}$$

$$\sum_{q:12h \leqslant |q-\dot{\mathsf{G}}| \wedge R_1^{-1} \leqslant |\phi(q)| \leqslant 1} h^2\Psi(q)\,|r-q'(q)|^{-3} \leqslant C\delta^{-1}|\log h| \quad \text{for all} \quad h>0 \quad \text{and} \quad r \in \mathsf{G}_h^* \cup \partial\mathsf{G}_h \tag{7.6.28}$$

where the constants γ *and* C *are independent of* δ, q, t, r *and* h.

Proof The first three inequalities follow from (7.6.24) and (7.4.8) with the triangle inequality. (7.6.28) follows with (7.6.27) and

$$|r'(r)-q| \leqslant k\,|r-q'(q)|$$

with a constant $k>0$ independent of $r \in \mathsf{G}_h^* \cup \partial\mathsf{G}_h$, q and $h \leqslant h_0$. Then we have

$$\sum_{q:12h \leqslant |q-\dot{\mathsf{G}}| \wedge R_1^{-1} \leqslant |\phi(q)| \leqslant 1} h^2\Psi(q)\,|r-q'(q)|^{-3}$$
$$\leqslant \gamma\delta^{-1}k^{-2} \iint_{8h \leqslant |r'(r)-q| \leqslant 2\,\mathrm{diam}\,(\mathsf{G})} |r'(r)-q|^{-2}[\mathrm{d}\xi, \mathrm{d}\eta]$$
$$\leqslant \delta^{-1}C\,|\log h|,$$

the desired estimate, since $|r'-q|^{-2}$ is monotonic.

Theorem 7.6.9 *Let be* $\dot{\mathsf{G}} \in C^{4+\alpha}$, $\alpha>0$. *Then there exists a constant* C

independent of h, δ and p such that

$$\sum_{q\in G_h \cup G_h^* \wedge |p-q|\geq h} h^2\Psi(q)\,|N(p,q)-N_h(p,q)| \leq Ch^2\,|\log h|^2\cdot\delta^{-1} \tag{7.6.29}$$

for all grid points $p\in\bar{G}_h$.

Remarks This inequality corresponds to Scott's estimate in [26]. Scott's estimate is sharper since it does not contain $\Psi(q)$ and is of order $h^2\,|\log h|$. One reason for this is on one hand that the difference approximation of $d_n v$ on $\dot{G}$ in [26] is implicitly given by Ritz equations to the corresponding bilinear form on piecewise linear functions and, hence, is *different* from our δ_n operator (7.5.8). Moreover, Scott considers the Neumann problem with *homogeneous* boundary data to the differential equation

$$-\Delta v + v = f \quad \text{in } G, \quad d_n v_{|\dot{G}} = 0$$

which is different from our problem (7.5.24) containing the point condition and also inhomogeneous boundary data.

For the above mentioned difference equations defined by the Ritz equations and the corresponding natural discrete homogeneous Neumann boundary condition to (7.5.24) *with the point condition*, Höhn's thesis [19] contains estimates yielding a bound (7.6.29) of order $h^2\,|\log h|$—the same as Scott's. But again, the boundary operator is implicitly given and different from δ_n and the boundary data are zero.

Proof of Theorem 7.6.9 According to our estimates in Theorems 7.6.4, 7.6.6 and Lemma 7.6.7 we break the sum up,

$$\begin{aligned}
&\sum_{q\in G_h\cup G_h^*\wedge|p-q|\geq h} h^2\Psi(q)\,|N(p,q)-N_h(p,q)| \\
&\leq \sum_{h\leq|p-q|<12h<|q-\dot{G}|} \Psi(q)h^2\,|N(p,q)-N_h(p,q)| \\
&\quad + \sum_{|q-p|\geq 12h} c_1'h^4\left\{\frac{1}{|p-q|^2}+\frac{1}{|p_0-q|^2}+|\log h|\right\} \\
&\quad + \sum_{12h\leq|p-q|\wedge(1/R_1)<|\phi(q)|\wedge|q-\dot{G}|\geq 12h} c_2'h^5\Psi(q)\left\{\sum_{r\in G_h^*} N_h(p,r)\,|r-q'|^{-3}\right. \\
&\quad \left. + \sum_{r\in\partial G_h} N_h(p,r)\,|r-q'|^{-3}\right\} \\
&\quad + \sum_{|q-\dot{G}|\leq 12h\wedge|q-p|\geq h} h^2\Psi(q)\,|N(p,q)-N_h(p,q)|.
\end{aligned} \tag{7.6.30}$$

For the first sum containing at most 441 points it suffices to prove an order of $h^2 |\log h|^2$ for each term. Since $N(p, q)$ grows at most logarithmically, according to (1.1.14) it remains to consider $\Psi(q)N_h(p, q)$ using Theorem 7.6.6,

$$\Psi(q)N_h(p, q) \leq \Psi(q)C |\log h|$$
$$+h \sum_{r\in\partial G_h} N_h(p, r)\left\{\Psi(q) |r-q|^{-1} + \sum_{t\in G_h^*} G_h(r, t)\frac{h}{h_t}|t-q'|^{-2}h\Psi(q)\right.$$
$$\left.+\sum_{t\in\partial G_h} hG_h(r, t)|t-q'|^{-2}h\Psi(q)\right\}. \tag{7.6.31}$$

Now Lemma 7.6.8. implies

$$\Psi(q)|r-q|^{-1} \leq \delta^{-1}, \qquad |t-q'|^{-2}h\Psi(q) \leq \gamma^2\delta^{-1}. \tag{7.6.32}$$

Inserting (7.6.32) into (7.6.31) and using (7.3.8) and (7.3.10) we find

$$\Psi(q)N_h(p, q) \leq C' |\log h|.$$

The second sum in (7.6.30) can be estimated as in (7.4.37).

In the third sum of (7.6.30) we use (7.6.28) and then (7.5.33) and (7.5.34).

For the last sum of (7.6.30) we use Lemma 7.6.7,

$$\sum_{|q-\dot{G}|\leq 12h \wedge |q-p|\geq h} h^2\Psi(q)|N(p, q)-N_h(p, q)|$$
$$\leq h^2 12h \sum_{|q-\dot{G}|\leq 12h} N_h(p, q) + \text{const}\cdot 12h^3 \sum_{|q-\dot{G}|\leq 12h} |\log h|$$
$$\leq \text{const}\cdot h^2 |\log h|.$$

Hence, all four different sums on the right hand side of (7.6.30) can be estimated as in (7.6.29).

7.7 Remarks on the Normalizing Condition

In (7.0.1) and (7.0.2) respectively we have used a point condition as a side condition whereas in Section 1.1 the side condition is an integral condition,

$$\kappa = \left\{\oint_{\dot{G}} \sigma \, ds\right\}^{-1} \oint_{\dot{G}} v\sigma \, ds. \tag{7.7.1}$$

Hence, it is also sensible to discretize the Neumann problem corresponding to (7.7.1) and to introduce a discrete Neumann function with (7.7.1) instead of the point condition.

If $\sigma = |\phi_z|_{|\dot{G}}$ with ϕ being the conformal mapping satisfying $\phi(p_0) = 0$ then

$$v(p_0) = \frac{1}{2\pi} \oint_{\dot{G}} v(p)\, |\phi_z| \, ds + \frac{1}{2\pi} \iint_{G} \Delta v \log |\phi(z)| \, [dx, dy] \tag{7.7.2}$$

and the point condition and (7.7.1) coincide for every v, i.e. for every solution of the Neumann problem where Δv is given. For more general σ and (7.7.1) let us consider the side condition

$$\begin{aligned} v(p_0) - \oint_{\dot{G}} \left\{ v(z(s)) \frac{1}{2\pi} |\phi_z| - \nu\sigma \right\} ds &= \mu \\ &\equiv \nu\kappa + \frac{1}{2\pi} \iint_{G} \Delta v \log |\phi(z)| \, [dx, dy] \end{aligned} \tag{7.7.3}$$

with

$$\nu \equiv (\max \sigma)^{-1} \cdot \min |\phi_z| > 0 \tag{7.7.4}$$

where the right hand side μ is given. This side condition makes the solution of the Neumann problem unique since for $v \equiv 1$ the left hand side of (7.7.3) becomes $\nu \oint \sigma \, ds > 0$.

Replacing the integration in (7.7.3) by a numerical quadrature with weights $w_q > 0$ (see e.g. Isaacson and Keller [20, Chap. 7]) we get

$$V(p_0) - \sum_{q \in \partial G_h} V(q) w_q \left\{ \frac{1}{2\pi} |\phi_z(q)| - \nu\sigma(q) \right\} = \mu \tag{7.7.5}$$

with given μ from (7.7.3) which replaces the last equation in (7.5.25). The corresponding system for the new Neumann function $\tilde{N}_h$ is now

$$\begin{aligned} &-\Delta_{hp} \tilde{N}_h(p, q) = h^{-2} \delta(p, q) \quad \text{for all} \quad p \in G_h \cup G_h^* - \{p_0\}, \\ &\delta_{np} \tilde{N}_h(p, q) = h^{-1} \delta(p, q) \quad \text{for} \quad p \in \partial G_h \quad \text{and} \\ &\tilde{N}_h(p_0, q) - \sum_{r \in \partial G_h} w_r \left(\nu\sigma(r) - |\phi_z(r)| \frac{1}{2\pi} \right) \tilde{N}_h(r, q) = 0 \end{aligned} \tag{7.7.6}$$

where $q \in \bar{G}_h$.

Corresponding to Lemma 7.5.1 we have again:

Lemma 7.7.1 *The matrix $\tilde{B}$ to (7.7.6) is of positive type.*

Proof In the last equation of (7.7.6) we have now on the left side

$$\tilde{b}_{p_0 p_0} = 1 \quad \text{and}$$

$$\tilde{b}_{p_0 q} = \begin{cases} 0 & \text{for} \quad q \in \partial G_h \quad \text{and} \quad q \neq p_0 \quad \text{and} \\ w_q \left(\nu\sigma(q) - \frac{1}{2\pi} |\phi_z(q)| \right) < 0 \end{cases}$$

according to (7.7.4). If h is small enough then the relation

$$\sum_{q\in\partial G_h} \tilde{b}_{p_0q} + \tilde{b}_{p_0p_0} = 1 - \sum_{q\in\partial G_h} w_q\left(\frac{1}{2\pi}|\phi_z(q)| - \nu\sigma(q)\right)$$

$$= 1 - \oint_{\dot{G}} \left(\frac{1}{2\pi}|\phi_z| - \nu\sigma\right) ds + \text{order}(h^2)$$

$$= \nu\oint_{\dot{G}} \sigma\, ds + \text{order}(h^2) > 0$$

is fulfilled since $\nu\oint\sigma\, ds > 0$. Thus, $p_0 \in J \neq \varnothing$.

The remaining part of the proof coincides with the proof of Lemma 7.5.1.

Since the equations (7.7.6) possess a unique (positive) solution $\tilde{N}_h$ we find:

Lemma 7.7.2 *$\tilde{N}_h$ is nonnegative,*

$$\tilde{N}_h(p, q) \geq 0, \tag{7.7.7}$$

and $\tilde{N}_h$ provides the representation formula

$$F(p) = -h^2 \sum_{q\in G_h\cup G_h^*} \tilde{N}_h(p, q)\, \Delta_h F(q)$$

$$+ h \sum_{q\in\partial G_h} \tilde{N}_h(p, q)\delta_n F(q) + b[F(p_0) - \sum_{q\in\partial G_h} \tilde{b}_{p_0q} F(q)] \tag{7.7.8}$$

for all $p \in \bar{G}_h$ where b is given by

$$b \equiv \left\{1 - \sum_{q\in\partial G_h} w_q\left(\frac{1}{2\pi}|\phi_z(q)| - \nu\sigma(q)\right)\right\}^{-1} > 0. \tag{7.7.9}$$

The discrete Neumann function $\tilde{N}_h$ is the counterpart to

$$\tilde{N}(p, q) \equiv G^{II}(p, q) - G^{II}(p, p_0) \tag{7.7.10}$$

where G^{II} denotes the Green function of the second kind corresponding to σ [17, Section 4.7]. Now it can be shown that *all the results of Section* 7.6 *are also true for $\tilde{N}_h - \tilde{N}$.*

But the disadvantage of the new side condition (7.7.3) involving (7.7.1) is that we need $|\phi_z|_{|\dot{G}}$, the derivative of the conformal mapping on the boundary. But if $|\phi_z|_{|\dot{G}}$ is known then $\phi(z)$ can be found explicitly (see Gaier [16]) by a simple boundary integral. With $\phi(z)$ one may either use the method of Chapter 6 or transform (7.0.1) and (7.0.2) respectively into a problem in the unit disk and then discretize the new problem with some method which profits from the special domain.

7.8 A Convergent Finite Difference Method for the First-Order System

Since the preceding analysis is available for the Laplacian we have replaced (7.0.1) by (7.0.2). Before we formulate the corresponding difference equations and asymptotic error estimates let us prove the equivalence of both problems.

Theorem 7.8.1 *The problems* (7.0.1) *and* (7.0.2) *are equivalent for* $w = u + iv \in C^2(\mathrm{G}) \cap C^{1+\alpha}(\bar{\mathrm{G}})$.

Proof Since (7.0.2) was obtained from (7.0.1) by differentiation every solution of (7.0.1) satisfies (7.0.2) trivially.

If w is a given solution of (7.0.2) then the complex valued function

$$f \equiv \{u_x - v_y - (Au + Bv + C)\} - i\{u_y + v_x - (\tilde{A}u + \tilde{B}v + \tilde{C})\} \tag{7.8.1}$$

is well defined in $\bar{\mathrm{G}}$, $f \in C^1(\mathrm{G}) \cap C^\alpha(\bar{\mathrm{G}})$. From (7.0.2) it follows that f is *holomorphic*,

$$f_{\bar{z}} = 0 \quad \text{in G and} \quad \operatorname{Re} f\dot{z} = 0 \quad \text{on } \dot{\mathrm{G}}. \tag{7.8.2}$$

Hence f solves a homogeneous boundary value problem (1.2.5) of characteristic

$$n = \frac{1}{2\pi} \oint_{\dot{\mathrm{G}}} \mathrm{d} \arg \dot{z} \,\mathrm{d}s = 1 > 0. \tag{7.8.3}$$

Thus, f vanishes identically according to (1.3.2), i.e. u and v satisfy the system (7.0.1).

For the difference equations to (7.0.2) we also need besides Δ_h first order differences for the right hand side terms which we choose to be

$$\begin{aligned} \partial_{hx} F(p) &\equiv \frac{1}{2h}\{F(x+h+iy) - F(x-h+iy)\}, \quad \text{for all} \quad p = (x+iy) \in \mathrm{G}_h \\ \partial_{hy} F(p) &\equiv \frac{1}{2h}\{F(x+i(y+h)) - F(x+i(y-h))\} \end{aligned} \tag{7.8.4}$$

and

$$\begin{aligned} \partial_{hx} F(p) &\equiv \frac{1}{h_1 + h_3}\{F(x+h_1+iy) - F(x-h_3+iy)\} \quad \text{for} \quad p = x+iy \in \mathrm{G}_h^*. \\ \partial_{hy} F(p) &\equiv \frac{1}{h_2 + h_4}\{F(x+i(y+h_2)) - F(x+i(y-h_4))\} \end{aligned} \tag{7.8.5}$$

where $\{x+h_1+iy,\ x+i(y+h_2),\ x-h_3+iy,\ x+i(y-h_4)\}$ denote the four neighbours of $x+iy$. Clearly, (7.8.4) is an approximation of order h^2 and

(7.8.5) of order h [14, p 186], i.e.

$$|\partial_{hx}F(p)-F_x(p)|\leqslant h^2\tfrac{1}{6}\max|F_{xxx}| \quad \text{for} \quad p\in \mathrm{G}_h \tag{7.8.6}$$

and

$$|\partial_{hx}F(p)-F_x(p)|\leqslant \tfrac{1}{2}(h_1+h_3)\max|F_{xx}| \quad \text{for} \quad p\in \mathrm{G}_h^*. \tag{7.8.7}$$

To (7.0.2) let us design the

Difference Equations for the System

$$\Delta_h U(p)=\partial_{hx}\{\Psi(AU+BV)+C\}+\partial_{hy}\{\Psi(\tilde{A}U+\tilde{B}V)+\tilde{C}\} \qquad \text{for all } p\in \mathrm{G}_h\cup \mathrm{G}_h^*,$$

$$\Delta_h V(p)=-\partial_{hy}\{\Psi(AU+BV)+C\}+\partial_{hx}\{\Psi(\tilde{A}U+\tilde{B}V)+\tilde{C}\} \qquad \text{for all } p\in \mathrm{G}_h\cup \mathrm{G}_h^*-\{p_0\}, \tag{7.8.8}$$

$$U(p)=\psi(p) \text{ for } \quad p\in\partial\mathrm{G}_h$$

and

$$\begin{aligned}\delta_n V(p)&=-\dot{\psi}(p)+C\dot{x}+\tilde{C}\dot{y}\\ &\quad-\sum_{j=1}^{3}\alpha_j(p)x_j(p)y_j(p)\{-\ddot{\psi}(p)+(C\dot{x})^{\cdot}+(\tilde{C}\dot{y})^{\cdot}\}\\ &\equiv\gamma(p) \quad \text{for} \quad p\in\partial\mathrm{G}_h,\end{aligned}$$

$$V(p_0)=v_0.$$

Remark For the special case

$$A=B=\tilde{A}=\tilde{B}\,|_{\dot{\mathrm{G}}}=0 \quad \text{on } \dot{\mathrm{G}} \text{ we choose } \quad \Psi\equiv 1 \text{ in (7.8.8).} \tag{7.8.9}$$

In the general case Ψ is a sequence of functions (7.6.24) belonging to a sequence δ_h which will be chosen later on. Although we shall need some further auxiliary estimates before we can prove the convergence of $U+iV$ to w let us state our main result:

Theorem 7.8.2 *There exist constants $C_j\geqslant 0$ $j=1,2,3$ and $h_0>0$ such that*

(i) *the difference equations* (7.8.8) *are uniquely solvable for every combination of $\delta_h>0$ and $h>0$ satisfying*

$$\lim_{h\to 0}\delta_h^{-1}h|\log h|^2=0 \quad \textit{and} \quad h\leqslant h_0, \tag{7.8.10}$$

(ii) *the error estimate*

$$\max_{p\in G_h} |U+iV-u-iv| \leq C_1\delta_h^{-3-\gamma}h^2|\log h|\,\|w\|_{4+\gamma}+C_2\delta_h|\log \delta_h|\,\|w\|_0 \tag{7.8.11}$$

holds for a solution $w\in C^3(\bar{G})$ of (7.0.1).†

In case of $A=B=\tilde{A}=\tilde{B}=0$ on $\dot{G}$, (7.8.11) is valid with $C_2=0$ and $\delta_h=1$.

Corollary 7.8.3 *In general we have an error estimate*

$$\max_{p\in \bar{G}_h} |W-w| \leq (C_1+C_2)h^{1/2-\varepsilon} \quad \|w\|_{4+10\varepsilon} \tag{7.8.12}$$

with any $\varepsilon>0$ if we choose a sequence of functions Ψ in (7.6.24) with $\delta_h=h^{1/2-\varepsilon}$.

This choice of δ is optimal with respect to the orders of h and $|\log h|$ and (7.8.11) if $C_1>0$ and $C_2>0$.

For the proof of Theorem 7.8.2 we shall use an auxiliary function $\tilde{w}=\tilde{u}+i\tilde{v}$ which is defined as the solution of the boundary value problem

$$\begin{aligned} &\tilde{w}_{\bar{z}}=\Psi(\bar{a}\tilde{w}+b\bar{\tilde{w}})+c,\\ &\tilde{u}\,|_{\dot{G}}=\psi,\ \tilde{v}(p_0)=v_0 \end{aligned} \tag{7.8.13}$$

where a, b, c are given by (1.1.6).

Lemma 7.8.4 *The auxiliary function $\tilde{w}$ satisfies*

$$\|w-\tilde{w}\|_0 \leq C\delta|\log \delta|\,\|w\|_0, \tag{7.8.14}$$

$$\|\tilde{w}\|_{4+\gamma} \leq C'\delta^{-3-\gamma}\|w\|_{4+\gamma} \tag{7.8.15}$$

for every $\gamma>0$ where the constants C, C' are independent of δ and w for all $\delta\leq\delta_0$ and all w with a suitable $\delta_0>0$.

Proof From (1.1.25) and with (7.6.1), $N(p,z)=G^{II}(z,p)-G^{II}(z,p_0)$ we find the integral equation for $w-\tilde{w}$,

$$\begin{aligned}(w-\tilde{w})(p)&=\mathsf{K}_\Psi(w-\tilde{w})+(\mathsf{K}_1-\mathsf{K}_\Psi)w\\
&=\iint_G \Psi\{(A(u-\tilde{u})+B(v-\tilde{v}))(G^I_x(p,z)-N_y(p,z))\\
&\quad+i(\tilde{A}(u-\tilde{u})+B(v-\tilde{v}))(G^I_y(p,z)+N_x(p,z))\}[dx,dy]\\
&\quad+\iint_G (1-\Psi)\{(Au+Bv)(G^I_x(p,z)-N_y(p,z))\\
&\quad+i(\tilde{A}u+\tilde{B}v)(G^I_y(p,z)+N_x(p,z))\}[dx,dy].\end{aligned} \tag{7.8.16}$$

† The norms here and in Section 7.9 are maximum norms or Hölder norms.

K_Ψ is essentially the operator in (1.1.25) with coefficients $\Psi A, \ldots, \Psi\tilde{B}$. Hence, according to Theorem 1.1.2, the operator $(\mathsf{I}-\mathsf{K}_\Psi)$ is continuously invertible in C^0. Since in

$$(\mathsf{K}_1-\mathsf{K}_\Psi)f$$

the integration takes only place in the strip $|z-\dot{\mathsf{G}}|\leq\delta$, it follows from (1.1.15) that an inequality of the form

$$\|\mathsf{K}_1-\mathsf{K}_\Psi\|\leq K\,\delta|\log\delta| \tag{7.8.17}$$

holds with respect to the operator norm induced by C^0 where the constant K is independent of δ. Hence, the inverses are uniformly bounded,

$$\|(\mathsf{I}-\mathsf{K}_\Psi)^{-1}\|\leq C_0 \tag{7.8.18}$$

for all $0<\delta\leq\delta_0$. Multiplication of (7.8.16) with $(\mathsf{I}-\mathsf{K}_\Psi)^{-1}$ and estimates (7.8.17) and (7.8.18) yield (7.8.14). The estimate (7.8.15) can be obtained by differentiating

$$(w-\tilde{w})_{\bar{z}}=\Psi\{\bar{a}(w-\tilde{w})+b(\bar{w}-\bar{\tilde{w}})\}+(1-\Psi)(aw+b\bar{w})$$

three times, using the *a priori* estimate (4.1.23) for the corresponding system together with

$$\mathrm{Re}\,(w-\tilde{w})=0 \quad \text{on } \dot{\mathsf{G}} \tag{7.8.19}$$

and its derivations along $\dot{\mathsf{G}}$ and the further use of

$$\|\tilde{w}\|_\gamma\leq k\|w\|_\gamma \tag{7.8.20}$$

with k independent of δ. (7.8.20) follows from (7.8.16).

In the following we shall compare $W=U+\mathrm{i}V$, the solution of the difference equations (7.8.8) with $\tilde{w}=\tilde{u}+\mathrm{i}\tilde{v}$, the auxiliary solution of (7.8.13) since the coefficients $\Psi\bar{a}, \Psi b$ of the differential equations vanish at $\dot{\mathsf{G}}$. To this end we consider both as solutions of corresponding integral equations. With the representation formulas (7.3.7) for U and (7.5.31) for V in connection with (7.8.8) the difference equations (7.8.8) become equivalent to the system

$$
\begin{aligned}
U(p) = &-h^2 \sum_{q \in \mathsf{G}_h \cup \mathsf{G}_h{}^*} G_h(p, q)\{\partial_{hx}(\Psi(AU+BV)) + \partial_{hy}(\Psi(\tilde{A}U+\tilde{B}V))\}(q) \\
&-h^2 \sum_{q \in \mathsf{G}_h \cup \mathsf{G}_h{}^*} G_h(p, q)\{\partial_{hx}C + \partial_{hy}\tilde{C}\}(q) \\
&+h \sum_{q \in \partial\mathsf{G}_h} G_h(p, q)\psi(q),
\end{aligned} \tag{7.8.21}
$$

$$
\begin{aligned}
V(p) = &-h^2 \sum_{q \in \mathsf{G}_h \cup \mathsf{G}_h{}^* - \{p_0\}} N_h(p, q)\{-\partial_{hy}(\Psi(AU+BV)) \\
&+\partial_{hx}(\Psi(\tilde{A}U+\tilde{B}V))\}(q) \\
&-h^2 \sum_{q \in \mathsf{G}_h \cup \mathsf{G}_h{}^* - \{p_0\}} N_h(p, q)\{-\partial_{hy}C + \partial_{hx}\tilde{C}\}(q) \\
&+h \sum_{q \in \partial\mathsf{G}_h} N_h(p, q)\delta_n V(q) + v_0
\end{aligned}
$$

for all $p \in \bar{\mathsf{G}}_h$ where $\delta_n V(q)$ is given by (7.8.8). As in Section 6.3, we extend the grid function $W(p)$ to a piecewise constant function $(\Pi W)(z) = W(z)$ with the prolongation operator Π defined by

$$
W(z) \equiv (\Pi W)(z) \equiv W(p) \quad \text{for all } (x, y) \in F_p \tag{7.8.22}
$$

where the surface pieces $F_p \ni p$ are defined in the following way:
If $p \in \mathsf{G}_h \cup \mathsf{G}_h^*$ then $p = \xi + \mathrm{i}\eta$ has four neighbours on the grid,

$$
\{(\xi + h_1 + \mathrm{i}\eta), (\xi + \mathrm{i}(\eta + h_2)), (\xi - h_3 + \mathrm{i}\eta), (\xi + \mathrm{i}(\eta + h_4))\}.
$$

Then we choose

$$
F_p \equiv \{z \in \bar{\mathsf{G}} \mid \xi - h_3/2 < x \leq \xi + h_1/2 \text{ and } \eta - h_4/2 < y \leq \eta + h_2/2\}. \tag{7.8.23}
$$

For boundary points p', e.g. if $p' = \xi_1 + h_1 + \mathrm{i}\eta \in \partial\mathsf{G}_h$ is neighbour of p, we define

$$
F_{p'} \equiv \{z \in \bar{\mathsf{G}} \mid \xi + h_1/2 < x \quad \text{and} \quad \eta - h_4/2 < y\} \tag{7.8.24}
$$

and correspondingly for the remaining cases. Now (7.8.21) can be interpreted as an *integral equation* in the Banach space B of the Baire functions of first category equipped with the supremum norm $\|\cdot\|_0$. Let us denote this equation by

$$
W(z) = (\Pi W(p)) = \mathsf{K}_h W + F_h. \tag{7.8.25}
$$

Correspondingly, we rewrite (7.8.13) into an integral equation for $\tilde{w}$, using the Green and Neumann functions G^I and $N(p, z)$ (7.6.1) (see

(1.1.25)) finding

$$\tilde{u}(p) = -\iint_G G^I(p,z)\{(\Psi(A\tilde{u}+B\tilde{v}))_x + (\Psi(\tilde{A}\tilde{u}+\tilde{B}\tilde{v}))_y\}[\mathrm{d}x,\mathrm{d}y]$$

$$-\iint_G G^I(p,z)(C_x+\tilde{C}_y)[\mathrm{d}x,\mathrm{d}y] - \oint_{\dot{G}} \psi\,\mathrm{d}_n G^I, \tag{7.8.26}$$

$$\tilde{v}(p) = -\iint_G N(p,z)\{-(\Psi(A\tilde{u}+B\tilde{v}))_y + (\Psi(\tilde{A}\tilde{u}+\tilde{B}\tilde{v}))_x\}[\mathrm{d}x,\mathrm{d}y]$$

$$-\iint_G N(p,z)(-C_y+\tilde{C}_x)[\mathrm{d}x,\mathrm{d}y]$$

$$+\oint_{\dot{G}} N(p,z)(-\mathrm{d}\psi + C\,\mathrm{d}x + \tilde{C}\,\mathrm{d}y) + v(p_0).$$

Let us write these equations in short form as

$$\tilde{w} = \mathsf{K}_\delta \tilde{w} + F. \tag{7.8.27}$$

The operator K_δ in (7.8.26) is at first defined only for $\tilde{w} \in C^1$. But integration by parts yields:

Proposition 7.8.5 K_δ *in* (7.8.26) *coincides with* K_Ψ *defining a completely continuous operator from* B *into* C^0.

If we choose $\delta = \delta_h$, a sequence satisfying (7.8.10) then we can prove:

Theorem 7.8.6 *The operator families* $\{\mathsf{K}_h\}$ *and* $\{\mathsf{K}_{\delta_h}\}$ *are collectively compact in* B *and satisfy*

$$\lim_{h\to 0} \|(\mathsf{K}_h - \mathsf{K}_{\delta_h})\mathsf{K}_h\| = 0 \tag{7.8.28}$$

provided (7.8.10) *holds where* $\|\cdot\|$ *denotes the operator norm induced by* B.

Proof The collective compactness of $\{\mathsf{K}_{\delta_h}\}$ follows from that of $\{\mathsf{K}_\Psi\}$. The latter can be obtained from the weak singularity of the corresponding kernels and their continuity properties. From (1.1.27) and the corresponding estimates for G_y^I and the first derivatives of N it follows that the set of functions $\{\mathsf{K}_\Psi f\}$ with $\|f\|_0 \leq 1$ is uniformly Hölder continuous independently of δ. Hence this set is relatively compact in C^0 due to the Arzela–Ascoli Theorem.

For the remaining properties it suffices to consider only one pair of corresponding terms in (7.8.21) and (7.8.26) since all remaining pairs can be treated in exactly the same manner. Let us choose e.g.

$$\mathsf{T}_h f \equiv h^2 \sum_{q \in \mathsf{G}_h \cup \mathsf{G}_h^* - \{p_0\}} N_h(p, q) \partial_{hy} (\Psi A f)(q),$$

$$\mathsf{T}_{\delta_h} f \equiv \iint_{\mathsf{G}} N(p, z)(\Psi A f)_y [\mathrm{d}x, \mathrm{d}y]. \tag{7.8.29}$$

For technical reasons let us further introduce the auxiliary operators

$$\mathsf{T}'_h f \equiv h^2 \sum_{q \in \mathsf{G}_h \cup \mathsf{G}_h^* - \{p_0\} \wedge |q-p| \geq h} N(p, q) \partial_{hy} (\Psi A f)(q),$$

$$\mathsf{T}''_h f \equiv \sum_{q \in \mathsf{G}_h \cup \mathsf{G}_h^*} \iint_{F_q} N(p, z) \Pi \partial_{hy} (\Psi A f)[\mathrm{d}x, \mathrm{d}y]. \tag{7.8.30}$$

Proposition 7.8.7

$\{\mathsf{T}''_h\}$ *with* $\mathsf{B} \to C^0$ *is collectively compact*
i.e. *the set* $\{\mathsf{T}''_h f \mid f \in \mathsf{B} \wedge \|f\|_0 \leq 1\}$ *is relatively compact in* C^0; (7.8.31)

$$\lim_{h \to 0} \|(\mathsf{T}''_h - T_{\delta_h}) f\|_0 = 0 \text{ for every } f \in C^0. \tag{7.8.32}$$

To prove this proposition we use summation by parts for T''_h which yields the representation

$$T''_h f(p) = \sum_{q \in \mathsf{G}_h^1} \iint_{F_q} \frac{1}{2h} \{N(p, z - \mathrm{i}h) - N(p, z + \mathrm{i}h)\}[\mathrm{d}x, \mathrm{d}y](\Psi A f)(q)$$
$$+ \frac{1}{h} \sum_{q \in (\mathsf{G}_h \cup \mathsf{G}_h^*) \setminus \mathsf{G}_h^1} \iint_{F_q} N(p, z)[\mathrm{d}x, \mathrm{d}y] \sum_{l=-1}^{1} \beta_{ql} (\psi A f)(q + ilh_l) \tag{7.8.33}$$

where $\mathsf{G}_h^1 \equiv \{q \in \mathsf{G}_h \mid \text{all neighbours of } q \text{ belong to } \mathsf{G}_h\}$ and where the β_{ql} are suitable coefficients which satisfy $|\beta_{ql}| \leq 1$. If we use with (7.8.33) the estimates (1.1.27), (7.8.10),

$$\left| \log \left| \frac{z - p_1}{z - p_2} \right| \right| \leq C(\alpha) |p_1 - p_2|^\alpha \{|z - p_1|^{-\alpha} + |z - p_2|^{-\alpha}\} \left| \log \left| \frac{z - p_1}{z - p_2} \right| \right|$$

and $\Psi \leq c\delta^{-1}h$ in the boundary strip then we obtain

$$|(\mathbf{T}_h''f)(p_1)-(\mathbf{T}_h''f)(p_2)|$$

$$\leq C|p_1-p_2|^\alpha \max_{p_1,p_2\in\bar{G}} \iint_G \{|z-p_1|^{-1}+|z-p_2|^{-1}\}[dx, dy]$$

$$+C\iint_G |p_1-p_2|\,|z-p_1|^{-1}|z-p_2|^{-1}[dx, dy]$$

$$+C|p_1-p_2|^\alpha \max_{p_1,p_2\in G} \delta^{-1} \iint_{|z-\dot{G}|\leq 4h} \{|z-p_1|^{-\alpha}+|z-p_2|^{-\alpha}\}$$

$$\times\left|\log\left|\frac{z-p_1}{z-p_2}\right|\right|[dx, dy]$$

$$\leq C|p_1-p_2|^\alpha\{1+\delta_h^{-1}h|\log h|\} \tag{7.8.34}$$

for $\|f\|_0 \leq 1$ and with $0<\alpha<1$.
Hence, the functions $\{T_h''f(p)\mid \|f\|_0\leq 1\}$ are uniformly Hölder continuous due to (7.8.10) and form a relatively compact subset in C^0 according to the well known Arzela–Ascoli Theorem.

Concerning (7.8.32) it is clear that the difference approximation $\Pi\partial_{hy}(\Psi Af)$ converges uniformly to $(\Psi Af)_y$. Since $C^1 \subset C^0$ is densely embedded and the sequence $\mathbf{T}_h''-\mathbf{T}_{\delta_h}$ is uniformly bounded due to (7.8.31) and (7.8.17) we find (7.8.32) is also valid for each $f\in C^0$.

As a consequence of Proposition 7.8.6 we have

$$\lim_{h\to 0} \|(\mathbf{T}_h''-\mathbf{T}_{\delta_h})\mathbf{T}_h''\|=0 \tag{7.8.35}$$

with respect to the operator norm induced by $\mathbf{B}$ [1, Proposition 1.7, p. 7].

For $\mathbf{T}_h$ and $\mathbf{T}_h'$, (7.8.29) and (7.8.30), we observe

Proposition 7.8.8

$$\lim_{h\to 0}\|\mathbf{T}_h-\mathbf{T}_h'\|=0, \qquad \lim_{h\to 0}\|\mathbf{T}_h'-\mathbf{T}_h''\|=0. \tag{7.8.36}$$

The first limit follows essentially from our main Theorem 7.6.9 together with Theorem 7.6.6, Lemma 7.6.8 and Lemma 7.5.3 and (7.8.10).

For the second limit we use the fact that $\mathbf{T}_h'$ can be considered as the numerical evaluation of $\mathbf{T}_h''$ by using the rectangular rule. For the error of this quadrature formula [20, p. 300] we have the estimate

$$\left|h^2N(p,q)-\iint_{F_q} N(p,z)[\mathrm{d}x,\mathrm{d}y]\right|$$

$$\leqslant Ch^2\iint_{F_q}\{|z-p|^{-2}+|z-p_0|^{-2}\}[\mathrm{d}x,\mathrm{d}y] \tag{7.8.37}$$

if we restrict p, q to $|q-p|\geqslant h$, $q\neq p_0$ and $|q-\dot{\mathsf{G}}|>2h$ using (1.1.15) for the second derivatives of N. Hence, for $|f|\leqslant 1$ we have

$$|T_h'f(p)-T_h''f(p)|\leqslant Ch \sum_{\substack{|q-p|\geqslant h\\ \wedge q\neq p_0\wedge|q-\dot{\mathsf{G}}|>2h}} \iint_{F_q}\{|z-p|^{-2}+|z-p_0|^{-2}\}[\mathrm{d}x,\mathrm{d}y]$$

$$+Ch^2|\log h|+C\sum_{|q-\dot{\mathsf{G}}|\leqslant 2h} h|\log h|\Psi(q)$$

$$\leqslant C'\{h|\log h|+h|\log h|\delta_h^{-1}\}. \tag{7.3.38}$$

The right hand side tends to zero for $h\to 0$ due to (7.8.10).

The Propositions 7.8.7 and 7.8.8 yield the collective compactness of $\{\mathsf{K}_h\}$ and (7.8.28).

Proof of Theorem 7.8.2 Repeating Anselone's arguments [1, Theorem 11.1, p. 10] it turns out that (7.8.28) together with the inverse stability (7.8.18) provides the existence of $(\mathsf{I}-\mathsf{K}_h)^{-1}$ and the inverse stability

$$\|(\mathsf{I}-\mathsf{K}_h)^{-1}\|\leqslant C \tag{7.8.39}$$

uniformly for all $h\leqslant h_0$ where $h_0>0$ is a suitable meshwidth. Hence the equations for W and $\tilde{w}$ yield

$$\|W-\tilde{w}\|_0\leqslant\|(\mathsf{I}-\mathsf{K}_h)^{-1}\{(\mathsf{K}_h-\mathsf{K}_{\delta_h})\tilde{w}+F_h-F\}\|_0\leqslant C\|H_h\|_0 \tag{7.8.40}$$

where

$$H_h=(\mathsf{K}_h-\mathsf{K}_{\delta_h})\tilde{w}+F_h-F.$$

According to equations (7.8.26) with $\mathsf{K}_{\delta_h}\tilde{w}$ and (7.8.21) we find for H_h the following system of difference equations:

$$\begin{aligned}\Delta_h H_h &= (\Delta-\Delta_h)(\mathsf{K}_{\delta_h}\tilde{w}+F)+\Delta_h\mathsf{K}_h\tilde{w}-\Delta\mathsf{K}_{\delta_h}\tilde{w}\\ &=(\Delta-\Delta_h)\tilde{w}+(\partial_{hx}-\partial/\partial x)\Psi((A\tilde{u}+B\tilde{v})+\mathrm{i}(\tilde{A}\tilde{u}+\tilde{B}\tilde{v})) \qquad (7.8.41)\\ &\equiv\chi_h \quad \text{for} \quad p\in\mathsf{G}_h\cup\mathsf{G}_h^*-\{p_0\}\end{aligned}$$

and on the boundary

$$\operatorname{Re} H_h(p)=0, \qquad \delta_n \operatorname{Im} H_h(p)=\delta_n\tilde{w}-\frac{\partial \tilde{w}}{\partial n}+\sum_{j=1}^{3}\alpha_j x_j y_j \frac{d}{ds}\left(\frac{\partial \tilde{w}}{\partial n}\right)\equiv \varepsilon_h$$

$$\text{for} \quad p\in\partial G_h \quad \text{and} \quad \operatorname{Im} H_h(p_0)=0. \tag{7.8.42}$$

Since from (7.8.41), (7.8.42) we have

$$|\chi_h|\leq\begin{cases}\dfrac{h^2}{12}\{\|\tilde{w}\|_4+\delta^{-3}\|w\|_0\} & \text{for} \quad p\in G_h,\\[2mm] \dfrac{h}{3}\{\|\tilde{w}\|_3+\delta^{-2}\|w\|_0\} & \text{for} \quad p\in G_h^*\end{cases},$$

$$|\varepsilon_h|\leq ch^2\|\tilde{w}\|_3 \qquad \text{for} \quad p\in\partial G_h \tag{7.8.43}$$

[14, p. 200 ff.] the Theorems 7.3.4 and 7.5.4 imply

$$\|W-\tilde{w}\|_0\leq Ch^2|\log h|\{\|\tilde{w}\|_4+\delta^{-3}\|\tilde{w}\|_0\}. \tag{7.8.44}$$

Triangle inequality and Lemma 7.8.3 yield the desired estimate (7.8.11).

In case $A=B=\tilde{A}=\tilde{B}\,|_{\dot{G}}=0$ we can use $w=\tilde{w}$ and $\delta=$ const. independently of h.

7.9 Remarks on the Treatment of the Difference Equations

Since for $A=B=\tilde{A}=\tilde{B}\equiv 0$ the matrices in (7.8.8) are of positive type one should use the *overrelaxation method* for solving (7.8.8) iteratively [20, p. 463 ff.]. But in general the first-order terms on the right hand side of (4.8.8) might destroy the convergence properties since the terms containing V in $\Delta_h U$ and those containing U in $\Delta_h V$ destroy the positive type of the coefficient matrix. But in all cases discrete versions of the method in Section 6.4 converge.

Let us first consider the case of small coefficients $A,\ldots,\tilde{B}$. Here we use the iteration

$$\begin{aligned}\Delta_h U^{(N+1)}(p)&=\partial_{hx}\{\Psi(AU^{(N)}+BV^{(N)})+C\}\\&\quad+\partial_{hy}\{\Psi(\tilde{A}U^{(N)}+\tilde{B}V^{(N)})+\tilde{C}\} \quad \text{for} \quad p\in G_h\cup G_h^*,\\ \Delta_h V^{(N+1)}(p)&=-\partial_{hy}\{\Psi(AU^{(N)}+BV^{(N)}+C\}\\&\quad+\partial_{hx}\{\Psi(\tilde{A}U^{(N)}+\tilde{B}V^{(N)})+\tilde{C}\} \quad \text{for} \quad p\in G_h\cup G_h^*-\{p_0\},\end{aligned} \tag{7.9.1}$$

$$U^{(N+1)}(p)=\psi(p) \quad \text{and} \quad \delta_n V^{(N+1)}=\gamma(p)$$

for $p\in\partial G_h$, $V^{(N+1)}(p_0)=v_0$ with $U^{(-1)}\equiv 0$ and $V^{(-1)}\equiv v_0$, *where every single step $N=0,1,\ldots$ can be solved by overrelaxation.* Let us indicate why this method converges. Since for $A,\ldots,\tilde{B}$ the operator K_1 (7.8.16) has norm smaller than 1, its spectral radius ρ is also smaller than 1. Then the operators K_Ψ possess spectral radii ρ_δ with $\rho_\delta \to \rho$ for $\delta \to 0$ [1, 4.3 p. 60]. Hence

$$\rho_\delta \leqslant \rho_0 < 1 \quad \text{for all} \quad \delta \leqslant \delta_0$$

with a suitable $\delta_0 > 0$. Since we have Theorem 7.8.5 for T_h, Anselone's arguments [1, Theorem 4.8 p. 65] remain valid for T_h, assuring that the spectral radii ρ_h of T_h also converge to ρ provided (7.8.10) is valid. Therefore there is a meshwidth $h_0 > 0$ such that

$$\rho_h \leqslant \rho' < 1 \quad \text{for all} \quad h \leqslant h_0. \tag{7.9.2}$$

Therefore the successive approximation (7.9.1) converges with a rate of convergence depending only on $A,\ldots,\tilde{B}$ but *not* on $h \leqslant h_0$.

For bigger coefficients $A,\ldots,\tilde{B}$ let us choose $\lambda(\eta)$ in (6.4.12) with the properties defined in Section 6.4 and let us define the corresponding iteration:

$$\Delta_h U^{(N+1)}(p) = \partial_{hx}\left\{\sum_{\nu=1}^{\min\{N,M\}} c_\nu \eta_0^\nu \Psi(AU^{(N-\nu)} + BV^{(N-\nu)}) + C\right\}$$
$$+\partial_{hy}\left\{\sum_{\nu=1}^{\min\{N,M\}} c_\nu \eta_0^\nu \Psi(\tilde{A}U^{(N-\nu)} + \tilde{B}V^{(N-\nu)}) + \tilde{C}\right\}$$
$$\text{for} \quad p \in \mathsf{G}_h \cup \mathsf{G}_h^*,$$

$$\Delta_h V^{(N+1)}(p) = -\partial_{hy}\left\{\sum_{\nu=1}^{\min\{N,M\}} c_\nu \eta_0^\nu \Psi(AU^{(N-\nu)} + BV^{(N-\nu)}) + C\right\}$$
$$+\partial_{hx}\left\{\sum_{\nu=1}^{\min\{N,M\}} c_\nu \eta_0^\nu \Psi(\tilde{A}U^{(N-\nu)} + \tilde{B}V^{(N-\nu)}) + \tilde{C}\right\}$$
$$\text{for} \quad p \in \mathsf{G}_h \cup \mathsf{G}_h^* - \{p_0\},$$

$$U^{(N+1)}(p) = \psi(p), \qquad \delta_n V^{(N+1)} = \gamma(p) \quad \text{for} \quad p \in \partial\mathsf{G}_h,$$

$$V^{(N+1)}(p_0) = v_0 \quad \text{with} \quad U^{(-1)} \equiv 0, \qquad V^{(-1)} \equiv 0, \qquad N = 0,1,\ldots.$$

Again, *every single step $N=0,1,\ldots$ requires the solution of difference equations e.g. with overrelaxation.*

Thanks to Theorem 7.8.5 all the arguments of Section 6.4 can here be repeated implying that this iteration converges like (6.4.25), i.e. with a rate of convergence independent of $h \leqslant h_0$.

7.10 Remarks on Semilinear Problems

Nonlinear elliptic difference equations allow monotone convergent iteration methods if they have monotonicity properties which are similar to linear problems of positive type. Törnig characterized in [30, 31] Dirichlet problems which can be treated by nonlinear implicit or Newton overrelaxation methods. For nonlinear problems defined by the Euler equation to a variational problem and the difference equations defined by their Ritz equations established by piecewise linear finite elements, Frehse [15] for homogeneous Dirichlet conditions and Höhn [19] for the corresponding homogeneous natural Neumann conditions proved the solvability and convergence of the difference solutions.

For nonlinear first-order problems, however, the coupling of u and v seems to prevent a direct application of these methods. Nevertheless the conversion of the difference equations into discrete integral equations can be applied to problems as e.g. (1.4.1) with $\tau \equiv 0$,

$$\begin{aligned} &w_{\bar{z}} = \tfrac{1}{2}(u_x - v_y + \mathrm{i}(u_y + v_x)) = H = \tfrac{1}{2}(H_1(z, u, v) + \mathrm{i}H_2(z, u, v)) \quad \text{in } \mathrm{G} \\ &u\,|_{\dot{\mathrm{G}}} = \psi, \quad v(p_0) = v_0. \end{aligned} \tag{7.10.1}$$

Here the corresponding difference equations are given by

$$\begin{aligned} &\Delta_h U = \partial_{hx}\{\Psi H_1\} + \partial_{hy}\{\Psi H_2\} \quad \text{for} \quad p \in \mathrm{G}_h \cup \mathrm{G}_h^*, \\ &\Delta_h V = -\partial_{hy}\{\Psi H_1\} + \partial_{hx}\{\Psi H_2\} \quad \text{for} \quad p \in (\mathrm{G}_h \cup \mathrm{G}_h^*) - \{p_0\}, \\ &\qquad\qquad\qquad\qquad\qquad\qquad V(p_0) = v_0 \end{aligned} \tag{7.10.2}$$

$$U(p) = \psi(p) \quad \text{and}$$

$$\delta_n V(p) = \gamma(p) \equiv -\dot{\psi}(p) + \sum_{j=1}^{3} \alpha_j(p) x_j y_j \ddot{\psi}(p) \quad \text{for} \quad p \in \partial \mathrm{G}_h.$$

$\{\Psi H_j\}$ denote the composed functions

$$\{\Psi H_j\}(z) \equiv \Psi(z) H_j(z, u(z), v(z)), \quad j = 1, 2. \tag{7.10.3}$$

These nonlinear difference equations are equivalent to the *discrete Urysohn integral equations*

$$\begin{aligned} U(p) = &-h^2 \sum_{q \in \mathrm{G}_h \cup \mathrm{G}_h^*} G_h(p, q)(\partial_{hx}\{\Psi H_1\} + \partial_{hy}\{\Psi H_2\})(q) \\ &+ h \sum_{q \in \partial \mathrm{G}_h} G_h(p, q)\psi(q) \\ V(p) = &-h^2 \sum_{q \in \mathrm{G}_h \cup \mathrm{G}_h^* - \{p_0\}} N_h(p, q)(-\partial_{hy}\{\Psi H_1\} + \partial_{hx}\{\Psi H_2\})(q) \\ &+ h \sum_{q \in \partial \mathrm{G}_h} N_h(p, q)\gamma(q) + v_0 \end{aligned} \tag{7.10.4}$$

abbreviated as

$$W = \mathsf{K}_h(W) + F_h. \tag{7.10.5}$$

The corresponding *Urysohn integral equation* for the equation approximating (7.10.1),

$$\tilde{w}_{\bar{z}} = \Psi \cdot H \quad \text{in } \mathsf{G}, \qquad u = \psi \quad \text{on } \dot{\mathsf{G}}, \qquad v(p_0) = v_0, \tag{7.10.6}$$

is given by

$$\tilde{u}(p) = -\iint\limits_{\mathsf{G}} G^I(p, z)(\{\Psi H_1\}_x + \{\Psi H_2\}_y)[\mathrm{d}x, \mathrm{d}y] - \oint\limits_{\dot{\mathsf{G}}} \psi \, \mathrm{d}_n G^I,$$

$$\tilde{v}(p) = -\iint\limits_{\mathsf{G}} N(p, z)(-\{\Psi H_1\}_y + \{\Psi H_2\}_x)[\mathrm{d}x, \mathrm{d}y] \tag{7.10.7}$$

$$+ \oint\limits_{\dot{\mathsf{G}}} \dot{\psi} N(p, z(s)) \, \mathrm{d}s + v_0$$

or abbreviated,

$$\tilde{w} = \mathsf{K}_\Psi(\tilde{w}) + F. \tag{7.10.8}$$

The equations (7.10.5) can be considered as approximations to (7.10.8), and both together as approximations to

$$\tilde{w} = \mathsf{K}_1(w) + F. \tag{7.10.9}$$

For smooth H satisfying (6.4.3), (1.4.4) these approximations provide many of the properties underlying the convergence theories by Stummel [29, Section 10] or by Anselone [1, Chapter 6]. For example, after estimates similar to those in the proof of Theorem 7.8.5 it turns out that the Frechet derivatives satisfy an equation corresponding to (7.8.28),

$$\lim_{h \to 0} \|(K'_{\psi h}(w) - K'_\psi(w))K'_h(w)\| = 0 \tag{7.10.10}$$

uniformly if $\|w\| \leq R < \infty$ and $w \in C^0$ provided (7.8.10) holds. This indicates that by a slight variation of Anselone's approach [1, Theorem 6.8 p. 109] it can be proved that the nonlinear difference system (7.10.2) is uniquely solvable if $h \leq h_0$ for a certain $h_0 > 0$. But this is yet to be done.

Since the strong properties (1.4.3), (1.4.4) also imply equicontinuity properties of $\mathsf{K}_1(w)$ at all $w \in \mathbf{B}$ the embedding Newton method (1.4.8) can be applied to (7.10.2), i.e.

$$\Delta_h U^{(N+1)}(p) = t_j \partial_{hx}(\{\Psi H_{1u}^{(N)}\}(U^{(N+1)} - U^{(N)}) + \{\Psi H_{1v}^{(N)}\}(V^{(N+1)} - V^{(N)})) + t_j \partial_{hy}(\{\Psi H_{2u}^{(N)}\}(U^{(N+1)} - U^{(N)}) + \{\Psi H_{2v}^{(N)}\} \times (V^{(N+1)} - V^{(N)})) \quad \text{for} \quad p \in \mathsf{G}_h \cup \mathsf{G}_h^*,$$

$$\Delta_h V^{(N+1)}(p) = -t_j \partial_{hy}(\{\Psi H_{1u}^{(N)}\}(U^{(N+1)} - U^{(N)}) + \{\Psi H_{1v}^{(N)}\}(V^{(N+1)} - V^{(N)})) + t_j \partial_{hx}(\{\Psi H_{2u}^{(N)}\}(U^{(N+1)} - U^{(N)}) + \{\Psi H_{2v}^{(N)}\} \times (V^{(N+1)} - V^{(N)})) \quad \text{for} \quad p \in \mathsf{G}_h \cup \mathsf{G}_h^* - \{p_0\}, \quad (7.10.11)$$

$$V^{(N+1)}(p_0) = v_0 \quad \text{and}$$

$$U^{(N+1)}(p) = \psi(p), \qquad \delta_h V^{(n+1)}(p) = \gamma(p) \quad \text{for} \quad p \in \partial \mathsf{G}_h,$$

where $\{\Psi H_{1u}^{(N)}\} \equiv \Psi(z) H_{1u}(z, U^{(N)}, V^{(N)})$ etc., $j = 1, 2, \ldots, \tilde{N}$.
Note that for each $N = 1, 2, \ldots$, (7.10.11) is a *linear* problem of the form (7.8.8). The ‘initial’ solution is chosen as

$$U^{(0)} \equiv U(t_{j-1}, p), \qquad V^{(0)} \equiv V(t_{j-1}, p). \quad (7.10.12)$$

Of course, in practical computations, (7.10.12) will be replaced by a reasonable iterate $U^{(N)}(t_{j-1}, p)$, $V^{(N)}(t_{j-1}, p)$.

Also for (7.10.11) Anselone’s approach [1, p. 100 ff.] can be applied but the details are yet to be done.

References

1 Anselone, P. M., *Collectively compact operator approximation theory.* Prentice-Hall, London, 1971.

2 Bramble, J. and Hubbard, B., A priori bounds on the discretization error in the numerical solution of the Dirichlet problem. *In* La Salle (Ed.), *Contributions to differential equations.* John Wiley, New York, 1963.

3 Bramble, J. H. and Hubbard, B. E., A theorem on error estimation for finite difference analogues of the Dirichlet problem for elliptic equations. *In* La Salle (Ed.), *Contributions to differential equations,* pp. 319–340. John Wiley, New York, 1963.

4 Bramble, J. and Hubbard, B., On a finite difference analogue of an elliptic boundary problem which is neither diagonally dominant nor of non negative type. *J. Math. Phys.* **43,** 117–132, 1964.

5 Bramble, J. and Hubbard, B., A finite difference analog of the Neumann problem for Poisson’s equation. *Siam J. Num. Anal.* **2,** 1–14, 1964.

6 Bramble, J. H., and Hubbard, B. E., Approximation of solutions of mixed boundary value problems for Poisson’s equations by finite differences. *J. Ass. Computing Machinery* **12,** 114–123, 1965.

7 Bramble, J. Hubbard, B. and Thomée, V., Convergence estimates for essentially positive type discrete Dirichlet problems. *Math. Comp.* **23,** 695–709, 1969.
8 Bramble, J. and Thomée, V., Pointwise bounds for discrete Green's functions. *SIAM J. Num. Anal.* **6,** 583–590, 1969.
9 Ciarlet, G., Discrete maximum principle for finite difference operators. *Aequ. math.* **4,** 338–353, 1970.
10 Collatz, L., *The numerical treatment of differential equations.* Springer, Berlin, 1960.
11 Courant, R. Friedrichs K. and Lewy, H., Über die partiellen Differentialgleichungen der Mathematischen Physik. *Math. Ann.* **100,** 32–74, 1928.
12 Deeter, C. R. and Springer, G., Discrete harmonic kernels. *J. Math. Mech.* **14,** 413–438, 1965.
13 Dieudonné, G., *Foundations of modern analysis.* Academic Press, New York, London, 1960.
14 Forsythe, G. and Wasow, W., *Finite difference methods for partial differential equations.* John Wiley, New York, 1960.
15 Frehse, J., Eine gleichmäßige asymptotische Fehlerabschätzung zur Methode der finiten Elemente bei quasilinearen elliptischen Randwertproblemen. *In Theory of nonlinear operators. Constructive aspects.* Tagungsband der Akademie der Wissenschaften DDR, Berlin, 1976.
16 Gaier, D., *Konstruktive Methoden der konformen Abbildung.* Springer, Berlin, 1969.
17 Haack, W. and Wendland, W., *Lectures on partial and pfaffian differential equations.* Pergamon Press, Oxford, New York, 1972.
18 Hainer, K. Lösung des Dirichletproblems und Konvergenz der Differenzenapproximation für elliptische Differentialgleichungen zweiter Ordnung. *Math. Zeitschr.* **131,** 189–211, 1973
19 Höhn, W., Numerische Behandlung nichtlinearer Neumann-Probleme mit finiten Elementen. *Dissertation* T. H. Darmstadt, Germany, D 17, 1976.
20 Isaacson, E. and Keller, H. B., *Analysis of numerical methods.* John Wiley, New York, 1966.
21 Laasonen, P., On the solution of Poisson's difference equation. *J. Ass. Computing Machinery* **5,** 370–382, 1958.
22 McAllister, G. T. and Sabotka, E. F., Discrete Green's Functions. *Math. Comp.* **27,** 59–80, 1973.
23 McCrea, W. H. and Whipple, F. H. W., Random paths in two and three dimensions. *Proc. Roy. Soc. Edinburgh* Sect. A, **60,** 281–298, 1940.
24 Natterer, F., Über die punktweise Konvergenz finiter Elemente. *Num. Math.*, to appear.
25 Scott, R., Finite element convergence for singular data. *Num. Math.* **21,** 317–327, 1973.
26 Scott, R., Optimal L^∞-estimates for the finite element method on irregular meshes. *Math. Comp.* **30,** 681–697, 1976.
27 Stummel, F., Elliptische Differenzenoperatoren unter Dirichletrandbedingungen. *Math. Zeitschr.* **97,** 169–221, 1967.
28 Stummel, F., Diskrete Konvergenz linearer Operatoren. I. *Math. Annalen*

190, 45–92, 1970; II *Math. Zeitschr.* **120,** 231–264, 1971; III Proc. Oberwolfach Conference on Linear Operators and Approximation 1971. *Int. Ser. Num. Math.* **20,** pp. 231–264, Birkhäuser, Basel, Stuttgart, 1972.

29 Stummel, F., Approximation methods in analysis. *Lecture Notes* Ser. No. **35,** Mat. Inst. Aarhus Universitet, 1973.

30 Törnig, W., Monoton konvergente Iterationsverfahren zur Lösung nichtlinearer Differenzen-Randwertprobleme. *Beiträge Num. Math.* **4,** 245–257, 1975.

31 Törnig, W., Monotonie- und Randmaximumsätze bei Diskretisierungen des Dirichletproblems allgemeiner nichtlinearer elliptischer Differentialgleichungen. *Computing* **11,** 391–401, 1973.

32 Varga, R., *Matrix iterative analysis.* Prentice Hall, Englewood Cliffs, N. J., 1962.

8

Continuous Approximations by Minimizing the Defect

Here we mainly consider approximations of the general BVP (4.1.1),

$$\begin{aligned} \mathsf{D}\mathbf{u} \equiv \mathbf{u}_x + \mathbf{B}\mathbf{u}_y + \mathbf{C}\mathbf{u} = \mathbf{f} \quad & \text{in } \mathsf{G} \quad \text{and} \\ \mathsf{r}\mathbf{u} = \boldsymbol{\psi} \quad & \text{on } \dot{\mathsf{G}} \end{aligned} \tag{8.0.1}$$

provided the Lopatinski condition (4.1.2) is satisfied. If (8.0.1) has nullity $N > 0$ according to Theorem 4.3.2 then we add to (8.0.1) N linearly independent requirements

$$\Lambda_l \mathbf{u} = c_l, \qquad l = 1, \ldots, N \tag{8.0.2}$$

with given $c_l \in \mathbb{R}$ and linear functionals Λ_l such that the extended problem has at most a unique solution. For (1.1.1) e.g. Λ corresponds to (1.1.36) or the point condition (7.0.1). We further assume throughout this chapter that the given $\mathbf{f}$, $\boldsymbol{\psi}$ possibly satisfy the solvability conditions (4.3.25) such that (8.0.1), (8.0.2) admits a unique solution $\mathbf{u}$.

The approximations will be chosen in a given scale of finite dimensional subspaces $\tilde{H}(h)$ of the domain of (8.0.1), (8.0.2) depending on a parameter $h > 0$ and approximating the domain of (8.0.1) for $h \to 0$. Inserting any $\boldsymbol{\chi} \in \tilde{H}$ into (8.0.1), (8.0.2), the defect

$$\mathsf{D}\boldsymbol{\chi} - \mathbf{f} \quad \text{in} \quad \mathsf{G}, \qquad \mathsf{r}\boldsymbol{\chi} - \boldsymbol{\psi} \quad \text{on } \dot{\mathsf{G}} \text{ and } \quad (\Lambda_l \boldsymbol{\chi} - c_l) \text{ in } \mathbb{R}^N \tag{8.0.3}$$

will not vanish in general. Throughout this chapter (except (8.6.7)), we shall minimize the defect (8.0.3) over $\tilde{H}$ in various ways. The corresponding minimal solutions $\mathbf{U} \in \tilde{H}$ will then depend on the metric chosen for (8.0.3); different norms for the defect yield different approximations. The size of the minimal defect gives estimates for $\mathbf{U} - \mathbf{u}$ if *a priori* estimates for the solution of (8.0.1), (8.0.2) are available. Besides the Hölder estimates (4.1.23) we shall also use the *a priori* estimates in H_s^p spaces by Agmon, Douglis and Nirenberg [1] for $s \geq 1$ and their extensions for smooth solutions and smaller s by Schechter [43] and Dikanskij [18].

In Section 8.1 we begin with the Hölder norms yielding Chebyshev approximations. In Sections 8.2–8.6 we use Hilbert space norms for

(8.0.3) which yield least squares approximations. These are variational problems for which a rich theory including approximation theory is available; we refer to the key work by Mikhlin [30] and his practical treatment [31].

Sections 8.1 and 8.2 have merely introductory character; there we use spaces $\tilde{H}$ without special properties (except polynomials). In Sections 8.3–8.6 we specialize $\tilde{H}$ to spaces of regular finite element functions on triangulations. Here we follow essentially the presentations by Fix and Strang [20]; for more information on these topics we refer to Aubin [5], Babuška and Aziz [6], Raviart [42], Zienkiewicz [55] and the references given there. For finite elements the least squares approximation subordinate (8.0.3) does not converge with optimal rate unless one uses weights in the corresponding norms which depend on h. This idea was used by Bramble and Schatz [10–12] and Aubin [5] who proved optimal order of convergence for even order BVPs. Following the simplified presentation of Baker [7] in Section 8.4 this approach is carried over to the first order problems (8.0.1), (8.0.2) as a special case of [47]. The optimal order of convergence follows with Nitsche's trick [35] and using Dikanskij's results [18].

In Section 8.5 we replace the required integrations by quadrature formulas. For the general problems (8.0.1), (8.0.2) it turns out that the least squares method in Section 8.4 combines the advantages of second-order Galerkin methods with the most general linear boundary conditions. The relation between the original Gauss equations and their discretized versions is presented following Fix and Strang [20]. A more efficient method based on the interpolation with finite element functions can be found in the work of Ciarlet and Raviart [16] which provides the properties for the general setting in the framework of discrete approximation theory by Stummel [50]. For our problems we formulate a condition related to the 'patch test' [20]. If this is fulfilled and if the right hand sides are evaluated with high accuracy—and this is the payment for the rather wide range of applicability—we find the same convergence rate for the 'numerical' solutions as in Section 8.4 for the 'theoretical' least squares approximations.

In Section 8.6 we specialize (8.0.1), (8.0.2) to the simple standard problem (1.1.1) restricting the boundary conditions to zero and using finite elements satisfying the homogeneous boundary condition. It turns out that the corresponding Ritz equations coincide with the Galerkin equations of a certain second-order elliptic system differing from the Galerkin equations to (7.0.2) only by a compact bilinear form. In both cases Mikhlin's approach [30] is available yielding uniform bounds for the corresponding Gauss respectively, Galerkin operators. This yields optimal convergence in the Sobolev spaces $H^s(G)$. For Dirichlet and for

Neumann problems corresponding to these second-order systems maximum norm error estimates have also been developed recently. Using L^1 estimates for the gradient of the differences of the finite element approximation to the Neumann function obtained by Scott [44, 45] and the corresponding estimates for the Green function of Frehse and Rannacher [22] we also find pointwise error estimates for the above Ritz and Galerkin methods. Finally we indicate the treatment of corresponding nonlinear problems.

Throughout the chapter we use the notation c, c', $c_1, \cdots$ for a generic constant which might change its size in different inequalities.

8.1 Chebyshev Approximations

According to the properties (4.1.23) and Theorem 4.3.2, for (8.0.1) the simplest approximation involving uniform convergence seems to be: *Minimize*

$$\rho(\boldsymbol{\chi}) \equiv \|\mathbf{D}\boldsymbol{\chi}-\mathbf{f}\|_{C^\alpha(\bar{G})} + |\mathbf{r}\boldsymbol{\chi}-\boldsymbol{\psi}|_{C^{1+\alpha}(\dot{G})} + \sum_{l=1}^{N} |\Lambda_l \boldsymbol{\chi} - c_l| \tag{8.1.1}$$

for $\boldsymbol{\chi} \in \tilde{H}$ $(0<\alpha<1,\ \alpha$ is fixed).

Let $M = M(h) < \infty$ be the dimension of $\tilde{H}$. Then (8.1.1) is a problem of *Chebyshev approximation.* Since (8.1.1) involves images of three mappings, it is connected with simultaneous approximations (see e.g. [3], [14]). For the approximating spaces $\tilde{H}$ we may choose e.g. *two-dimensional polynomials*

$$\boldsymbol{\chi} \equiv \sum_{j,k=0}^{L} \boldsymbol{\alpha}_{jk} x^j y^k \tag{8.1.2}$$

with fixed L, $\boldsymbol{\alpha}_{jk} \in \mathbb{R}^{2n}$. Here the dimension of $\tilde{H}$ is $M = 2n(L+1)^2$. (Another choice for $\tilde{H}$ will be finite element spaces in Section 8.3–8.5.)

Lemma 8.1.1 *The finite dimensional problem* (8.1.1) *has at least one solution* $\mathbf{U} \in \tilde{H}$.

This follows from the continuity of (8.1.1) on $\tilde{H}$ which is finite dimensional and the Weierstrass Theorem for continuous functions on a compact set. For the details see e.g. [53, p. 17].

If $h \to 0$ and $M(h) \to \infty$ then the approximations $\mathbf{U}$ form a sequence which should converge to $\mathbf{u}$, the unique solution of (8.0.1), (8.0.2).

Theorem 8.1.2 *Let* $\bigcup_{h\leq h_0}\tilde{H}(h)$ *be dense in* $C^{1+\alpha}(\bar{G})$ *for every* $h_0>0$ *and let* Λ_l *be continuous on* $C^{1+\alpha}(\bar{G})$.

Then

$$\lim_{h\to 0}\|\mathbf{u}-\mathbf{U}\|_{1+\alpha}=0. \tag{8.1.3}$$

Proof Since (8.0.1), (8.0.2) has a *unique* solution, the *a priori* estimate (4.1.23) together with the Fredholm property (Theorem 4.3.2) implies an *a priori* estimate

$$\gamma^{-1}\|\mathbf{v}\|_{1+\alpha}\leq\left\{\|\mathbf{Dv}\|_\alpha+|\mathbf{rv}|_{1+\alpha}+\sum_{l=1}^{N}|\Lambda_l\mathbf{v}|\right\}\leq\gamma\|\mathbf{v}\|_{1+\alpha} \tag{8.1.4}$$

for all $\mathbf{v}\in C^{1+\alpha}(\bar{G})$ where $\gamma>0$ is a suitable constant independent of $\mathbf{v}$. Since the spaces $\bigcup_{h\leq h_0}\tilde{H}(h)$ are dense we can find a sequence of functions $\mathbf{u}_h\in\tilde{H}(h)$ satisfying

$$\lim_{h\to 0}\|\mathbf{u}_h-\mathbf{u}\|_{1+\alpha}=0 \tag{8.1.5}$$

Hence (8.1.4) with (8.1.1) yields

$$\|\mathbf{u}_h-\mathbf{u}\|_{1+\alpha}\geq\frac{1}{\gamma}\rho(\mathbf{u}_h)\geq\frac{1}{\gamma}\rho(\mathbf{U})\geq\gamma^{-2}\|\mathbf{U}-\mathbf{u}\|_{1+\alpha}\to 0$$

according to (8.1.5).

The above method has at least one disadvantage, i.e. the use of the complicated Hölder norms in (8.1.1) which lead to weighted maximum norms subordinated to the product sets $\bar{G}\times\bar{G}$ and $\dot{G}\times\dot{G}$. This difficulty can be overcome by using *a priori* estimates other than (8.1.4).

Lemma 8.1.3 *For* (8.0.1), (8.0.2) *there holds an a priori estimate*

$$\gamma^1\|\mathbf{v}\|_\alpha\leq\left\{\|\mathbf{Dv}\|_{C^0(\bar{G})}+|\mathbf{rv}|_{C^1(\dot{G})}+\sum_{l=1}^{N}|\Lambda_l\mathbf{v}|\right\} \tag{8.1.6}$$

where $0<\alpha<1$ *and where* $\gamma^1>0$ *is independent of* $\mathbf{v}$, *provided the* Λ_l *are continuous on* $W_p^1(\Omega)$ *for every* $p>1$.

Proof Let us indicate the proof of (8.1.6) omitting the details. Agmon, Douglis and Nirenberg proved in [1, II p. 77 and p. 74, Remark 2] the *a priori* estimate

$$\gamma(\|\mathbf{v}_x\|_{L^p(G)}+\|\mathbf{v}_y\|_{L^p(G)}+\|\mathbf{v}\|_{L^p(G)})\leq\left\{\|\mathbf{Dv}\|_{L^p(G)}+|\mathbf{rv}|_{1-1/p}+\sum_{l=1}^{N}|\Lambda_l\mathbf{v}|\right\} \tag{8.1.7}$$

where $1<p<\infty$ and where the $1-1/p$ boundary norm is defined by

$$|\boldsymbol{\psi}|_{1-1/p} \equiv g.l.b.\{\|\boldsymbol{\phi}\|_{L^p(G)} + \|\boldsymbol{\phi}_x\|_{L^p(G)} + \|\boldsymbol{\phi}_y\|_{L^p(G)}\}$$

where $\boldsymbol{\phi}$ traces all functions with $\boldsymbol{\phi}|_{\dot{G}} = \boldsymbol{\psi}$, if we specialize their general elliptic problems to our problem (8.0.1), (8.0.2) using the uniqueness. If $D\mathbf{v} \in C^0(\bar{G})$ and $\mathbf{r}\mathbf{v} \in C^1(\dot{G})$ then the right hand side of (8.1.7) can surely be estimated by the right hand side of (8.1.6) with any $p>1$. For estimating the left hand side in (8.1.7) below by the left hand side of (8.1.6) we use the Sobolev embedding theorem [46, p. 84] choosing $p>1$ in (8.1.7) appropriately.

The estimate (8.1.6) suggests replacing (8.1.1) by the new Chebyshev problem:
Minimize

$$\rho_0(\boldsymbol{\chi}) \equiv \|\mathbf{D}\boldsymbol{\chi} - \mathbf{f}\|_{C^0(\bar{G})} + |\mathbf{r}\boldsymbol{\chi} - \boldsymbol{\psi}|_{C^1(\dot{G})} + \sum_{l=1}^{N} |\Lambda_l \boldsymbol{\chi} - c_l| \tag{8.1.8}$$

for $\boldsymbol{\chi} \in \tilde{H}$.

Of course, Lemma 8.1.1 remains valid. Instead of Theorem 8.1.2 we can easily prove the corresponding theorem.

Theorem 8.1.4 *Let* $\bigcup_{h\le h_0} \tilde{H}(h)$ *be dense in* $C^1(\bar{G})$ *for every* $h_0>0$ *and let* Λ_l *be continuous with respect to the norm on the left hand side of* (8.1.7) *for every* $p>1$.

Then

$$\lim_{h\to 0} \|\mathbf{u} - \mathbf{U}\|_\alpha \to 0. \tag{8.1.9}$$

Proof The proof repeats all arguments of the proof of Theorem 8.1.2:

Begin with $\mathbf{u}_h \in \tilde{H}(h)$ providing $\|\mathbf{u}_h - \mathbf{u}\|_1 \to 0$. Then the continuity properties of $\mathsf{D}, \mathsf{r}, \Lambda_l$ imply with (8.1.8) and (8.1.6) the inequality

$$\|\mathbf{u}_h - \mathbf{u}\|_1 \geq \gamma' \rho_0(\mathbf{u}_h) \geq \gamma'' \rho_0(\mathbf{U}) \geq \gamma''\gamma' \|\mathbf{u} - \mathbf{U}\|_\alpha$$

which yields (8.1.9).

Although (8.1.8) contains much more practical norms than (8.1.1), the solution of the finite dimensional problem (8.1.8) is by no means trivial. Since in $\|\mathbf{D}\boldsymbol{\chi} - \mathbf{f}\|_{C^0(\bar{G})}$ we have two independent variables, the Haar condition will be violated [29] and hence, (8.1.8) can not be solved by the ordinary Remez algorithm anymore [53]. Here only the more complicated version by Töpfer [51] is available. There one needs to determine the points of expressions $\rho_0(\boldsymbol{\chi})$ where ρ_0 takes its maximal values. This requires another discretization of (8.1.8) by restricting all functions to a suitable fine net of grid points on G and $\dot{G}$. The corresponding finite dimensional Chebyshev approximation can be treated either by linear programming [8] or by Töpfer's version of the Remez algorithm [15], [51].

All the more complicated procedures above result from using the $C^0(\bar{G})$ and $C^1(\dot{G})$ norms for minimizing the defect (8.0.3). Clearly, the corresponding minimal problem will become much simpler if we use Hilbert space norms for (8.0.3) as we shall do in the following sections.

8.2 Least Squares Approximations without Weighted Norms

The basic inequalities for least squares approximations in connection with (8.0.3) are the *a priori* estimates by Agmon, Douglis, Nirenberg in the Sobolev spaces $H^t(G)$ and $H^r(\dot{G})$ [1] with $t \geq 1$ and by Dikanskij [18] who extended corresponding estimates for even order equations by Schechter [43] to elliptic boundary value problems with pseudodifferential operators on one hand, and continuity of D, r, Λ_l on the other hand.

Lemma 8.2.1 *Let* $\dot{G}$ *be smooth enough and let* Λ_l *be continuous on* $H^t(G)$, $t \geq 0$. *Then to* $t \geq 0$ *there exists a constant* $\gamma > 0$ *such that*

$$\gamma^{-1} \|\mathbf{v}\|^2_{H^t(G)} \leq \|D\mathbf{v}\|^2_{H^{t-1}(G)} + |\mathbf{r}\mathbf{v}|^2_{H^{t-1/2}(\dot{G})} + \sum_{l=1}^{N} |\Lambda_l \mathbf{v}|^2 \tag{8.2.1}$$

holds for all $\mathbf{v} \in H^t(G) \cap H^1(G)$.

This follows from (8.1.7) for $t \geq 1$ and from [18, Theorems 1.1 and 2.3] for $0 \leq t < \frac{1}{2}$ by interpolation.

Moreover, for $t \neq \frac{1}{2}$, $D: H^t(G) \to H^{t-1}(G)$ is continuous [28, p. 85] and for $t > \frac{1}{2}$ the trace theorem from Lions and Magenes holds [28, p. 41]. Hence we also have:

Lemma 8.2.2 *For* $t > \frac{1}{2}$, *both sides of* (8.2.1) *define equivalent norms on* $H^t(G)$ *provided the* Λ_l *are continuous on* $H^t(G)$.

(See also Hörmander [25, p. 267 ff.].)

For $t = 1$ Lemma 8.2.2 provides

$$\gamma^{-1} \|\mathbf{v}\|^2_{H^1(G)} \leq \|D\mathbf{v}\|^2_{H^0(G)} + |\mathbf{r}\mathbf{v}|^2_{H^{1/2}(\dot{G})} + \sum_{l=1}^{N} |\Lambda_l \mathbf{v}|^2 \leq \gamma \|\mathbf{v}\|^2_{H^1(G)}. \tag{8.2.2}$$

These inequalities suggest the choice $\mathbf{v} = \mathbf{u} - \boldsymbol{\chi}$ and the minimization of the quadratic middle expression relative to the defect. But, unfortunately, the $H^{1/2}(\dot{G})$ norm cannot be calculated numerically without enormous difficulties. Hence, one way for overcoming this difficulty would be to weaken

(8.2.2) to

$$\gamma^{-1}\|\mathbf{v}\|^2_{H^1(G)}\le\|\mathbf{D}\mathbf{v}\|^2_{L^2(G)}+|\mathbf{r}\mathbf{v}|^2_{H^1(\dot G)}+\sum_{l=1}^{N}|\Lambda_l\mathbf{v}|^2 \tag{8.2.3}$$

and minimizing the corresponding quadratic functional

$$\rho_1^2(\boldsymbol{\chi})\equiv\|\mathbf{D}\boldsymbol{\chi}-\mathbf{f}\|^2_{L^2(G)}+|\mathbf{r}\boldsymbol{\chi}-\boldsymbol{\psi}|^2_{H^1(\dot G)}+\sum_{l=1}^{N}|\Lambda_l\boldsymbol{\chi}-c_l|^2 \tag{8.2.4}$$

over $\tilde H$. It is well known that

$$\rho_1^2(\mathbf{U})=\min_{\boldsymbol{\chi}\in\tilde H}\rho_1^2(\boldsymbol{\chi}) \tag{8.2.5}$$

is equivalent to the Gauss equations

$$(\mathbf{D}\mathbf{U}-\mathbf{f},\mathbf{D}\boldsymbol{\chi})_0+\langle\mathbf{r}\mathbf{U}-\boldsymbol{\psi},\mathbf{r}\boldsymbol{\chi}\rangle_0+\left\langle\frac{\mathrm{d}}{\mathrm{d}s}(\mathbf{r}\mathbf{U}-\boldsymbol{\psi}),\frac{\mathrm{d}}{\mathrm{d}s}\mathbf{r}\boldsymbol{\chi}\right\rangle_0 + \sum_{l=1}^{N}(\Lambda_l\mathbf{U}-c_l)\cdot\Lambda_l\boldsymbol{\chi}=0 \tag{8.2.6}$$

for all $\boldsymbol{\chi}\in\tilde H$ with $\mathbf{U}\in\tilde H$. The scalar products are defined by

$$\begin{aligned}(\mathbf{V},\boldsymbol{\phi})_0&\equiv\sum_{j=1}^{2n}\iint_G V_j\phi_j(x,y)[\mathrm{d}x,\mathrm{d}y],\\ \langle\mathbf{g},\boldsymbol{\psi}\rangle_0&\equiv\sum_{k=1}^{n}\oint_{\dot G} g_k\psi_k\,\mathrm{d}s.\end{aligned} \tag{8.2.7}$$

The equations (8.2.6) are equivalent to a system of M linear equations for the M coefficients γ_i of

$$\mathbf{U}=\sum_{i=1}^{M}\gamma_i\chi_i(x,y) \tag{8.2.8}$$

where the $\boldsymbol{\chi}_i$ form a basis of $\tilde H$. The matrix of coefficients defined by (8.2.6) is positive definite providing the application of corresponding methods for solving these linear equations, e.g. Cholesky's method [19]. Note that the linearity of (8.2.6) and uniqueness of (8.0.1), (8.0.2) yield *unique solvability* of (8.2.6). If the coefficients in $\mathbf{B}$ and $\mathbf{C}$ are polynomials of x and y (and they can always be approximated in this way according to the Weierstrass approximation theorem) and if we choose $\tilde H$ with (8.1.2) also as polynomials then the surface integrals in (8.2.6) can all be evaluated by boundary integrals due to the Gauss–Green theorem,

$$\iint_G x^j y^k[\mathrm{d}x,\mathrm{d}y]=\frac{1}{j+1}\oint_{\dot G} x^{j+1}y^k\,\mathrm{d}y. \tag{8.2.9}$$

In this case all the integrations in (8.2.6) can be reduced to one-dimensional integrals such that the numerical expense becomes acceptable. Using (8.2.3) and the continuity properties of D, r and Λ_l we can prove the following theorem word for word as Theorem 8.1.4.

Theorem 8.2.3 *Let* $\bigcup_{h\leq h_0} \tilde{H}(h)$ *be dense in* $H^1(\mathsf{G})$ *and let the corresponding functions on* $\dot{\mathsf{G}}$ *be dense in* $H^1(\dot{\mathsf{G}})$ *for every* $h_0>0$. *Let* Λ_l *be continuous on* $H^1(\mathsf{G})$. *Then*

$$\lim_{h\to 0} \|\mathbf{u}-\mathbf{U}\|_{H^1(\mathsf{G})}=0. \tag{8.2.10}$$

Clearly, in (8.2.3) and the following formulas we could use the $H^{t-1}(\mathsf{G})$ and $H^t(\dot{\mathsf{G}})$ scalar products corresponding to integers ≥ 2. Then (8.2.6) becomes more complicated. Nevertheless, for polynomials given in $\tilde{H}$, B and C one would have to evaluate only boundary integrals of the form (8.2.9). For this case with $t\geq 2$ the Sobolev embedding theorem would provide uniform convergence of $\mathbf{U}$ to $\mathbf{u}$ including their derivatives for t big enough.

For higher M, however, the presented method leads to numerically badly conditioned Gauss equations (8.2.6) for the γ_i in (8.2.8) and, moreover, the corresponding matrix of coefficients is *completely filled* in general. The latter can be eliminated if one uses suitable finite element spaces for $\tilde{H}$ which lead to sparse matrices in (8.2.6). This will be done in the following section.

8.3 Finite Element Approximations

In the following let us choose for $\tilde{H}(h)$ a sequence of *regular finite element spaces* $\tilde{H}\subset H^m(\mathsf{G})$, $m\geq 1$, defined over the region $\bar{\mathsf{G}}$. Although such spaces and many of their properties have been studied intensively during the last decade this field of applied mathematics is by no means completed yet. It was Courant [17] who already used two-dimensional finite elements for approximating the Dirichlet problem using most of their basic properties. For recent surveys we refer the reader to Aubin [5], Babuška and Aziz [6], Fix and Strang [20], Raviart [42] and Zienkiewicz [55]. In the following, we shall introduce triangular elements following essentially Fix and Strang [20]. (The whole analysis can also be performed for quadrilateral elements.)

Let the region $\bar{\mathsf{G}}$ be covered by a family of 'regular' triangular nets; i.e. if $h>0$ denotes the maximal side length of the triangles of one of these nets then we require that every triangle of the particular net contains a disk of radius $c\cdot h$ where c is a constant independent of h, $0<c<\frac{1}{2}$. The family corresponds to a sequence in h with $h \to 0$. We further assume that each triangle T_k containing points of the boundary curve $\dot{\mathsf{G}}$ allows the placement of a disk with radius $c \cdot h$ into the intersection $T_k' \equiv \bar{T}_k \cap \bar{\mathsf{G}}$. The choice of the integer $m \geq 1$ depends on the smoothness and accuracy desired. To each of these triangulations with $h > 0$ we define $\tilde{H}(h)$ to be the piecewise polynomials of fixed degree $\tilde{m}$, i.e. on each triangle T_k these functions are polynomials

$$\mathbf{p}(x, y) = \sum_{0 \leq j,k}^{j+k \leq \tilde{m}} \alpha_{jk} x^j y^k. \tag{8.3.1}$$

Moreover, the additional assumption

$$\tilde{H}(h) \subset H^m(\mathsf{G}) \cap C^{m-1}(\bar{\mathsf{G}}) \tag{8.3.2}$$

is required which restricts the possible $\tilde{m}(m)$. For $m = 1$, the Courant elements are piecewise linear, continuous and $\tilde{m} = m = 1$. For further definitions of elements see [20, p. 85] (Cubics: $m = 1$, $\tilde{m} = 3$; Quintics: $m = 2$, $\tilde{m} = 5$) and [13] ($m \geq 1$, $\tilde{m} = 4m - 3$) and the references in [6]. But the above requirements are still not satisfactory enough; we further assume that $\tilde{H}(h)$ possesses a basis of trial functions $\chi_1, \ldots, \chi_{M(h)}$ satisfying a 'uniformity condition' [20, p. 137] of order m implying with the uniform triangulations the following convergence (C) and stability (S) (or 'inverse assumption') properties:

(C) *For $k \leq r$ with $-m-1 \leq k \leq m$, $-m \leq r \leq m+1$ and for any $\mathbf{v} \in H^r(\mathsf{G})$ there exists a $\mathbf{V} \in \tilde{H}$* such that

$$\|\mathbf{V} - \mathbf{v}\|_{H^k(\mathsf{G})} \leq c_{rk} h^{r-k} \|\mathbf{v}\|_{H^r(\mathsf{G})} \tag{8.3.3}$$

holds where $\mathbf{V}$ is independent of k and c_{rk} is independent of $\mathbf{V}$, $\mathbf{v}$ and h. (See [10, p. 659], [20, p. 145], [6, Theorem 4.1.5], [5, p. 153].

(S) *For $k \leq r$ with $|r|$, $|k| \leq m$ there exists $M_{rk} > 0$ independent of h and $\chi \in \tilde{H}$ such that*

$$\|\chi\|_{H^r(\mathsf{G})} \leq M_{rk} h^{k-r} \|\chi\|_{H^k(\mathsf{G})} \quad \textit{for all} \quad \chi \in \tilde{H} \tag{8.3.4}$$

(See [34], [20, p. 167], [6 Theorem 4.1.3], [5. p. 157].) The uniformity properties allow estimates between maximum norm and Sobolev norms on the spaces $\tilde{H}(h)$ which will be of use for error estimates and numerical integration.

Lemma 8.3.1 *Let $\mathcal{P}_l$ be the class of all polynomials of degree $\leq l$ on T_k. Then there exists a constant $c_l > 0$ independent of T_k and h such that*

$$|p(z)| \leq c_l h^{-1} \left\{ \iint_{T_k'} |p(z)|^2 \, [\mathrm{d}x, \mathrm{d}y] \right\}^{1/2} \tag{8.3.5}$$

holds for all $p \in \mathcal{P}_l$.

Proof Let us consider first the standard (or reference) triangle T_0 with $h=1$ having vertices $z=0;\ 1;\ \frac{1}{2}(1+i\sqrt{3})$. Since the dimension of $\mathcal{P}_l$ is $\frac{1}{2}(l+1)(l+2)$ and on a finite dimensional space all norms are equivalent, there holds

$$|\tilde{p}(\zeta)| \leqslant \tilde{c}_l \left\{ \iint_{K_0} |\tilde{p}(\zeta)|^2 \,[d\xi, d\eta] \right\}^{1/2} \quad \textit{for all} \quad \tilde{p} \in \mathcal{P}_l, \tag{8.3.6}$$

where $K_0 \equiv \{\zeta \mid |\zeta - \zeta_0| \leqslant c\}$ is any disk with $K_0 \subset T_0$. Mapping T_k onto T_0 by an affine linear transformation, the uniformity guarantees a uniform estimate

$$|\mathcal{J}| = \left\| \frac{\partial(\xi, \eta)}{\partial(x, y)} \right\| \leqslant c' \frac{1}{h^2}$$

for the Jacobian $\mathcal{J}$ which implies (8.3.5) from (8.3.6).

Using similar arguments as in the proof of the preceding lemma the following stability properties can be shown [20, p. 167].

For $0 \leqslant k \leqslant r \leqslant \tilde{m}$ there exist M_{rk} and $M'_{rk} > 0$ independent of h and $\chi \in \tilde{H}$ such that

$$\|\chi_{|T_l}\|_{C^r(T_l')} \leqslant h^{k-r} M_{rk} \, \|\chi_{|T_l}\|_{C^k(T_l')} \tag{8.3.7}$$

and

$$\|\chi_{|T_l}\|_{H^r(T_l')} \leqslant h^{k-r} M'_{rk} \, \|\chi_{|T_l}\|_{H^k(T_l')} \tag{8.3.8}$$

hold for all $\chi \in \tilde{H}$. For $k \leqslant m-1$ on the right hand side of (8.3.7) *and for $r \leqslant m-1$ on both sides the norms can be replaced by the corresponding norms over $\bar{\mathsf{G}}$. The corresponding replacement is possible in* (8.3.8) *for $k \leqslant m$* [33]. For further estimates we use the following lemma from Il'in:

Lemma 8.3.2 [26] *Let $\dot{\mathsf{G}}$ satisfy a cone condition with radius $d > 0$. Then there exists a constant c such that*

$$\|g\|_{C^0(\bar{\mathsf{G}})} \leqslant c \cdot \|g\|_{H^1(\mathsf{G})} \cdot \begin{cases} 1 + |\log q|^{1/2} & \textit{for} \quad q \leqslant d \\ 1 & \textit{for} \quad q > d \end{cases} \tag{8.3.9}$$

holds for all $g \in H^1(\mathsf{G})$ with $g_z,\ g_{\bar{z}} \in L^\infty(\mathsf{G})$ where

$$q \equiv \|g\|_{H^1(\bar{\mathsf{G}})} \left\{ \operatorname*{ess.\,sup}_{z \in \mathsf{G}} \left(|g_z| + |g_{\bar{z}}|\right) \right\}^{-1}.$$

Since $\dot{\mathsf{G}}$ is assumed to be smooth, it satisfies the required cone condition in the above lemma. Now we can prove the announced further stability condition on $\tilde{H}$:

Lemma 8.3.3 *For $m \geq 1$ there exists a constant M independent of h and $\chi \in \tilde{H}$ such that*

$$\|\boldsymbol{\chi}\|_{C^0(G)} \leq M |\log h|^{1/2} \|\boldsymbol{\chi}\|_{H^1(G)} \quad \textit{for all} \quad \boldsymbol{\chi} \in \tilde{H}. \tag{8.3.10}$$

Proof Since all entries of (grad $\boldsymbol{\chi}$) are polynomials of degree $\leq m-1$ on every T_k we can use Lemma 8.3.1 yielding

$$\max_{T_k} (|x_{jz}| + |x_{j\bar{z}}|) \leq ch^{-1} \|\boldsymbol{\chi}_j\|_{H^1(G)} \tag{8.3.11}$$

for every component χ_j of $\boldsymbol{\chi}$. Hence,

$$q(\chi_j) \geq c^{-1}h$$

and Il'in's lemma provides the desired estimate

$$\|\chi_j\|_{C^0(G)} \leq c |\log h|^{1/2} \|\chi_j\|_{H^1(G)}$$

for every component.

Let us return to the least squares method of Section 8.2 and the corresponding Gauss equations (8.2.6). Since many of the finite element spaces provide a basis $\boldsymbol{\chi}_l$ for (8.2.8) where each $\boldsymbol{\chi}_l$ has a support of diameter $\leq c \cdot h$ [20, p. 81, 102] the corresponding matrix of coefficients to (8.2.6) is a sparse matrix. But the integrations (8.2.7) have to be done numerically in (8.2.6) and for suitable quadrature formulas the class of finite element functions with bases providing a sparse matrix is even bigger [20, p. 136].

Using the *a priori* estimates (8.2.1) and (8.2.3) we find (8.2.5)

$$\|\mathbf{U}-\mathbf{u}\|^2_{H^1(G)} \leq \gamma\rho_1^2(\mathbf{U}) \leq \gamma\rho_1^2(\boldsymbol{\chi}) \leq c\|\boldsymbol{\chi}-\mathbf{u}\|^2_{H^{3/2}(G)} \quad \text{for all} \quad \boldsymbol{\chi} \in \tilde{H}. \tag{8.3.12}$$

Using (8.3.3) we observe:

Proposition 8.3.4 *The least squares method* (8.2.5), (8.2.6) *converges for $m \geq 2$ and $\frac{3}{2} \leq s \leq m+1$ as*

$$\|\mathbf{U}-\mathbf{u}\|_{H^1(G)} \leq ch^{s-\frac{3}{2}} \|\mathbf{u}\|_{H^s(G)}. \tag{8.3.13}$$

Note that the order of convergence is of order $\frac{1}{2}$ less than the corresponding finite element approximation to $\mathbf{u}$ according to (8.3.3) and, moreover, in (8.3.13) the Courant elements are not admissible. Both disadvantages disappear if we replace (8.2.5) by an even simpler least squares method described in the next section.

8.4 A Least Squares Method with Optimal Order of Convergence

Bramble and Schatz used weighted norms in [10–12] in order to improve the least squares method for even order problems. (See also Aubin's book [5, p. 279 ff.].) Baker [7] simplified the original proof for Dirichlet problems by using an explicit version of Ehrling's inequality due to Thomée. This approach yields (for $\mathbf{u} \in \tilde{H}$) convergence of the least squares solutions with the same order as the best approximation (8.3.3), hence the convergence is of optimal order. Following Baker's presentation, Stephan and Wendland showed the same result for a rather large class of regular elliptic problems including pseudodifferential equations [47]. Here we shall follow their representation and shall find that for the first-order problems (8.0.1), (8.0.2) this least squares method combines the advantages of Galerkin methods to second-order problems with the treatment of the most general boundary conditions (8.0.1). The convergence proof, however, hinges essentially on Nitsche's trick (see e.g. [20, p. 49]) for which we shall need, that either (8.0.1) admits solutions for *every* $\mathbf{f}$, $\boldsymbol{\psi}$ smooth enough, i.e. the nullity of R^* (Theorem 4.3.2) is zero or we shall have to modify the approximation by adjoining a suitable finite dimensional additional space (independently of $\tilde{H}$). This last modification seems to be of a technical nature but the corresponding proof for the simpler first method is yet to be done. The corresponding numerical integrations will be introduced in Section 8.5.

Let us consider first the case that (8.0.1), (8.0.2) is uniquely solvable *for all* $\mathbf{f}$, $\boldsymbol{\psi}$. Following [47] the weighted least squares problem reads:
Minimize

$$\rho_h^2(\boldsymbol{\chi}) \equiv \|\mathsf{D}\boldsymbol{\chi} - \mathbf{f}\|^2_{L^2(\mathsf{G})} + h^{-1}\,|\mathsf{r}\boldsymbol{\chi} - \boldsymbol{\psi}|^2_{L^2(\dot{\mathsf{G}})} + \sum_{l=1}^{N} |\Lambda_l\boldsymbol{\chi} - c_l|^2 \tag{8.4.1}$$

for $\boldsymbol{\chi} \in \tilde{H}$.

For fixed $h > 0$ (8.4.1) has exactly one solution $\mathbf{U} \in \tilde{H}$ due to the uniqueness of (8.0.1), (8.0.2), and $\mathbf{U} \in \tilde{H}$ is the unique solution of the Gauss equations to (8.4.1),

$$\begin{aligned}[\mathbf{U} - \mathbf{u}, \boldsymbol{\chi}] &\equiv (\mathsf{D}(\mathbf{U} - \mathbf{u}), \mathsf{D}\boldsymbol{\chi})_0 + h^{-1}\langle \mathsf{r}(\boldsymbol{U} - \boldsymbol{u}), \mathsf{r}\boldsymbol{\chi}\rangle_0 + \sum_{l=1}^{N} (\Lambda_l(\boldsymbol{U} - \boldsymbol{u})) \cdot \Lambda_l\boldsymbol{\chi} \\ &= [\mathbf{U}, \boldsymbol{\chi}] - \mathcal{L}(\boldsymbol{\chi}) = 0 \quad \text{for all} \quad \boldsymbol{\chi} \in \tilde{H} \end{aligned} \tag{8.4.2}$$

where

$$\mathcal{L}(\boldsymbol{\chi}) \equiv \iint_{\mathsf{G}} \mathbf{f}^T \mathsf{D}\boldsymbol{\chi}[dx, dy] + h^{-1} \oint_{\dot{\mathsf{G}}} \boldsymbol{\psi}^T \mathsf{r}\boldsymbol{\chi}\, ds + \sum_{l=1}^{N} c_l \Lambda_l \boldsymbol{\chi}.$$

Here $[\cdot, \cdot]$ defines a scalar product for every $h > 0$. Note that (8.4.2) is equivalent to a system of M linear equations for the coefficients γ_j in (8.2.8). For a basis of elements with minimal supports this system will have a sparse coefficient matrix. Moreover, since D is of first order, (8.4.2) can be used for $m=1$, i.e. the piecewise linear Courant elements corresponding to the *Galerkin* method for second-order equations (and *not* the least squares method for second-order equations) [20, p. 134] although we handle rather general boundary conditions. This is a very valuable advantage of using first-order systems.

For the solutions $\mathbf{U}$ of the Gauss equations (8.4.2) we find optimal order of convergence:

Theorem 8.4.1 *For $t \leqslant s$ and*

$$1 \leqslant s \leqslant m+1, \qquad 0 \leqslant t \leqslant m \tag{8.4.3}$$

there exists a constant c which is independent of h and $\mathbf{u}$ *such that the asymptotic error estimate*

$$\|\mathbf{U}-\mathbf{u}\|_{H^t(\mathsf{G})} \leqslant c h^{s-t} \|\mathbf{u}\|_{H^s(\mathsf{G})} \tag{8.4.4}$$

holds provided the Λ_l *are continuous on* $H^0(\mathsf{G})$. *For* $1 \leqslant s \leqslant m+1$ *and* $\mathbf{u} \in H^s(\mathsf{G}) \cap C^{[s]}(\bar{\mathsf{G}})$ *the least squares solutions converge uniformly with*

$$\|\mathbf{U}-\mathbf{u}\|_{C^0(\bar{\mathsf{G}})} \leqslant c\{h^{s-1} |\log h|^{1/2} \|\mathbf{u}\|_{H^s(\mathsf{G})} + h^{[s]} \|\mathbf{u}\|_{C^{[s]}(\bar{\mathsf{G}})}\} \tag{8.4.5}$$

provided $\tilde{H}$ *satisfies the stronger conditions of Theorem* 3.3 *in the book by Fix and Strang* [20, p. 144].

Remark A simple generalization of (8.3.10) to higher derivatives also yields estimates (8.4.5) for derivatives of the error.

The proof of (8.4.4) follows [47]; let us begin with some preliminary estimates. Since to every given $\mathbf{u}$ there corresponds exactly one $\mathbf{U} \in \tilde{H}$ solving (8.4.2), the mapping

$$\underset{\sim}{G}\mathbf{u} \equiv \mathbf{U}, \qquad \mathbf{u} \mapsto \mathbf{U}$$

is well defined. $\underset{\sim}{G}$ is called the *Gauss projection* since

$$\underset{\sim}{G}\boldsymbol{\chi} = \boldsymbol{\chi} \tag{8.4.6}$$

holds for every $\boldsymbol{\chi} \in \tilde{H}$.

Lemma 8.4.2 *There exists a constant c independent of h and* $\mathbf{v}$ *such that for all* $1 \leqslant s \leqslant m+1$ *and* $\mathbf{v} \in H^s(\mathsf{G})$ *the estimate*

$$\inf_{\boldsymbol{\chi} \in \tilde{H}} [\boldsymbol{\chi}-\mathbf{v}, \boldsymbol{\chi}-\mathbf{v}]^{1/2} \leqslant c h^{s-1} \|\mathbf{v}\|_{H^s(\mathsf{G})} \tag{8.4.7}$$

holds.

Proof For every $\chi \in \tilde{H}$ we have from continuity

$$[\chi-\mathbf{v}, \chi-\mathbf{v}] = \|\mathsf{D}(\chi-\mathbf{v})\|_{L^2(\mathrm{G})} + h^{-1}\,|\mathsf{r}(\chi-\mathbf{v})|^2_{L^2(\dot{\mathrm{G}})} + \sum_{l=1}^{N} |\Lambda_l(\chi-\mathbf{v})|^2$$
$$\leq c\{\|\chi-\mathbf{v}\|^2_{H^1(\mathrm{G})} + h^{-1}\,|\chi-\mathbf{v}|^2_{L^2(\dot{\mathrm{G}})}\}.$$

For the last expression we use an explicit version of Ehrling's inequality due to Thomée (which can be found in [7, Lemma 2.1] or [28, p. 41 and p. 102]) and which can easily be obtained from the Green Theorem:

$$\oint_{\dot{\mathrm{G}}} |\chi-\mathbf{v}|^2\,ds \leq c\{\varepsilon^{-1}\,\|\chi-\mathbf{v}\|^2_{L^2(\mathrm{G})} + \varepsilon\,\|\chi-\mathbf{v}\|^2_{H^1(\mathrm{G})}\} \tag{8.4.8}$$

for every $\varepsilon > 0$.

Choosing $\varepsilon = \|\chi-\mathbf{v}\|_{H^0}/\|\chi-\mathbf{v}\|_{H^1}$, the above inequality becomes

$$\oint_{\dot{\mathrm{G}}} |\chi-\mathbf{v}|^2\,ds \leq 2c\,\|\chi-\mathbf{v}\|_{H^0}\,\|\chi-\mathbf{v}\|_{H^1}. \tag{8.4.9}$$

This yields the estimate

$$[\chi-\mathbf{v}, \chi-\mathbf{v}] \leq c\,\|\chi-\mathbf{v}\|_{H^1(\mathrm{G})}\{\|\chi-\mathbf{v}\|_{H^1(\mathrm{G})} + h^{-1}\,\|\chi-\mathbf{v}\|_{L^2(\mathrm{G})}\}.$$

If we chose χ according to the convergence property (8.3.3), this becomes the desired estimate (8.4.7).

Now we execute Nitsche's trick using the *a priori* estimate (8.2.1) for the error

$$\mathbf{e} \equiv \mathbf{u} - \mathbf{U}$$

together with orthogonality in (8.4.2).

Lemma 8.4.3 *For* $1 \leq s \leq m+1$, $0 \leq p \leq m$ *and* $-\frac{1}{2} \leq q \leq m$ *we have the following estimates*:

$$[\mathbf{e}, \mathbf{e}]^{1/2} \leq ch^{s-1}\,\|\mathbf{u}\|_{H^s(\mathrm{G})}, \tag{8.4.10}$$

$$\|\mathsf{D}\mathbf{e}\|_{H^{-p}(\mathrm{G})} \leq ch^{s+p-1}\,\|\mathbf{u}\|_{H^s(\mathrm{G})}, \tag{8.4.11}$$

$$|\mathsf{r}\mathbf{e}|_{H^{-q-1/2}(\dot{\mathrm{G}})} \leq ch^{s+q}\,\|\mathbf{u}\|_{H^s(\mathrm{G})}, \tag{8.4.12}$$

$$|\Lambda_l\mathbf{e}| \leq ch^{s+p-1}\,\|\mathbf{u}\|_{H^s(\mathrm{G})}. \tag{8.4.13}$$

Proof The estimate (8.4.10) follows from the definition of $\mathbf{U}$ according to (8.4.1) and (8.4.7),

$$[\mathbf{e}, \mathbf{e}] = \inf_{\chi \in \tilde{H}} \rho_h^2(\chi) = \inf_{\chi \in \tilde{H}} [\chi-\mathbf{u}, \chi-\mathbf{u}] \leq c^2 h^{2s-2}\,\|\mathbf{u}\|^2_{H^s(\mathrm{G})}. \tag{8.4.14}$$

If $\mathbf{g} \in H^p(\mathrm{G})$, $\boldsymbol{\phi} \in H^{p+1/2}(\dot{\mathrm{G}})$ and $c_1, \ldots, c_N$ are chosen arbitrarily then we find corresponding solutions $\mathbf{w} \in H^{p+1}(\mathrm{G})$ of

$$\mathsf{D}\mathbf{w} = \mathbf{g}, \qquad \mathsf{r}\mathbf{w} = \boldsymbol{\phi}, \qquad \Lambda_l\mathbf{w} = c_l. \tag{8.4.15}$$

Here we need the full space $H^p(\mathsf{G})\times H^{p+1/2}(\dot{\mathsf{G}})\times \mathbb{R}^N$ with solvability conditions. Orthogonality (8.4.2) and (8.4.7) (with $p+1=s$) yields

$$[\mathbf{e},\mathbf{w}]=[\mathbf{e},\mathbf{w}-\boldsymbol{\chi}]\leq[\mathbf{e},\mathbf{e}]^{1/2}[\mathbf{w}-\boldsymbol{\chi}]^{1/2},\quad \text{for}\quad \boldsymbol{\chi}\in\tilde{H};$$
$$[\mathbf{e},\mathbf{w}]\leq[\mathbf{e},\mathbf{e}]^{1/2}ch^p\,\|\mathbf{w}\|_{H^{p+1}(\mathsf{G})} \tag{8.4.16}$$

and with the definition of [,] and *a priori* estimate (8.2.1),

$$(\mathbf{De},\mathbf{g})_0+h^{-1}\langle\mathbf{re},\boldsymbol{\phi}\rangle_0+\sum_{l=1}^{N}\Lambda_l\mathbf{e}\,\mathrm{d}_l=[\mathbf{e},\mathbf{w}]$$
$$\leq ch^p[\mathbf{e},\mathbf{e}]^{1/2}\left\{\|\mathbf{g}\|_{H^p(\mathsf{G})}+|\boldsymbol{\phi}|_{H^{p+1/2}(\dot{\mathsf{G}})}+\sum_{l=1}^{N}|\mathrm{d}_l|\right\}. \tag{8.4.17}$$

Hence, in particular for $\boldsymbol{\phi}=0$, $\mathrm{d}_l=0$ we have

$$(\mathbf{De},\mathbf{g})_0\leq ch^p[\mathbf{e},\mathbf{e}]^{1/2}\,\|\mathbf{g}\|_{H^p}\quad\text{for every}\quad g\in H^p. \tag{8.4.18}$$

If $\mathbf{g}$ traces $\|\mathbf{g}\|_{H^p}\leq 1$ then the supremum of the left hand side coincides with $\|\mathbf{De}\|_{H^{-p}(\mathsf{G})}$, and hence

$$\|\mathbf{De}\|_{H^{-p}(\mathsf{G})}\leq ch^p[\mathbf{e},\mathbf{e}]^{1/2}\leq c'h^{p+s-1}\,\|\mathbf{u}\|_{H^s(\mathsf{G})}$$

with (8.4.7).

For (8.4.12) we proceed in the same way; (8.4.17) with $\mathbf{g}=0$ and $d_l=0$ yields

$$h^{-1}\langle\mathbf{re},\boldsymbol{\phi}\rangle_0\leq ch^p[\mathbf{e},\mathbf{e}]^{1/2}\,|\boldsymbol{\phi}|_{H^{p+1/2}(\dot{\mathsf{G}})}$$

and with duality and supremum over $|\boldsymbol{\phi}|_{p+1/2}\leq 1$ and (8.4.7),

$$|\mathbf{re}|_{H^{-p-1/2}(\dot{\mathsf{G}})}\leq ch^{p+1}[\mathbf{e},\mathbf{e}]^{1/2}\leq c'h^{p+s}\,\|\mathbf{u}\|_{H^s(\mathsf{G})}. \tag{8.4.19}$$

Moreover, the definition of [,] yields

$$|\mathbf{re}|_{H^0(\dot{\mathsf{G}})}\leq h^{1/2}[\mathbf{e},\mathbf{e}]^{1/2}\leq ch^{s-1/2}\,\|\mathbf{u}\|_{H^s(\mathsf{G})}. \tag{8.4.20}$$

(8.4.19) is already (8.4.12) for $q=p\geq 0$ whereas (8.4.20) corresponds to $q=-\frac{1}{2}$. Interpolation between (8.4.19) and (8.4.20) leads to (8.4.12) [5, p. 83].

The inequality (8.4.13) follows from (8.4.17) with $\mathbf{g}=0$, $\boldsymbol{\phi}=0$, $d_l=\delta_{lj}$, and (8.4.7),

$$|\Lambda_j\mathbf{e}|\leq ch^p[\mathbf{e},\mathbf{e}]^{1/2}\leq c'h^{p+s-1}\,\|\mathbf{u}\|_{H^s(\mathsf{G})}.$$

Lemma 8.4.4 *For $0\leq r\leq\frac{1}{2}$ and $1\leq s\leq m+1$ the error satisfies*

$$\|\mathbf{e}\|_{H^r(\mathsf{G})}\leq ch^{s-r}\,\|\mathbf{u}\|_{H^s(\mathsf{G})}. \tag{8.4.21}$$

The Gauss projection $\underset{\sim}{G}$ satisfies

$$\|\underset{\sim}{G}\mathbf{v}\|_{H^0(\mathsf{G})}\leq\|\mathbf{v}\|_{H^0(\mathsf{G})}+ch\,\|\mathbf{v}\|_{H^1(\mathsf{G})}. \tag{8.4.22}$$

Proof For $\mathbf{u} \in H^s(\mathrm{G})$ we have $\mathbf{e} \in H^1(\mathrm{G})$ (at least) and (8.2.1) is available for $\mathbf{e}$ providing

$$\|\mathbf{e}\|_{H^r(\mathrm{G})} \leqslant \gamma \left\{ \|\mathbf{De}\|_{r-1} + |\mathbf{re}|_{r-1/2} + \sum_{l=1}^{N} |\Lambda_l \mathbf{e}| \right\}. \tag{8.4.23}$$

Inserting (8.4.11)–(8.4.13) we find (8.4.21).

For the Gauss projection we use (8.4.21) with $r=0$, $s=1$,

$$\|\underset{\sim}{G}\mathbf{u}\|_{H^0(\mathrm{G})} = \|\mathbf{U}\|_{H^0(\mathrm{G})} \leqslant \|\mathbf{u}\|_{H^0(\mathrm{G})} + \|\mathbf{e}\|_{H^0(\mathrm{G})} \leqslant \|\mathbf{u}\|_{H^0(\mathrm{G})} + ch\, \|\mathbf{u}\|_{H^1(\mathrm{G})}.$$

Proof of Theorem 8.4.1 For $0 \leqslant t \leqslant \frac{1}{2}$ the proposition (8.4.4) coincides with (8.4.21). For the remaining t, $\frac{1}{2} < t \leqslant m$ we use the approximation property and inverse assumption (8.3.3), (8.3.4) in connection with (8.4.6) and (8.4.22) [34]:

$$\begin{aligned}
\|\mathbf{u}-\mathbf{U}\|_{H^t} &\leqslant \|\mathbf{u}-\boldsymbol{\chi}\|_{H^t} + \|\underset{\sim}{G}(\mathbf{u}-\boldsymbol{\chi})\|_{H^t} \\
&\leqslant \|\mathbf{u}-\boldsymbol{\chi}\|_{H^t} + mh^{-t}\, \|\underset{\sim}{G}(\mathbf{u}-\boldsymbol{\chi})\|_{H^0} \\
&\leqslant \|\mathbf{u}-\boldsymbol{\chi}\|_{H^t} + Mh^{-t}\, \{\|\mathbf{u}-\boldsymbol{\chi}\|_{H^0} + ch\, \|\mathbf{u}-\boldsymbol{\chi}\|_{H^1}\} \\
&\leqslant c' h^{s-t}\, \|\mathbf{u}\|_{H^s}.
\end{aligned}$$

For the remaining proposition (8.4.5) we use the interpolate $\mathbf{u}_I \in \tilde{H}$ to $\mathbf{u}$ since $\mathbf{u}$ is continuous for $s > 1$ due to the Sobolev embedding theorem. It follows from [20, pp. 136–143] that for integers $k = [s]$

$$\|\boldsymbol{u} - \boldsymbol{u}_I\|_{C^0} \leqslant ch^{[s]}\, \|\mathbf{u}\|_{C^{[s]}}$$

and from [20, Theorem 3.3 p. 144] that

$$\|\mathbf{u} - \mathbf{u}_I\|_{H^1} \leqslant ch^{s-1}\, \|\mathbf{u}\|_{H^s}.$$

Hence, with (8.3.10), (8.4.4) and the above estimates we find the desired inequality,

$$\begin{aligned}
\|\mathbf{u}-\mathbf{U}\|_{C^0} &\leqslant \|\mathbf{u}-\mathbf{u}_I\|_{C^0} + \|\mathbf{u}_I - \mathbf{U}\|_{C^0} \\
&\leqslant ch^{[s]}\, \|\mathbf{u}\|_{C^{[s]}} + c'\, |\log h|^{1/2}\, \{\|\mathbf{u}_I - \mathbf{u}\|_{H^1} + \|\mathbf{u}-\mathbf{U}\|_{H^1}\} \\
&\leqslant ch^{[s]}\, \|\mathbf{u}\|_{C^{[s]}} + c''\, |\log h|^{1/2}\, h^{s-1}\, \|\mathbf{u}\|_{H^s}.
\end{aligned}$$

In the most general case, if R in (8.0.1) has smaller range than $H^{s-1}(\mathrm{G}) \times H^{s-1/2}(\dot{\mathrm{G}})$ $(s > \frac{1}{2})$ let N^* denote the codimension of the range of R in this space. Since the eigenspace of R^*, the adjoint problem (4.3.8) is in $\bigcap_s H^s(\mathrm{G})$ thanks to the corresponding *a priori* estimate like (8.2.1), and since the solvability conditions for R are given by (4.3.25) for defining the range with the help of the L_2 scalar product over G, it can easily be shown that we can find (nonunique) linearly independent functions

$$\mathbf{l}_j(x, y),\ \mathbf{l}_j \in H^m(\mathrm{G}) \backslash \{\mathbf{g} = \mathbf{Dv} \quad \text{where } \mathbf{rv} = 0 \text{ on } \dot{\mathrm{G}} \quad \text{and}$$
$$\Lambda_l \mathbf{v} = 0, \quad l = 1, \ldots, N, \quad \mathbf{v} \in H^{m+1}\}, \quad j = 1, \ldots, N^* \tag{8.4.24}$$

such that the new, extended problem

$$\begin{aligned} \mathsf{D}\mathbf{v}+\sum_{j=1}^{N^*}\alpha_j\mathbf{l}_j&=\mathbf{f}\quad\text{in}\quad \mathsf{G},\\ \mathsf{r}\mathbf{v}&=\boldsymbol{\psi}\quad\text{on}\quad \dot{\mathsf{G}},\\ \Lambda_l\mathbf{v}&=c_l,\qquad l=1,\ldots,N \end{aligned}\tag{8.4.25}$$

admits exactly one solution $\mathbf{v}\in H^s$, $\boldsymbol{\alpha}=(\alpha_1,\ldots,\alpha^*_N)\in\mathbb{R}^{N^*}$ *for every given* $\mathbf{f}\in H^{s-1}(\mathsf{G})$, $\boldsymbol{\psi}\in H^{s-1/2}(\dot{\mathsf{G}})$, $c_l\in\mathbb{R}(s>\frac{1}{2})$. Hence, the mapping defined by (8.4.25) becomes a bijection $H^s(\mathsf{G})\times\mathbb{R}^{N^*}\to H^{s-1}(\mathsf{G})\times H^{s-1/2}(\dot{\mathsf{G}})\times\mathbb{R}^N$ due to the Fredholm properties, Theorem 4.3.2. Consequently, by Banach's Theorem on the bounded inverse (see also the closed graph theorem [54]) the *a priori* estimate (8.2.1) can be changed to

$$\begin{aligned}\gamma^{-1}\{\|\mathbf{v}\|^2_{H^t(\mathsf{G})}+|\boldsymbol{\alpha}|^2\}&\le\left\{\|\mathsf{D}\mathbf{v}\|^2_{H^{t-1}(\mathsf{G})}+|\mathsf{r}\mathbf{v}|^2_{H^{t-1/2}(\dot{\mathsf{G}})}+\sum_{l=1}^{N}|\Lambda_l\mathbf{v}|^2\right\}\\ &\le\gamma\{\|\mathbf{v}\|^2_{H^t(\mathsf{G})}+|\boldsymbol{\alpha}|^2\}\end{aligned}\tag{8.4.26}$$

where the left inequality holds for all $\mathbf{v}\in H^t\cap H^1(\mathsf{G})$ $0\le t\le m$ (with a new constant γ) and the right inequality holds for $\frac{1}{2}<t\le m$ provided that the Λ_l are continuous on $H^t(\mathsf{G})$, $|\boldsymbol{\alpha}|^2\equiv\sum_{j=1}^{N^*}\alpha_j^2$. Let $\mathbf{u}$ be the unique solution of (8.0.1), (8.0.2) with $\mathbf{f}$, $\boldsymbol{\psi}$ appropriately given in the range.

The new minimal problem corresponding to (8.4.25) is defined by: *Minimize the quadratic functional*

$$\begin{aligned}\rho_h'^2(\boldsymbol{\chi},\boldsymbol{\beta})\equiv&\left\|\mathsf{D}\boldsymbol{\chi}+\sum_{j=1}^{N^*}\beta_j\mathbf{l}_j-\mathbf{f}\right\|^2_{L^2(\mathsf{G})}\\ &+h^{-1}|\mathsf{r}\boldsymbol{\chi}-\boldsymbol{\psi}|^2_{L^2(\dot{\mathsf{G}})}+\sum_{l=1}^{N}|\Lambda_l\boldsymbol{\chi}-c_l|^2\end{aligned}\tag{8.4.27}$$

for $\boldsymbol{\chi}\in\tilde{H}$ *and* $\boldsymbol{\beta}\in\mathbb{R}^{N^*}$.

The Gauss equations to (8.4.27) and the corresponding scalar product are given by

$$\begin{aligned}[(\mathbf{V}-\mathbf{u},\boldsymbol{\alpha}),(\boldsymbol{\chi},\boldsymbol{\beta})]'\equiv&\left(\mathsf{D}(\mathbf{V}-\mathbf{u})+\sum_{j=1}^{N^*}\alpha_j\mathbf{l}_j,\mathsf{D}\boldsymbol{\chi}+\sum_{j=1}^{N^*}\beta_j\mathbf{l}_j\right)_0\\ &+h^{-1}\langle\mathsf{r}(\mathbf{V}-\mathbf{u}),\mathsf{r}\boldsymbol{\chi}\rangle_0+\sum_{l=1}^{N}(\Lambda_l(\mathbf{V}-\mathbf{u}))\Lambda_l\boldsymbol{\chi}\\ =&\left(\mathsf{D}\mathbf{V}+\sum_{j=1}^{N^*}\alpha_j\mathbf{l}_j-f,\mathsf{D}\boldsymbol{\chi}+\sum_{j=1}^{N^*}\beta_j\mathbf{l}_j\right)_0\\ &+h^{-1}\langle\mathsf{r}\mathbf{V}-\boldsymbol{\psi},\mathsf{r}\boldsymbol{\chi}\rangle_0+\sum_{l=1}^{N}(\Lambda_l\mathbf{V}-c_l)\Lambda_l\boldsymbol{\chi}=0\end{aligned}\tag{8.4.28}$$

for all $(\boldsymbol{\chi}, \boldsymbol{\beta}) \in \tilde{H} \times \mathbb{R}^{N^*}$ with $(\mathbf{V}, \boldsymbol{\alpha}) \in \tilde{H} \times \mathbb{R}^{N^*}$. Clearly, (8.4.28) and (8.4.27) have exactly one solution $(\mathbf{V}, \boldsymbol{\alpha}) \in \tilde{H} \times \mathbb{R}^{N^*}$.

Replacing the trial functions, the error $\mathbf{e}$, etc., by pairs $(\boldsymbol{\chi}, \boldsymbol{\beta})$, $(\mathbf{V}-\mathbf{u}, \boldsymbol{\alpha}-\mathbf{0})$ etc., the lemmata 8.4.2–8.4.4 remain valid word by word yielding:

Theorem 8.4.5 *Under the same assumptions as in Theorem* 8.4.1 *we have corresponding error estimates for*

$$\|\mathbf{V}-\mathbf{u}\|+|\boldsymbol{\alpha}|$$

on the left hand sides of (8.4.4) *and* (8.4.5) *where* $\|\cdot\|$ *stands for the* $H^t(\mathrm{G})$ *respectively* $C^0(\bar{\mathrm{G}})$ *norm.*

We leave the details to the reader.

8.5 The Discrete Equations for the Least Squares Method

In order to solve the equations (8.4.2) computationally, all the integrations must be replaced by quadrature formulas on each element T_k' and on $\dot{\mathrm{G}}_k$ respectively. As is pointed out in [20], the coefficients $[\boldsymbol{\chi}_j, \boldsymbol{\chi}_r]$ of the matrix and $\mathscr{L}(\boldsymbol{\chi}_r)$ of the given right hand sides have to be evaluated with high accuracy to preserve both definiteness and error estimates. For the matrix elements we shall see that quadratures providing a property corresponding to the patch test of Galerkin methods preserve the definiteness. This is the already mentioned advantage of the least squares method for first-order systems with finite elements which yields sparse matrices for practically all the finite element spaces [20, p. 101 ff.]. Using a quadrature formula on T_k' of high enough order this test will succeed. A very thorough investigation of this question has been done by Ciarlet and Raviart [16]; our assumptions here will be much cruder. Since we have gained so much for the general BVP we must pay for it with a higher accuracy of the quadratures in $\mathscr{L}(\boldsymbol{\chi}_r)$. The reason for this trouble is the penalty term $h^{-1}\langle \mathrm{r}\mathbf{U}, \mathrm{r}\boldsymbol{\chi}\rangle_0$ in (8.4.2) spoiling uniform continuity and definiteness ((independent of $h>0$) over $H^1(\mathrm{G})$. This will imply either a loss of order two in our error estimates or e.g. a recursive refinement of the numerical quadratures.

For the matrix elements $[\boldsymbol{\chi}_j, \boldsymbol{\chi}_r]$ we use a quadrature formula

$$\iint_{T_k'} F(z)[\mathrm{d}x, \mathrm{d}y] = \sum_{j=1}^{L} \Omega_{kj} F(z_{kj}) + R_k(F) \tag{8.5.1}$$

with L nodal points $z_{kj} \in T_k'$ where L is independent of h. Correspondingly we use for the boundary integrals a quadrature formula

$$\oint_{\dot{G}_{k'}} F(z)\,ds_z = \sum_{j=1}^{L'} \omega_{k'j} F(z'_{k'j}) + R'_{k'}(F) \tag{8.5.2}$$

with L' nodal points on $\dot{G}_{k'}$, $z'_{k'j} \in \dot{G}_{k'}$. R_k and $R'_{k'}$ denote the error terms.

The bilinear form [,] also contains the linear functionals Λ_l which are continuous over $H^0(G)$. Hence, they have an integral representation

$$\Lambda_l \mathbf{v} = \iint_G \mathbf{\Lambda}_l^T(z)\mathbf{v}(z)[dx, dy] + \oint_{\dot{G}} \boldsymbol{\lambda}_l^T \mathbf{v}(z)\,ds_z \tag{8.5.3}$$

with $\mathbf{\Lambda}_l \in H^0(G)$ and $\boldsymbol{\lambda}_l \in H^0(\dot{G})$. For the following let us assume that $\mathbf{\Lambda}_l$ and $\boldsymbol{\lambda}_l$ are as smooth as we need. (Without loss of generality, they can always be chosen as smooth functions.)

Replacing all integrals in $[\boldsymbol{\chi}_j, \boldsymbol{\chi}_r]$ by quadratures (8.5.1), respectively (8.5.2), we define a new computable quadratic form by

$$\begin{aligned}[\boldsymbol{\psi}, \boldsymbol{\chi}] \equiv {} & \sum_{k=1}^{K}\sum_{j=1}^{L} \Omega_{kj}(\mathsf{D}\boldsymbol{\psi})^T(\mathsf{D}\boldsymbol{\chi})|_{z_{kj}} + h^{-1}\sum_{k'=1}^{K'}\sum_{j=1}^{L'} \omega_{k'j}(\boldsymbol{\psi}^T\mathbf{r}^T\mathbf{r}\boldsymbol{\chi})|_{z'_{k'j}} \\ & + \sum_{l=1}^{N}\left\{\sum_{k=1}^{K}\sum_{j=1}^{L} \omega_{kj}\mathbf{\Lambda}_l^T\boldsymbol{\psi}|_{z_{kj}} + \sum_{k'=1}^{K'}\sum_{j=1}^{L'} \omega_{k'j}\boldsymbol{\lambda}_l^T\boldsymbol{\psi}|_{z'_{k'j}}\right\} \\ & \times\left\{\sum_{k=1}^{K}\sum_{j=1}^{L} \Omega_{kj}\mathbf{\Lambda}_l^T\boldsymbol{\chi}|_{z_{kj}} + \sum_{k'=1}^{K'}\sum_{j=1}^{L'} \omega_{k'j}\boldsymbol{\lambda}_l^T\boldsymbol{\chi}|_{z'_{k'j}}\right\}. \end{aligned} \tag{8.5.4}$$

The given right hand side in (8.4.2) expressed by $\mathcal{L}(\boldsymbol{\chi})$ can be replaced by using the same refined quadratures. In the latter case every T'_k, respectively $\dot{G}_k$, is again broken up subordinate to a second partition with a new mesh width $h_0 \leqslant c \cdot h^2$ and $T'_k = \bigcup_{p=1}^{L''} T_{kp}$ where the T_{kp} are intersections of subtriangles with $\bar{G}$; for the boundary part we use a respective refinement. Here we do not need any uniformity. Since L'' will be of order h^{-2} this is a recursive refinement. The corresponding quadrature formulas are denoted by

$$\iint_{T_{kp}} F(z)[dx, dy] = \sum_{j=1}^{L^{01}} \Omega_{kpj} F(z_{kpj}) + R^0_{kp}(F) \tag{8.5.5}$$

with positive weights, $z_{kpj} \in T_{kp}$ and fixed L^0, and

$$\oint_{\dot{G}_{k'p}} F(z)\,ds_z = \sum_{j=1}^{L^{01}} \omega_{k'pj} F(z'_{k'pj}) + R^{01}_{k'p}(F), \tag{8.5.6}$$

$$L^{01} = 0(h^{-1}).$$

The computable replacement of $\mathscr{L}$ will be

$$\mathscr{L}_*(\chi) \equiv \sum_{k=1}^{K} \sum_{p=1}^{L''} \sum_{j=1}^{L^0} \Omega_{kpj} \mathbf{f}^T \mathsf{D}\chi|_{z_{kpj}} + \sum_{k'=1}^{K'} \sum_{p=1}^{L'''} \sum_{j=1}^{L^0} \omega_{k'pj} \boldsymbol{\psi}^T \mathbf{r}\chi|_{z'_{k'pj}}, \quad L'' = 0(h^{-2}). \tag{8.5.7}$$

In case of the modified equations (8.4.28) we have to add the corresponding numerical integrals involving $\mathbf{l}_j$. But since all the aspects in the following are the same regardless of this modification let us consider here only the first simpler case.

The computational equations are

$$\sum_{j=1}^{M} \gamma_j^* [\chi_j, \chi_r]_* - \mathscr{L}_*(\chi_r) = 0, \quad r = 1, \ldots, M(h) \tag{8.5.8}$$

providing a solution

$$\mathbf{U}^* = \sum_{j=1}^{M} \gamma_j^* \chi_j \tag{8.5.9}$$

if (8.5.8) has essentially the same properties as (8.4.2). For this purpose let us impose the following conditions on the quadrature formulas (8.5.1), (8.5.2), and (8.5.5), (8.5.6), respectively, [20, Sec. 4.3]:
Patch condition (P):

$$|[\boldsymbol{\psi}, \chi] - [\boldsymbol{\psi}, \chi]_*| \leq ch^{m+1/2} \|\boldsymbol{\psi}\|_{H^1(G)} \|\chi\|_{H^{1/2}(G)} \tag{8.5.10}$$

for all $\boldsymbol{\psi}, \chi \in \tilde{H}$ *where* c *is independent of* h, $\boldsymbol{\psi}$, χ.
Accuracy condition (A):

$$|\mathscr{L}(\chi) - \mathscr{L}_*(\chi)| \leq ch^{\sigma} \|\mathbf{u}\|_{H^s(G)} \|\chi\|_{H^{1/2}(G)} \tag{8.5.11}$$

for all $\chi \in \tilde{H}$ *with* $0 \leq \sigma \leq s - \frac{1}{2}$, $s \geq 1$, $\mathbf{u}$ *the solution of the original problem* (8.0.1), (8.0.2), *and where* c *is independent of* h, $\mathbf{u}$ *and* χ.

As we have already mentioned, Ciarlet and Raviart [16] show how to choose quadrature formulas with nodal points corresponding to the finite element interpolation in order to find optimal estimates providing patch and accuracy. They also consider isoparametric elements.

Let us give here simpler but much stronger assumptions guaranteeing (P) and (A).
(I) (8.5.1) *is exact for all polynomials of degree* $\leq m + 2\tilde{m} - 2$ *and* (8.5.2) *is exact for all polynomials* (in s) *of degree* $\leq m + 2\tilde{m}$.

Note, that in general (I) requires a higher degree than the patch for Galerkin's method in [20, p. 182]. But this again is some payment for the general boundary conditions. For Courant elements, however, we have $m + 2\tilde{m} - 2 = 1$ and the conditions coincide. Here (I) can be achieved with just 3 nodal points, e.g. the vertices of the triangles in (8.5.1) and with

Simpson's rule or the two point Gaussian formula in the boundary integrals (8.5.2).

For proving (*P*) or (*A*) with (*I*) we need well known estimates for the errors of integration. If the quadrature (8.5.1) is exact for all polynomials of degree $\leq\lambda$ then the error satisfies

$$|R_k(F)|\leq h^{\lambda+1}\cdot c_\lambda \sum_{\alpha+\beta=\lambda+1} \|(\partial^{\lambda+1}/\partial x^\alpha \partial y^\beta)F\|_{C^0(T_k')} \times \cdot \iint_{T_k'} [dx, dy]. \tag{8.5.12}$$

For (8.5.2) we have the estimate

$$|R_{k'}(F)|\leq h^{\lambda+1} c_\lambda' \oint_{\dot{G}_{k'}} \left|\frac{d^{\lambda+1}}{ds^{\lambda+1}} F\right| ds. \tag{8.5.13}$$

The constants c_λ, c_λ' are independent of h, F and k.

Lemma 8.5.1 *Let all coefficients of* $\mathbf{D}$, $\mathbf{r}$ *and* Λ_l *be smooth enough. If* (*I*) *is satisfied then the patch condition* (8.5.10) *is valid. Using* (8.5.1) *also for* $\mathcal{L}_*$ *then the accuracy in* (8.5.11) *is* $\sigma=s-\frac{5}{2}-\varepsilon$ *with any* $\varepsilon>0$ *arbitrarily small.* (The constant in (8.5.11) depends on ε.)

Proof For the patch we observe first that $\boldsymbol{\psi}$ and $\boldsymbol{\chi}$ on T_k' are polynomials of degree $2\tilde{m}$. Since all terms in the difference (8.5.10) can be handled similarly let us prove the desired estimate just for one term of the surface integrals and for one boundary integral. With (8.5.12) we have

$$I_1 \equiv \left| \iint_G \boldsymbol{\psi}_y^T \mathbf{B}^T \mathbf{B} \boldsymbol{\chi}_y [dx, dy] - \sum_{k=1}^{K} \sum_{j=1}^{L} \Omega_{kj} \boldsymbol{\psi}_y^T \mathbf{B}^T \mathbf{B} \boldsymbol{\chi}_{y|z_{kj}} \right|$$

$$\leq \sum_{k=1}^{K} |R_k(\boldsymbol{\psi}_y \mathbf{B}^T \mathbf{B} \boldsymbol{\chi}_y)|$$

$$\leq h^{\lambda+1} \cdot c \cdot \sum_{k=1}^{K} \sum_{|\alpha|=\lambda+1} \|\partial^\alpha \boldsymbol{\psi}_y^T \mathbf{B}^T \mathbf{B} \boldsymbol{\chi}_y\|_{C^0(T_k')} \iint_{T_k'} [dx, dy] \tag{8.5.14}$$

$$(\partial^\alpha = (\partial/\partial x)^{\alpha_1} (\partial/\partial y)^{\alpha_2}, \quad |\alpha| = \alpha_1 + \alpha_2).$$

Here we use the product rule and observe that all derivatives of $\boldsymbol{\psi}$ or $\boldsymbol{\chi}$, respectively, of degree higher than $\tilde{m}$ vanish identically. The derivatives of $\mathbf{B}^T\mathbf{B}$ are bounded. This gives rise to the estimate

$$I_1 \leq c \cdot h^{\lambda+3} \sum_{k=1}^{K} \sum_{1\leq|\alpha|,|\beta|\leq\tilde{m}} \|\partial^\alpha \boldsymbol{\psi}\|_{C^0(T_k')} \|\partial^\beta \boldsymbol{\chi}\|_{C^0(T_k')}.$$

Here we use the stability property (8.3.7) yielding

$$I_1 \leqslant c \cdot h^{\lambda+5-2\tilde{m}} \sum_{k=1}^{K} \sum_{|\alpha|=1} \|\partial^\alpha \boldsymbol{\psi}\|_{C^0(T_k')} \sum_{|\beta|=1} \|\partial^\beta \boldsymbol{\chi}\|_{C^0(T_k')}.$$

Now Lemma 8.3.1 can be applied which leads to

$$I_1 \leqslant ch^{\lambda+3-2\tilde{m}} \sum_{k=1}^{K} \|\boldsymbol{\psi}\|_{H^1(T_k')} \|\boldsymbol{\chi}\|_{H^1(T_k')}.$$

$\lambda = m + 2\tilde{m} - 2$. The Schwarz inequality and (8.3.4) yield the desired estimate (8.5.10).

For the boundary integrals we find from (8.5.13) that

$$I_2 \equiv h^{-1} \left| \oint_{\dot{G}} \boldsymbol{\psi}^T \mathbf{r}^T \mathbf{r} \boldsymbol{\chi} \, ds - \sum_{k'=1}^{K'} \sum_{j=1}^{L'} \omega_{k'j} \boldsymbol{\psi}^T \mathbf{r}^T \mathbf{r} \boldsymbol{\chi}_{|z'_{k'j}} \right|$$

$$\leqslant h^\lambda \cdot c \sum_{k'=1}^{K'} \left\| \frac{d^{\lambda+1}}{ds^{\lambda+1}} \boldsymbol{\psi}^T \mathbf{r}^T \mathbf{r} \boldsymbol{\chi} \right\|_{C^0(T')} \oint_{\dot{G}_{k'}} ds.$$

Here we use the chain rule and again the fact that the $\tilde{m}+1+k$ derivations of $\boldsymbol{\psi}$, $\boldsymbol{\chi}$ on T_k' vanish identically. Hence we find

$$I_2 \leqslant c' h^{\lambda+1} \sum_{k'=1}^{K'} \|\boldsymbol{\psi}\|_{C^{\tilde{m}}(T_{k'}')} \|\boldsymbol{\chi}\|_{C^{\tilde{m}}(T_{k'}')}.$$

Stability (8.3.7) and (8.3.5) for the first derivatives yield

$$I_2 \leqslant c'' h^{\lambda+1-2\tilde{m}} \sum_{k'=1}^{K'} \|\boldsymbol{\psi}\|_{H^1(T_{k'}')} \|\boldsymbol{\chi}\|_{H^1(T_{k'}')}.$$

Schwarz inequality, $\lambda = 2\tilde{m} + m$ and (8.3.4) finally imply the desired estimate

$$I_2 \leqslant c''' h^{m+1/2} \|\boldsymbol{\psi}\|_{H^1(G)} \|\boldsymbol{\chi}\|_{H^{1/2}(G)}.$$

For the Λ_l of the form (8.5.3) we have the numerical approximation

$$\Lambda_{l*}\boldsymbol{\chi} \equiv \sum_{k=1}^{K} \sum_{j=1}^{L} \Omega_{kj} \boldsymbol{\Lambda}_l^T \boldsymbol{\chi}_{|z_{kj}} + \sum_{k'=1}^{K'} \sum_{j=1}^{L'} \omega_{k'j} \boldsymbol{\lambda}_l^T \boldsymbol{\chi}_{|z'_{k'j}}. \tag{8.5.16}$$

Hence, with the same arguments as above we find

$$I_3 \equiv |\Lambda_l \boldsymbol{\chi} - \Lambda_{l*} \boldsymbol{\chi}| \leqslant ch^{m+2\tilde{m}+1} \sum_{k=1}^{K} \|\boldsymbol{\chi}\|_{C^{\tilde{m}}(T_k')}$$

and with (8.3.7), (8.3.5), the Schwarz inequality, (8.3.4) and $\tilde{m} \geqslant 1$,

$$I_3 \leqslant c' h^{m+1} \|\boldsymbol{\chi}\|_{H^1(G)} \leqslant ch^{\tilde{m}+1/2} \|\boldsymbol{\chi}\|_{H^{1/2}(G)}. \tag{8.5.17}$$

Since Λ_l is continuous on H^1, (8.5.17) shows that Λ_{l*} there is also continuous. This together with (8.5.17) finally implies that

$$|\Lambda_l \boldsymbol{\psi} \Lambda_l \boldsymbol{\chi} - \Lambda_{l*} \boldsymbol{\psi} \Lambda_{l*} \boldsymbol{\chi}| \leqslant h^{m+1/2} c'''(2 + c'' h^{m+1}) \, \|\boldsymbol{\psi}\|_{H^1(\mathsf{G})} \, \|\boldsymbol{\chi}\|_{H^{1/2}(\mathsf{G})},$$

the last contribution to (8.5.10).

For proving the second proposition for $\mathscr{L}_*$ in Lemma 8.5.1 replace $\mathbf{f}$ in $\mathscr{L}$ and $\mathscr{L}_*$ by a finite element approximation e.g. by $\mathbf{g} = P_h \mathbf{f}$. Then with the same arguments as above we find after some simple computations

$$\begin{aligned} I_4 &= \left| \iint_{\mathsf{G}} \mathbf{f}^T \mathsf{D} \boldsymbol{\chi} [dx, dy] - \sum_{k=1}^{K} \sum_{j=1}^{L} \Omega_{kj} \mathbf{f}^T \mathsf{D} \boldsymbol{\chi}_{|z_{kj}} \right| \\ &\leqslant c \, \|\boldsymbol{\chi}\|_{H^1(\mathsf{G})} \, h^{s-1} \, \|\mathbf{f}\|_{H^{s-1}(\mathsf{G})} + c' h^2 \sum_{k=1}^{K} \|\mathbf{f} - \mathbf{g}\|_{C^0(T_k')} \, \|\boldsymbol{\chi}\|_{C^1(T_k')}. \end{aligned} \qquad (8.5.18)$$

In the last term we use the Sobolev embedding theorem [46, p. 84] for $\|\mathbf{f} - \mathbf{g}\|_{C^0}$ and (8.3.5) with the Schwarz inequality for the remaining sum,

$$I_4 \leqslant c'' \, \|\mathbf{f} - \mathbf{g}\|_{H^{1+\varepsilon}(\mathsf{G})} \, \|\boldsymbol{\chi}\|_{H^1(\mathsf{G})} \leqslant c''' h^{s-5/2-\varepsilon} \, \|\mathbf{f}\|_{s-1} \, \|\boldsymbol{\chi}\|_{H^{1/2}(\mathsf{G})}.$$

The boundary integral,

$$I_5 \equiv \frac{1}{h} \left| \oint_{\dot{\mathsf{G}}} \boldsymbol{\psi}^T \mathsf{r} \boldsymbol{\chi} \, ds - \sum_{k'=1}^{K'} \sum_{j=1}^{L'} \omega_{k'j} \boldsymbol{\psi}^T \mathsf{r} \boldsymbol{\chi}_{|z'_{k'j}} \right| \qquad (8.5.19)$$

can be handled similarly by the use of a $H^s(\mathsf{G})$ prolongation of the $H^{s-1/2}(\dot{\mathsf{G}})$ boundary values $\mathsf{r}^T \boldsymbol{\psi} = \mathbf{H}$ on $\dot{\mathsf{G}}$ to $\bar{\mathsf{G}}$ which can be shown to satisfy

$$c_2 \, \|\mathbf{H}\|_{H^s(\mathsf{G})} \leqslant |\mathsf{r}^T \boldsymbol{\psi}|_{H^{s-1/2}(\dot{\mathsf{G}})} \leqslant c_1 \, \|\mathbf{H}\|_{H^s(\mathsf{G})}, \qquad (8.5.20)$$

acçording to Lions–Magenes [28, p. 42]. Approximating $\mathbf{H}$ by a finite element function, e.g. by $P_h \mathbf{H}$ we find with a similar analysis as for I_2 an estimate

$$I_5 \leqslant c h^{s-3/2} \, \|\mathbf{H}\|_{H^s(\mathsf{G})} \, \|\boldsymbol{\chi}\|_{H^{1/2}(\mathsf{G})} + c' h^{-1} \, \|(P_h - 1)\mathbf{H}\|_{C^0(\bar{\mathsf{G}})} \, \|\boldsymbol{\chi}\|_{C^0(\bar{\mathsf{G}})}.$$

Using (8.3.10), (8.3.4), the Sobolev embedding theorem and (8.3.3) we finally find

$$I_5 \leqslant c h^{s-3/2} \, \|\mathbf{H}\|_{H^s(\mathsf{G})} \, \|\boldsymbol{\chi}\|_{H^{1/2}(\mathsf{G})} + c'' h^{s-5/2-\varepsilon'} \, |\log h| \, \|\mathbf{H}\|_{H^s(\mathsf{G})} \, \|\boldsymbol{\chi}\|_{H^{1/2}(\mathsf{G})}.$$

Inserting (8.5.20) and the *a priori* estimate (8.2.1) we arrive at an estimate (8.5.11) with $\sigma = s - \frac{5}{2} - \varepsilon$.

The remaining difference $\Lambda_l \boldsymbol{\chi} - \Lambda_{l*} \boldsymbol{\chi}$ can be treated in the same manner as (8.5.18) and (8.5.19).

For satisfying the accuracy condition (A) with $\sigma = s - \frac{1}{2}$ using the recursive refinement in (8.5.7) we can prove the following *sufficient* condition.

Lemma 8.5.2 *Let* (8.5.5) *be exact for all polynomials of degree* $\leqslant \max\{2, \tilde{m}+[m/2]\}$ *and let* (8.5.6) *be exact for all polynomials* (in s) *of degree* $\leqslant \max\{2, \tilde{m}+[m/2]-3\}$. *Then the accuracy condition* (A) *holds with* $\sigma = s-\frac{1}{2}$ *for* $s>4$.

For the surface integrals, the proof is based on the refinement $h_0 \leqslant ch^2$ and similar estimates as for (8.5.18), (8.5.19). In the surface integrals approximated by sums (8.5.5) replace $\mathbf{f}$ by finite elements $\mathbf{g}=P_h\mathbf{f}$ and use the corresponding error terms with different λ; with maximal λ for the terms containing only finite element functions and with $\lambda'=[s]-2$ for the terms containing $(\mathbf{f}-\mathbf{g})$. These estimates have the form

$$\left| \iint\limits_{\mathsf{G}} \mathbf{f}^T \mathsf{D}\boldsymbol{\chi}[dx, dy] - \sum_{k=1}^{K} \sum_{p=1}^{L''} \sum_{j=1}^{L^0} \Omega_{kpj} \mathbf{f}^T \mathsf{D}\boldsymbol{\chi}_{|z_{kpj}} \right|$$

$$\leqslant \sum_{k=1}^{K} \sum_{p=1}^{L''} \{ |R^0_{kp}((P_h\mathbf{f}-\mathbf{f})^T \mathsf{D}\boldsymbol{\chi})| + |R^0_{kp}((P_h\mathbf{f})^T \mathsf{D}\boldsymbol{\chi})| \}$$

$$\leqslant c_1 \sum_{k=1}^{K} \sum_{p=1}^{L''} h_0^{\lambda'+3} \{ \|P_h\mathbf{f}-\mathbf{f}\|_{C^0(\bar{\mathsf{G}})} \cdot \|\boldsymbol{\chi}\|_{C^{\lambda'+2}(T_{kp})}$$

$$+ \cdots + \|P_h\mathbf{f}-\mathbf{f}\|_{C^{\lambda'+1}(\bar{\mathsf{G}})} \cdot \|\boldsymbol{\chi}\|_{C^1(T_{kp})} \}$$

$$+ c_2 \sum_{k=1}^{K} \sum_{p=1}^{L''} h_0^{\tilde{m}+[m/2]+1} \|P_h\mathbf{f}\|_{C^{\tilde{m}}(T_{kp})} \cdot \|\boldsymbol{\chi}\|_{C^{\tilde{m}}(T_{kp})}.$$

Summing up the terms over T_{kp} with respect to p first, then using (8.3.7), (8.3.5), the Schwarz inequality and Sobolev's embedding theorem, one finally arrives at (8.5.11).

In the boundary integrals one can again use the prolongation $\mathbf{H}$ and its approximation by a finite element function. With different error terms as for the preceding surface integrals and summing first up $p=1, \ldots, L''$, then using (8.3.7), (8.3.5), (8.3.10) and Sobolev's embedding theorem, one again finds an estimate (8.5.11) for the boundary integrals (8.5.6). The remaining differences $\Lambda_l\boldsymbol{\chi}-\Lambda_{l*}\boldsymbol{\chi}$ can be estimated in the same manner. We shall omit further details.

$s>4$ is a very high regularity condition requiring $m \geqslant 4$. Since then $\mathbf{u}$ is in $C^3(\bar{\mathsf{G}})$ one might rather use maximum estimates instead of our Sobolev space estimates.

If the patch condition (8.5.10) is available and the right hand sides represented by $\mathcal{L}_*(\boldsymbol{\chi})$ are accurate enough then the numerical solutions (8.5.9) computed from (8.5.8) will converge to $\mathbf{u}$. In order to show this we first prove some definiteness and continuity properties for $[\ ,\ ]$ and $[\ ,\ ]_*$.

Lemma 8.5.3 *For $h \leq 1$ the bilinear form* [,] (8.4.2) *satisfies*

$$k\,\|\chi\|^2_{H^{1/2}(G)} \leq [\chi, \chi] \leq k' h^{-1}\,\|\chi\|_{L^2(G)} \cdot \|\chi\|_{H^1(G)} \tag{8.5.21}$$

where $k > 0$ and k' are independent of h and $\chi \in \tilde{H}$.

Proof The left inequality follows from the continuous embedding of L^2 into $H^{-1/2}$ and the *a priori* estimate (8.2.1) with $t = \frac{1}{2}$ since $\chi \in H^m \supset H^1$:

$$\begin{aligned}[\chi, \chi] &\geq \|\mathbf{D}\chi\|^2_{L^2(G)} + |\mathbf{r}\chi|^2_{L^2(\dot{G})} + \sum_{l=1}^{N} |\Lambda_l \chi|^2 \\ &\geq \|\mathbf{D}\chi\|^2_{H^{-1/2}(G)} + |\mathbf{r}\chi|_{L^2(\dot{G})} + \sum_{l=1}^{N} |\Lambda_l \chi|^2 \\ &\geq \gamma\,\|\chi\|^2_{H^{1/2}(G)}.\end{aligned}$$

The right inequality follows from (8.4.9), the continuity of $\mathbf{D}$, Λ_l and the inverse assumption (8.3.4):

$$\begin{aligned}[\chi, \chi] &\leq c\,\|\chi\|^2_{H^1(G)} + c' h^{-1}\,\|\chi\|_{L^2(G)}\,\|\chi\|_{H^1(G)} \\ &\leq c'' h^{-1}\,\|\chi\|_{L^2(G)}\,\|\chi\|_{H^1(G)}.\end{aligned}$$

Lemma 8.5.4 *If the patch condition* (8.5.10) *holds then* $[\ ,\ ]_*$ *is also uniformly definite:*

$$k''\,\|\chi\|^2_{H^{1/2}(G)} \leq [\chi, \chi]_* \tag{8.5.22}$$

where $k'' > 0$ is independent of $h \leq h_0$ and χ.

Proof (8.5.22) follows from (8.5.21) and (8.5.10) just by the triangle inequality and (8.3.4) since $m \geq 1$:

$$\begin{aligned}[\chi, \chi]_* &\geq [\chi, \chi] - |[\chi, \chi]_* - [\chi, \chi]| \\ &\geq k\,\|\chi\|^2_{H^{1/2}(G)} - c h^{3/2}\,\|\chi\|_{H^1(G)}\,\|\chi\|_{H^{1/2}(G)} \\ &\geq (k - c'h)\,\|\chi\|^2_{H^{1/2}(G)}.\end{aligned}$$

This implies (8.5.22) for $h \leq 0.5k/c'$.

Theorem 8.5.5 (See also [20, p. 186]) *If the patch condition (P) and accuracy condition (A) are fulfilled then we have the asymptotic error estimates*

$$\|\mathbf{U}^* - \mathbf{U}\|_{H^t(G)} \leq c h^{\sigma+1/2-t}\,\|\mathbf{u}\|_{H^s(G)}, \tag{8.5.23}$$

$$\|\mathbf{U}^* - \mathbf{u}\|_{H^t(G)} \leq c' h^{\sigma+1/2-t}\,\|\mathbf{u}\|_{H^s(G)} \tag{8.5.24}$$

for $1 \leq s \leq m+1$, $\frac{1}{2} \leq t \leq m$ and $t \leq \sigma + \frac{1}{2} \leq s$. If in addition the assumptions of Theorem 3.3 in [20, p. 144] are fulfilled then for $\sigma > \frac{1}{2}$ we also have *uniform convergence,*

$$\|\mathbf{U}^*-\mathbf{u}\|_{C^0(\bar{G})} \leqslant c''\{|\log h|^{1/2}\, h^{\sigma-1/2}\, \|\mathbf{u}\|_{H^s(\bar{G})} + h^{[s]}\, \|\mathbf{u}\|_{C^{[s]}(\bar{G})}\}. \tag{8.5.25}$$

Remark The corresponding estimates are also valid for the numerical equations if we replace (8.4.28) with $\|\mathbf{V}^*-\mathbf{u}\|+|\boldsymbol{\alpha}^*|$ and the corresponding norms in (8.5.23–25).

Proof With the definiteness (8.5.22) and the equations (8.5.8) for $\mathbf{U}^*$ and (8.4.2) for $\mathbf{U}$ we find

$$\begin{aligned}\|\mathbf{U}^*-\mathbf{U}\|^2_{H^{1/2}(G)} &\leqslant k''^{-1}[\mathbf{U}-\mathbf{U}^*, \mathbf{u}-\mathbf{U}^*]_* \\ &= k''^{-1}\{[\mathbf{U}, \mathbf{U}-\mathbf{U}^*]_* - [\mathbf{U}, \mathbf{U}-\mathbf{U}^*] - \mathscr{L}_*(\mathbf{U}-\mathbf{U}^*) \\ &\qquad + \mathscr{L}(\mathbf{U}-\mathbf{U}^*)\}.\end{aligned}$$

For the right hand side we use the patch condition (8.5.10) and accuracy (8.5.11) yielding after division by $\|\mathbf{U}^*-\mathbf{U}\|_{H^{1/2}}$:

$$\|\mathbf{U}^*-\mathbf{U}\|_{H^{1/2}(G)} \leqslant c_1 h^{m+1/2}\, \|\mathbf{U}\|_{H^1(G)} + c_2 h^{\sigma}\, \|\mathbf{u}\|_{H^s(G)}. \tag{8.5.26}$$

Since $1 \leqslant s \leqslant m+1$, $0 \leqslant \sigma \leqslant s-\frac{1}{2}$ and $\|\mathbf{U}\|_{H^1} \leqslant c\, \|\mathbf{u}\|_{H^1}$ from (8.4.4) (with $t=s=1$) the first term on the right hand side of (8.5.26) can be estimated by $ch^{\sigma}\, \|u\|_{H^s}$ i.e. (8.5.23) for $t=\frac{1}{2}$. For other t use the stability property (8.3.4).

The estimate (8.5.24) is an immediate consequence of (8.5.23), (8.4.4) and $\sigma \leqslant s-\frac{1}{2}$.

For (8.5.25) we use the stability property (8.3.10) and (8.4.5), (8.5.23):

$$\begin{aligned}\|\mathbf{U}^*-\mathbf{u}\|_{C^0(G)} &\leqslant \|\mathbf{U}^*-\mathbf{U}\|_{C^0(G)} + \|\mathbf{U}-\mathbf{u}\|_{C^0(G)} \\ &\leqslant c\, |\log h|^{1/2}\, \|\mathbf{U}^*-\mathbf{U}\|_{H^1(G)} + \|\mathbf{U}-\mathbf{u}\|_{C^0(G)} \\ &\leqslant c''\{|\log h|^{1/2}\, h^{\sigma-1/2}\, \|\mathbf{u}\|_{H^s(G)} + h^{[s]}\, \|\mathbf{u}\|_{C^{[s]}(\bar{G})}\}.\end{aligned}$$

As we have seen, the general boundary conditions required the penalty term in (8.4.1) with the weight h^{-1}. Therefore the whole method becomes much more efficient if the trial function itself satisfies the required boundary conditions. This can be built into the Ritz–Galerkin equations for simple boundary conditions such as Dirichlet and Neumann data.

8.6 A Ritz and a Galerkin Method for the Standard Problem (1.1.1)

In this section we shall specialize (8.0.1) (8.0.2) to the first boundary value problem for a system in Hilbert's normal form (1.1.1) with normalizing condition (1.1.36) (or point condition) and with homogeneous boundary data $\psi = 0$. Using the function space corresponding to $\psi = 0$, we shall see that the least squares method (8.4.2) and the Galerkin method for the equivalent second-order system (7.0.2) differ only by a compact bilinear form. The bilinear equations belong to a well known class of variational equations which have been investigated extensively; e.g. by Mikhlin [30] who also investigated their approximations [31]. We shall indicate how collective compact approximations yield asymptotic error analysis following Anselone's approach [2]. For our equations much more is available such as numerical discretization using numerical quadrature and finite differences which can now be handled (especially for the variational problem) with the discrete convergence theory of Stummel [50]. This approach allows further to weaken the required zero-Dirichlet condition for the first component χ_1 of the trial functions $\boldsymbol{\chi}$ which was started by Nitsche [36]; for this problem see also the survey by Bramble [9]. Moreover, for nonlinear problems these methods yield formulations combining the advantages of the Galerkin method with those of variational problems. But all this opens so many new questions for more complicated systems with similar properties that we confine ourselves to the optimal rate of convergence and the formulation of some corresponding problems.

If one uses Courant's piecewise linear elements which are continuous then the discrete equations correspond to certain difference equations. (This connection has been studied especially by Strang in [48].) Therefore the variational formulation turns out to be no restriction for pointwise error estimates which have been developed at a rapid rate during recent years. Using corresponding finite element approximations of the Green and Neumann functions, the Galerkin equations can be considered as approximations of corresponding integral equations. Since Scott [44, 45] and Frehse and Rannacher [22] provided just the right error estimates for the differences of the Green and Neumann functions respectively we find optimal uniform convergence of order $h^2 |\log h|$. We finally formulate the corresponding equations for nonlinear problems for which Anselone's [2, Chapter 6] and Stummel's [49] results should be applicable. It seems that the Galerkin type equations provide both solvability and error estimates corresponding to the results by Frehse and Rannacher [57, 58]. Finally, Newton's method with embedding (1.4.8) can be applied to the finite element equations. The real numerical work begins when we stop; i.e. at the computational treatment of the discrete problems. Törnig [52] investigated which discretizations of elliptic problems have to be chosen for providing iterative computational methods. How big the computational expense for relaxation procedures (for related higher order problems) will be was estimated by Astrahancev and Ruhovec [4].

If we formulate (8.4.2) for

$$\mathbf{Du} = \begin{cases} u_x - v_y - (Au + Bv) = C, & \mathbf{u} = \{u, v\}^T, \\ u_y + v_x - (\tilde{A}u + \tilde{B}v) = \tilde{C} & \text{in } \mathsf{G}, \end{cases} \tag{8.6.1}$$

$$\mathbf{ru} = u = 0 \quad \text{on } \dot{\mathsf{G}} \quad \text{and}$$

$$\Lambda u \equiv \oint_{\dot{\mathsf{G}}} v\sigma \, \mathrm{d}s = \kappa\Sigma$$

using only smooth trial functions $\boldsymbol{\chi} \in C^0(\bar{\mathsf{G}}) \cap \mathring{H}^1(\mathsf{G}) \times H^1(\mathsf{G})$ i.e.

$$\chi_{1|\dot{\mathsf{G}}} = 0 \tag{8.6.2}$$

then the trouble maker $h^{-1} \oint_{\dot{\mathsf{G}}} \mathbf{u}^T \mathbf{r}^T \mathbf{r} \boldsymbol{\chi} \, \mathrm{d}s$ vanishes and (8.4.2) becomes (after applying the Gauss–Green theorem):

$$[\mathbf{u}, \boldsymbol{\chi}] = \mathrm{d}[\mathbf{u}, \boldsymbol{\chi}] + k[\mathbf{u}, \boldsymbol{\chi}] = l(\boldsymbol{\chi}) \quad \text{for all } \boldsymbol{\chi} \in \mathring{H}^1 \times H^1. \tag{8.6.3}$$

Here d denotes the Dirichlet bilinear form,

$$\mathrm{d}[\mathbf{u}, \boldsymbol{\chi}] = \iint_{\mathsf{G}} \{u_x \chi_{1x} + u_y \chi_{1y} + v_x \chi_{2x} + v_y \chi_{2y} + u\chi_1 + v\chi_2\} [\mathrm{d}x, \mathrm{d}y]$$

$$= (\mathbf{u}, \boldsymbol{\chi})_{H^1}, \tag{8.6.4}$$

and k and l are given by

$$k[\mathbf{u}, \boldsymbol{\chi}] \equiv -\iint_{\mathsf{G}} (\{Au + Bv, \tilde{A}u + \tilde{B}v\} \mathbf{D}\boldsymbol{\chi}$$

$$+ (u_x - v_y)(A\chi_1 + B\chi_2) + (u_y + v_x)(\tilde{A}\chi_1 + \tilde{B}\chi_2) + u\chi_1 + v\chi_2)$$

$$\times [\mathrm{d}x, \mathrm{d}y] + \oint_{\dot{\mathsf{G}}} v\sigma \, \mathrm{d}s \oint_{\dot{\mathsf{G}}} \chi_2 \sigma \, \mathrm{d}s, \tag{8.6.5}$$

$$l(\boldsymbol{\chi}) \equiv \iint_{\mathsf{G}} \{C, \tilde{C}\} \mathbf{D}\boldsymbol{\chi} [\mathrm{d}x, \mathrm{d}y] + \kappa\Sigma \oint_{\dot{\mathsf{G}}} \chi_2 \sigma \, \mathrm{d}s.$$

Since in the following we use only spaces $H^s(\mathsf{G})$, $\mathring{H}^s(\mathsf{G})$ and $C^0(\mathsf{G})$ we shall omit the notation of G. ($\mathring{H}^s \equiv$ closure of $C_0^\infty(\mathsf{G})$ in H^s).

Note that the bilinear forms and the linear form l are all continuous on $\mathring{H}^1 \times H^1$, hence by continuity (8.6.3) holds for all $\boldsymbol{\chi} \in \mathring{H}^1 \times H^1$, if $\mathbf{u} \in \mathring{H}^1 \times H^1$ solves (8.6.1). If we consider (8.6.3) on this Sobolev space then we cannot use the point condition $v(p_0) = c_0$ but (1.1.36). Since (8.6.3) are the Ritz equations to (8.4.1) we call it the *Ritz method.* But obviously with (8.6.4, 8.6.5) the bilinear equations (8.6.3) correspond to the weak formulation of a corresponding second-order boundary value problem for $\mathbf{u}$ [30].

Another weak formulation appears if we multiply $\mathbf{Du}$ in (8.6.1) by

$\{\chi_{1x}-\chi_{2y},\ \chi_{1y}+\chi_{2x}\}$ scalarly and integrate over G—using again the Gauss–Green theorem and $\chi \in \mathring{H}^1 \times H^1 \cap C^0$. Here we find

$$d[\mathbf{u}, \chi]+c[\mathbf{u}, \chi]=\mathscr{L}(\chi) \quad \text{for all } \chi \in \mathring{H}^1 \times H^1 \tag{8.6.6}$$

with

$$c[\mathbf{u}, \chi] \equiv -\iint\limits_{G} \{(\chi_{1x}-\chi_{2y})(Au+Bv)+(\chi_{1y}+\chi_{2x})(\tilde{A}u+\tilde{B}v)$$
$$+u\chi_1+u\chi_2\}[dx, dy]+\oint\limits_{\dot{G}} v\sigma\, ds \cdot \oint\limits_{\dot{G}} \chi_2\sigma\, ds \tag{8.6.7}$$

and

$$\mathscr{L}(\chi) \equiv \iint\limits_{\dot{G}} \{C(\chi_{1x}-\chi_{2y})+\tilde{C}(\chi_{1y}+\chi_{2x})\}[dx, dy]+\kappa\Sigma \oint\limits_{\dot{G}} \chi_2\sigma\, ds.$$

This bilinear equation (8.6.7) for $\mathbf{u} \in \mathring{H}^1 \times H^1$ is just the weak formulation of the second-order system (7.0.2). The treatment of such 'variational' problems can be found in many presentations, e.g. Mikhlin [30] who also used Galerkin's method for their solution.

It follows from Rellich's theorem [54] that the bilinear forms k and c are *compact* defining via the Riesz representation theorem completely continuous linear operators k and c in $\mathring{H}^1 \times H^1$ by

$$k[\mathbf{u}, \chi]=(k\mathbf{u}, \chi)_1, \quad c[\mathbf{u}, \chi]=(c\mathbf{u}, \chi)_1. \tag{8.6.8}$$

Clearly,

$$d[\chi, \chi]=(\chi, \chi)_1=\|\chi\|_1^2$$

is definite and corresponds to the identity.

In the special case $k=c=0$, (8.6.3) and (8.6.6) coincide giving the weak formulation of the complete decomposite system

$$\begin{aligned} &\Delta u-u=F_1 \equiv C_x+\tilde{C}_y \quad \text{in G}, \quad u|_{\dot{G}}=0 \text{ on } \dot{G},\\ &\Delta v-v=F_2 \equiv -C_y+\tilde{C}_x \quad \text{in G},\\ &\qquad\qquad d_n v|_{\dot{G}}=C\, dx+\tilde{C}\, dy+\kappa\Sigma\sigma\, ds \quad \text{on } \dot{G}. \end{aligned} \tag{8.6.9}$$

Hence, for approximating the bilinear equations we have to decide how to choose $\tilde{H}$ subject to the boundary conditions; the Dirichlet condition and the Neumann condition in (8.6.9). Since the Neumann condition is the natural condition belonging to (8.6.3) or (8.6.6) there is no problem, in contradiction to the Dirichlet condition which presents problems. If we choose $\tilde{H}\subset \mathring{H}^1\times H^1$—as we shall do—only the Courant elements can satisfy $\chi_1|_{\dot{\mathsf{G}}}=0$ without difficulties. For higher order elements, however, and curved $\dot{\mathsf{G}}$ this condition is too strong. There are several methods available which violate the Dirichlet condition for $h>0$ but approximate it for $h\to 0$. A survey on these methods was given by Bramble [9], see also Aubin [5], Babuška and Aziz [6], King [27], Nitsche [36], Stummel [50]. Most of these methods are based on penalty terms as in (8.4.2). Nitsche [36] uses additional properties of $\tilde{H}$.

We require here $\tilde{H}\subset \mathring{H}^1\times H^1$ and, hence, the triangulation in Section 8.3 should be adjusted to $\dot{\mathsf{G}}$ [20, Sections 1.9 and 4.4]. (Moreover, the domains of integration in (8.6.3), (8.6.6) should be replaced by a suitable scale of polygonal domains G_h [50].) The Ritz and Galerkin methods read as:

Find $\mathbf{U}\in\tilde{H}$ *and* $\mathbf{V}\in\tilde{H}$ *respectively such that the Ritz equations*

$$d[\mathbf{U},\boldsymbol{\chi}]+k[\mathbf{U},\boldsymbol{\chi}]=l(\boldsymbol{\chi}) \tag{8.6.10}$$

and the Galerkin equations

$$d[\mathbf{V},\boldsymbol{\chi}]+c[\mathbf{V},\boldsymbol{\chi}]=\mathscr{L}(\boldsymbol{\chi}) \tag{8.6.11}$$

respectively are satisfied for all trial functions $\boldsymbol{\chi}\in\tilde{H}$.

Note that these equations correspond to finite difference approximations of the corresponding second-order systems; i.e. for (8.6.11) the system (7.0.2) [48].

Since (8.6.3) corresponds to

$$\mathbf{u}+k\mathbf{u}=\mathbf{l}, \tag{8.6.12}$$

an equation of the second kind with completely continuous k where $\mathbf{l}\in\mathring{H}^1\times H^1$ is given by

$$(\mathbf{l},\boldsymbol{\chi})_1=\mathscr{L}(\boldsymbol{\chi}) \quad \text{for all} \quad \boldsymbol{\chi}\in\mathring{H}^1\times H^1, \tag{8.6.13}$$

and since (8.6.10) corresponds to

$$\mathbf{U}+P_1kP_1\mathbf{U}=P_1\mathbf{l} \tag{8.6.14}$$

where $P_1:\mathring{H}^1\times H^1\to\tilde{H}$ denotes the H^1 orthogonal projection onto $\tilde{H}$ with

$$P_1\mathbf{v}\to \mathsf{I}\mathbf{v} \quad \text{in } \mathring{H}^1\times H^1 \quad \text{for every} \quad \mathbf{v},\ \|P_1\|_{H^1,H^1}=1 \tag{8.6.15}$$

the family P_1kP_1 is *collectively compact* converging to k with respect to the operator norm [2, p. 8]. Due to the unique solvability of (8.6.1) and the Fredholm alternative [54] the inverse $(\mathsf{I}+k)^{-1}$ exists and is bounded. Hence we deduce with well known arguments [2, p. 10], [31, Section 14]:

Proposition 8.6.1 *The approximating inverses to* (8.6.10) *and* (8.6.11) *exist, they are uniformly bounded and converge to the corresponding* $(\mathsf{I}+k)^{-1}$ *and* $(\mathsf{I}+c)^{-1}$ *respectively,*

$$\|(\mathsf{I}+P_1kP_1)^{-1}\|_{H^1,H^1}\leqslant c \quad \text{for all } h \quad \text{and}$$
$$\|(\mathsf{I}+P_1cP_1)^{-1}\|_{H^1,H^1}\leqslant c' \quad \text{for all} \quad h\leqslant h_0, \tag{8.6.16}$$

with constants independent of h.
With the Gauss projection $\underset{\sim}{G}:\mathbf{u}\mapsto\mathbf{U}$ satisfying

$$\mathbf{U}=\underset{\sim}{G}\mathbf{u}=(\mathsf{I}+P_1kP_1)^{-1}P_1(\mathsf{I}+k)\mathbf{u} \tag{8.6.17}$$

and the corresponding relation for the Galerkin projection $\underset{\sim}{G}'$ to (8.6.11) the Proposition 8.6.1 implies

Proposition 8.6.2 *The Gauss and Galerkin projections* $\underset{\sim}{G}$ *and* $\underset{\sim}{G}'$ *with*

$$\mathbf{U}=\underset{\sim}{G}\mathbf{u} \quad \text{and} \quad \mathbf{V}=\underset{\sim}{G}'\mathbf{u} \tag{8.6.18}$$

are uniformly bounded:

$$\|\underset{\sim}{G}\|_{H^1,H^1}\leqslant c, \qquad \|\underset{\sim}{G}'\|_{H^1,H^1}\leqslant c \quad \text{and} \tag{8.6.19}$$

$$\lim_{h\to 0}\mathbf{U}=\lim_{h\to 0}\underset{\sim}{G}\mathbf{u}=\mathbf{u} \quad \text{and} \quad \lim_{h\to 0}\mathbf{V}=\lim_{h\to 0}\underset{\sim}{G}'\mathbf{u}=\mathbf{u} \tag{8.6.20}$$

in $\mathring{H}^1\times H^1$.
As in the proof of Theorem 8.4.1, this implies an optimal rate of convergence

$$\|\mathbf{u}-\mathbf{W}\|_{H^t}\leqslant ch^{s-t}\|\mathbf{u}\|_{H^s}$$

for $1\leqslant t\leqslant s\leqslant m+1$, $t\leqslant m$ for both methods, $\mathbf{U}=\mathbf{W}$ or $\mathbf{V}=\mathbf{W}$. For second-order systems (7.0.2) and that corresponding to (8.6.3) the *a-priori* estimates by Schechter [43] provide the application of Nitsche's trick [35] extending the above estimate to $t\geqslant 0$:

Theorem 8.6.3 *The methods* (8.6.10) *and* (8.6.11) *both converge with optimal order, i.e. for* $0\leqslant t\leqslant s\leqslant m+1$, $t\leqslant m$, $1\leqslant s$.

$$\|\mathbf{u}-\mathbf{W}\|_{H^t}\leqslant ch^{s-t}\|u\|_{H^s} \tag{8.6.21}$$

where $\mathbf{W}=\mathbf{U}$ *for* (8.6.10) *and* $\mathbf{W}=\mathbf{V}$ *for* (8.6.11).
We shall omit the proof. From (8.6.21) with Lemma 8.3.3 we also have:

Corollary 8.6.4 *If in addition the finite elements satisfy the assumptions of* [20, Theorem 3.3] *then we have for both methods and* $1\leqslant s\leqslant m+1$

$$\|\mathbf{u}-\mathbf{W}\|_{C^0}\leqslant c'\{h^{s-1}|\log h|\,\|\mathbf{u}\|_{H^s}+h^{[s]}\|\mathbf{u}\|_{C^{[s]}}\}. \tag{8.6.22}$$

Since all the bilinear equations and their approximations (8.6.10), (8.6.11) operate in $\mathring{H}^1 \times H^1$ continuously and since the inverses are uniformly bounded, we find exactly the same situation for the numerical replacement as for the finite element methods for (8.6.9). Hence, the treatment by Fix and Strang [20, p. 181 ff.] is also valid word for word for (8.6.10) and (8.6.11). Here also the assumptions by Stummel [50, especially Section 3.2] are satisfied and the 'numerical' solutions converge to the same order as **U** and **V** respectively, if the quadrature formulas satisfy the patch condition. The numerical equations to (8.6.10) provide the big advantage that they have a *positive definite symmetric matrix* of coefficients which is also sparse. Hence, here the overrelaxation methods are recommended, their computational expense corresponds to that in [4].

For the Galerkin method (8.6.11) these advantages can only be expected asymptotically for $h \to 0$. Although the methods (8.6.10) and (8.6.11) seem to be entirely linked to the Hilbert space structure of $\mathring{H}^1 \times H^1$, the correspondence to difference approximations shows that they also have strong connections with pointwise estimates as e.g. the maximum principle for the discrete equations to (8.6.9). Recently the pointwise estimates of $u - U$ for finite element methods have been improved up to optimal orders of convergence $h^2 |\log h| \, \|F\|_{C^0}$ by Nitsche [39] for the Poisson equation with homogeneous Dirichlet condition (and in [37] for problems with restrictions $u \leqslant z$) and in [38] for more general differential equations, and by Natterer [32]. For Neumann problems (8.6.9) with homogeneous boundary data the corresponding estimates have been proved by Scott [45] and Höhn [24]. Rannacher [40] and Scott [45] use a method which is of key importance for our systems: they approximate the Green and Neumann functions by finite elements and use W_1^1 estimates for these approximations which have been proved for the Green functions by Frehse and Rannacher [22] and for the Neumann functions by Scott [44]. This is just the approach which we have used for finite difference approximations in Section 7.8. Let $\Gamma_h(\zeta; z)$ with $\zeta = \xi + i\eta$, $z = x + iy$ be a 2×2 matrix of finite element functions of z for every fixed $\zeta \in \bar{G}$,

$$\Gamma_{hk}(\zeta; \cdot) \in \tilde{H} \subset \mathring{H}^1 \times H^1, \quad k = 1, 2,$$

which is defined by the solution of

$$d[\boldsymbol{\chi}, \Gamma_h(\zeta; \cdot)] = -\boldsymbol{\chi}(\zeta) \quad \text{for all} \quad \boldsymbol{\chi} \in \tilde{H} \tag{8.6.23}$$

and for every fixed $\zeta \in \bar{G}$. The component Γ_{h11} corresponds to the discrete Green function and Γ_{h22} to the discrete Neumann function; $\Gamma_{h12} = \Gamma_{h21} = 0$.

Hence, let us define the matrix

$$\Gamma(\zeta; z) \equiv \begin{pmatrix} G(\zeta, z), & 0 \\ 0 & N(\zeta, z) \end{pmatrix}$$

corresponding to (8.6.9) by

$$d[\boldsymbol{\chi}, \Gamma(\zeta; \cdot)] = -\boldsymbol{\chi}(\zeta) \quad \text{for all} \quad \boldsymbol{\chi} \in \mathring{H}^1 \times H^1. \tag{8.6.24}$$

G, the Green function to (8.6.9) is different from G^I (1.1.7) and N, the Neumann function to (8.6.9) is different from G^{II} (1.1.11) but both have the same logarithmic singularity at $z=\zeta$ and both satisfy (1.1.15). (They both can be expressed in terms of G^I, G^{II} and suitable smooth solutions of Fredholm integral equations of the second kind. We leave these details to the reader.)
For Γ_h and Γ we shall use the following:

Theorem 8.6.5 [22][45, Theorem 2]

$$\sup_{\zeta \in G} \iint_G |\Gamma_h(\zeta; z) - \Gamma(\zeta, z)| \, [dx, dy] \leq c_1 h^2 |\log h| \tag{8.6.25}$$

$$\sup_{\zeta \in G} \iint_G \{|\Gamma_{hx}(\zeta; z) - \Gamma_x(\zeta; z)| + |\Gamma_{hy}(\zeta; z) - \Gamma_y(\zeta; z)|\}[dx, dy] \leq c_2 h |\log h| \tag{8.6.26}$$

$$\sup_{\zeta \in G} \oint_{\dot{G}} |N_h(\zeta; z) - N(\zeta; z)| \, ds_z \leq c_3 h |\log h| \tag{8.6.27}$$

where c_1, c_2, c_3 are independent of h.

Remark (8.6.27) is not explicitly proved in [45] but it can be obtained with the same methods and additional estimates at the boundary.

Due to (8.6.23) and (8.6.24) we see that $\mathbf{u}$ and $\mathbf{U}$ satisfy the Fredholm integral equations of the second kind,

$$\mathbf{u}(\zeta) = -k[\mathbf{u}, \Gamma(\zeta; \cdot)] + \mathbf{l}(\Gamma(\zeta, \cdot)), \tag{8.6.28}$$

$$\mathbf{U}(\zeta) = -k[\mathbf{U}, \Gamma_h(\zeta; \cdot)] + \mathbf{l}(\Gamma_h(\zeta; \cdot)). \tag{8.6.29}$$

Using Theorem 8.6.5, the application of perturbation arguments to these integral equations in $C^0(G)$ yields pointwise error estimates. Note that

$$k[\mathbf{u}, \Gamma(\zeta; \cdot)] = k\mathbf{u}, \qquad k[\mathbf{u}, \Gamma_h(\zeta; \cdot)] = P_1 k P_1 \mathbf{u} \quad \text{for all} \quad \mathbf{u} \in \mathring{H}^1 \times H^1.$$

Hence, (8.6.28), (8.6.29) are extensions from (8.6.12) and (8.6.14) to $C^0(G)$.

Lemma 8.6.6 *The integral operators in* (8.6.28)(8.6.29) *satisfy*

$$\|k[\mathbf{u}, \Gamma(\zeta;\cdot)] - k[\mathbf{u}, \Gamma_h(\zeta;\cdot)]\|_{C^0} \leq ch\,|\log h|\,\|\mathbf{u}\|_{C^0} \tag{8.6.30}$$

for all $\mathbf{u} \in C^0(\bar{G})$ *with c independent of h and* $\mathbf{u}$.

Proof From (8.6.5) we distinguish three different expressions
(i) integrals containing u, v;
(ii) integrals containing $u_x, \ldots, v_y$;
(iii) boundary integrals.
For the first and the last terms (8.6.25) and (8.6.27) yield immediately the desired estimate. For the second kind of terms we use the Gauss–Green theorem; e.g.

$$\iint_G u_x A(G-G_h)[dx, dy] = -\iint_G u\{A_x(G-G_h) + A(G_x - G_{hx})\}[dx, dy]$$

$$\iint_G u_x B(N-N_h)[dx, dy] = -\iint_G u\{B_x(N-N_h) + B(N_x - N_{hx})\}[dx, dy]$$

$$+ \oint_{\dot{G}} uB(N-N_h)\dot{y}\, ds$$

and then apply (8.6.25–27).
Hence, since $(I+k)^{-1}$ exists defining a bounded linear operator in C^0 (this can be proved in exactly the same way as Theorem 1.1.2) and since (8.6.30) shows the norm convergence of the approximating operators in (8.6.29) we have from [2, p. 10]:

Proposition 8.6.7 *The approximating inverses to* (8.6.29) *exist, they are uniformly bounded, and converge to the corresponding* $(I+k)^{-1}$,

$$\|(I + k[\cdot, \Gamma_h(\zeta;\cdot)])^{-1}\|_{C^0, C^0} \leq c \quad \text{for all} \quad h \to 0$$

$$\text{where} \quad c \text{ is independent of } h. \tag{8.6.31}$$

Since all the above conclusions are valid word for word for the Galerkin equations (8.6.11) with the corresponding bilinear form c, we finally find:

Theorem 8.6.8 *Both the solutions of* (8.6.10) *and* (8.6.11) *converge uniformly and satisfy error estimates*

$$\|\mathbf{u}-\mathbf{W}\|_{C^0}\leqslant c_1 h\,|\log h|\,(\|C\|_{C^0}+\|\tilde{C}\|_{C^0}+\|\mathbf{u}\|_{C^0}) \tag{8.6.32}$$

$$\|\mathbf{u}-\mathbf{W}\|_{C^0}\leqslant c_2 h^2\,|\log h|(\|C\|_{C^{1+\alpha}}+\|\tilde{C}\|_{C^{1+\alpha}}+\|\mathbf{u}\|_{C^{1+\alpha}}) \tag{8.6.33}$$

where $\alpha>0$ *can be chosen arbitrarily small.*
($\mathbf{W}=\mathbf{U}$ for (8.6.10) and $\mathbf{W}=\mathbf{V}$ for (8.6.11)).

Remark The theorem remains valid if we replace the side condition $\Lambda\mathbf{u}=\kappa\Sigma$ in (8.6.1) by the point condition

$$\Lambda\mathbf{u}\equiv v(z_0)=c_0,\quad z_0\in \mathsf{G}\quad \text{fixed,} \tag{8.6.34}$$

since this new mapping (8.6.34) is continuous on $C^0(\bar{\mathsf{G}})$. In this case for the finite element methods, cancel the boundary integrals in (8.6.5) and in (8.6.7) respectively, and add the point condition (8.6.34) to (8.6.10) and to (8.6.11) respectively.

Proof From (8.6.31) it follows in the usual manner [2, p. 12] that

$$\|\mathbf{u}-\mathbf{U}\|_{C^0}\leqslant c\sup_{\zeta}|k[\mathbf{u},\Gamma(\zeta;\cdot)-\Gamma_h(\zeta;\cdot)]+l(\Gamma(\zeta;\cdot)-\Gamma_h(\zeta;\cdot))|. \tag{8.6.35}$$

Inserting (8.6.25)–(8.6.27) into the definitions of k and l, (8.6.5) we find from (8.6.35) the desired estimate (8.6.32).

For the second estimate we integrate by parts; i.e. apply the Gauss–Green theorem to all terms in (8.6.5) containing derivatives of $\Gamma-\Gamma_h$. For the new surface integrals apply (8.6.25) which already has the desired form (8.6.33). For the remaining boundary integrals; e.g. for

$$w(\zeta)\equiv\oint_{\dot{\mathsf{G}}} Bv\dot{x}N(\zeta;z)\,\mathrm{d}s_z,\qquad w_h(\zeta)\equiv\oint_{\dot{\mathsf{G}}} Bv\dot{x}N_h(\zeta;z)\,\mathrm{d}s_z \tag{8.6.36}$$

we compute that $w-w_h$ solves

$$(w-w_h,\chi_2)_1=0\quad\text{for all finite elements}\quad \chi_2,\{0,\chi_2\}\in\tilde{H}. \tag{8.6.37}$$

These are the Ritz equations to $\|w-w_h\|^2_{H^1}\to\min$. Hence, we can use the error estimates by Höhn [24] for which we need $w\in C^2(\bar{\mathsf{G}})$. In general, this we can only ensure if the boundary data are in $C^{1+\alpha}$. Höhn's estimates yield

$$\|w-w_h\|_{C^0}\leqslant c'\,\|w\|_{C^2}h^2\,|\log h|\leqslant c''\,\|Bv\dot{x}\|_{C^{1+\alpha}(\dot{\mathsf{G}})}h^2\,|\log h|$$

and corresponding estimates for the remaining boundary integrals. This implies (8.6.33).

We now finally consider the semilinear problem (1.4.1),

$$\begin{aligned} &u_x - v_y = H_1(z, \mathbf{u}),\\ &u_y + v_x = H_2(z, \mathbf{u}) \quad \text{in} \quad \mathsf{G},\\ &u = 0 \quad \text{on} \quad \dot{\mathsf{G}} \quad \text{and} \quad \oint_{\dot{\mathsf{G}}} v\sigma \, ds = \kappa. \end{aligned} \tag{8.6.38}$$

(This last condition can be replaced by the point condition $v(z_0) = c_0$ with $z_0 \in \mathsf{G}$ fixed.)

Let us assume that (8.6.38) admits a solution, e.g. by conditions (1.4.3), (1.4.4).

For nonlinear Dirichlet problems, most of the convergence results with an order $h^2 |\log h|$ have already been proved by Frehse [21] and Rannacher [41], for nonlinear Neumann problems by Höhn [24].

Let us relate the following variational problem:
Minimize the functional

$$\begin{aligned} I(\mathbf{u}) \equiv &\iint_{\mathsf{G}} (u_x - v_y - H_1(z, \mathbf{u}))^2 + (u_y + v_x - H_2(z, \mathbf{u}))^2 [dx, dy]\\ &+ \left(\oint_{\dot{\mathsf{G}}} v\sigma \, ds - \kappa \right)^2\\ = &\iint_{\mathsf{G}} \{u_x^2 + u_y^2 + v_x^2 + v_y^2 - 2H_1(u_x - v_y) - 2H_2(u_y + v_x) + H_1^2 + H_2^2\}\\ &\times [dx, dy] + \left(\oint_{\dot{\mathsf{G}}} v\sigma \, ds - \kappa \right)^2 \end{aligned} \tag{8.6.39}$$

over $\mathbf{u} \in \mathring{H}^1 \times H^1 \cap C^0(\bar{\mathsf{G}})$.
Under the assumptions of Theorem 1.4.1, this variational problem has a unique solution.

In the case that (8.6.39) can be decomposed into two separate problems for u and v such that for u belongs to the class in [21] and [23] and that for v to the class in [24], there $h^2 |\log h|$ convergence has been proved. Hence the same can be expected for (8.6.39). But the proof is yet to be done.

If the gradient of $I(\mathbf{u})$ defines a strictly convex functional [20, p. 100]

then the Ritz method converges. The Ritz equations to (8.6.39) take the form [31, Section 70]:
Find $\mathbf{U} \in \tilde{H}$ *satisfying*

$$d[\boldsymbol{\chi}, \mathbf{U}] + p[\boldsymbol{\chi}, \mathbf{U}] = 0 \quad \textit{for all} \quad \boldsymbol{\chi} \in \tilde{H} \tag{8.6.40}$$

where the form $p[\ ,\ ]$ is given by

$$\begin{aligned} p[\boldsymbol{\chi}, \mathbf{U}] \equiv \iint_{G} \{&(\chi_{1x} - \chi_{2y}) H_1(z, \mathbf{U}) + (\chi_{1y} + \chi_{2x}) H_2(z, \mathbf{U}) \\ &+ (\chi_1 H_{1u}(z, \mathbf{U}) + \chi_2 H_{1v}(z, \mathbf{U}))(U_{1x} - U_{2y} - H_1(z, \mathbf{U})) \\ &+ (\chi_1 H_{2u}(z, \mathbf{U}) + \chi_2 H_{2v}(z, \mathbf{U}))(U_{1y} + U_{2x} - H_2(z, \mathbf{U}))\}[\mathrm{d}x, \mathrm{d}y] \\ &+ \oint_{\dot{G}} \chi_2 \sigma \,\mathrm{d}s \left(\oint_{\dot{G}} U_2 \sigma \,\mathrm{d}s - \kappa \right). \end{aligned} \tag{8.6.41}$$

Note that $p[\cdot, \cdot]$ is linear with respect to the first argument but nonlinear with respect to the second.

(8.6.39) provides essentially just the assumptions under which Törnig [52] constructs suitable numerical equations.

Another approach uses Galerkin's method for (8.6.38) directly:
Find $\mathbf{V} \in \tilde{H}$ *such that the equations*

$$d[\boldsymbol{\chi}, \mathbf{V}] = q[\boldsymbol{\chi}, \mathbf{V}] \tag{8.6.42}$$

are satisfied for all $\boldsymbol{\chi} \in \tilde{H}$ *where the nonlinear form* q *is given by*

$$\begin{aligned} q[\boldsymbol{\chi}, \mathbf{V}] \equiv \iint_{G} &\{V_1 \chi_1 + V_2 \chi_2 + (\chi_{1x} - \chi_{2y}) H_1(z, \mathbf{V}) + (\chi_{1y} + \chi_{2x}) H_2(z, \mathbf{V})\} \\ &\times [\mathrm{d}x, \mathrm{d}y] \\ &+ \oint_{\dot{G}} \chi_2 \sigma \,\mathrm{d}s \left(\kappa \Sigma - \oint_{\dot{G}} V_2 \sigma \,\mathrm{d}s \right). \end{aligned} \tag{8.6.43}$$

If we use instead of $\boldsymbol{\chi}$ the columns of the matrix $\Gamma_h(\zeta, z)$ then we find that (8.6.43) is equivalent to the nonlinear Urysohn integral equation

$$\mathbf{V}(\zeta) = q[\Gamma_h(\zeta, \cdot), \mathbf{V}]. \tag{8.6.44}$$

This equation is an approximation to

$$\mathbf{u}(\zeta) = q[\Gamma(\zeta, \cdot), \mathbf{u}], \tag{8.6.45}$$

a nonlinear Urysohn equation which is equivalent to (8.6.38). Note that

(8.6.45) has a weakly singular kernel. Under the assumptions (1.4.3), (1.4.4) this equation *has a unique solution.* For (8.6.44) the linear dependence of q on the first argument and Theorem 8.6.5 provide:

Lemma 8.6.9 *If H satisfies* (1.4.3) *and* (1.4.4) *then we have the estimate*

$$\|q[\Gamma_h(\zeta,\cdot)-\Gamma(\zeta,\cdot),\mathbf{v}]\|_{C^0}\leq c\,\|\mathbf{v}\|_{C^0}h\,|\log h|. \tag{8.6.46}$$

With this result we can prove:

Theorem 8.6.10 *If $h\leq h_0$ with $h_0>0$ suitably small then the Galerkin equations* (8.6.42) *can be solved uniquely. The finite element solutions converge with*

$$\|\mathbf{u}-\mathbf{V}\|_{C^0}\leq ch^2\,|\log h|\,\|\mathbf{u}\|_{C^{1+\alpha}} \tag{8.6.47}$$

where $\alpha>0$ is arbitrarily small and c is independent of h and $\mathbf{u}$ *but might depend on α.*

The proofs of Lemma 8.6.9 and Theorem 8.6.10 will be published elsewhere. (See [74].)

Since the linearized problems to (8.6.38) are of the form (1.1.1) they always admit a unique solution. Hence, under weaker assumptions, the theory by Anselone [2] and Stummel [49] should be applicable to (8.6.44) and (8.6.45).

Although the method (8.6.42) converges, the finite element equations do not provide the nice convexity properties of (8.6.39). Hence, we have to worry about a method for solving the nonlinear finite dimensional equation (8.6.42). Here the Newton method in connection with embedding can be used as in (1.4.8). This yields the following iterative procedure which involves solving sequences of linear problems (8.6.10) or (8.6.11):

$$\begin{aligned}&d[\mathbf{V}_{n+1},\boldsymbol{\chi}]+t_jk_n[\mathbf{V}_{n+1},\boldsymbol{\chi}]=t_jl_n(\boldsymbol{\chi}),\ n=0,1,\ldots,\\&\mathbf{V}_0(t_j;z)\equiv\mathbf{V}_\infty(t_{j-1};z)\end{aligned} \tag{8.6.48}$$

or

$$\begin{aligned}&d[\mathbf{W}_{n+1},\boldsymbol{\chi}]+t_jc_n[\mathbf{W}_{n+1},\boldsymbol{\chi}]=t_j\mathscr{L}_n(\boldsymbol{\chi}),\ n=0,1,\ldots,\\&\mathbf{W}_0(t_j;z)\equiv\mathbf{W}_\infty(t_{j-1};z)\end{aligned} \tag{8.6.49}$$

where t_j is chosen as in (1.4.8). The forms k_n, l_n, and c_n, $\mathscr{L}_n$ respectively are recursively defined by (8.6.5) by choosing there the coefficients for D by

$$\begin{aligned}
&A \equiv H_{1u}(z, \mathbf{V}_n), \qquad B \equiv H_{1v}(z, \mathbf{V}_n),\\
&\tilde{A} \equiv H_{2u}(z, \mathbf{V}_n), \qquad \tilde{B} \equiv H_{2v}(z, \mathbf{V}_n),\\
&C \equiv -\mathbf{V}_n^T\{H_{1u}, H_{1v}(z, \mathbf{V}_n)\}^T + H_1(z, \mathbf{V}_n),\\
&\tilde{C} \equiv -\mathbf{V}_n^T\{H_{2u}, H_{2v}(z, \mathbf{V}_n)\}^T + H_2(z, \mathbf{V}_n)
\end{aligned} \tag{8.6.50}$$

and in (8.6.7) correspondingly.

Thanks to the fact that $\Gamma_h \to \Gamma$ in Theorem 8.6.5 here we have:

Theorem 8.6.11 *There exist* $h_0 > 0$ *and* $\delta > 0$ *such that the embedding Newton methods* (8.6.48) *and* (8.6.49) *with* $t_0 = 0$, $t_{j+1} \equiv t_j + \delta$, $j = 0, 1, \ldots$ *converge.*

The proof will be published elsewhere. (See [74].)

Note that for (8.6.48) the finite element equations are always positive definite and symmetric.

8.7 Remarks on the Additional References

Since an enormous amount of work has been done on finite element methods, the following remarks can by no means be complete or general. These references are of interest in connection with the preceding presentations and they might be of some use for further error analysis for the indicated methods besides the already given references.

Maximum norm estimates for second and higher order problems have been obtained in [56], [70], [71]. The use of spaces without boundary conditions was made in [60], [68]. The effect of corner singularities on the Galerkin approximation was investigated in [69], [72].

For nonlinear problems and their approximation the hard part is to prove global *a priori* estimates for the approximations [62]. Further estimates for nonlinear second-order problems and isoparametric elements are in [65–67] and with a remarkable use of the Schauder continuity method in [57], [58]; for higher order problems see [59], [60]. The hard numerical work begins with solving the discrete problems for which monotonicity properties and convergent iterations are of invaluable importance [61], [63], [64], [73].

References

1 Agmon, S., Douglis, A. and Nirenberg, L., Estimates near the boundary for solutions of elliptic partial differential equations satisfying general boundary conditions. I *Comm. Pure Appl. Math.* **12,** 623–727, 1959; II *Comm. Pure Appl. Math.* **17,** 35–92, 1964.

2 Anselone, P. M., *Collectively compact operator approximation theory*. Prentice Hall, London, 1971.

3 Arndt, D., Näherungsweise Lösung partieller Differentialgleichungen durch simultane Defektminimierung. *Diplomarbeit* T. H. Darmstadt, Germany, 1973.

4 Astrahancev, G. P. and Ruhovec, L. A., Konvergenzgeschwindigkeit der Methode der oberen Gruppenrelaxation der Lösung von Variations—Differenzenschemata für eine elliptische Gleichung der Ordnung 2m in einem beliebigen zweidimensionalen Gebiet. *Z. vyc. Mat. nat. Fiz.* **13,** 1425–1440, 1973.

5 Aubin, J. P., *Approximation of elliptic boundary-value problems*. Wiley-Interscience, New York, 1972.

6 Babuška, I. and Aziz, A. K., Survey lectures on the mathematical foundations of the finite element method. *In* K. Aziz (Ed.), *The mathematical foundations of the finite element method with applications to partial differential equations*, pp. 3–359, Academic Press, New York, 1972.

7 Baker, G., Simplified proofs of error estimates of the least squares method for Dirichlet's problem. *Math. Comp.* **27,** 229–235, 1973.

8 Bittner, L., Das Austauschverfahren der linearen Tschebyscheff-Approximation bei nicht erfüllter Haarscher Bedingung. *ZAMM* **41,** 238–256, 1961.

9 Bramble, J. H., A survey of some finite element methods proposed for treating the Dirichlet problem. *Adv. Math.* **16,** 187–196, 1975.

10 Bramble, J. and Schatz, A., Rayleigh-Ritz-Galerkin-methods for Dirichlet's problem using subspaces without boundary conditions. *Comm. Pure Appl. Math.* **23,** 653–675, 1970.

11 Bramble, J. and Schatz, A., On the numerical solution of elliptic boundary value problems by least squares approximation of the data. *Num. Sol. of Partial Differential Equations*, Synspade, 1970, pp. 107–131, Academic Press, New York, 1971.

12 Bramble, J. and Schatz, A., Least squares methods for 2mth order elliptic boundary-value problems. *Math. Comp.* **25,** 1–32, 1971.

13 Bramble, J. and Zlámal, M., Triangular elements in the finite element method. *Math. Comp.* **24,** 809–819, 1970.

14 Bredendiek, E., Simultanapproximationen. *Arch. Rat. Mech. Anal.* **33,** 307–330, 1969.

15 Bruhn, G., Approximation mit Maximumnorm. Preprint *Fachbereich Mathematik*. T. H., Darmstadt, Germany, 1970.

16 Ciarlet, P. G. and Raviart, P. A., The combined effect of curved boundaries and numerical integration in isoparametric finite element methods. *In* K. Aziz (Ed.), *The mathematical foundations of the finite element method with applications to partial differential equations*, pp. 409–474, Academic Press, New York, 1972.

17 Courant, R., Variational methods for the solution of problems of equilibrium and vibrations. *Bull. Am. Math. Soc.* **49,** 1–23, 1943.

18 Dikanskij, A. S., Conjugate problems of elliptic differential and pseudo-differential boundary value problems in a bounded domain. *Math. USSR Sbornik* **20,** 67–83, 1973.

19 Faddejev, L. and Faddejeva, V., *Computational methods of linear algebra.* Dover, New York, 1959.

20 Fix, G. J. and Strang, G., *An analysis of the finite element method.* Prentice Hall, Englewood Cliffs, N. J., 1973.

21 Frehse, J., Eine gleichmäßig asymptotische Fehlerabschätzung zur Methode der finiten Elemente bei quasilinearen elliptischen Randwertproblemen. *In Theory of nonlinear operators. Constructive Aspects.* Tagungsband der Akademie der Wissenschaften DDR, Berlin, 1976.

22 Frehse, J. and Rannacher, R., Eine L^1-Fehlerabschätzung für diskrete Grundlösungen in der Methode der finiten Elemente. *In Finite Elemente.* Tagungsband, Bonn. Math. Schriften, 1976.

23 Frehse, J. and Rannacher, R., Punktweise Konvergenz der Methode der finiten Elemente bei n-dimensionalen quasilinearen Randwertaufgaben, to appear.

24 Höhn, W., Numerische Behandlung nichtlinearer Neumann-Probleme mit finiten Elementen. *Dissertation.* T. H., Darmstadt, 1976.

25 Hörmander, L., *Linear partial differential operators.* Springer, Berlin, New York, Heidelberg, 1964.

26 Il'in, V. P., Some inequalities in function spaces and their application to the study of the convergence of variational processes. *Trudy Math. Inst. Steklov, Am. Math. Soc. Transl.* **81,** 1–66, 1969.

27 King, J. T., New error bounds for the penalty method and extrapolation. *Num. Math.* to appear.

28 Lions, J. L. and Magenes, E., *Non-homogeneous boundary-value problems and applications,* Vol. I. Springer, Berlin, 1972.

29 Mairhuber, J. C., On Haar's theorem concerning Chebysheff problems having unique solutions. *Proc. Am. Math. Soc.* **7,** 609–615, 1956.

30 Mikhlin, S. G., *Variational methods in mathematical physics.* Pergamon Press, New York, 1964.

31 Mikhlin, S. G., *The numerical performance of variational methods.* Wolters-Noordhoff, Groningen, The Netherlands, 1971.

32 Natterer, F., Über die punktweise Konvergenz finiter Elemente, to appear.

33 Nitsche, J. A., Umkehrsätze für Spline-Approximationen. *Comp. Math.* **21,** 400–416, 1969.

34 Nitsche, J. A., Zur Konvergenz von Näherungsverfahren bezüglich verschiedener Normen. *Num. Math.* **15,** 224–228, 1970.

35 Nitsche, J. A., Lineare Spline-Funktionen und die Methode von Ritz für elliptische Randwertprobleme. *Arch. Rat. Mech. Anal.* **36,** 348–355, 1970.

36 Nitsche, J. A., On Dirichlet problems using subspaces with nearly zero boundary conditions. *In* K. Aziz (Ed.), *The mathematical foundation of the finite element method with applications to partial differential equations,* pp. 603–627, Academic Press, New York, 1972.

37 Nitsche, J. A., L_∞-convergence of finite element approximations. *In Mathematical aspects of the finite element methods.* Rome, Italy, 1975, to appear.

38 Nitsche, J. A., On L_∞-convergence of finite element approximations to the solution of a nonlinear boundary value problem. *In Proc. Conf. Numerical Analysis,* Dublin, 1976. Springer Lecture Notes, to appear.

39 Nitsche, J. A., L_∞-convergence of the Ritz-method with linear finite elements for second order elliptic boundary value problems. *Acad. Sci. USSR, ded. 70th birthday I. N. Vekua*, N. Bogolubov and M. Lavrentiev (Eds.).
40 Rannacher, R., Zur L^∞-Konvergenz linearer finiter Elemente. Preprint, University of Bonn, 1975.
41 Rannacher, R., Some asymptotic error estimates for finite element approximation of minimal surfaces. *Revue Franc. Automatique Inf. Rech. Operationnelle*, to appear.
42 Raviart, P. A., Methode des êlêments finis. *Lecture Notes*, Université de Paris VI, Laboratoire Associê 189, 1973.
43 Schechter, M., On L^p estimates and regularity II. *Math. Scand.* **13,** 47–69, 1963.
44 Scott, R., Finite element convergence for singular data. *Num. Math.* **21,** 317–327, 1973.
45 Scott, R., Optimal L^∞-estimates for the finite element method on irregular meshes. *Math. Comp.* **30,** 681–697, 1976.
46 Sobolev, S. L., *Applications of functional analysis in mathematical physics.* Transl. Amer. Math. Soc., Providence, 1963.
47 Stephan, E. and Wendland, W., Remarks to Galerkin and least squares methods with finite elements for general elliptic problems. *In Proc. Conf. on Ord. and Partial Differential Equations*, Dundee, 1976, Springer Lecture Notes, No. 564, 1977; and in *Manuscripta Geodaetica* **1,** 93–123, 1976.
48 Strang, G., The finite element method and approximation theory. *In* B. Hubbard (Ed.), Synspade 1970, *Numerical Solution of Partial Differential Equations* II. pp. 547–83, Academic Press, New York, 1971.
49 Stummel, F., *Approximation methods in analysis.* Lecture Notes Ser. No. 35, Mat. Inst. Aarhus Universitet, 1973.
50 Stummel, F., *Näherungsmethoden für Variationsprobleme elliptischer Differentialgleichungen.* Fachbereich Mathematik der Johann Wolfgang Goethe Universität, Robert-Mayer-Straße 10, 6 Frankfurt/Main, Germany, 1976.
51 Töpfer, H. J., Tschebyscheff-Approximation und Austauschverfahren bei nicht erfüllter Haarscher Bedingung. *In* L. Collatz and H. Unger (Eds.), *Funktionalanalysis, Approximationstheorie, Numerische Mathematik, ISNM* **7,** Birkhauser, Basel, Stuttgart, 1967.
52 Törnig, W., Monotonieeigenschaften von Diskretisierungen des Dirichletproblems quasilinearer elliptischer Differentialgleichungen. *In Springer Lecture Notes* No. **333,** 274–291, 1973.
53 Werner, H., Vorlesung über Approximationstheorie. *Springer Lecture Notes* No. **14,** Berlin, Heidelberg, New York, 1966.
54 Wloka, J., *Funktionalanalysis und Anwendungen.* de Gruyter, 1971.
55 Zienkiewicz, O. C., *The finite element method in engineering science* McGraw-Hill, London, 1971.

Additional References

56 Bramble, J. H., Nitsche, J. A. and Schatz, A. H., Maximum-norm interior estimates for Ritz–Galerkin methods. *Math. Comp.* **29,** 677–688, 1975.

57 Frehse, J. and Rannacher, R., Asymptotic L^∞-error estimates for linear finite element approximations of quasilinear boundary value problems. *SIAM J. Numer. Anal.* to appear.

58 Frehse, J. and Rannacher, R., Optimal uniform convergence for the finite element approximation of a quasilinear elliptic boundary value problem. *In Proc. U.S.-Germ. Symp. Formulations and Comp. Algorithms in Finite Element Analysis.* M.I.T., Cambridge, Mass., 1976, to appear.

59 Gentzsch, W., *Zur Diskretisierung quasilinearer elliptischer Differentialgleichungen vierter Ordnung.* Fachbereich Math., T. H., Darmstadt. Preprint No. 193, 1975.

60 Gentzsch, W., *Penalty-Methode bei der numerischen Behandlung von quasilinearen elliptischen Differentialgleichungen vierter und höherer Ordnung.* Fachbereich Math., T. H. Darmstadt, Preprint No. 243, 1976.

61 Höhn, W. and Törnig, W., *Ein Maximum-Minimum-Prinzip für Lösungen von finite-Element-Approximationen für quasilineare elliptische Randwertprobleme.* Fachbereich Math., T. H., Darmstadt, Preprint No. 280, 1976.

62 Johnson, C. and Thomée, V., Error estimates for a finite element approximation of a minimal surface. *Math. Comp.* **29,** 343–349, 1975.

63 Meis, T. and Törnig, W., Diskretisierung des Dirichletproblems nichtlinearer elliptischer Differentialgleichungen. *In* B. Brosowski and E. Martensen (Eds.), *Methoden und Verfahren der Mathematischen Physik.* BI, Mannheim, 8, 1973.

64 Merten, K., Zur Theorie der Diskretisierung von Variationsproblemen. *Dissertation.* T. H., Darmstadt, Germany, 1975.

65 Mittelmann, H. D., *On pointwise estimates for a finite element solution of nonlinear boundary value problems.* Fachbereich Math., T. H., Darmstadt, Preprint No. 207, 1975.

66 Mittelmann, H. D., Numerische Behandlung des Minimalflächenproblems mit finiten Elementen. *ISNM* **28,** 91–108, Birkhauser, Basel, Stuttgart, 1975.

67 Mittelmann, D., Die Methode der finiten Elemente zur numerischen Lösung von Randwertproblemen quasilinearer elliptischer Differentialgleichungen. *Habilitationsschrift,* T. H., Darmstadt, 1976.

68 Nitsche, J. A., Über ein Variationsprinzip zur Lösung von Dirichletproblemen bei Verwendung von Teilräumen, die keinen Randbedingungen unterworfen sind. *Abh. d. Hamb. Math. Sem.*, **36,** 9–15, 1971.

69 Nitsche, J. A., Zur lokalen Konvergenz von Projektionen auf finite Elemente, to appear.

70 Nitsche, J. A. and Schatz, A. H., Interior estimates for Ritz-Galerkin methods. *Math. Comp.* **28,** 937–958, 1974.

71 Rannacher, R., Punktweise Konvergenz der Methode der finiten Elemente beim Plattenproblem. *Manuscripta Math.* **19,** 401–416, 1976.

72 Stephan, E., A finite element scheme for the biharmonic equation in domains with corners, to appear (Fachbereich Math., T. H., Darmstadt, Preprint 363, 1977).

73 Törnig, W., Monoton konvergente Iterationsverfahren zur Lösung nichtlinearer Differenzen-Randwertprobleme. *Beiträge z. Numer. Math.* **4,** 245–257, 1975.

74 Wendland, W., On the imbedding method for semilinear first order elliptic systems and related finite element methods, to appear (Fachbereich Math., T. H., Darmstadt, Preprint 410, 1978).

Appendix

Applications of Systems of the First Order

In this appendix some applications where plane elliptic equations arise are indicated. Since such problems come from so many different fields it is impossible to give a reasonable presentation or a survey here. Therefore we present only a few examples and the corresponding references can neither be complete nor systematic.

The examples stem from stationary or time harmonic problems in elasticity and fluid dynamics. Besides these, the classical application of the homogeneous and inhomogeneous Cauchy Riemann equations, i.e. (1.0.1) with $A=B=\tilde{A}=\tilde{B}$, is the two dimensional theory of electrostatics (see e.g. [61]). Another classical field of application is acoustic and electromagnatic waves which are governed by the Helmholtz equation (see e.g. [40, 41]). In two dimensions, it can be replaced by a first order system, as is shown in section A.1. But for these equations, as everybody knows, much more specific theories are available. Hence, these applications are not dealt with here.

The elasticity problems have a rather long history and their mathematical theory is well developed in the framework of strongly even order elliptic systems or equations (e.g. [17, 32, 42]). Hence, one should not expect to produce completely new results just by using the approach via first order systems. But some of the advantages, such as representation of solutions, function theoretic properties, integral equations etc., might become more apparent. Besides the classical problems we also refer to some newer mechanical structures whose elastic equations could be formulated as a first order system with $n \geqslant 2$.

In fluid dynamics we meet essentially the same situation. Here the relationships of first order systems to the well known mathematical theory of rotational free flow problems [4, 50, 51, 61], and to the linearizations of the Navier-Stokes equations, are briefly indicated, adding a few references to more complicated fluid models.

In order to indicate the transformation of higher order equations to first order systems let us begin with one second order equation. Higher order problems can be handled in a similar way.

A.1 One Second Order Equation

In general the equation

$$\Delta u + g u_x + h u_y + cu = f \tag{A.1.1}$$

with real valued functions can only be replaced by a system of (four real or) two complex first order equations in the Pascali normal form:

$$\begin{aligned} u_{\bar z} &= \tfrac{1}{2}\bar w, \\ w_{\bar z} &= -\tfrac{1}{2}(g+ih)w - \tfrac{1}{2}(g-ih)\bar w - cu + f. \end{aligned} \tag{A.1.2}$$

The Dirichlet condition $u = \psi$ for (A.1.1) becomes

$$\operatorname{Re}\begin{pmatrix}1 & 0\\ 0 & \dot z\end{pmatrix}\begin{pmatrix}u\\ w\end{pmatrix} = \begin{pmatrix}\psi\\ \dot\psi\end{pmatrix} \quad \text{on } \dot{\mathsf{G}} \tag{A.1.3}$$

and the Neumann condition $u_x\dot y - u_y\dot x = \psi$ becomes

$$\operatorname{Re} i\begin{pmatrix}1 & 0\\ 0 & \dot z\end{pmatrix}\begin{pmatrix}u\\ w\end{pmatrix} = \begin{pmatrix}0\\ -\psi\end{pmatrix} \quad \text{on } \dot{\mathsf{G}}. \tag{A.1.4}$$

Both boundary matrices P satisfy the Šapiro-Lopatinski condition and have index $\nu = 0$, (3.4.3).

A.2 Plane Elastic Deformations

Let us consider an infinitely long cylindrical elastic rod with plane cross section G (perpendicular to the generating direction) and let us assume that all the displacements and forces take place only within the cross section plane. Let us consider first an isotropic material. Then the equations for the displacement vector $\mathbf{u} = (u_1, u_2, u_3)^T$ (see e.g. [30, pp. 8–14]),

$$\sum_{j,k,l=1}^{3} \frac{\partial}{\partial x_j}\left\{(\lambda\delta_{pj}\delta_{kl} + \mu(\delta_{pk}\delta_{jl} + \delta_{pl}\delta_{jk}))\frac{\partial}{\partial x_k}u_l\right\} = f_p,\ p = 1, 2, 3 \tag{A.2.1}$$

reduce with $u_3 = 0$, $x_1 = x$, $x_2 = y$ and

$$w_1 \equiv (2\mu+\lambda)(u_{1x} + u_{2y}) + i\mu(u_{2x} - u_{1y}),$$

$$w_2 \equiv 2\mu\frac{2\mu+\lambda}{\mu+\lambda}\{-u_{1x} + u_{2y} + i(u_{1y} + u_{2x})\},\ \mathbf{w} \equiv (w_1, w_2)^T,$$

to a first order system. If the Lamé coefficients λ and μ are constant then the system takes the Douglis normal form [19]

$$\mathbf{w}_{\bar z}-\begin{pmatrix}0 & 0\\ -1 & 0\end{pmatrix}\mathbf{w}_z=\mathbf{f}. \tag{A.2.2}$$

In the case of an isotropic but inhomogeneous material with variable λ and μ, the equation (A.2.2) is extended by linear terms without derivatives [19, (4.9)]. Clearly, system (A.2.2) satisfies condition (5.3.30) providing the unique continuation property (Theorem 5.3.7). In the case of dynamic problems, the functions f_p and, hence $\mathbf{f}$ in (A.2.2) also contain terms of the form $\rho\mathbf{u}_{tt}$. If the state is time harmonic, then $\mathbf{u}$ is of the form $\mathbf{u}_h e^{i\omega t}$ and the equations for the time independent amplitude with corresponding $\mathbf{w}$ again take the form (A.2.2) with additional terms without derivatives [60]. Clearly, these systems correspond to (A.2.2),

$$\mathbf{w}_{\bar z}-\begin{pmatrix}0 & 0\\ -1 & 0\end{pmatrix}\mathbf{w}_z=A\mathbf{w}+B\bar{\mathbf{w}}+\mathbf{f} \tag{A.2.3}$$

They satisfy the condition (5.3.30) of Hile and Protter providing the unique continuation (Theorem 5.3.7) but in the case of variable λ and μ they do *not* satisfy the conditions for the Carleman property except if A and B are analytic matrices.

The simplest boundary conditions correspond to the fundamental boundary value problems of elasticity [42, Section 20], [54, 32]:

1 The displacement

$u_1=\psi_1$, $u_2=\psi_2$ is given on $\dot{\mathsf{G}}$.

Hence, also their derivatives are given along $\dot{\mathsf{G}}$. This with (A.2.1) implies

$$\operatorname{Re}\begin{pmatrix}2\dot x(2\mu+\lambda)^{-1}+2i\dot y\mu^{-1}, & -(\lambda+\mu)(2\mu+\lambda)^{-1}(\dot x+i\dot y\mu^{-1})\\ 2\dot y(2\mu+\lambda)^{-1}-2i\dot x\mu^{-1}, & \mu^{-1}(2\mu+\lambda)^{-1}(\dot y-i\dot x(\lambda+\mu))\end{pmatrix}\mathbf{w}$$

$$=4\begin{pmatrix}\dot\psi_1\\ \dot\psi_2\end{pmatrix}\text{ on }\dot{\mathsf{G}}. \tag{A.2.4}$$

The boundary matrix P in (A.2.4) satisfies the Lopatinski condition (3.4.3), and $\nu=-2$. The two solvability conditions correspond to

$$\oint_{\dot{\mathsf{G}}}\dot\psi_1\,ds=0,\qquad \oint_{\dot{\mathsf{G}}}\dot\psi_2\,ds=0. \tag{A.2.5}$$

Since the solution is unique [30, chapter 5], these are the only conditions.

2. The stress,

$$T\mathbf{u}=\begin{pmatrix}\{(\lambda+2\mu)u_{1x}+\lambda u_{2y}\}\dot{y}-\{\mu(u_{1y}+u_{2x})\}\dot{x}\\ \mu(u_{1y}+u_{2x})\dot{y}-\{\lambda u_{1x}+(\lambda+2\mu)u_{2y}\}\dot{x}\end{pmatrix}=\boldsymbol{\psi}$$

is given on $\dot{G}$ [32]. This yields

$$\text{Re}\begin{pmatrix}\dot{y}(2\mu+\lambda)-i\dot{x}\mu, & -(\lambda+\mu)(\dot{x}+i\dot{y})\\ -\dot{x}\lambda+i\dot{y}\mu, & -i(\bar{\lambda}+\mu)(\dot{x}+i\dot{y})\end{pmatrix}\mathbf{w}=(2\mu+\lambda)\boldsymbol{\psi}. \tag{A.2.6}$$

The boundary matrix P in (A.2.6) satisfies the Lopatinski condition (3.4.3),

$$\text{Det}\,\mathsf{P}=-(\lambda+\mu)^2(2\lambda+\mu)^{-2}\dot{z}^2\neq 0,\quad \text{and}\quad \nu=-2.$$

The two solvability conditions correspond to the vanishing of total forces:

$$\oint_{\dot{G}}\psi_1\,ds=0,\qquad \oint_{\dot{G}}\psi_2\,ds=0. \tag{A.2.7}$$

Again, uniqueness [30, Chapter 5] implies that (A.2.7) are the only conditions.

Since (A.2.2), (A.2.3) are of the Douglis normal form, their solution $\mathbf{w}$ can be constructed by the use of the generating solution and hyperanalytic functions (see Section 5.2, (5.2.63) ff.).

If the body is *anisotropic* and homogeneous then for an orthotropic medium with the coordinate axes as principal directions one finds in [35, section 6] a different first order system. If the given body forces can be obtained from a potential then the stresses and strains can be derived from an elastic potential F [35, p. 26]. With $\mu_1=\alpha_1+i\beta_1$, $\mu_2=\alpha_2+i\beta_2$, $\beta_1\neq 0$, $\beta_2\neq 0$ and $w_1=g_1$, $w_2=\bar{g}_3$, $w_3=g_2$, $w_4=\bar{F}$ one finds from [35, p. 27 ff.] the system in the normal form (3.1.30),

$$\begin{aligned}
&w_{1\bar{\zeta}_1}=0,\\
&w_{2\bar{\zeta}_1}=\frac{i}{2\beta_1}\bar{w}_3, \qquad \zeta_1=x+\mu_1 y, \qquad \bar{\zeta}_1=x+\bar{\mu}_1 y,\\
&w_{3\bar{\zeta}_2}=\frac{i}{2\beta_2}w_1, \qquad \zeta_2=x+\mu_2 y, \qquad \bar{\zeta}_2=x+\bar{\mu}_2 y,\\
&w_{4\bar{\zeta}_2}=\frac{i}{2\beta_2}w_2.
\end{aligned} \tag{A.2.8}$$

(see also [37, Section 1.62]).

In case $\mu_1\neq\mu_2$ the elastic potential F corresponding to the general solution of (A.2.8) is of the form

$$F=\text{Re}\left\{\int_{\zeta_{10}}^{\zeta_1}\phi_1(\xi)\,d\xi+\int_{\zeta_{20}}^{\zeta_2}\phi_2(\eta)\,d\eta\right\} \tag{A.2.9}$$

with two holomorphic functions $\phi_1(\zeta_1)$, $\phi_2(\zeta_2)$. In this case (A.2.8) can be replaced by

$$\phi_{1\bar{\zeta}_1}=0,$$
$$\phi_{2\bar{\zeta}_2}=0.$$

The fundamental boundary conditions correspond to

$$\operatorname{Re}\begin{pmatrix} p_1 & p_2 \\ q_1 & q_2 \end{pmatrix}\begin{pmatrix} \phi_1 \\ \phi_2 \end{pmatrix}=\begin{pmatrix} \psi_1 \\ \psi_2 \end{pmatrix} \quad \text{on } \dot{G}$$

for given displacements [35, (8.3)] where $p_1, \ldots, q_2$ are given constants and

$$\operatorname{Re}\begin{pmatrix} 1 & 1 \\ \mu_1 & \mu_2 \end{pmatrix}\begin{pmatrix} \phi_1 \\ \phi_2 \end{pmatrix}=\begin{pmatrix} \psi_1 \\ \psi_2 \end{pmatrix} \quad \text{on } \dot{G}$$

for given forces on $\dot{G}$ [35, (8.7)]. We observe that the Šapiro-Lopatinski condition is only satisfied if $\mu_1 \neq \mu_2$. Hence, for general inhomogeneous anisotropic plane problems, the formulation with equations (A.2.8) seems to be the most reasonable including also isotropy points $\mu_1=\mu_2$.

In the general case of plane anisotropic problems the equilibrium equations and Hooke's law yield first order elliptic systems with a principal part of the form (A.2.8) if one follows the presentation in [35].

For further plane problems and complex methods let us refer to [38, 32, 15]. To such problems belong the torsion problems which lead to holomorphic or generalized analytic functions whose real and imaginary parts define the components of the displacements [23, p. 49 ff.], [42, Chapter 22].

A.3 Plate and Shell Problems

For plates and shells the geometry of the elastic body implies that all elastic quantities depend mainly on two variables within the middle surface. By using a formal expansion with respect to the thickness of the plate, in [18] it was shown that for homogeneous isotropic plates the equilibrium states can be described by the superposition of a plane problem as in A.2 and a purely bending problem corresponding to the lowest order terms in the formal expansion. For shells a corresponding analysis has been made in [45]. A complex formulation of the differential equations for shells can also be found in [46]. (Introducing the complex derivative of the trace P, the equations [46, (20)] form a first order elliptic system.) The corresponding moment free shell theory yields first order systems with $n=1$, which were thoroughly studied in [58, Chapter VI]. For the plane elasticity theory for plates we refer to the books [32, 35,

39, 42, 54, 57] and the survey [53]. The fundamental solutions for the equations of elasticity and the use of boundary layer potentials and singular integral equations are developed in [42, 32] and [16] (see also [53, 22, 44]). The theory of weak solutions of the elastic boundary value problems by the use of coercive bilinear forms is developed in [17]. All these classical problems correspond to $n=1$ or 2. Problems with more unknowns i.e. $n>2$ arise for more complicated materials as micromorphic continua [14, (18.12) (18.13)], Cosserat shells [21, (5.12)], micropolar plates [9, 10, 11, 12] etc. For further two dimensional elasticity problems see e.g. [1, 6, 7, 20, 28, 29, 36, 47, 48, 52, 55, 56].

Of course, for time harmonic vibrations of plates and shells the corresponding equations arise for the time independent amplitudes. These equations defer from the stationary equations just by additional terms without spatial derivatives

Here we indicate the formulation with first order systems only for the simplest cases, the bending of homogeneous isotropic and orthotropic plates.

For the homogeneous isotropic plate one finds the biharmonic equation

$$\Delta\Delta u = 0$$

for the normal displacement. Then since the function

$$v = \tfrac{1}{4}\Delta u = u_{z\bar{z}} \tag{A.3.1}$$

must be harmonic, we may introduce a holomorphic function $\phi(z)$ by

$$v = \operatorname{Re} \phi. \tag{A.3.2}$$

The most common boundary conditions are:

1. *The rigidly deformed edge,*

$$u = \psi_1 \quad \text{and} \quad \mathrm{d}_n u = \psi_2 \, \mathrm{d}s \quad \text{on } \dot{\mathrm{G}} \tag{A.3.3}$$

are given. Here we introduce with

$$w_{1z} \equiv \phi, \qquad w_{1\bar{z}} = 0, \tag{A.3.4}$$

the complex strain

$$u_{\bar{z}} = \tfrac{1}{2}(w_1 + \bar{w}_2). \tag{A.3.5}$$

Then inserting (A.3.5) into (A.3.1), the relations (A.3.2) – (A.3.4) yield a system in the Douglis normal form,

$$\begin{aligned} w_{1\bar{z}} &= 0, \\ w_{2\bar{z}} - w_{1z} &= 0. \end{aligned} \tag{A.3.6}$$

The boundary conditions (A.3.3) (with $du = \dot{\psi}_1\, ds$) take the form

$$\operatorname{Re}\begin{pmatrix} \bar{\dot{z}}, & \dot{z} \\ i\bar{\dot{z}}, & -i\dot{z} \end{pmatrix}\begin{pmatrix} w_1 \\ w_2 \end{pmatrix} = \begin{pmatrix} \dot{\psi}_1 \\ \psi_2 \end{pmatrix} \quad \text{on } \dot{\mathsf{G}}. \tag{A.3.7}$$

Since Det $\mathsf{P} = -2i$, we have from (3.3.1) the Fredholm index $\nu = 2$. The only three eigensolutions are

$$\mathbf{w}_h = \begin{pmatrix} -1 \\ 1 \end{pmatrix}, \quad \mathbf{w}_h = i\begin{pmatrix} 1 \\ 1 \end{pmatrix} \quad \text{and} \quad \mathbf{w}_h = i\begin{pmatrix} z \\ \bar{z} \end{pmatrix}. \tag{A.3.8}$$

They correspond to translations of the plate. The condition $\oint_{\dot{\mathsf{G}}} \dot{\psi}_1\, ds = 0$ corresponds to the solvability condition.

2. *The loaded edge*

Here the bending moments and the transverse forces are given by

$$\frac{1+\tilde{\nu}}{2}\Delta u - (1-\tilde{\nu})\{\tfrac{1}{2}(\dot{x}^2 - \dot{y}^2)(u_{xx} - u_{yy}) + 2\dot{x}\dot{y}u_{xy}\} = \psi_1,$$

$$d_n \Delta u + (1-\tilde{\nu})\, d\{\dot{x}\dot{y}(u_{xx} - u_{yy}) + (\dot{y}^2 - \dot{x}^2)u_{xy}\} = \psi_2\, ds \quad \text{on } \dot{\mathsf{G}}. \tag{A.3.9}$$

Introducing

$$w_1 \equiv \phi \quad \text{and} \quad w_2 \equiv 2u_{zz} = \tfrac{1}{2}(u_{xx} - u_{yy}) - iu_{xy}, \tag{A.3.10}$$

one again finds the hyperanalytic system

$$w_{1\bar{z}} = 0$$

$$w_{2\bar{z}} - w_{1z} = 0.$$

The boundary conditions (A.3.9) take the form

$$\operatorname{Re}\begin{pmatrix} \dfrac{1+\tilde{\nu}}{2} & -(1-\tilde{\nu})\dot{z}^2 \\ -\dfrac{i}{2} & -i(1-\tilde{\nu})\dot{z}^2 \end{pmatrix}\begin{pmatrix} w_1 \\ w_2 \end{pmatrix} = \begin{pmatrix} \psi_1 \\ \displaystyle\oint_0^s \psi_2(\sigma)\, d\sigma \end{pmatrix}. \tag{A.3.11}$$

Since $\operatorname{Det}\mathsf{P} = -(i/2)\nu(1-\tilde{\nu})\dot{z}^2$ holds, the Šapiro-Lopatinski condition is satisfied and we have from (3.3.1) $\nu = -2$. The solvability conditions correspond to the vanishing of the total forces and the total moments,

$$\oint_{\dot{\mathsf{G}}} \psi_1\, ds = 0, \qquad \oint_{\dot{\mathsf{G}}} \psi_2\, ds = 0. \tag{A.3.12}$$

For both problems the systems (A.3.6), (A.3.10) correspond to the Douglis hyperanalytic function theory which actually has been used in mechanics for a long time (see the references in [31]). Using the representation formula (5.2.34), the boundary value problem (A.3.7) and (A.3.11) can be solved by integral equations on the boundary curve $\dot{G}$. This approach appears in a number of variations and has also been used for many numerical computations [31]. For the connection of first kind integral equations, such as in (6.5.41), with these problems see also the references {[16, 17, 59] to Chapter 5} and {[10, 30, 34, 38, 40] to Chapter 6}. Using finite elements in G the method in Sections 8.4 and 8.5 for (A.3.6), (A.3.7) and (A.3.10) (A.3.11), respectively, correspond to hybrid element methods in elasticity which also are extensively developed.

In the case of anisotropic, in particular orthotropic plates, one again finds from [35 Chapter 9] a system such as (A.2.8). In the case $\mu_1=\mu_2$ one proceeds as in the isotropic case above. For $\mu_1\neq\mu_2$ the displacement is of the form

$$u=2\,\mathrm{Re}\left\{\int_{\zeta_{10}}^{\zeta_1} w_1(\xi)\,d\xi+\int_{\zeta_{20}}^{\zeta_2} w_2(\sigma)\,d\sigma\right\} \tag{A.3.13}$$

where w_1, w_2 are solutions of a system in normal form,

$$w_{1\bar{\zeta}_1}=0,$$
$$w_{2\bar{\zeta}_2}=0,$$

subject to the boundary conditions for

1. *the rigidly deformed edge,*

$$\mathrm{Re}\begin{pmatrix}1 & 1\\ \mu_1 & \mu_2\end{pmatrix}\begin{pmatrix}w_1\\ w_2\end{pmatrix}=\begin{pmatrix}\psi_1\\ \psi_2\end{pmatrix}\quad \text{on } \dot{G},$$

and

2. *the loaded edge,*

$$\mathrm{Re}\begin{pmatrix}\dfrac{p_1}{\mu_1}, & \dfrac{p_2}{\mu_2}\\ q_1, & q_2\end{pmatrix}\begin{pmatrix}w_1\\ w_2\end{pmatrix}=\begin{pmatrix}\psi_1\\ \psi_2\end{pmatrix}\quad \text{on } \dot{G}.$$

The systems for time harmonic vibrations are of similar form. Their formulation and investigation with even order strongly elliptic equations and Hilbert space methods can be found in [33, 34, 60].

A.4 Flow Problems

Most of the systems and equations in flow problems correspond to $n=1$. For example it is well known that incompressible irrotational plane steady flows are governed by the Cauchy Riemann equations

$$w_{\bar{z}}=0 \quad \text{for} \quad w=u-iv \tag{A.4.1}$$

[39, Chapter II], [43]. The most natural problem is the flow around an obstacle G corresponding to the Neumann problem

$$-\mathrm{Re}\, i\dot{z}w = u\dot{y} - v\dot{x} = \psi \quad \text{on } \dot{G} \tag{A.4.2}$$

in the *exterior* domain $\mathbb{C}\backslash\bar{G}$ with $w \to 0$ for $|z| \to \infty$. Interior problems with (A.4.1), often with mixed boundary conditions, arise for flows in porous media [49, p. 638]. For *compressible* steady plane irrotational fluid flows one has a nonlinear system [4, p. 14]

$$\begin{aligned}(c-u^2)u_x - uv(u_y+v_x)+(c^2-u^2)v_y &= 0,\\ u_y - v_x &= 0\end{aligned} \tag{A.4.3}$$

which corresponds for subsonic flows to a system (2.6.3) where $w = u - iv$. Again, the principal problem is the flow in the exterior of G subject to the boundary condition (A.4.2). The existence of this flow corresponds to the existence of a quasiconformal homeomorphism (Theorem 2.6.2). But for the flow problem additional difficulties arise, such as for example, the uniformity of the subsonic flow and the unboundedness of the domain. These flow problems have started many extensive investigations on elliptic plane problems ($n = 1$), especially the work by Bers on pseudoanalytic functions. The complete solution of these flow problems can be found in [4, 50, 51, 62]. A similar system for four unknowns ($n = 2$) can be found in [8, (38.1)] governing the subsonic flows in domains with shocks.

For flows with a small Reynolds number, i.e. viscous flows, a formal expansion with respect to the Reynolds number yields a sequence of Stokes and Oseen flow problems. These are governed by the Bilaplacian [3, p. 204], [39, p. 214] corresponding to (A.3.6) and the Dirichlet conditions (A.3.7).

The Oseen flow [3, (78.11), (78.12)] corresponds to a system in Pascali's normal form,

$$\begin{aligned}&w_{1\bar{z}} = 0\\ &w_{2\bar{z}} = \frac{i\lambda}{2}(w_2 - \bar{w}_2), \quad \text{in } G\\ &w_{3\bar{z}} = \bar{w}_2,\end{aligned} \tag{A.4.4}$$

(with $w_1 = \frac{1}{2}(\phi_y + i\phi_z)$, $w_2 = \frac{1}{2}(\chi_y - i\chi_z)$, $w_3 = \chi$ in the notations of [3]) subject to the boundary condition

$$\mathrm{Re}\begin{pmatrix} 1, & (2\lambda)^{-1}, & 0\\ -1, & i(2\lambda)^{-1}, & 0\\ 0, & 0\;\;, & i\end{pmatrix}\begin{pmatrix} w_1\\ w_2\\ w_3\end{pmatrix} = \begin{pmatrix}\psi_1\\ \psi_2\\ 0\end{pmatrix}. \tag{A.4.5}$$

(see also [39, p. 214 ff.]).

These boundary value problems play important roles for both, interior and exterior domains.

For such flows past an obstacle in exterior domains, the correct formulation of the singular perturbation problem can be found in [24, 25, 26].

In [13] one finds a system of equations corresponding to the Navier Stokes equations for polar fluids. Here the approximations similar to Stokes and Oseen approximations yield systems in normal form of $n=4$, respectively of $n=6$ complex unknowns. Similarly, the linearizations of stationary higher order non Newtonian fluid flows yield elliptic systems with many unknowns, see for example [27 (4.1.26) and Section 42], [5].

A different kind of application in flow problems is based on the *analytic theory* for elliptic systems which has been worked out by Garabedian. This is related to Section 5.4. In [2] the analytic continuation of the flow equations (A.4.3) into $\mathbb{C}^4$ even provides a suitable continuation across the sonic manifold into the domain of supersonic flow. Integrating along complex characteristic curves, shockless transonic airfoils are designed in an ingenious way.

References

1 Abramyan, K. G., General equations of nonsymmetrical sandwich shells with a light filler. *In Theory of Shells and Plates*, S. M. Durgar'yan (Ed.), Israel Program Sc. Transl., Jerusalem, pp. 133–141, 1966.

2 Bauer, F., Garabedian, P. and Korn, D., Supercritical wing sections. *Lecture Notes in Economics and Mathematical Systems* No. 66, Springer, Berlin, Heidelberg, New York, 1972.

3 Berker, R., Intégration des équations du mouvement d'un fluide visqueux incompressible. *In* S. Flügge (Ed.), *Handbuch der Physik*, Vol. VIII/2, pp. 1–384, Springer, Berlin, Göttingen, Heidelberg, 1963.

4 Bers, L., *Mathematical Aspects of Subsonic and Transonic Gas Dynamics.* John Wiley, London 1958.

5 Böhme, G., *Eine Theorie für sekundäre Strömungserscheinungen in nicht-Newtonschen Fluiden.* Habilitationsschrift Technische Hochschule, Darmstadt, 1974.

6 Bryukker, L. E., Bending of sandwich plates allowing far temperature stresses. *In* S. M. Durgar'yan (Ed.), *Theory of Shells and Plates*, pp. 241–253, Israel Program Sc. Transl. Jerusalem, 1966.

7 Burmistrov, E. F. and Mel'nichechko, A. A., Stability of structurally orthotropic cylindrical panels under the action of shearing and normal forces and internal pressure. *In* S. M. Durgar'yan (Ed.), *Theory of Shells and Plates*, pp. 246–253, Israel Progr. Sc. Transl. Jerusalem, 1966.

8 Cabannes, H., Theorie des ondes de choc. *In* S. Flügge (Ed.), *Handbuch der Physik*, vol. IX, pp. 162–224, Springer, Berlin, Göttingen, Heidelberg, 1960.

9 Constanda, C., On the bending of micropolar plates. *Letters in Appl. and Eng. Sc.*, **2,** 329–339, 1974.
10 Constanda, C., Sur la flexion des plaques élastiques micropolaires. *C. R. Acad. Sc. Paris,* **278,** 1267–1269, 1974.
11 Constanda, C. Asupra unor probleme de existentă si unicitate in teoria elasticităţii micropolare. *St. Cerc. Mat.*, **25,** 1075–1093, Bucuresti 1974.
12 Constanda, C., Complex variable treatment of bending of micropolar plates. *Int. J. Eng. Sci.*, **15,** 666–669, 1977.
13 Cowin, S. C., The characteristic length of a polar fluid. *In* E. Kröner (Ed.), *Mechanics of Generalized Continua,* pp. 90–94, (IUTAM Symp. Freudenstadt · Stuttgart 1967), Springer, Berlin, Heidelberg, New York, 1968.
14 Ehringen, A. C., Mechanics of micromorphic continua. *In* E. Kröner (Ed.), *Mechanics of Generalized Continua,* pp. 18–35, (IUTAM Symp. Freudenstadt, Stuttgart 1967), Springer, Berlin, Heidelberg, New York, 1968.
15 England, A. H., *Complex Variable Methods in Elasticity.* J. Wiley-Interscience, London, New York, Sydney, Tokyo, 1971.
16 Fichera, G., Linear elliptic equations of higher order in two independent variables and singular integral equations, with applications to anisotropic inhomogeneous elasticity. *In* R. Langer (Ed.), *Partial Differential Equations and Continuum Mechanics,* pp. 55–80, University Wisconsin Press, 1961.
17 Fichera, G., Existence theorems in elasticity. *In Handbuch der Physik,* S. Flügge (Ed.), vol. VI a/2, Springer, Berlin, Heidelberg, New York, pp. 347–389, 1972.
18 Friedrichs, K. O. and Dressler, F. R., A boundary-layer theory for elastic plates. *Comm. Pure Appl. Math.*, **14,** 1–33, 1961.
19 Gilbert, R. P. and Wendland, W., Analytic, generalized, hyperanalytic function theory and an application to elasticity. *Proc. Royal Soc. Edinburgh,* **73A,** 317–331, 1975.
20 Godzevich, V. G., Free vibrations of circular conic shells. *In* S. M. Durgar'yan (Ed.), *Theory of Shells and Plates,* pp. 339–343, Israel Program Sc. Transl., Jerusalem, 1966.
21 Green, A. E. and Naghdi, P. M., The Cosserat surface. *In* E. Kröner (Ed.), *Mechanics of Generalized Continua,* pp. 36–48, (IUTAM Symp. Freudenstadt, Stuttgart 1967), Springer, Berlin, Heidelberg, New York, 1968.
22 Hansen, E., Numerical solution of integro differential and singular integral equations for plate bending problems. *J. Elasticity,* **6,** 39–56, 1976.
23 Hearmon, R. F. S., *An Introduction to Applied Anisotropic Elasticity.* Oxford Univ. Press, London, 1961.
24 Hsiao, G. and MacCamy, R. C., Solution of boundary value problems by integral equations of the first kind. *SIAM Rev.*, **15,** 687–705, 1973.
25 Hsiao, G. and MacCamy, R. C., *On Slow Viscous Flows Past Cylinders.* University of Delaware, Dept. Mathematics, Newark, Delaware, Technical Report 19711, 1973.
26 Hsiao, G. C., Singular perturbations of an exterior Dirichlet problem. *SIAM J. Math. Anal.*, to appear. (Preprint 255, *Fachbereich Mathematik.* T.H. Darmstadt, Germany 1976).

27 Huilgol, R. R., *Continuum Mechanics of Viscoelastic Liquids.* John Wiley, New York, 1975.

28 Imenitov, L. B., Stress concentration in thin elastic spherical shells. *In* S. M. Durgar'yan (Ed.), *Theory of Shells and Plates,* pp. 431–434, Israel Program Sc. Transl., Jerusalem, 1966.

29 Kalinin, V. S., On the calculation of nonlinear vibrations of flexible plates and shallow shells by the small parameter method. *In* S. M. Durgar'yan (Ed.), *Theory of Shells and Plates,* pp. 435–443, Israel Program Sc. Transl., Jerusalem, 1966.

30 Knops, R. J. and Payne, L. E., *Uniqueness Theorems in Linear Elasticity.* Springer, Berlin, Heidelberg, New York, 1971.

31 Krawietz, A., Ein Beitrag zur Behandlung ebener Elastizitätsprobleme. TU Berlin Publ. No. 546, 1975, (ISBN 3 7983 0546 3).

32 Kupradze, V. D., *Potential Methods in the Theory of Elasticity.* Israel Program Sc. Transl., Jerusalem, 1965.

33 Leis, R., Zur Theorie elastischer schwingungen in inhomogenen Medien. *Archive Rat. Mech. Anal.,* **39,** 158–168, 1970.

34 Leis, R., Zur Theorie elastischer Schwingungen. *Ber. d. Ges. Mathematik u. Datenverarbeitung Bonn* No. 72, 1973.

35 Lekhnitskii, S. G., *Anisotropic Plates.* Gordon and Breach, New York, 1968.

36 Malkina, R. L., Application of the method of asymptotic integration to problems of shell vibrations. *In* S. M. Durgar'yan (Ed.), *Theory of Shells and Plates,* pp. 614–619, Israel Program Sc. Transl., Jerusalem, 1966.

37 Milne-Thomson, L. M., *Plane Elastic Systems.* Springer, Berlin, Göttingen, Heidelberg, 1960.

38 Milne-Thomson, L. M., *Antiplane Elastic Systems.* Springer, Berlin, Göttingen, Heidelberg, 1962.

39 Mises, R. v. and Friedrichs, K. O., *Fluid Dynamics.* Springer, New York, Heidelberg, Berlin, 1971.

40 Morse, P. M. and Ingard, K. U., Linear acoustic theory. *In* S. Flügge (Ed.), *Handbuch der Physik.* vol. XI/1, pp. 1–128, Springer, Berlin, Göttingen, Heidelberg, 1961.

41 Müller, C., *Grundprobleme der Mathematischen Theorie Elektromagnetischer Schwingungen.* Springer, Berlin, 1957.

42 Muskhelishvili, N. I., *Some Basic Problems of the Mathematical Theory of Elasticity.* Noordhoff Ltd., Groningen, 1953.

43 Oswatitsch, K., Physikalische Grundlagen der Strömungslehre. *In* S. Flügge (Ed.), *Handbuch der Physik,* vol. VIII/1, pp. 1–124, Springer, Berlin, Göttingen, Heidelberg, 1959.

44 Rieder, G., Iterationsverfahren und Operatorgleichungen in der Elastizitätstheorie. *Abh. d. Braunschweigischen Wiss. Ges.,* **14,** 109–342, 1962.

45 Rutten, H. S., Asymptotic Approximation in the Three Dimensional Theory of Thin and Thick Elastic Shells. Thesis Technical University Delft 1971 (printed in s'Hertogenbosch). A short survey on this work can be found in: *Theory of Thin Shells,* (IUTAM-Conf., Copenhagen, 1967), F. I. Niordson (Ed.), Springer, Berlin, Heidelberg, New York, pp. 115–134, 1969.

46 Sanders, J. L., On the shell equations in complex form. *In Theory of Thin*

Shells, (IUTAM-Conf., Copenhagen, 1967), F. I. Niordson (Ed.), Springer, Berlin, Heidelberg, New York, pp. 135–156, 1969.

47 Sarkisyan, V. S., On the bending of lengthy anisotropic plates moving in a gas at constant supersonic velocity. *In* S. M. Durgar'yan (Ed.), *Theory of Shells and Plates*, pp. 787–796, Israel Program Sc. Transl., Jerusalem, 1966.

48 Savin, G. N., Nonlinear problems of stress concentration near holes in plates. *In* S. M. Durgar'yan (Ed.), *Theory of Shells and Plates*, pp. 97–118, Israel Program Sc. Transl., Jerusalem, 1966.

49 Scheidegger, A. E., Hydrodynamics in porous media. *In* S. Flügge (Ed.), *Handbuch der Physik*, vol. VIII/2, pp. 625–662, Springer, Berlin, Göttingen, Heidelberg, 1963.

50 Schiffer, M., Analytic theory of subsonic and supersonic flows. *In* S. Flügge (Ed.), *Handbuch der Physik*, vol. IX, pp. 1–161, Springer, Berlin, Göttingen, Heidelberg, 1960.

51 Serrin, J., Mathematical principles of classical fluid mechanics. *In* S. Flügge (Ed.), *Handbuch der Physik*, vol. VIII/1, pp. 125–263, Springer, Berlin, Göttingen, Heidelberg, 1959.

52 Skomorovskii, Ya. G., On the influence of creep on the state of thin sandwich shells. *In* S. M. Durgar'yan (Ed.), *Theory of Shells and Plates*, pp. 812–817, Israel Program Sc. Trans., Jerusalem, 1966.

53 Sneddon, I. N. and Berry, D. S., The classical theory of elasticity. *In* S. Flügge (Ed.), *Handbuch der Physik*, vol. VI, pp. 1–126, Springer, Berlin, Göttingen, Heidelberg, 1958.

54 Sokolnikoff, I. S., The elastic boundary value problems. *In* E. F. Beckenbach (Ed.), *Modern Mathematics for the Engineer*, pp. 145–164, McGraw-Hill, New York, 1956.

55 Sternberg, E., Couple-stresses and singular stress concentrations in elastic solids. *In* E. Kröner (Ed.), *Mechanics of Generalized Continua*, pp. 95–108, (IUTAM Symp. Freudenstadt, Stuttgart 1967), Springer, Berlin, Heidelberg, New York, 1968.

56 Teodosiu, C., Continuous distributions of dislocations in hyperelastic materials of grade 2. *In* E. Kröner (Ed.), *Mechanics of Generalized Continua*, pp. 279–282, (IUTAM Symp. Freudenstadt, Stuttgart, 1967) Springer, Berlin, Heidelberg, New York, 1968.

57 Timoshenko, S., *Theory of Plates and Shells*. McGraw-Hill, New York and London, 1940.

58 Vekua, I. V., *Generalized Analytic Functions*. Pergamon Press, Oxford, 1962.

59 Vekua, I. N., On one version of the consistent theory of elastic shells. *In* F. I. Niordson (Ed.), *Theory of Thin Shells* (IUTAM-Conf., Copenhagen, 1967), pp. 59–84, Springer, Berlin, Heidelberg, New York, 1969.

60 Weck, N,, Außenraumaufgaben in der Theorie stationärer Schwingungen inhomogener elastischer Körper. *Math. Zeitschr.*, **111,** 387–398, 1969.

61 Wendt, G., Statische Felder und stationäre Ströme. *In* S. Flügge (Ed.), *Handbuch der Physik*, vol. XVI, pp. 1–164, Springer, Berlin, Göttingen, Heidelberg, 1958.

62 Bojarski, B., Subsonic flow of compressible fluid. *Math. Probl. Fluid Mech.*, pp. 9–23, 1967.

Index